Gut Microbiome and Brain Ageing

Surajit Pathak • Antara Banerjee
Editors

Gut Microbiome and Brain Ageing

Brain Aging

Springer

Editors
Surajit Pathak
Faculty of Allied Health Sciences
Chettinad Hospital & Research Institute,
Chettinad Academy of Research and Education
Chennai, Tamil Nadu, India

Antara Banerjee
Faculty of Allied Health Sciences
Chettinad Hospital & Research Institute,
Chettinad Academy of Research and Education
Chennai, Tamil Nadu, India

ISBN 978-981-99-8802-0 ISBN 978-981-99-8803-7 (eBook)
https://doi.org/10.1007/978-981-99-8803-7

This Springer imprint is published by the registered company Springer Nature Singapore Pte Ltd.
The registered company address is: 152 Beach Road, #21-01/04 Gateway East, Singapore 189721, Singapore

Paper in this product is recyclable.

Dedicated to our beloved mother the late Mrs. Bela Rani Pathak.

Preface

The book Gut Microbiome and Brain Ageing deals with the mechanisms and phenomena that relate the gut microbiome to the ageing of the brain. This book brings forth the connection between good gut and brain ageing and includes some cutting-edge research topics dealing with gut physiology and its importance as a future therapeutic option in neurological disorders.

Recent research has shed light on the crucial role of the gut-brain axis (GBA) in brain health and disease. This interplay, along with the diverse gut microbiome, appears to play a crucial role not only in brain development and function but also in the underlying mechanisms of neurodegenerative disorders, including Alzheimer's disease, Parkinson's disease, depression, and multiple sclerosis. As life expectancy increases, our society faces a significant demographic shift, necessitating effective measures to address the rising demand for the diagnosis and management of neurological and neurodegenerative conditions.

The complex and dynamic interplay of the gut-brain axis and its potential for preventing and treating a range of neuropsychiatric diseases are explored in this book. The discovery of the intestinal flora as a critical regulator, providing a flawless two-way communication route between the colon and the brain, is the core of this book. Unravelling the complexities of this interaction offers a promising avenue for advancing healthcare practices.

Advancements in gut microbiome research have given rise to numerous emerging strategies, such as microfluidic assays, high-throughput culturing, and engineered organoids derived from human stem cells, which are discussed in this volume, demonstrating improved efficiency and quality of research.

The gut microbiome's impact on neuronal function, mediated by neurotransmitters, vitamins, and neuroactive microbial metabolites, is an exciting area of investigation that this book thoroughly explores. It also explores the expanding understanding of the importance of the early-life gut microbiota and its implications for long-term health consequences. Environmental influences, stress, antibiotic exposure, and host genetics are all factors that can disrupt this delicate balance and leave enduring imprints on the host's behaviour and physiology.

This book aspires to be a comprehensive guide that explores the latest research, unravels ongoing debates, and ignites curiosity for further exploration. We hope that it inspires researchers, clinicians, and curious minds alike to collaborate in this exciting field and, together, unlock the potential of the gut-brain axis to transform the landscape of neurological health and well-being.

We sincerely thank Springer Nature Publishers for the help in publishing this book.

Chennai, Tamil Nadu, India Surajit Pathak
Chennai, Tamil Nadu, India Antara Banerjee

Contents

1 **Gut Microbiome and Brain Aging** 1
Anjana Suresh, Pravi Prasad, and Sreejith Parameswara Panicker

2 **The Scenario of Emerging Gut Microbiome Cross Talks in Brain Aging** 25
Swetha Panneerselvam, Selvaraj Jayaraman, and Ponnulakshmi Rajagopal

3 **The Gut Microbiome and the Central Nervous System (CNS)** 43
Amulya Vijay, Faraz Ahmad, Sudesh Ravi, and Anandan Balakrishnan

4 **Gut-Brain Interplay and Cognitive Degeneration** 71
Kalaiselvi Krishnamoorthy, Ponnulakshmi Rajagopal, and Selvaraj Jayaraman

5 **Neurological Disorders: The Impact of Alteration in Gut Microbiome and Inflammation on Disease Development** 93
Jeane Rebecca Roy, Coimbatore Sadagopan Janaki, Ponnulakshmi Rajagopal, and Selvaraj Jayaraman

6 **Influence of Altered Gut Microbiota in Cellular Senescence** 125
Bhuvaneswari Ponnusamy and Selvaraj Jayaraman

7 **Molecular Interplay of Oxidative Stress and Gut Microbiome in Aging** 143
Ahamed Basha Abdul Bari, Madhan Krishnan, and Shyamaladevi Babu

8 **The Gut Microbiome, Microbial-Produced Pro-inflammatory Neurotoxins, and Neurological Disorders** 161
Monisha Prasad, Ponnulakshmi Rajagopal, and Selvaraj Jayaraman

9 **Gut Microbiota and Altered Behaviour: Target on Neuroimmune Interplays** 185
Selvaraj Jayaraman and Shiny Paul

10 Customised Microbiome Restoration Approaches in Older People: Perspectives and Therapeutic Prospects 209
Rajalakshmi Manikkam, Vijayalakshmi Periyasamy, and Indu Sabapathy

11 Gut-Microbiota as a Therapeutic Intervention for Cognitive Damage 227
Hadeer M. Hamdalla, Manal Abdul-Hamid, and Mohamed Y. Zaky

12 Pharmacological and Nutritional Approaches to Modulate Microglial Polarization in Cognitive Senescence 243
Bhuvaneswari Ponnusamy, Ponnulakshmi Rajagopal, and Selvaraj Jayaraman

13 Role of Gut Brain and Gut Oral Axis in Progression of Parkinson's Disease with Special Focus on Gut Microbes 261
Ashwin Rajeev and Indranil Chattopadhyay

14 The Gut Microbiome-Induced Changes in Brain Immune System and Their Role in Epilepsy 287
Ramajayam Govindan, Maheshkumar Poomarimuthu, S. N. Arumugam, Senthilkumar Kalimuthu, Selvaraj Jayaraman, and C. S. Christudass

15 Exploring the Interplay Between Gut Microflora and Parkinson's 307
M. Swathi, Avipsa Hazra, A. S. Dhanu, and Kanthesh M. Basalingappa

16 Gut Microbiome: A Key Factor in Ageing and an Aim for Anti-senescence Intervention 335
Hanaa R. Abbas, Mohamed Y. Zaky, and Manal Abdul-Hamid

Editors and Contributors

About the Editors

Surajit Pathak, PhD, has been working as a Professor at Chettinad Academy of Research and Education, Kelambakkam, Chennai, since 2015. He worked as a Principal Research Scientist in Linkoping University, Linkoping, Sweden (2022 June–Dec 2022). He has received his Ph.D. degree from India in 2007, and completed his Post-Doctoral training (2007–2015) at the University of Alabama, USA, University of Padova, Italy, and University of Linkoping, Sweden.

His current research focuses on the role of intracellular and extracellular microRNAs for early detection of colorectal cancer. He has published more than **125 research and review articles** in peer-reviewed international journals of repute with 3238 citations and 32 h-index. **He has been granted with 3 patents from the Govt of India on his research works.** He received various research grants from national and international funding agencies. He has active research collaboration with top research laboratories in Sweden, Italy, Norway, Turkey, Mexico, and India.

He is an editorial board member of various renowned international high-impact journals. Prof. Pathak has edited several books for **Springer Nature:** ***Cancer Stem Cells: New Horizons in Cancer Therapies*****; MRW book** ***Handbook of Animal Models and Its Uses in Cancer Research*****; and** ***Evidence-Based Functional Foods for Prevention of Age-Related Diseases***. He is a member of various professional research bodies of India including **Indian Association of Cancer Research, Indian Science Congress.**

Antara Banerjee has been working as an Associate Professor at Chettinad Academy of Research and Education, Chennai, since 2016. She completed her Ph.D. in 2008 and pursued her Post-Doctoral research at the University of Padova, Italy (2009–2012), and later she joined as Senior Research Associate at the University of Linkoping, Sweden (2013–2015), and worked in collaboration with Karolinska Institute, Sweden.

She has published over **105 research and review articles** in peer-reviewed international and national journals with citation-**2294, i10-index 55 and h-index-25** and has active collaboration with various research laboratories in Sweden, Italy, Norway, Turkey, Mexico, and India. **She has been granted with 2 patent on her research works**. She serves as an editorial board member and reviewer of several international and national journals. She received various research grants from national and international funding agencies. She is an editor of books with Springer Publishers, namely ***Cancer Stem Cells: New Horizons in Cancer Therapies***; ***Handbook of Animal Models and Its Uses in Cancer Research***, **and** ***Evidence-Based Functional Foods for Prevention of Age-Related Diseases***. She is the lifetime member of Indian Association of Cancer Research and Indian Science Congress.

Contributors

Hanaa R. Abbas Cell Biology, Histology and Genetics Division, Department of Zoology, Faculty of Science, Beni-Suef University, Beni-Suef, Egypt

Manal Abdul-Hamid Cell Biology, Histology and Genetics Division, Department of Zoology, Faculty of Science, Beni-Suef University, Beni-Suef, Egypt

Faraz Ahmad School of Bioscience and Technology, Vellore Institute of Technology, Vellore, India

S. N. Arumugam DHR-Multidisciplinary Research Unit, Dr. ALM PGIBMS, University of Madras, Taramani Campus, Chennai, India

Shyamaladevi Babu Research, Chettinad Hospital and Research Institute, Chettinad Academy of Research and Education, Chengalpattu, India

Anandan Balakrishnan Department of Genetics, Dr ALM PG Institute of Basic Medical Sciences, University of Madras, Chennai, India

Ahamed Basha Abdul Bari Physiology, Chettinad Hospital and Research Institute, Chettinad Academy of Research and Education, Chengalpattu, India

Kanthesh M. Basalingappa Division of Molecular Biology, School of Life Sciences, JSS Academy of Higher Education and Research, Mysore, Karnataka, India

Indranil Chattopadhyay Department of Biotechnology, Central University of Tamil Nadu, Thiruvarur, India

C. S. Christudass Unit of Neurochemistry, Department of Neurological Sciences, Christian Medical College and Hospital, Ranipet Campus, Vellore, Tamil Nadu, India

A. S. Dhanu Division of Molecular Biology, School of Life Sciences, JSS Academy of Higher Education and Research, Mysore, Karnataka, India

Ramajayam Govindan DHR-ICMR, Multidisciplinary Research Unit, Madurai Medical College, Government Rajaji Hospital, Madurai, Tamil Nadu, India

Hadeer M. Hamdalla Cell Biology, Histology and Genetics Division, Department of Zoology, Faculty of Science, Beni-Suef University, Beni-Suef, Egypt

Avipsa Hazra Division of Molecular Biology, School of Life Sciences, JSS Academy of Higher Education and Research, Mysore, Karnataka, India

Coimbatore Sadagopan Janaki Department of Anatomy, Bhaarath Medical College and Hospital, Bharath Institute of Higher Education and Research (BIHER), Chennai, India

Selvaraj Jayaraman Department of Biochemistry, Centre of Molecular Medicine and Diagnostics (COMManD), Saveetha Dental College and Hospital, Saveetha Institute of Medical and Technical Sciences, Saveetha University, Chennai, India

Senthilkumar Kalimuthu Central Research Laboratory, Swamy Vivekanandha Medical College Hospital and Research Institute, Namakkal, Tamil Nadu, India

Kalaiselvi Krishnamoorthy Department of Biochemistry, Centre of Molecular Medicine and Diagnostics (COMManD), Saveetha Dental College and Hospital, Saveetha Institute of Medical and Technical Sciences, Saveetha University, Chennai, India

Madhan Krishnan Research, Chettinad Hospital and Research Institute, Chettinad Academy of Research and Education, Chengalpattu, India

Rajalakshmi Manikkam PG & Research Department of Zoology, Holy Cross College (Autonomous), Tiruchirappalli, Tamil Nadu, India

PG & Research Department of Biotechnology & Bioinformatics, DBT-BIF Centre, Holy Cross College (Autonomous), Affiliated to Bharathidasan University, Tiruchirappalli, Tamil Nadu, India

Sreejith Parameswara Panicker Department of Zoology, Advanced Centre for Regenerative Medicine and Stem Cell in Cutaneous Research (AcREM-Stem), University of Kerala, Thiruvananthapuram, Kerala, India

Swetha Panneerselvam Department of Biochemistry, Centre of Molecular Medicine and Diagnostics (COMManD), Saveetha Dental College and Hospital, Saveetha Institute of Medical and Technical Sciences, Saveetha University, Chennai, India

Shiny Paul Department of Biochemistry, Centre of Molecular Medicine and Diagnostics (COMManD), Saveetha Dental College and Hospital, Saveetha Institute of Medical and Technical Sciences, Saveetha University, Chennai, India

Vijayalakshmi Periyasamy PG & Research Department of Biotechnology & Bioinformatics, DBT-BIF Centre, Holy Cross College (Autonomous), Affiliated to Bharathidasan University, Tiruchirappalli, Tamil Nadu, India

Bhuvaneswari Ponnusamy Department of Biochemistry, Centre of Molecular Medicine and Diagnostics (COMManD), Saveetha Dental College and Hospital, Saveetha Institute of Medical and Technical Sciences, Saveetha University, Chennai, India

Maheshkumar Poomarimuthu Multidisciplinary Research Unit, Madurai Medical College, Government Rajaji Hospital, Madurai, Tamil Nadu, India

Monisha Prasad Department of Biochemistry, Centre of Molecular Medicine and Diagnostics (COMManD), Saveetha Dental College and Hospital, Saveetha Institute of Medical and Technical Sciences, Saveetha University, Chennai, India

Pravi Prasad Department of Zoology, Advanced Centre for Regenerative Medicine and Stem Cell in Cutaneous Research (AcREM-Stem), University of Kerala, Thiruvananthapuram, Kerala, India

Ponnulakshmi Rajagopal Department of Central Research Laboratory, Meenakhsi Ammal Dental College and Hospitals, Meenakhsi Academy of Higher Education and Research, Chennai, India

Central Research Laboratory, Meenakshi Ammal Dental College and Hospitals, Meenakshi Academy of Higher Education and Research, Chennai, India

Ashwin Rajeev Department of Biotechnology, Central University of Tamil Nadu, Thiruvarur, India

Sudesh Ravi School of Bioscience and Technology, Vellore Institute of Technology, Vellore, India

Jeane Rebecca Roy Department of Anatomy, Bhaarath Medical College and Hospital, Bharath Institute of Higher Education and Research (BIHER), Chennai, India

Indu Sabapathy PG & Research Department of Biotechnology & Bioinformatics, DBT-BIF Centre, Holy Cross College (Autonomous), Affiliated to Bharathidasan University, Tiruchirappalli, Tamil Nadu, India

Anjana Suresh Department of Zoology, Advanced Centre for Regenerative Medicine and Stem Cell in Cutaneous Research (AcREM-Stem), University of Kerala, Thiruvananthapuram, Kerala, India

M. Swathi Division of Molecular Biology, School of Life Sciences, JSS Academy of Higher Education and Research, Mysore, Karnataka, India

Amulya Vijay Department of Genetics, Dr ALM PG Institute of Basic Medical Sciences, University of Madras, Chennai, India

Mohamed Y. Zaky Molecular Physiology Division, Department of Zoology, Faculty of Science, Beni-Suef University, Beni-Suef, Egypt

Gut Microbiome and Brain Aging

1

Anjana Suresh, Pravi Prasad,
and Sreejith Parameswara Panicker

Abstract

The concept that the gut microbiota plays a significant role in maintaining physiological status in the gastrointestinal (GI) system is supported by both qualitative and quantitative transformation of the intestinal flora in various physiological and pathological conditions, as revealed in various research studies. A relationship between the gastrointestinal system and cognitive functioning is clearly suggested by the evidence of neuroinflammation that is seen in neurodegenerative illnesses like Parkinson's and Alzheimer's. Many factors will affect or influence the gut microbiota. Dysbiosis in the gut microbiome is linked to a decreased immune response, which encourages the growth of skin diseases and causes hair-related disturbances. Gut microbiome alteration is related to many skins and hair-related disorders. Gut-brain cross-talk has significantly improved the interaction between gut microbiota and gut hormones. Ayurveda can be applied to cosmetics to balance or symbiosis the gut microbial ecology using organic plant botanicals. According to this perspective, a physiological and non-pathological aging process may be facilitated by modulating gut microbiota, which considers the interdependence of microbiota products, inflammation mediators, and the immune system. It may also help to contrast the advancement of degenerative mechanisms. Certain investigations, with encouraging outcomes, have already characterized the gut microbiota of elderly people. The connection between gut microbiota, aging, and the degenerative diseases that plague older people should be better understood through future research.

A. Suresh · P. Prasad · S. P. Panicker (✉)
Department of Zoology, Advanced Centre for Regenerative Medicine and Stem Cell Research in Cutaneous Biology (AcREM-Stem), University of Kerala, Thiruvananthapuram, Kerala, India
e-mail: psreejith@keralauniversity.ac.in

S. Pathak, A. Banerjee (eds.), *Gut Microbiome and Brain Ageing*,
https://doi.org/10.1007/978-981-99-8803-7_1

Keywords

Gut microbiome · Aging · Gut-brain axis · Ayurveda · Homeostasis

1.1 Introduction

Age-related physiological changes can cause a steady reduction in cognitive function as well as overall brain health. Patients' quality of life is reduced by this decline in brain health, which is shown in a number of age-related disorders, including moderate cognitive impairment (MCI), Alzheimer's disease (AD), Parkinson's disease (PD), and multiple sclerosis (MS) (Murman 2015). One's chronological age (time since birth) and biologic age differ greatly from one another, and variations in the rate of aging, particularly around midlife, can have a big impact on the future age-related decline. Furthermore, dementia development in healthy elderly people can be predicted by biological age. Trillions of bacteria inhabit the GI tract, all continually observing, adapting to, and reacting to their surroundings. These microorganisms, collectively referred to as the gut microbiota, have coexisted and developed with their host, and they are now more commonly recognized for their contributions to preserving that host's health throughout their lifetime, including the health of the brain. The gut microbiota can communicate with the brain in both directions through a number of methods, including endocrine, neurological, immunological, and microbial metabolite-driven pathways. Thanks to studies into these pathways, it is now possible to more clearly understand how the gut microbiota affects biological processes in its host, including the brain (Morais et al. 2021). These studies also utilized direct gut microbiota manipulation techniques such as antibiotic therapy, fecal microbiota transplantation, and microbiota injection. Germ-free (GF) mice are those that have no bacteria. Importantly, the gut microbiota all impact the blood-brain barrier, neurochemistry, and cellular processes in the brain, including immunology and neuroplasticity. When taken as a whole, the gut microbiota is essential for maintaining the host's healthy neurological and cognitive processes, including active participation in important aspects of brain aging. However, it is widely acknowledged that the structure of the gut microbiota remains rather constant during early- and mid-adulthood, despite community stability being disrupted in the later stages of aging (Boehme et al. 2023). However, as people age, their gut microbiome's diversity changes in diverse ways, maybe as a function of factors like diet and health. The general population of centenarians showed a notable increase in alpha diversity, which was explained by greater relative abundances of subdominant taxa and decreased relative abundances of core taxa. By contrast, elderly patients in nursing homes showed a general decline in the diversity of fecal microbes, which was also correlated with a worsening of the patient's health status. These contradicting results suggest that aging-related alterations in the gut microbiome depend on the community being sampled and likely lead to an increase in microbiome uniqueness. The possibility that the gut microbiota will influence health is especially important for elderly people. This is due to the possibility that changes in innate immunity, sarcopenia, and cognitive function brought on by aging, and all components of frailty may be modulated by the microbiota. It has been shown through studies employing cell culture-dependent and culture-independent methodologies that the gut microbiota of older persons differs from that of younger adults. The composition of the microbiota does not change abruptly at any particular age or point in time; instead, changes take place over time. Our thorough assessments have classified the microbiota into categories according to age, prolonged residential care, dietary habits, and the degree of retention of a core microbiome. We are starting to comprehend how these groups alter as people age and how they connect to clinical characteristics.

1.2 Gut Microbiota and Homeostasis

Numerous human diseases are linked to alterations in the gut microbiota's makeup. However, based on the existence or absence of specific microbial species, still unable to differentiate between homeostasis and dysbiosis (Ipci et al. 2017). Diet and host variables that control and guide microbial growth determine the makeup and function of the adult gut microbiota. The small intestine's lumen receives oxygen and nitrate from the host, favoring bacteria that employ respiration to produce energy. By contrast, the host restricts the availability of nitrate and oxygen in the colon, creating a bacterial community that depends on fermentation for growth. Although nutrition affects the composition of the microbiota, a poor diet inhibits the host regulatory systems that manage the microbiota. Therefore, assessing host factors that influence microbial development may assist in distinguishing between homeostasis and dysbiosis and provide alternate methods for treating dysbiosis. Numerous human diseases, including cancer, obesity, and even neurological disorders, are correlated with alterations in the makeup of the gut-associated microbial populations (gut microbiota) (Scott et al. 2013). Identifying the factors that affect the gut microbiota is, therefore, one of the key objectives of microbiome research. Diet is one factor that affects the gut microbiota's makeup. Weaning causes a sudden change in food that alters the phylum-level makeup of the fecal microbiota. However, the fecal microbiota of healthy people and the community present in breastfed children before weaning are homeostatic. Therefore, alterations in the gut microbiota caused by nutrition are not always linked to disease.

Furthermore, due to the significant variation in species composition of the gut microbiota between individuals, it is impossible to discriminate between homeostasis and dysbiosis simply on the presence or absence of certain microorganisms. It is challenging to transfer studies into medical interventions due to the lack of knowledge regarding the ecological causes of dysbiosis and the causal impact of dysbiosis on disease. Energy in the form of adenosine triphosphate (ATP) is necessary for bacterial development. Microbes that can produce the most energy in a certain environment will expand more quickly and take control of microbial communities. When respiratory electron acceptors, such as oxygen, are present in the environment, bacteria that respire will predominate a microbial community, because respiration produces more energy than fermentation (Siddiqui et al. 2022). The host molds its gut microbiota based on these bacterial community composition principles. To encourage the composition of the microbial community to be dominated by bacteria that employ respiration for energy production, our body maintains a high luminal oxygen content in the small intestine. Contrarily, by keeping the epithelium in a condition of physiological hypoxia, the host restricts the amount of oxygen that can enter the large intestine lumen, which promotes the predominance of bacteria that use fermentation as a source of energy. According to recent findings on the ecological causes of dysbiosis, a breakdown of homeostasis is typically accompanied by a deterioration of the host's control mechanisms for managing the microbiota. Mouse models of colorectal cancer, ulcerative colitis, antibiotic therapy, or enteric infection result in compromised host control mechanisms that restrict the amount of oxygen available for bacteria in the colon. The abundance of oxygen-rebreathing bacteria in the colonic microbiota is increased due to increased oxygen availability (Widhiati et al. 2022). Similarly, a Western-style high-energy, low-fiber diet has unfavorable health impacts and is linked to weakened host factors that regulate the microbial ecosystem in the gut. According to these mechanistic discoveries, dysbiosis is connected to a condition of diminished host control over the microbial environment. In contrast, gut homeostasis is a condition in which host activities regulating microbial development are regular (i.e., those characterizing a healthy or properly functioning person).

Dysbiosis in an environment caused by a loss of microbial homeostasis may result from altered metabolic processes, lower alpha diversity, a lack of keystone taxa, a decrease in metabolic function, or an increase in pathobiont abundance. A unique microbial ecological state can be seen there. Taxonomic and functional dysbiosis are the two basic classifications of dysbiosis (Sun et al. 2023). Taxonomic dysbiosis is an ecosystem's imbalance of microbial species, characterized by abnormal constituents, perturbations, altered composition, and decreased diversity and richness. Pathogen abundance may increase, and keystone taxa may disappear, or alpha diversity may decrease. A decline in microbial diversity may be seen at several taxonomic levels, such as phylum, class, genus, or even species level (Tsai et al. 2023). Obesity is associated with a phylum level imbalance, with a higher proportion of Firmicutes than the other prominent phylum Bacteroidetes. Firmicutes have enriched genomes that enable them to produce obesity and energy from food components more effectively. IBD, on the other hand, is connected to a decline in microbial diversity. Taxonomic dysbiosis is the imbalance of microbial species in an ecosystem, characterized by anomalous constituents, perturbations, altered composition, and decreased diversity and richness (Riaz Rajoka et al. 2021). Pathogen abundance may increase as a result, and keystone taxa may disappear, or alpha diversity may decrease. A decrease in microbial diversity may be seen at several taxonomic levels, such as phylum, class, genus, or species level. A phylum level imbalance, with a higher proportion of Firmicutes than the other significant phylum Bacteroidetes, is linked to obesity. Firmicutes have enriched genomes that make it easier for them to manufacture components of obesity and food-based energy. IBD, however, has been linked to a decrease in microbial diversity.

The homeostasis state of microbial ecology may be impacted by factors that affect microbial growth, colonization effectiveness, metabolic processes, and communication between microbial species (Turroni et al. 2020). Diet and xenobiotics significantly impact gut microbial ecology's short-term and long-term homeostasis. The three nutrients that people consume most frequently are carbohydrates, proteins, and lipids. Diets high in fat and poor in fiber diminish the variety of microbes in the gastrointestinal tract. *Prevotella copri* and *Xylanobacter* grow best on diets high in fiber, while *Proteobacteria* thrive on diets high in energy-dense simple sugars. Bacteroides thrive in diets high in protein. In healthy subjects, our recent study also showed a relationship between dietary practices and the makeup of the GIT microbiota.

1.3 The Gut Microbiome During Aging

Modern medicine has greatly increased life expectancy, and most people in affluent nations may today anticipate spending at least 4 score years on earth (Lee et al. 2022). The current challenge is to increase life expectancy and decrease the length and intensity of morbidity that occurs before death. In addition, an increasing number of elderly persons prefer or are expected to be competent in independent living due to changes in family and societal structures. Frailty is the biggest obstacle to independent living. Frailty does not result from chronological aging but rather from the accumulation of illnesses. The "organ" of the human body made up of the bacteria in the gut may not follow the typical trajectory of physiological deterioration (Ballini et al. 2020). Although gut bacteria do not age per se, comorbid conditions linked to the stomach and gut bacteria may start to affect older persons. Thus, whether the microbiota in the human gut can influence aging or if it merely adapts with advancing age naturally emerges.

Homeostasis gradually disappears as we become older, and we become less functional and more susceptible to death. Among age-related diseases are infectious, neoplastic, metabolic, and degenerative illnesses associated with frailty and cognitive loss (Guang et al. 2020). The fundamental signs of aging in mammals have been recognized, including their molecular and cellular characteristics; however, these signs are accompanied by changes in the microbiome, which in turn influence the speed of age-related decline. The microbiome alterations associated with aging are extremely varied and regulated by internal and external environmental variables. For instance, the gradual degradation in the alimentary tract's physiology would inevitably impact the gut microbiota. Because the newborn comes into contact with the mother's feces, vaginal, skin microbiota, and other microbial communities, the human body becomes more colonized with bacteria both before and after delivery (Quaglio et al. 2022). In turn, factors like age, sex, immune system development, and environmental factors affect the composition of the gut microbiota, which stabilizes between the ages of 6 and 36 months. A continuous endogenous flora, referred to as the "core microbiota," can be distinguished from bacteria that are still provisional and vulnerable to external disruptions. After a genetically and environmentally established age, the gut flora changes with age, depending on a person's racial or ethnic background, drug use, way of life, and dietary preferences. Although Bacteroidetes bacteria dominate in the elderly and Firmicutes bacteria in adults, most members of the Firmicutes and Bacteroidetes genera continue to be dominant.

Reduced diversity, a lack of species that produce butyrate, and potential pathogens in the gut microflora of centenarians are the main characteristics of the aged population (Hajishengallis and Lamont 2016). The link between the host and the gut microbiota has been seen as centered on the generation of SCFAs, which helps explain the connection between changes in the flora and a rise in frailty in the elderly. According to Biagi et al., the gut microbiota of older individuals has an unusual composition. Four groups were compared in a small area of Italy in this study. Twenty participants aged 25–40, 21 subjects aged 99–104, and 21 subjects aged 59–78 are centenarians' offspring. *Bifidobacteria* levels dropped, and the mucin-degrading bacteria rose. Compared to young individuals, *Akkermansia muciniphila* has also been seen in older people (İnceoğlu et al. 2011). A group of centenarian patients' descendants was also included in the study, even though the gut microbiota's general organization was identical to that of other subjects. Although the gut microbiota's overall composition matched that of subjects of a comparable age (70 years old on average), children of centenarian parents who shared a home with them showed a higher prevalence of opportunistic or potentially pathogenic bacterial groups than children who did not. As a result, cohabitation may impact these participants' gut microbiota's composition.

1.4 Elderly-Specific Factors Affecting the Gut Microbiota

One of the most well-known elements that may be able to alter the nature of the gut microbiota is diet, which has been proven in older adults. Dietary changes brought on by aging include loss of taste and smell as well as difficulty chewing (Chiu 2022). People who suffer from these conditions usually prefer diets heavy in carbohydrates and fats while consuming less foods with a plant origin. The Nutritious Food Diversity (HFD), which emphasizes the importance of eating nutritious foods, was recently developed. The investigation by the ELDERMET cooperation validated the link between microbial elements and a diet with a low HFD score. Certain researchers discovered a connection between microbial diversity, the functional independence measure (FIM), and the elderly patients' capacity to carry out daily activities as determined by the Barthel index, depending on whether the elderly patients were receiving long-term residential care, day hospital care, or rehabilitation (Das and Nair 2019). Particularly, it was discovered that dietary differences between institutionalized seniors and seniors living in the community impacted the makeup of the gut microbiota: A less varied diet was associated with a less diverse population of gut bacteria. But in addition to food, other factors may affect the microbiota of the elderly. In addition to food, Claesson et al. found that the loss in microbial diversity was linked to increased frailty, inflammatory markers, and worse health indices. The modification of gut microbiota was also influenced by residence and antibiotic use.

Last but not least, it should be emphasized that some bacterial composition changes are less vulnerable to external influences and may likely reflect the fundamental components of the gut microbial community in the elderly. *Ruminococcus* and *Blautia* spp., as well as other butyrate-producing bacteria (*Clostridium cluster* XIVa and *Clostridium cluster* IV), have been seen to decline in studies involving senior persons compared to younger participants. *Escherichia coli* and other facultative anaerobes are on the rise due to aging and inflammation, which is another typical finding. Most likely, the variability caused by dietary patterns outweighs changes in immune system function linked to immunosenescence in predicting these alterations.

According to recent research, the gut microbiota has a role in initiating and maintaining the inflammatory process. Zhang et al. studied how aged mice's gut microbiota was transferred to young germ-free mice to see how those organs responded. These basic lymphopoietic organs include the spleen, lymph nodes, and small intestines. (Veenhoven 2005). Strong links between gut microbiota structure and inflammatory aging have been found; in particular, the absence of *Akkermansia* and the rise in TM7 bacteria and *Proteobacteria* abundance were linked to the beginning of the local and systemic inflammatory response. The microbial population in the gut encouraged T-cell activation in the systemic compartment and small intestine inflammation. Specifically, the activation of the microglia in the central nervous system can be modulated by the modulation of inflammatory cytokines linked to changes in the gut microbiota. However, either total or partial gut cleansing can impact microglial function. Germ-free mice have abnormal maturation and a general malfunction of the microglial system. SCFAs mostly regulate the integrity of the microglia. This is very important for elderly people and causes cognitive deterioration. Additionally, there is evidence that substances from the gut microbiota that enter the bloodstream and are either components of bacterial structure or metabolism can activate macrophages into a pro-inflammatory state responsible for atherosclerosis. This mechanism may bring on both vascular dementia and cardiovascular illness. Consequently, it has been proposed that altering the gut microbiome can treat neurodegenerative diseases. There is proof that the gut microbiota's regulation significantly affects the inflammatory process. Indeed, it has been demonstrated that giving *Lactobacillus brevis* OW38 orally to elderly mice strengthens the tight junctions of the intestinal barrier, lowers the levels of circulating LPS, and inhibits the release of pro-inflammatory cytokines through suppressing NF-κB activation (Candela et al. 2014). Additionally, p16, p53, and SAMHD1 senescence markers in the colon and hippocampus decreased, whereas *Bacteroidetes*. Amyloid-β (Aβ) peptide and tau are deposited in the brain during Alzheimer's disease, which causes cognitive performance to deteriorate. It looks like the intestinal microbiota might be contributing to this procedure, albeit the exact cause is not yet known. Studies conducted in vivo and in vitro have revealed a correlation between certain factors and the development of lipopolysaccharide (LPS), one of the microbiota's most significant products. Strongly regarded to be amyloidogenic are these proteins.

On the other hand, research using overexpressed mouse models of the amyloid precursor protein and presenilin 1 (APPPS1) has demonstrated that changes in the inflammatory cytokine-dependent pattern brought on by prolonged antibiotic therapy affect the microbiota, which in turn affect the formation and deposition of the amyloid protein. Studies using antibiotic medication administered at birth and examining the best time for the microbiota's influence on neurodegeneration further supported these findings.

1.5 Gut-Brain Axis

One of the most involved areas of biology is the gut-brain axis. Simply, it connects our brain and gut (Drago et al. 2012). The microbiome in this region is having a huge impact on our overall health and well-being. The enteric nervous system and the central nervous system are strongly linked by the GBA axis. The brain and spinal cord are both parts of the central nervous system. The brain controls most of the body's vital functions (Boris and Vanessa 2023). It is very complex in nature. It is mainly divided into three major parts: cerebrum, cerebellum, and medulla. The largest and highest region, the cerebrum, is split into two hemispheres. It mainly controls voluntary actions (Garehdaghi and Sarbaz 2023). The cerebellum is a small part of the brain located at the back of the head. It plays a major role in regulating motor movement and balance. Medulla oblongata is a tale-like structure forming the brain's bottommost part (Park 2023). It is the chief site of many important reactions, such as cardiovascular, respiration, vasomotor, and involuntary functions. It connects the brain stem and spinal cord. The spinal cord is a tubular structure tightly packed by the column of nerve tissues. It extends downward from the brain stem through the central column of the spine. It mainly transmits the signals to-and-fro from the brain and other body parts. On the other hand, the gut nervous system, also called the enteric nervous system, controls digestion, absorption, and secretion and is known as the second brain (Nakamura et al. 2022). Enteric nervous system is a large division of peripheral nervous system. It autonomously controls gastrointestinal function. Its mesh-like structure consists of millions of neurons in the gastrointestinal tract, mainly motor, sensory, and interneurons. The neurons in the ENS can be grouped into three categories mainly: intrinsic primary afferent neurons, interneurons, and motor neurons. Ingested food and drink items primarily cause intrinsic primary afferent neurons to sense mechanical and chemical inputs. The motor neurons synchronize peristalsis and muscular contractions in the GI tract. Interneurons process information from sensory neurons and give instructions to motor nerves, to act accordingly on the effector system.

Numerous studies conducted in the past years found that the gut and brain are always in constant communication. The cognitive center of the brain and peripheral gastrointestinal tract is connected via gut-brain axis and linked directly through each other. This signaling is done in neuronal, endocrinal, immunological, and humoral ways. These transduction pathways allow the brain to influence the gut directly and vice versa (Matsumoto et al. 2023). That is, physiological concerns like stress, anxiety, depression, sadness, anger, etc. can be directly worse the gut condition. Similarly, any alterations or disorders in the gastrointestinal tract can affect the cognitive behavior in a serious way. The gut microbiota controls healthy aging, which mediates this bidirectional transmission of nerve signals between the central nervous and enteric nervous systems through the gut-brain axis (Tarride 2023).

Microbiota in the human gastrointestinal tract regulates metabolic and physiological functions, and their relative abundance and distribution are similar under normal conditions (Carabottia et al. 2015). In addition to affecting intestine function, the central nervous system is significantly impacted by gut microbes. Evidence from the clinical and experimental studies validated this, and the concept of gut-brain microbiome is one of the most sought areas of biology. It is a large community of symbiotic microorganisms. Among 1000 different bacterial species, the most prominent are bacterial phyla *Firmicutes*, *Bacteroidetes*, *Actinobacteria*, and *Proteobacteria* (Dahiya and Nigam 2023). The harmonious existence of host and microbiome is essential, because it has immense control over digestion and absorption of nutrients, vitamin synthesis, angiogenesis, cell maturation, development and defense mechanism. Any distortions in the gut microbiome compositions alter the CNS mechanism and lead to neurological diseases (Naseribafrouei et al. 2014). Gut microorganisms have a substantial impact on the central nervous system, in addition to influencing intestinal function (Rutsch et al. 2020). These changes might be due to genetic, lifestyle, and environmental factors.

The microbiome in the GBA acts as an effective modulator of healthy aging. Aging is a serious concern worldwide due to its physical, mental, and socioeconomical features. Elder people need keen attention from society in every aspect, because aging demands it. One of the major reasons behind brain aging is the lack of renewal of degenerative neuronal cell. Since the gut microbiome is a key controller of brain function and regulation, a rise in the study of GBA axis microbiota occurs to confirm its linking property with CNS (Yousefi et al. 2022). Healthy aging and gut microbiome are closely allied. The balanced composition with host microbiome is essential to maintain less detrimental effects of unhealthy aging. As ages increase, there is a high chance of gut microbiome composition alterations worsening conditions. Gut microbiota dysbiosis also contributes to unhealthy aging. So, to keep the nutritional profile of the host homeostasis, it is apparent to maintain the balance between microbial community. It can be achieved through corrections in lifestyle, physical exercises, following an appropriate diet, and meditation. Maintaining a harmonious association leads to a well-being life of the host. Tracking the microbial community composition and brain alterations paves the way to innovative microbial drug therapy (Zeng et al. 2023).

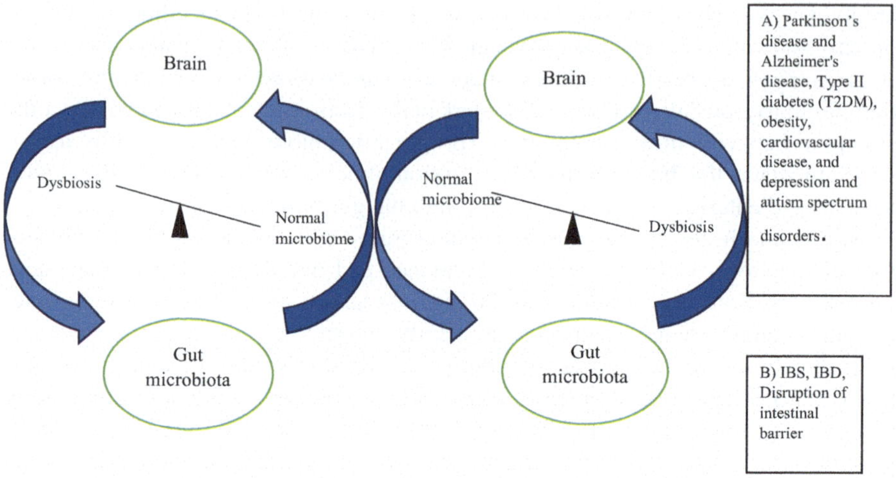

Fig. 1.1 Relationship between dysbiosis in the gut microbiota and disease. (**a**) Dysbiosis in the gut microbiota affects several neurological and gut-related diseases (**a**) Parkinson's disease and Alzheimer's disease, type II diabetes (T2DM), obesity, cardiovascular disease, and depression and autism spectrum disorders. (**b**) Disease caused in the gut due to dysbiosis: IBS, IBD, disruption of intestinal barrier

1.6 Gut-Brain Axis in Aging

GBA axis is the connecting link of the cognitive behavior center of the brain with peripheral intestine functions which holds a significant role in aging (Li et al. 2023). The degeneration of neuronal cells without renewal and repair leads to brain aging. It results in severe and detrimental consequences for the patients, their families, and society as a whole. Numerous studies validated that the alterations in the gut-brain axis's microbial composition contribute to various neurodegenerative disorders. GBA axis consists of beneficial bacteria which help in human physiological and metabolic activities. At the same time, it also comprises harmful bacteria species which impart infections, inflammations, and lethal effects to humans (Dos Santos et al. 2022). There is evidence tying the gut microbiota to degenerative diseases like Parkinson's disease and Alzheimer's disease as well as chronic problems like type II diabetes (T2DM), obesity, cardiovascular disease, and neuropsychiatric disorders including depression and autism spectrum disorders (Fig. 1.1.).

1.7 Gut Microbiome and Alzheimer's Disease

The excessive buildup of abnormal extracellular amyloid plaques in Alzheimer's disease (AD), a form of dementia, is brought on by the degradation of the amyloid precursor protein (APP), tau protein hyperphosphorylation, and protein misfolding, which results in the formation of intra-neurofibrillary tangles inside neurons

(Toricelli et al. 2021). Gut microbiome dysbiosis contributes to this dreadful disease. The microbial population secretes negative metabolites in addition to neuroactive substances like serotonin, γ-aminobutyric acid (GABA), acetylcholine, tryptophan, and catecholamines (Pearce et al. 2018). This makes the gut barrier more permeable, which eventually allows these toxic substances to be transported to the brain *via* the gut-brain axis, where they cause AD to develop. Gut microbiome alterations cause numerous toxins produced by bacteria and inflammatory product formation. This neurotoxin will accumulate in the CNS system. According to studies, the gut microbiota of people with AD stimulated C/EBP/AEP signaling, which in turn triggered the brain's polyunsaturated fatty acid (PUFA) metabolism in a pro-inflammatory manner, driving microglial maturation to AD pathologies and cognitive deficits. It confirms the causal relationship between GM changes and the pathophysiology of AD. Considering all these supporting documents, the GM community can be used as the predictive tool for diagnosing AD. It will assist in halting the development of AD and monitoring its prognosis (van Olst et al. 2021). So, from all the available clinical and experimental data, it can be concluded that gut microbial alterations lead to the pathogenesis of AD.

1.8 Gut Microbiome and Parkinson's Disease

A prevalent neurodegenerative condition is Parkinson's disease, affecting motor neurons in the CNS. The neurons in the substantia nigra are lost, where the production of dopamine takes place (Wasén et al. 2022). So, this will lead to tremors in one hand, leading to slow movement causing stiffness and loss of balance. The central, autonomous, and enteric nervous systems' neurons are affected by the accumulation of improperly folded alpha-synuclein (α-syn) proteins, which create intracellular inclusions within the neurons. The gut microbiome alteration causes these changes, distorting the gut's permeability and functions of the intestinal barrier. This will produce inflammation, promoting injury and degeneration of neurons. All these changes contribute to the misfolding of alpha-synuclein (α-syn), and gut microbiome, which can be the risk factor for diagnosis and prognosis of Parkinson's disease (Kalia and Lang 2015).

1.9 Gut Microbiome and Schizophrenia

Schizophrenia is a serious psychiatric disorder which is characterized by delusions, hallucinations, apathy, withdrawal, slowness, and many more concerns (Vizcarra et al. 2015). It is terribly affecting the lives of the patients in numerous ways. Dysbiosis in gut microbe bacteria is linked with this dreadful condition; still, future studies are evident for confirmation. Recent studies noted that in schizophrenia, the elevation of certain species in the gut microbiome is prominent. It was found that the significant increases of *Lactobacilli* somehow contribute to this disease and are used as a marker. In reaction to changes in the gut microbiome, inflammatory substances such interferons, tumor necrosis factor, and antibodies are produced more frequently. Also, higher levels of the bacterial translocation marker sCD14 have been observed in schizophrenia (Scheurink et al. 2023). All these potential toxic compounds lead to neurons' inflammation and harmfully affect the central nervous system.

1.10 Gut Microbiome and Multiple Sclerosis

Multiple sclerosis is a neurodegenerative disease in which the immune system destroys the myelin sheath that protects nerve fibers, thereby leading to miscommunication between the brain and the rest of the body (Hassanzadeh et al. 2023). The major symptoms include demyelination, axonal damage, gliosis, and neuronal loss. This disabling disease causes terrible effects on patients and society. The dysbiosis in GM triggers this condition. Any variations in microbial community composition cause an inflammatory response and induce neuronal injury and degeneration. The changes in the permeability of gut and blood-brain barriers lead to the transport of immune cells, CD4+ and CD8+ T lymphocytes, to the central nervous system. Multiple interferons will be produced as a result, activating macrophages to release reactive oxygen and reactive nitrogen species (ROS and RNS), which inflict damage on neighboring tissues.

Additionally, they generate interleukin 12 (IL-12), which raises INF and increases the production of TNF. All these factors contribute to inflammation, eventually leading to demyelinating plaque deposition. Studies found that the elevation of *Streptococcus mitis* can induce the production of Th17 cells and causes neuronal tissue damage. It is also validated that the significantly higher amount of pro-inflammatory microbial species Methanol *Brevibacter* (phylum *Euryarchaeota*) and *Akkermansia* (phylum *Verrucomicrobia)* contributes to MS. The subsequent reduction of anti-inflammatory microbe *Butyricimonas* is also documented. So, together they contributed to inflammation and, thereby, neuronal damage in the central nervous system (Sadek et al. 2023).

1.11 Gut Microbiome, Depression, and Anxiety

Clinical evidence and laboratory studies have shown a tight connection between gut microbiome dysbiosis and psychological disorders. Stress, rage, melancholy, bipolar illness, anxiety, etc. directly impact gut health, which in turn worsens our mood patterns. Thus, study on this cross-talk is constantly focused. Alterations in the gut microbiome initiate inflammatory mechanisms, leading to the disruption of central nervous system neurons. In patients with psychiatric illnesses, a sizable drop in microorganisms that produce butyrate has been observed (Nikolova et al. 2021). As a result, the gut microbiome's composition can be used as a diagnostic indicator for sickness, opening up new possibilities for the study of psychobiotics.

The imbalance in the GM composition will distort the intestinal epithelial barrier. As a result, inflammatory bacterial products (such lipopolysaccharides) are transported across the intestinal mucosa and into the bloodstream. The neurological system's ability to operate properly can then be hampered by inflammatory by-products, which can also cause inflammation to worsen and provoke unfavorable immunological reactions that might worsen the symptoms of psychiatric disease. Therefore, healthy maintenance of well-balanced GBA axis microbiome is essential for the effective association between the central and enteric nervous systems (Collins and Bercik 2009).

1.12 Skin Pathophysiology and Gut Microbiome

By tolerating the antigens in healthy flora, the digestive tract, which is important for immune system defense, keeps the balance with commensal microorganisms. Dysbiosis may result in tissue damage or autoimmune reactions after an inflammatory response (Tlaskalová-Hogenová et al. 2004; Bourlioux et al. 2003). The pathophysiology of cancer has been connected to changes in the gut microbiota and allergy, cardiovascular, gastrointestinal, metabolic, neurodevelopmental, and psychiatric illnesses. Additionally, there is growing proof that the skin-gut axis, sometimes known as the gastrointestinal microbiome, is related to skin conditions (Polkowska-Pruszyńska et al. 2020; Bosman et al. 2019). It has long been known that the intestine and skin have a strong link since gastrointestinal (GI) problems frequently present with skin symptoms. Numerous studies have demonstrated a reciprocal association between the stomach and skin related to GI health, skin homeostasis, and allostasis. Firmicutes, Bacteroidetes, Proteobacteria, and Actinobacteria are the four phyla that predominately comprise the gut microbiota (Bai et al. 2019). These phyla make up 93.5% of isolated human fecal material, and their composition reflects their regional physiological characteristics. As a result, they exhibit distinctive traits in inflammatory skin diseases as chronic urticaria, psoriasis, atopic dermatitis, and acne vulgaris.

The term "chronic urticaria" (CU) refers to persistent urticaria that lasts for more than 6 weeks without a known cause (Nettis et al. 2004). Changes in the bacteria *Akkermansia muciniphila*, *Faecalibacterium prausnitzii*, *Clostridium leptum*, and *Enterobacteriaceae* have been observed in CU patients (Nabizadeh et al. 2017). Additionally, serum metabolome study demonstrated that butanoate metabolic pathway and gut microbiota-associated changes in unsaturated fatty acids may contribute to CSU. These findings may be linked to inflammation caused by an imbalance of the cytokines Th1/Th2/Th17, which may have a role in the pathophysiology of CSU. More studies in this area could lead to more effective clinical, diagnostic, and therapeutic methods for treating CSU patients.

A prevalent and persistent skin condition known as psoriasis is regarded as a systemic inflammatory disorder. Psoriasis and psoriatic arthritis' subclinical gut inflammation and microbiome exhibit a distinctive pattern of dysbiosis (Myers et al. 2019). Numerous studies look at various theories on the role of the gut microbiome in the pathogenesis of psoriasis and psoriatic arthritis, such as compromised immune homeostasis, increased intestinal permeability, and an imbalance of bacteria that produce short- and medium-chain fatty acids (Scher et al. 2015). Recent scientific studies have revealed that people with psoriasis have significant gut and skin dysbiosis. Microbiome alpha-diversity (the abundance of several bacterial taxa assessed in one sample) and beta-diversity (microbial diversity in different samples) were significantly reduced in psoriasis-affected skin. It has been demonstrated that when compared to healthy skin, the presence of *Cutibacterium*, *Burkholderia* spp., and *Lactobacilli* is decreased. In contrast, *Corynebacterium kroppenstedtii*, *Corynebacterium simulans*, *Neisseria* spp., and *Finegoldia* spp. are more prevalent. Psoriasis patients' gut microbiomes had changes comparable to those seen in inflammatory bowel disease patients. In those two disorders, the prevalence of *Salmonella* spp., *Campylobacter* spp., *Helicobacter* spp., *Escherichia coli*, *Alcaligenes* spp., *Mycobacterium* spp., *F. prausnitzii*, *Bifidobacterium* spp., *Lactobacillus* spp., *Parabacteroides*, and *Coprobacillus* were underrepresented (Olejniczak-Staruch et al. 2021).

Dryness and itching of the skin are symptoms of the chronic, multifactorial inflammatory skin condition known as atopic dermatitis (Henley et al. 2014). After childhood, the gastrointestinal microbiota appears to stabilize. For the gut microbiota, there is also a sizable heritable component. There are notable differences in the microbial taxa, relative percentages, and amounts between the various digestive tract regions. Early gut bacteria colonization may be crucial in stopping AD from progressing. Topical medications that target the skin barrier aid in the development of a neo-microbiome from deeper compartments. Although probiotics, prebiotics, and symbiotic have been studied for the treatment of AD, further research is required (Petersen et al. 2019).

Propionibacterium acnes (formerly known as *Cutibacterium acnes*) proliferation, hyperkeratinization, and inflammation work together to generate acne, a dermatological condition that commonly presents and leaves behind debilitating psychological scars (Jafferany and Davari 2019). The plethora of scientific evidence that suggests the gut microbiome is involved in the host's overall health and physiology is consistent with the findings of numerous studies demonstrating the gut microbiome of people with acne is different and exhibits less microbial diversity than that of persons without acne. The phrases "acne," "*C. acnes*," "IGF-1," "sebum," and "gut microbiome" are used in this article (O'Neill and Gallo 2018).

1.13 Gut Microbiome and Skin Aging

The gut and skin are the human system's chief organs and play an important role in the defense mechanism (Dalmo et al. 1997). They share a structural and functional unique relationship with each other. The gut's inner surface and the skin's outer surface are protected by epithelial cells (Gallo and Hooper 2012). The EC cells prevent the entry of microorganisms into the host cells. The squamous epithelium contains keratin which makes skin resistant to toxic pathogens, and the mucosa's glycoprotein layer helps eliminate the bacteria (Coates et al. 2019).

Similarly, the gut system contains saliva, and lysozyme secretes compounds to prevent the entry of antigens. It forms a physical barrier between the host and microorganisms (Jena et al. 2023). Gut system is immensely rich with diverse species of microorganisms. Various studies reported a direct connection between the gut microbiome and skin health (Ngo et al. 2022). It is essential to maintain the GM composition balanced for skin homeostasis. Gut microbes are key in maintaining epithelial cell barrier texture and providing immunity (Madison 2003). Dysbiosis triggers the production of inflammatory pathways, which cause the formation of toxic inflammatory products. It will reach the skin through the gut-skin axis and manifest different skin disorders.

Aging leads to GM dysbiosis. Similarly, GM dysbiosis contributed to the early aging of skin (Kim and Ho 2010). Though both are bidirectionally connected to each other, the balanced association is very essential. Modern busy lifestyle, adequate nutritional diet, lack of physical exercise, unmanageable stress, non-hygienic environmental conditions, and genetics together contribute to the imbalance in microbial community composition in the gut (Anwar et al. 2019). It will alter changes, and as a result, significant addition or reduction of many microbial species is noted in the skin-associated disturbance. This will lead to unwanted immunological responses, toxic inflammatory products disrupting homeostasis, and, finally, skin aging.

1.14 Gut Microbiome in Skin and Hair Pathology

Evidence indicates that GM dysbiosis manifests pathological skin conditions. Even the slightest variation from the normal composition triggers adverse effects. In *Acne vulgaris*, patients showed decreased diversity in GM with reduced *Firmicutes* and increased levels of *Bacteroides (*Widhiati et al. 2022*)*. The inflammatory disease atopic dermatitis manifests due to increased production of *Faecalibacterium prausnitzii*, which produce harmful compounds that damage the epithelial barrier's integrity, reduce the production of butyrate and propionate, and exhibit anti-inflammatory property. Patients with severe chronic inflammatory disease psoriasis have more *Streptococcus* and *Staphylococcus* species than healthy individuals (Lewis et al. 2019). Experimental data provide insight into the link between GM dysbiosis and skin cancer. Due to the prognosis, cancer patients are exposed to various chemical and immune therapies, causing a disturbance in the gut microbiome. It is revealed that colorectal cancer (CRC) is associated with an increase in *Bacteroides fragilis* (Gribble and Reimann 2016). All this evidence and data confirm the direct association between gut microbiota and skin pathology.

Clinical analysis and experimental results documented Gm dysbiosis causes hair-related disturbance. Relevant studies from the past few years conducted research on the topic, and results were validated. The consumption of *Lactobacillus paracasei* strain in seborrheic dermatitis is found to be effective in lowering the severity of the disease (Thomas et al. 2016). The solid association between hair follicle disruption in alopecia and GM dysbiosis is verified in clinical cases. Alopecia manifestation triggers an alteration in the gut and produces interferons in the system. This leads to the abnormal development hair follicles and, ultimately, destruction.

1.15 Gut Skin Brain Axis

The human system's microbial ecology is crucial for properly regulating and sustaining homeostasis (Woźniak et al. 2021). Slight variations in composition may lead to inflammatory pathways, which cause the formation of toxic compounds damages the integrity of cells in any part of the body. The direct connection of gut microbiota with other distant sites, skin, and brain is essential in diagnosing and prognosis chronic diseases. Any disturbance in the cognitive center will disturb the normal composition of GM, and exposure to harmful components also alters the gut microbiota. In neurological and endocrinological diseases, gastrointestinal tract microbe's alteration directs toxic inflammatory products and carries them through gut-brain axis. They reach the central nervous system and accumulate. Ultimately it will cause neuronal injury and unrepairable neuronal degeneration. Similarly, GM dysbiosis in the gut system will impair the integrity of epithelial cells. So, the skin might unable to perform as a physical barrier to foreign bodies and fail to perform defense mechanisms. Therefore, it is an undeniable fact that to maintain a healthy well-being state, we have to keep the microbial community composition in the gut system balanced (Zoppi 2021).

1.16 Hormone Interaction and Gut-Brain Axis

The equilibrium of the gut-brain axis has been linked to several effects on both physical and mental health. Important signaling molecules within the gut-brain axis include the gut hormones generated by enteroendocrine cells dispersed throughout the gastrointestinal tract (Thomas et al. 2016). Gut-brain cross talk has considerably valued the connection between gut bacteria and gut hormones. The microbiota is crucial in regulating numerous gut-brain axis-related diseases, from gastrointestinal problems to psychiatric conditions. Similar to this, gut hormones are crucial signals in the gut-brain axis that play a variety of vital and pleiotropic roles in maintaining health. More critically, the release and activities of gut hormones can be impacted by gut bacteria. Since some bacteria may create or metabolize these hormones, the gut microbiota may directly impact the host's hormones. Neurohormone production by gut bacteria has been discovered; some bacteria, such as *Bacillus* and *Serratia*, can create dopamine, for instance. According to reports, numerous hormones in GF mice were significantly different from those in mice grown conventionally. Norepinephrine, 5-HT, and dopamine levels decreased, whereas GLP-1, corticosterone, and adrenocorticosterone levels were all raised. Following antibiotic therapy, the plasma leptin level dropped (Woźniak et al. 2021). Studies have uncovered crucial gut microbiota, host hormones, and gut-brain axis information. For 16 weeks, prebiotics were given to children who were overweight or obese. In the gut microbiome, In the gut microbiota, *Bifidobacterium* increased, whereas *Bacteroides vulgatus* significantly decreased. Despite making up only 1% of the intestinal epithelial cells, enteroendocrine cells (EECs) are found dispersed throughout the digestive system (Zoppi 2021). They serve crucial functions in regulating gut motility, hunger, and hormone release and can release a range of gut hormones in response to diet-related stimuli. EECs are typically divided into ten categories based on the main hormone they generate. These EECs are sensory cells that coordinate signal changes between the host responses, such as controlling food intake, secreting insulin, and adapting behavior and gut content (Kubrak et al. 2022). In order to develop new treatments for mental and gastrointestinal diseases like obesity, anxiety, and depression, future research must concentrate on the interaction between the microbiota, hormones, and gut-brain axis.

1.17 Ayurveda and Gut-Brain Axis: Role of Ayurveda in Gut-Brain Skin Axis

Ayurveda, which integrates individualized science and conventional practices to achieve general wellness, is a conventional medicinal practice still used today. Science has gained a deeper understanding of the fundamental principles of Ayurveda due to the role that microbiota plays in health and disease. We now know that a wide range of bacteria that live in different parts of our bodies, notably those in our gut microbiome, affect nearly every aspect of our physiology and health. The function of the microbiome in health and disease has led science to a clearer comprehension of the fundamental concepts of Ayurveda (Beri 2018). Ayurveda and other traditional health systems have long strongly emphasized diet and digestion. Numerous studies have connected the prakriti concept in Ayurveda to the human gut, mouth, and skin microbiota. The presence or high abundance of a small number of bacterial taxa in the human gut, mouth, and skin microbiomes was revealed by three prakriti categories, highlighting their special physiological significance. A plant-based diet is good for the gut microbiota, but there is less data on how certain botanicals affect the skin microbiome. Given the close relationship between the skin and gut microbiomes and the known effects of plant-based diets on gut health, we can infer that plant-based topicals in cosmetics may also serve as a healthy source of nutrients for the skin microbiome and possibly general health (Daniluk 2022). Ayurveda and Siddha medicine both use potent natural herbs to treat various diseases. These antiquated medical systems, which date back at least 5000 years and may have their origins in India, maintain that harmony can only exist when the Dosha, or individual constitution, is in tune with its surroundings. It is considered that a person's individual makeup, or "Dosha," and their surroundings are related. Considering the connections between this concept and the host and microbiome theories is intriguing. Despite its limits, new research on herbal medicine has shown how plant botanicals can promote wound healing, reduce inflammation, and increase keratinocyte proliferation (Budovsky et al. 2015).

1.18 Neutraceuticals and Probiotic

Neutraceuticals provide both nutrition and pharmaceutics. It can be used as a nutritional supplement and medicine for various disorders (Rajat et al. 2012). Neutraceuticals are prominent in regulating normal physiology, improving the gut system, delaying aging, preventing severe diseases, and maintaining homeostasis of the human body system (Nagpal et al. 2018). Neutraceuticals exhibit both dietary and therapeutic properties. Widely used neutraceutical consists of carotenoids; vitamin and mineral supplements; herbal products like ginger, turmeric, curcumin, omega-3-fatty acids; etc. It has a significant role in the modern era. Relevant studies indicate that neutraceuticals strongly impact various chronic diseases: allergy, Alzheimer's disease, cardiovascular diseases, cancer, diabetes, eye disorders, neurodegenerative diseases, obesity, inflammation, and many more (Nasri et al. 2014). Food in this category has health benefits. It reduces inflammation and lower the production of unwanted toxic inflammatory products, thereby maintaining homeostasis and promotes well-being of the human body. Changes in the gut flora may be how dietary supplementation with joint-protective nutraceuticals works.

On the other hand, probiotics are living microorganisms that have several beneficial effects on humans (Yaqoob et al. 2022). It normally resembles the microbial community living in our body. The main probiotic sources are yogurt, buttermilk, kombucha, kimchi, cheese, kefir, etc. Major probiotic bacteria are *Lactobacillus* and *Bifidobacterium*; sometimes *Saccharomyces boulardii* also may be used (Pais et al. 2020). Probiotics have gained recognition over the past 20 years for their role in boosting immunity and intestinal health and treating metabolic illnesses. The emerging idea of a gut-heart-brain axis has sparked several developments and approaches for adding probiotics to food and nutrition. Probiotics have a direct impact on brain neurotransmitter pathways as well as gut microbiota profiles, inflammation, and diseases. The idea that probiotics are good for the aging brain has garnered a lot of traction and emphasis in both study and product development, because brain health frequently declines with age. Probiotics work on the human body system in different aspects: to maintain the normal balanced composition of the gut microbiome, regulate the inflammatory pathway positively, prevent the accumulation of toxic compounds and diseases, etc. (Hou et al. 2022). The consumption of probiotics improves severe gastrointestinal conditions, oral disorders, allergic reactions, and cardiovascular, respiratory, and urinary tract infections. Besides, it boosts immunity and enhances the defense mechanism (Ashaolu 2020).

1.19 Conclusion

In general, studies on the host-microbiota dynamics brought on by aging can aid in discovering novel molecular and physiological mechanisms that assist the preservation of homeostasis. On the other hand, aging is linked to a decline in microbial diversity, but healthy aging is associated with diverse microbiome. Recent studies have connected the gut microbiome to a variety of diseases, including Parkinson's, autism, schizophrenia, and depression. It is still debatable whether the changes in the gut microbiota are essential to the pathogenesis of such diseases or are merely epiphenomenal. The microbiota can then be specifically targeted by potent medicines created using the insights supplied, because it is a novel fundamental player involved in regulating host aging processes. The precise processes underlying the connection between the microbiota and aging must be clarified by future research. Future research should explore how age-related changes in the gut microbiome might contribute to dysbiosis, inflammation, and ultimately, various age-related diseases. Investigating the key bacterial species, metabolites, and potential of prebiotics, probiotics, or FMT offers promising avenues for promoting gut health and potentially delaying the onset of age-related conditions. By unravelling the intricate relationship between the gut and aging, we can pave the way for novel therapeutic strategies to improve health and quality of life for older adults.

References

Anwar H, Irfan S, Hussain G, Faisal MN, Muzaffar H, Mustafa I et al (2019) Gut microbiome: a new organ system in body. Parasitol Microbiol Res 1:17–21

Ashaolu TJ (2020) Immune boosting functional foods and their mechanisms: a critical evaluation of probiotics and prebiotics. Biomed Pharmacother 130:110625

Bai J, Hu Y, Bruner DW (2019) Composition of gut microbiota and its association with body mass index and lifestyle factors in a cohort of 7–18 years old children from the American Gut Project. Pediatr Obes 14(4):e12480

Ballini A, Scacco S, Boccellino M, Santacroce L, Arrigoni R (2020) Microbiota and obesity: where are we now? Biology 9(12):415

Beri K (2018) Perspective: stabilizing the microbiome skin-gut-brain axis with natural plant botanical ingredients in cosmetics. Cosmetics 5(2):37

Boehme M, Guzzetta KE, Wasén C, Cox LM (2023) The gut microbiota is an emerging target for improving brain health during ageing. Gut Microbiome 4:e2

Boris V, Vanessa V (2023) Molecular systems biology approaches to investigate mechanisms of gut-brain communication in neurological diseases. Eur J Neurol 30(11):3622–3632

Bosman ES, Albert AY, Lui H, Dutz JP, Vallance BA (2019) Skin exposure to narrow band ultraviolet (UVB) light modulates the human intestinal microbiome. Front Microbiol 10:2410

Bourlioux P, Koletzko B, Guarner F, Braesco V (2003) The intestine and its microflora are partners for the protection of the host: report on the Danone Symposium "The Intelligent Intestine," held in Paris, June 14, 2002. Am J Clin Nutr 78(4):675–683

Budovsky A, Yarmolinsky L, Ben-Shabat S (2015) Effect of medicinal plants on wound healing. Wound Repair Regen 23(2):171–183

Candela M, Biagi E, Brigidi P, O'Toole PW, De Vos WM (2014) Maintenance of a healthy trajectory of the intestinal microbiome during aging: a dietary approach. Mech Ageing Dev 136:70–75

Carabottia M, Sciroccoa A, Masellib MA, Severia C (2015) The gut-brain axis: interactions between enteric microbiota, central and enteric nervous systems. Ann Gastroenterol 28(1):1–7
Chiu O (2022) The influence of sample processing and storage methods on the feline fecal metabolome and canine fecal microbiota. Doctoral dissertation, University of Guelph
Coates M, Lee MJ, Norton D, MacLeod AS (2019) The skin and intestinal microbiota and their specific innate immune systems. Front Immunol 10:2950
Collins SM, Bercik P (2009) The relationship between intestinal microbiota and the central nervous system in normal gastrointestinal function and disease. Gastroenterology 136(6):2003–2014
Dahiya D, Nigam PS (2023) Antibiotic-therapy-induced gut dysbiosis affecting gut microbiota—brain Axis and cognition: restoration by intake of probiotics and synbiotics. Int J Mol Sci 24(4):3074
Dalmo RA, Ingebrigtsen K, Bøgwald J (1997) Non-specific defence mechanisms in fish, with particular reference to the reticuloendothelial system (RES). J Fish Dis 20(4):241–273
Daniluk J (2022) The hot detox plan: cleanse your body and heal your gut with warming, anti-inflammatory foods. Hay House, Inc.
Das B, Nair GB (2019) Homeostasis and dysbiosis of the gut microbiome in health and disease. J Biosci 44:1–8
Dos Santos JCC, Lima MPP, de Castro Brito GA, de Barros Viana GS (2022) Role of enteric glia and microbiota-gut-brain axis in parkinson disease pathogenesis. Ageing Res Rev 84:101812
Drago L, Toscano M, Rodighiero V, De Vecchi E, Mogna G (2012) Cultivable and pyrosequenced fecal microflora in centenarians and young subjects. J Clin Gastroenterol 46:S81–S84
Gallo RL, Hooper LV (2012) Epithelial antimicrobial defence of the skin and intestine. Nat Rev Immunol 12(7):503–516
Garehdaghi F, Sarbaz Y (2023) Analyzing global features of magnetic resonance images in wide-spread neurodegenerative diseases: new hope to understand brain mechanism and robust neurodegenerative disease diagnosis. Med Biol Eng Comput 61:1–12
Gribble FM, Reimann F (2016) Enteroendocrine cells: chemosensors in the intestinal epithelium. Annu Rev Physiol 78:277–299
Guang L, Koomson DA, Jingyu H, Ewusi-Mensah D, Miwornunyuie N (2020) Performance of exoelectrogenic bacteria used in microbial desalination cell technology. Int J Environ Res Public Health 17(3):1121
Hajishengallis G, Lamont RJ (2016) Dancing with the stars: how choreographed bacterial interactions dictate nososymbiocity and give rise to keystone pathogens, accessory pathogens, and pathobionts. Trends Microbiol 24(6):477–489
Hassanzadeh M, Banitaba SM, Azizi S (2023) Validation of Persian Multiple Sclerosis quality of life-29 (P-MSQOL-29) questionnaire. Acta Neurol Belg 123:1–11
Henley JB, Ing RMM, Hana Zelenkova MD (2014) Microbiome of affected and unaffected skin of patients with atopic dermatitis before and after emollient treatment. J Drugs Dermatol 13(11):1365–1372
Hou K, Wu ZX, Chen XY, Wang JQ, Zhang D, Xiao C et al (2022) Microbiota in health and diseases. Signal Transduct Target Ther 7(1):135
İnceoğlu Ö, Al-Soud WA, Salles JF, Semenov AV, van Elsas JD (2011) Comparative analysis of bacterial communities in a potato field as determined by pyrosequencing. PLoS One 6(8):e23321
Ipci K, Altıntoprak N, Muluk NB, Senturk M, Cingi C (2017) The possible mechanisms of the human microbiome in allergic diseases. Eur Arch Otorhinolaryngol 274:617–626
Jafferany M, Davari ME (2019) Itch and psyche: psychiatric aspects of pruritus. Int J Dermatol 58(1):3–23
Jena AB, Samal RR, Bhol NK, Duttaroy AK (2023) Cellular Red-Ox system in health and disease: the latest update. Biomed Pharmacother 162:114606
Kalia LV, Lang AE (2015) Parkinson's disease. Lancet 386(9996):896–912
Kim YS, Ho SB (2010) Intestinal goblet cells and mucins in health and disease: recent insights and progress. Curr Gastroenterol Rep 12:319–330

Kubrak O, Koyama T, Ahrentløv N, Jensen L, Malita A, Naseem MT et al (2022) The gut hormone Allatostatin C/Somatostatin regulates food intake and metabolic homeostasis under nutrient stress. Nat Commun 13(1):692

Lee JY, Tsolis RM, Bäumler AJ (2022) The microbiome and gut homeostasis. Science 377(6601):eabp9960

Lewis DJ, Chan WH, Hinojosa T, Hsu S, Feldman SR (2019) Mechanisms of microbial pathogenesis and the role of the skin microbiome in psoriasis: a review. Clin Dermatol 37(2):160–166

Li RD, Zheng WX, Zhang QR, Song Y, Liao YT, Shi FC et al (2023) Longevity-associated core gut microbiota mining and effect of mediated probiotic combinations on aging mice: case study of a long-lived population in Guangxi, China. Nutrients 15(7):1609

Madison KC (2003) Barrier function of the skin:"la raison d'etre" of the epidermis. J Investig Dermatol 121(2):231–241

Matsumoto K, Sugimoto F, Mizuno T, Hayashi T, Okamura R, Nishioka T et al (2023) Immunohistochemical characterization of transient receptor potential vanilloid types 2 and 1 in a trinitrobenzene sulfonic acid-induced rat colitis model with visceral hypersensitivity. Cell Tissue Res 391(2):287–303

Morais LH, Schreiber HL IV, Mazmanian SK (2021) The gut microbiota–brain axis in behaviour and brain disorders. Nat Rev Microbiol 19(4):241–255

Murman DL (2015) The impact of age on cognition. Semin Hear 36(3):111–121. Thieme Medical Publishers

Myers B, Brownstone N, Reddy V, Chan S, Thibodeaux Q, Truong A et al (2019) The gut microbiome in psoriasis and psoriatic arthritis. Best Pract Res Clin Rheumatol 33(6):101494

Nabizadeh E, Jazani NH, Bagheri M, Shahabi S (2017) Association of altered gut microbiota composition with chronic urticaria. Ann Allergy Asthma Immunol 119(1):48–53

Nagpal R, Mainali R, Ahmadi S, Wang S, Singh R, Kavanagh K et al (2018) Gut microbiome and aging: physiological and mechanistic insights. Nutr Healthy Aging 4(4):267–285

Nakamura K, Nakamura Y, Kataoka N (2022) A hypothalamomedullary network for physiological responses to environmental stresses. Nat Rev Neurosci 23(1):35–52

Naseribafrouei A, Hestad K, Avershina E, Sekelja M, Linløkken A, Wilson R, Rudi K (2014) Correlation between the human fecal microbiota and depression. Neurogastroenterol Motil 26(8):1155–1162

Nasri H, Baradaran A, Shirzad H, Rafieian-Kopaei M (2014) New concepts in nutraceuticals as alternative for pharmaceuticals. Int J Prev Med 5(12):1487

Nettis E, Colanardi MC, Paradiso MT, Ferrannini A (2004) Desloratadine in combination with montelukast in the treatment of chronic urticaria: a randomized, double-blind, placebo-controlled study. Clin Exp Allergy 34(9):1401–1407

Ngo PA, Neurath MF, López-Posadas R (2022) Impact of epithelial cell shedding on intestinal homeostasis. Int J Mol Sci 23(8):4160

Nikolova VL, Hall MR, Hall LJ, Cleare AJ, Stone JM, Young AH (2021) Perturbations in gut microbiota composition in psychiatric disorders: a review and meta-analysis. JAMA Psychiatry 78(12):1343–1354

O'Neill AM, Gallo RL (2018) Host-microbiome interactions and recent progress into understanding the biology of acne vulgaris. Microbiome 6:1–16

Olejniczak-Staruch I, Ciążyńska M, Sobolewska-Sztychny D, Narbutt J, Skibińska M, Lesiak A (2021) Alterations of the skin and gut microbiome in psoriasis and psoriatic arthritis. Int J Mol Sci 22(8):3998

Pais P, Almeida V, Yılmaz M, Teixeira MC (2020) Saccharomyces boulardii: what makes it tick as successful probiotic? J Fungi 6(2):78

Park KS (2023) Humans and electricity: understanding body electricity and applications. Springer Nature

Pearce SC, Coia HG, Karl JP, Pantoja-Feliciano IG, Zachos NC, Racicot K (2018) Intestinal in vitro and ex vivo models to study host-microbiome interactions and acute stressors. Front Physiol 9:1584

Petersen EB, Skov L, Thyssen JP, Jensen P (2019) Role of the gut microbiota in atopic dermatitis: a systematic review. Acta Derm Venereol 99(1):5–11

Polkowska-Pruszyńska B, Gerkowicz A, Krasowska D (2020) The gut microbiome alterations in allergic and inflammatory skin diseases—an update. J Eur Acad Dermatol Venereol 34(3):455–464

Quaglio AEV, Grillo TG, De Oliveira ECS, Di Stasi LC, Sassaki LY (2022) Gut microbiota, inflammatory bowel disease and colorectal cancer. World J Gastroenterol 28(30):4053

Rajat S, Manisha S, Robin S, Sunil K (2012) Nutraceuticals: a review. Int Res J Pharm 3(4):95–99

Riaz Rajoka MS, Thirumdas R, Mehwish HM, Umair M, Khurshid M, Hayat HF et al (2021) Role of food antioxidants in modulating gut microbial communities: novel understandings in intestinal oxidative stress damage and their impact on host health. Antioxidants 10(10):1563

Rutsch A, Kantsjö JB, Ronchi F (2020) The gut-brain axis: how microbiota and host inflammasome influence brain physiology and pathology. Front Immunol 11:604179

Sadek MA, Kandil EA, El Sayed NS, Sayed HM, Rabie MA (2023) Semaglutide, a novel glucagon-like peptide-1 agonist, amends experimental autoimmune encephalomyelitis-induced multiple sclerosis in mice: involvement of the PI3K/Akt/GSK-3β pathway. Int Immunopharmacol 115:109647

Scher JU, Ubeda C, Artacho A, Attur M, Isaac S, Reddy SM et al (2015) Decreased bacterial diversity characterizes the altered gut microbiota in patients with psoriatic arthritis, resembling dysbiosis in inflammatory bowel disease. Arthritis Rheumatol 67(1):128–139

Scheurink TAW, Borkent J, Gangadin SS, El Aidy S, Mandl R, Sommer IE (2023) Association between gut permeability, brain volume, and cognition in healthy participants and patients with schizophrenia spectrum disorder. Brain Behav 13(6):e3011

Scott KP, Gratz SW, Sheridan PO, Flint HJ, Duncan SH (2013) The influence of diet on the gut microbiota. Pharmacol Res 69(1):52–60

Siddiqui R, Makhlouf Z, Khan NA (2022) The increasing importance of the gut microbiome in acne vulgaris. Folia Microbiol 67:1–11

Sun L, Zhang Y, Cai J, Rimal B, Rocha ER, Coleman JP et al (2023) Bile salt hydrolase in non-enterotoxigenic Bacteroides potentiates colorectal cancer. Nat Commun 14(1):755

Tarride MI (2023) Towards an organizational model based on the human being seen as an allopoietic gray box information processor. Kybernetes Vol. ahead-of-print No. ahead-of-print.

Thomas AM, Jesus EC, Lopes A, Aguiar S Jr, Begnami MD, Rocha RM et al (2016) Tissue-associated bacterial alterations in rectal carcinoma patients revealed by 16S rRNA community profiling. Front Cell Infect Microbiol 6:179

Tlaskalová-Hogenová H, Štěpánková R, Hudcovic T, Tučková L, Cukrowska B, Lodinová-Žádníková R et al (2004) Commensal bacteria (normal microflora), mucosal immunity and chronic inflammatory and autoimmune diseases. Immunol Lett 93(2–3):97–108

Toricelli M, Pereira AAR, Abrao GS, Malerba HN, Maia J, Buck HS, Viel TA (2021) Mechanisms of neuroplasticity and brain degeneration: strategies for protection during the aging process. Neural Regen Res 16(1):58

Tsai WH, Fang YT, Huang TY, Chiang YJ, Lin CG, Chang WW (2023) Heat-killed Lacticaseibacillus paracasei GMNL-653 ameliorates human scalp health by regulating scalp microbiome. BMC Microbiol 23(1):121

Turroni F, Milani C, Duranti S, Lugli GA, Bernasconi S, Margolles A et al (2020) The infant gut microbiome as a microbial organ influencing host well-being. Ital J Pediatr 46(1):1–13

van Olst L, Roks SJ, Kamermans A, Verhaar BJ, van der Geest AM, Muller M et al (2021) Contribution of gut microbiota to immunological changes in Alzheimer's disease. Front Immunol 12:683068

Veenhoven R (2005) Apparent quality-of-life in nations: how long and happy people live. Soc Indic Res 71:61–86

Vizcarra JA, Wilson-Perez HE, Espay AJ (2015) The power in numbers: gut microbiota in Parkinson's disease. Mov Disord 30(3):296–298

Wasén C, Simonsen E, Ekwudo MN, Profant MR, Cox LM (2022) The emerging role of the microbiome in Alzheimer's disease. Int Rev Neurobiol 167:101–139. Academic

Widhiati S, Purnomosari D, Wibawa T, Soebono H (2022) The role of gut microbiome in inflammatory skin disorders: a systematic review. Dermatol Reports 14(1):9188

Woźniak D, Cichy W, Przysławski J, Drzymała-Czyż S (2021) The role of microbiota and enteroendocrine cells in maintaining homeostasis in the human digestive tract. Adv Med Sci 66(2):284–292

Yaqoob MU, Wang G, Wang M (2022) An updated review on probiotics as an alternative of antibiotics in poultry—a review. Anim Biosci 35(8):1109

Yousefi B, Kokhaei P, Mehranfar F, Bahar A, Abdolshahi A, Emadi A, Eslami M (2022) The role of the host microbiome in autism and neurodegenerative disorders and effect of epigenetic procedures in the brain functions. Neurosci Biobehav Rev 132:998–1009

Zeng J, Yang K, Nie H, Yuan L, Wang S, Zeng L et al (2023) The mechanism of intestinal microbiota regulating immunity and inflammation in ischemic stroke and the role of natural botanical active ingredients in regulating intestinal microbiota: a review. Biomed Pharmacother 157:114026

Zoppi J (2021) Impact of psychological stress on the remodeling of the host-microbiota interactome. Doctoral dissertation, Nantes

The Scenario of Emerging Gut Microbiome Cross Talks in Brain Aging

2

Swetha Panneerselvam, Selvaraj Jayaraman, and Ponnulakshmi Rajagopal

Abstract

Brain aging is natural aging process that occurs over the lifetime of every individual which depends on various factors including genetic, environmental factors, lifestyle changes, and age which are major risk factors for developing brain aging. Brain aging has nine characteristic hallmarks which are majorly categorized into three types, namely, primary, antagonistic, and integrative hallmarks. Some of the neurological conditions that are associated with brain aging are neurological disorders like Alzheimer's disease, Parkinson's disease, amyotrophic lateral sclerosis, traumatic brain injury, stroke, epilepsy, and dementia. Among which neurodegenerative diseases may have a greater impact among the individuals with a variety of risk factors. Microbiome is composed of variety of microorganisms that are integral to the host which maintains a healthy relationship between the host gut. There are innumerable gut microbiomes which are present in the human gut. Those microbes can play a crucial role in the host metabolic, physiologic, pathologic, and psychologic conditions by maintaining the homeostatic environment. Certain pathological, psychological, and physiological alterations can defect the homeostasis maintained by the gut microbiota which results in the development and progression of diseases especially age-

S. Panneerselvam · S. Jayaraman
Department of Biochemistry, Centre of Molecular Medicine and Diagnostics (COMManD), Saveetha Dental College and Hospitals, Saveetha Institute of Medical and Technical Sciences, Saveetha University, Chennai, India

P. Rajagopal (✉)
Department of Central Research Laboratory, Meenakshi Ammal Dental College and Hospital, Meenakshi Academy of Higher Education and Research (Deemed to be University), Chennai, India
e-mail: drponnulakshmi.researchscientist@madch.edu.in

S. Pathak, A. Banerjee (eds.), *Gut Microbiome and Brain Ageing*,
https://doi.org/10.1007/978-981-99-8803-7_2

related neuronal diseases. This cross talk between the causative factors and the microbiota can be modulated by various therapeutic approaches which show significant results which are achieved by mediating the signaling of gut to the brain or vice versa.

Keywords

Brain aging · Hallmarks · Neurodegenerative disorders · Gut microbiome · GBA

2.1 Introduction

A myriad of microorganisms include bacteria, fungi, parasites, and viruses that reside in our body. The gastrointestinal tract harbors approximately 95% of the microbes which are collectively referred as gut microbiota. A variety of microorganisms exist within a complex ecosystem in the human gut (so called "microbiota superorganism"). Every individual has a unique gut microbiota signature which is composed of 100–1000 microbial population. In recent decades, both clinical and scientific research prove that the intestinal microbiota acts as super organism which is essential for maintaining regular homeostasis. Microbial richness or bacterial diversity is generally considered as a hallmark of a healthy status. Reduced bacterial diversity has been correlated with obesity and immune-related and inflammatory diseases. Additionally, although a balanced microbiome composition is necessary for many physiological processes, qualitative modifications, particularly at the level of the core microbiome, might result in the onset of disease. Most research on these phenomena has focused on inflammatory bowel disorders (IBD), such as Crohn's disease and ulcerative colitis. The role of normal gut microbiota includes host nutrient metabolism, maintaining gut mucosal barrier, immunomodulation, and prevention against various antigens. The factors that play a crucial role in the shaping of gut microbiota are their mode of delivery, diet, and environmental factors involving commensal community (D'Argenio and Salvatore 2015). The host human and its gut microbes have an intense relationship. The enteric microbiota gets involved in host gut physiology and health by influencing gut-brain communication. Modification or instability of the gut microbiota and changes in its biodiversity may lead to metabolic and gastrointestinal disorders. The gut-brain communication is bidirectional. Signals from the brain can influence sensory, motor, and secretory function of gastrointestinal tract which is achieved either directly by the influence of release of neuronally signaling molecules or indirectly by effecting changes in their mobility and permeability. Brain functioning can also be influenced by GI tract, and this is achieved by multiple mechanisms involving direct stimulation of receptor-mediated signalling (Aziz et al. 2012). Microbiome and host immune response interaction is diverse, bidirectional, and complex (Zheng et al. 2020). Gut microbiome is integral to the host (human) digestion and nutrition and utilizes nutrients from the substrate which are indigestible by the host. Gut microbiome is majorly involved in the development and functioning of the immune system which is demonstrated by numerous research findings. Gut microbiome has an impact over the functioning of CNS by various methods including metabolites, hormones, and neuroactive molecules. The degradation of dietary fiber by gut microbiota produces large amount of short-chain fatty acids (SCFA)—acetate, butyrate, and propionate in the gut. These short-chain fatty acids reduce the level of inflammatory cytokines, enhance gut motility, and modulate immune tolerance, gut hormones levels, and neuropeptides. These molecules also modulate gastrointestinal cells in various conditions. The development and differentiation of variety of immune cells (dendritic cells, Tregs) are affected by SCFAs. These molecules can also affect the maturation and functioning of microglial cells. Another microbial metabolite involving gut-brain axis is the secondary bile acids. Tryptophan is essential amino acid obtained from the diet source which is metabolized into indoles by gut microbiota or by kynurenine pathway by the host. Tryptophan metabolites can alter the intestinal immune cell functioning via aryl hydrocarbon receptor. Other gut-microbiota-derived molecules have an influence on host immunity and neurological problems. One of the most notable instances is endotoxin lipopolysaccharide (LPS), a marker for chronic inflammatory disorders. Translocation of this polysaccharide is mediated by gut permeability, and it also induces strong inflammatory response that results in the disruption of blood-brain barrier leading to microglia activation (Benakis et al. 2020).

The gut-brain axis is a complex network which interconnects the brain and gut microbiome. Microglia, tissue-resident immune cells in the brain which are responsible for modulation of neurogenesis, can influence synaptic remodeling and also involved in the regulation of neuronal inflammation. Microglial cells maturation and functioning can be regulated by the gut microbiome which is achieved by cross-talking between gut microbiome and the brain. Microglial cells play a crucial role in customizing neural circuitry during postnatal developmental stages which imply cognitive and social behavior (Abdel-Haq et al. 2018). Thus, in the current book chapter, the role of microbiome cross-talking involved in brain aging process will be emphasized.

2.2 Brain Aging

Aging is an inevitable part of an organism's lifetime that is linked to physical degeneration and an elevated risk of disease. The body's inability to overcome the physical damage due to aging can lead to subsequent loss of physiological functions including sensory, motor, and cognitive function. Age is also considered as one of the significant contributors to a number of clinical conditions including cancer, cardiovascular diseases, and neurodegenerative diseases associated with genetic and environmental factors. Neurodegeneration is the most common age-related disorders that raises the incidence and suggested the link between neurodegenerative diseases and brain aging in association with age-related changes including genomic instability, loss of protein homeostasis, and epigenetic modifications. Even though, aging is a crucial causative factor for many neurodegenerative diseases, the exact mechanism of action of aging involving neurodegeneration remains to be elucidated. Some research findings suggested that brain aging in association with the raised risk for the development of neurodegenerative diseases can contribute to cognitive impairment at a later stage. Environmental exposure to toxins can also attribute to cognitive impairment in the later stage of the life. It's interesting to consider that healthy aging, or aging without the onset of disease, may not be linked to longevity genes but rather to the lack of AD risk factors. The cohort study of the genetic sequences on individuals above the age group 80 without the progression of chronic condition suggested that most of the single-nucleotide polymorphisms are associated with only cognition. From this study, they concluded that the brain health and cognition are interlinked and the development of neurodegenerative disorders among individuals is solely determined by genetic factors. Postmitotic cells, such as neurons and oligodendrocytes, comprise the majority of brain tissue. These cells are particularly vulnerable to age-related alterations, especially epigenetic modifications (DNA damage, methylation). DNA methylation process can be altered by aging which results in DNA damage that might pave way for the development of neurodegeneration. Neurodegeneration is also stimulated by mitochondrial damage and dysfunction in association with brain aging via the production of reactive oxygen species (ROS) and neuroinflammation (Azam et al. 2021).

2.3 Hallmarks of Brain Aging

Hallmarks of brain aging include decline in cognitive performance, decreased brain volume, increased inflammation, accumulation of deposits, and changes in the brain chemistry. Over the period of time, there is a decline in brain cognition which is due to the changes in the structure and functioning of brain. Overall decrease in the brain volume can lead to cognitive impairment by mediating the reduction of neuron levels and synapses in the brain. Changes in the brain chemistry mainly involve the decline in the secretion of neurotransmitters which induces cognitive decline as well as mood disorders. Abnormal deposition of proteins in the brain could interfere with the normal brain functioning, and it also provides way for the development and progression of neurodegenerative disorders. Chronic inflammation can stimulate the progression of various neurodegenerative diseases, and it reduces the overall cognitive performance. Inflammation can also lead to protein deposition in the brain. The three basic categories of biological evidence of aging are primary, antagonistic, and integrative hallmarks. Genomic instability, epigenetic alterations, loss of protein homeostasis, and telomere attrition are the primary hallmarks of aging. Mitochondrial malfunction, improper nutrition sensing, and cellular aging are examples of antagonistic hallmarks. Integrative hallmarks, which include disrupted intercellular communication and stem cell depletion and are linked to neurological conditions including Alzheimer's and Parkinson's disease, are the increased damage of primary and antagonistic hallmarks (Hou et al. 2019) (Fig. 2.1).

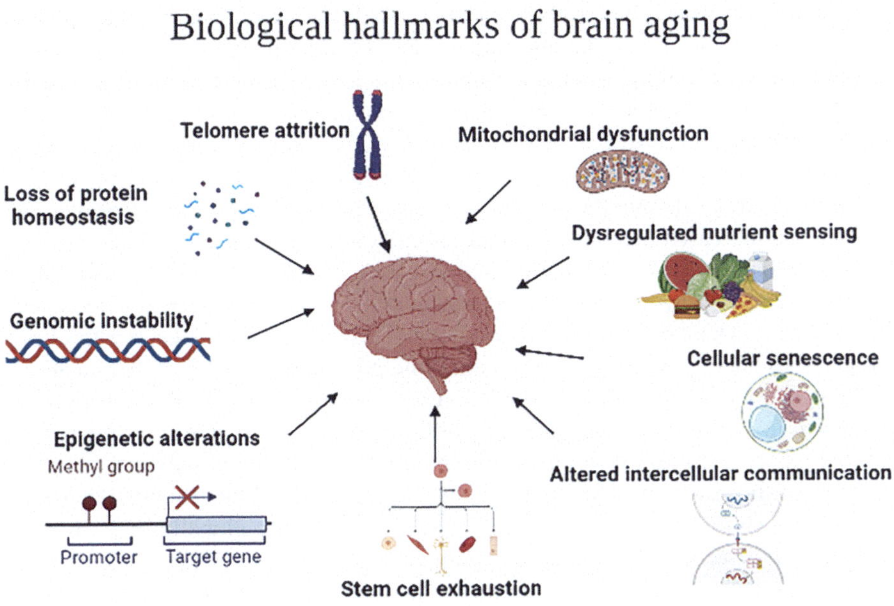

Fig. 2.1 Demonstrates the nine biological hallmarks of brain aging

2.4 Senescence

Senescence refers to the biological process of aging and the associated deterioration of cells, tissues, and organisms over time. It is a natural process that affects all living things and is characterized by a decline in physiological function and an increased susceptibility to disease and death. In cellular biology, senescence refers to a permanent and irreversible arrest of cell division that occurs in response to various stresses, including DNA damage, oxidative stress, and telomere shortening (McHugh and Gil 2018). This process is believed to play a role in the development of age-related diseases, such as cancer and neurodegenerative disorders. Overall, senescence is a complex biological process that involves multiple molecular pathways and is influenced by a variety of environmental and genetic factors. While it is a natural part of the aging process, researchers are studying ways to slow down or even reverse senescence in order to extend healthy life span and reduce the incidence of age-related diseases (van Deursen 2014).

2.5 Epidemiology of Neurological Diseases

Population aging continues to be a global phenomenon. According to reports from the UN Ageing Programme and the US Centres for Disease Control and Prevention, the global population of older people (65+ years) is expected to grow from 420 million in 2000 to nearly 1 billion by 2030, with the proportion of older people increasing from 7 to 12%. The absolute number of older people will rise most dramatically in developing countries. Consequently, the proportion of the world's aging population that comes from developing countries will rise from 59 to 71%. The overall prevalence of epilepsy is 7.6 in 1000 which may increase over age in association with social deprivation. Epilepsy has a bimodal distribution in correlation with age that gets into the peak in older-aged peoples as well as in younger age groups. The increased incidence of seizures and epilepsy in older-aged peoples can contribute to raise the risk of age-related diseases. An analysis of a national representative sample of Americans above the age of 70 revealed that the prevalence of AD was 9.7%. Dementia prevalence among individuals aged 60 and above has been estimated to be 3.9% globally, with geographic prevalence of 1.6% in Africa, 4.0% in China and the Western Pacific, 4.6% in Latin America, 5.4% in Western Europe, and 6.4% in North America. Over 25 million people worldwide are affected by approximately 5 million new cases of dementia that are diagnosed each year, the majority of those who had Alzheimer's disease. The second most significant cause of mortality (9 million) and the primary cause of disability globally in 2016 were neurological disorders (Beghi and Giussani 2018).

The prevalence of Alzheimer's disease increased steadily with age, from 0.6% in the 65–69 age group to 22.2% in the 90-and-older age group, according to European collaborative research of population-based cohorts. In both men and women, the prevalence of Alzheimer's disease rises rapidly with age, roughly doubling every 5 years between the ages of 50 and 80, after which the increase can be reduced in the oldest age groups. The WHO projects that by 2050, the estimate of 44 million people living with dementia worldwide would have increased to 135 million due to this extremely significant age-related incidence and prevalence (Dumurgier and Tzourio 2020).

2.6 Pathophysiology of Brain Aging

Brain aging is a complex process that involves multiple changes at the cellular, molecular, and structural levels. Overall, the pathophysiology of brain aging is a complex process that involves multiple factors. While there is no single cause of brain aging, it is likely that a combination of genetic, environmental, and lifestyle factors all play a role. Some of the pathophysiology of brain aging involves oxidative stress, inflammation, accumulation of toxic proteins, mitochondrial dysfunction, decreased blood flow, reduced neoplasticity, and genetic factors. Excessive production of ROS can lead to the accumulation of damaged molecules that interferes with the functioning of the brain (Kandlur et al. 2020; Floyd and Hensley 2002). Chronic inflammation is a hallmark of brain aging. Inflammation can be caused by a variety of factors, including infections, chronic diseases, and environmental toxins. Inflammatory cells and cytokines can cause damage to brain cells and promote the production of toxic proteins (Joseph et al. 2005; Yin et al. 2016; Garaschuk et al. 2018). The accumulation of toxic proteins, such as beta-amyloid and tau, is a characteristic feature of many neurodegenerative diseases, including Alzheimer's disease. These proteins can form aggregates that damage neurons and disrupt normal brain function (Bayer and Wirths 2010; Pickett et al. 2019). Mitochondria are the powerhouses of cells, and they play a critical role in energy production. Mitochondrial function declines during older age, leading to a decrease in energy production and an increase in oxidative stress (Müller et al. 2010). When we become older, the blood flow to the brain diminishes which results in the reduced supply of oxygen and other essential nutrients to the brain cells. This can impair the functioning of brain and also raises the incidence of cognitive decline (Aanerud et al. 2012). Neuroplasticity refers to the brain's ability to adapt to new situations and experiences. Diminished neuroplasticity can impair learning and memory (Toricelli et al. 2021). There are several genes that have been implicated in brain aging, including the APOE gene, which is associated with an increased risk of Alzheimer's disease (Mattson et al. 2002).

The pathophysiology of brain aging is explained by a few theories. Programmed and damage or error theories are the two primary subcategories of contemporary biology explanations for human aging (Table 2.1).

Table 2.1 Summarizes the list of theories explaining about brain aging

S. no	Theory	Definition	References
1	Programmed longevity	Genes that are responsible for controlling of aging and longevity and their activation and deactivation occur during lifetime in association with various factors including diet, exercise, and environmental factors	Davidovic et al. (2010)
2	Immunological theory	The immune system is designed to deteriorate over time, which increases sensitivity to infectious diseases, accelerates aging, and ultimately causes death	Cornelius (1972)
3	Endocrine theory	Endocrine system involves the synthesis of several hormones which act as a key regulator in diverse metabolic and physiological processes. And these hormones can also impart brain aging through various means (e.g., IGF-1 signaling controls the brain aging process)	van Heemst (2010)
4	Cross-linking theory	Cross-linked protein accumulation negatively impacts cells and tissues and slows down bodily functions, which causes aging	Bjorksten (1977)
5	Free radicals' theory	Harman stated that free radicals can cause the damage of cells leading to the accumulation of toxic substances resulting in the abnormal functioning. The primary process underlying the onset of cellular senescence and organismal aging is ROS	Afanas'ev (2010)
6	Somatic DNA damage theory	It is the cumulative damage of DNAs in the cells which is mediated by variety of factors that further lead to dysfunction. The major consequence of this damage is the accumulation of mutations in the DNA that results in improper functioning	Maynard Smith (1962)

2.7 Gut Microbiome and Its Functional Aspects

The exposure of microbiome-derived molecules to the gastrointestinal tract is continuous. Through this, homeostatic state is maintained by immune and epithelial cells in the gut via protecting host integrity and enhancing tolerogenic responses to microbes. In healthy conditions, there is a balance between the functioning of regulatory T cell (T_{regs}) and effector T cell (T_{eff}), whereas in diseased condition, the balance between modulatory and protective function of T cells is dysregulated. The molecular underlying principles of host-microbiome interaction are small compounds derived from the gut microbiome. A negligible amount of these chemicals, including bacterial metabolites, has been found to affect the host's neurophysiology through a variety of mechanisms, including blood flow, humoral pathways, the immune system, and neuronal pathways. The host gut microbiome ecology is also maintained by initiating intestinal defense mechanism. And this mechanism helps in the prevention of translocation of commensal bacteria into the tissues. The first line of this defense mechanism is mainly depending on the secretion of antimicrobial peptides (AMPs), and these molecules are otherwise referred as defensins which are secreted mainly by intestinal epithelial cells. The genes which encode for defensin molecule majorly depend on intestinal tract in response to gut microbiota, and during pathogenic condition, these molecules get secreted. The second line of defense mechanism involves the secretion of IgA molecules that can prevent the bacterial translocation by mediating the crossing over of epithelial barrier. The immunoglobulin molecule is derived from breast milk and transferred to the foetus at birth to serve as a defence mechanism until the intestinal defence system develops (Burcelin 2016). Numerous research findings provided evidence for the correlation between gut microbiome composition and the development of mental disorders like anxiety, depression, and so on. Morris and his coworkers revealed that the decreased synthesis of SCFA on the GAD patients leads to intestinal barrier dysfunction leading to brain dysfunction. Preclinical studies demonstrate that dysbiosis (microbiota dysregulation) affects anxiety and stress behaviors, indicating that the GBA may contribute to the risk of disease, including anxiety and mood disorders. The GBA axis refers to the bidirectional communication between the brain and gut bacteria. Through direct and indirect pathways, such as the limbic system; efferent, afferent, and sympathetic afferent systems; and immune (chemokines, cytokines), endocrine (hypothalamic-pituitary-adrenal [HPA] axis), and metabolic pathways, the bidirectional interactions between the gut microbiota and crucial parts of the central nervous system (CNS) and immune systems are maintained. The abovementioned correlation can conclude the relationship between gut microbiota and behavior through diverse systems. Some other factors that may have an impact on GBA are intestinal epithelial barrier, integrity of blood-brain barrier, and neurotransmitters synthesized by the gut microbiota.

2.8 Role of Gut Microbiome in CNS-Related Problems

Age-related cognitive decline is integrally dependent on the gut microbiome. Multiple pathways are affected by the dysbiotic state, a hallmark of the aging microbiome on cognition. It may trigger when proinflammatory bacteria are more prevalent than populations of immunoregulatory microbes; the release of proinflammatory cytokines and bacterial toxins interfere with the transmission of the regulatory neural signal via the vagus nerve and suppress the production and release of microbial metabolites and hormones. Globally, AD is the most typical factor in cognitive deterioration. It is characterized by the accumulation of Aβ, and tau proteins result in neuroinflammation, neuronal loss, and synaptic dysfunction. Pathogenesis of AD is linked with GBA dysfunction and the rise in intestinal inflammation. Parkinson's disease is a progressive disorder that leads to muscle stiffness, tremors, and other cognitive performances. Other risk factors for PD other than age include environmental toxins, obsessive uptake of caffeine, and smoking. Major pathophysiology of PD is frontal cortex atrophy and ventricular enlargement (Spires-Jones et al. 2017; Priyadarshi et al. 2001). Various key mechanisms that are involved in the progression of PD are misfolding and aggregation of α-synuclein, neuroinflammation, mitochondrial dysfunction, abnormal protein clearance system, impaired autophagy-lysosome system, and impaired ubiquitin-proteasome system (Moore et al. 2005).

2.9 Role of Gut Microbiota in Cognitive Decline and Brain Health

There is increasing attention on the potential impact of the intestinal microbiome on the susceptibility and weaknesses linked to the aging process. Numerous studies have demonstrated that the microbial makeup of the gut undergoes changes as individuals age, challenging the notion of a linear progression. Significantly, these alterations manifest later than initially anticipated, and centenarians, indicative of longevity, seem to exhibit a distinct microbiome (Biagi et al. 2010; Claesson et al. 2011, 2012). The predominant phyla in younger age groups, namely *Bacteroides* and *Firmicutes*, continue to be prominent as individuals age; however there is ongoing debate regarding potential changes in the ratio of these phyla. A shift is observed towards an increased prevalence of potentially harmful bacteria (pathobionts) at the cost of beneficial bacteria (symbionts). This shift is evident in the greater relative abundance of proteobacteria and a decrease in bifidobacteria species. A decline in the production of short-chain fatty acids is observed with aging, and the fatty acid butyrate holds particular significance due to its crucial role in preserving the integrity of the colonic epithelium and managing inflammation. The alterations in microbial composition and metabolism align with the concept of inflammaging, suggesting that chronic low-grade inflammation serves as a shared foundation for a wide range of age-related pathologies, including cognitive decline (Franceschi et al. 2000). Strong correlations were identified between microbiome profiles and indicators of frailty and compromised health in elderly individuals residing in long-term care facilities.

Inflammation is currently recognized as a leading candidate for contributing to cognitive decline, not only within the realm of typical aging but also in neurological conditions and sporadic Alzheimer's disease (Griffin 2013). A recent investigation led by vom Berg and colleagues demonstrated that the intraperitoneal administration of a neutralizing p40 antibody resulted in a reduction of amyloid accumulation in the brains of APPPS1 mice, a murine model of Alzheimer's disease (Vom Berg et al. 2012). The authors suggested a pivotal role for activated microglia in driving neuroinflammation in the APPPS1 model, positing that the activation of brain microglia might be influenced by the intestinal microbiome, as observed in a murine model of multiple sclerosis with a comparable neuroinflammatory profile (Berer et al. 2011). These findings, coupled with the previously discussed connections between age-related alterations in the intestinal microbiome and low-grade inflammation, underscore the need for meticulously controlled studies investigating the microbiome in individuals experiencing cognitive decline.

2.10 Gut-Brain Axis

A complex network of reciprocal connections between the gut microbiome and the central nervous system is referred as gut-brain axis (GBA). The GBA encompasses several biological systems and is essential for preserving the overall homeostasis of the body. Using the autonomic nervous system directly or indirectly through metabolites and chemical transmitters, signals move from the gut to the CNS and vice versa. The composition of the gut microbiota can vary and have an impact on each of these interactions (Morais et al. 2021). Due to its newly emerging function in mediating health and disease as well as its potential utility as a therapeutic target, the GBA has recently gained attention. Various aspects of brain growth and function, including blood-brain barrier (BBB) formation and permeability, neurogenesis, and myelination, are influenced by the gut microbiota. GBA disruption could be involved in the pathogenesis of an array of neurological conditions, such as neurodegenerative disorders (Diaz Heijtz et al. 2011). There is ongoing discussion over the precise mechanisms through which a changed gut microbiome may contribute to the development of CNS inflammation and degeneration. Clinical and experimental data both support the idea that enteric microbiota plays a significant role in the development of GBA, engaging with CNS directly through neuroendocrine and metabolic pathways, in addition to local interactions with intestinal cells and ENS. Dysbiosis can also arise in gastrointestinal conditions known as functional gastrointestinal disorders (FGID), which are closely related to mood disorders and also have an alteration in the GBA (Berrill et al. 2013). There is evidence that both gut-brain and brain-gut dysfunctions exist, with the former that dominates in diseases like irritable bowel syndrome (IBS). The alteration in the GBA will determine the changes in the motility and secretion of intestinal microbes that promotes visceral hypersensitivity, and it also leads to the modification of endocrine and immune system (Koloski et al. 2012).

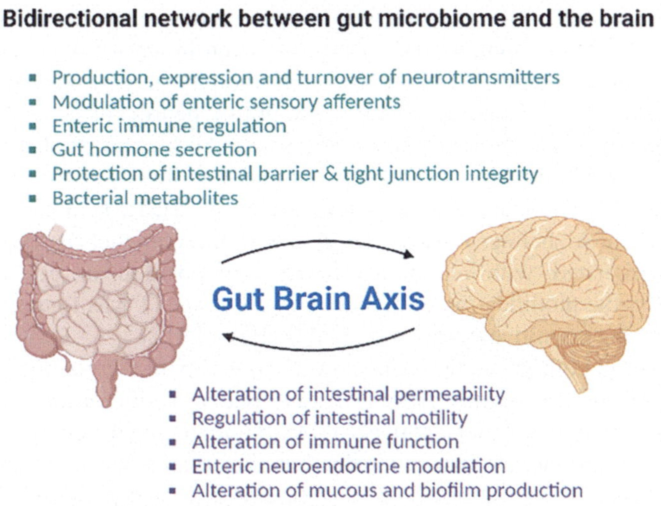

Fig. 2.2 Explains about the mechanisms involved in bidirectional communicational network between gut microbiome and brain

The principal mechanisms underlying the bidirectional network of brain-gut-microbiota axis mainly focus on the forward proceeding either from the gut to the brain or vice versa. The mechanism involving the signal from the gut to the brain includes production, expression, and turnover of neurotransmitters (such as serotonin, dopamine), modulation of enteric sensory afferents, enteric immune regulation, gut hormone secretion, protection of intestinal barrier and tight junction integrity, and bacterial metabolites (proteasome and metabolome effects). Brain to gut pathway involves alteration of intestinal permeability, regulation of intestinal motility, alteration of immune function, alteration of mucous and biofilm production, and enteric neuroendocrine modulation (Fig. 2.2).

The human gut microbiome profile has a strong correlation between the microstructure of the brain, brain FC, intrinsic neuronal activities, mood, and cognitive performances. A well-designed longitudinal study is required for proving the evaluations such as structure and metabolomics of gut microbial community in conjugation with neuroimaging results and behavioral testing data. To further understand the manner the gut and the brain interact, additional research is needed on inflammation, immunological activation, neurotransmitters, neuromodulators, microbial metabolomics, intestinal permeability, motility, and visceral sensitivity.

2.11 Therapeutic Approaches

Based on the recent findings revealing the potential of the microbial interventions in the regulation of intestinal dysbiosis-driven neurological disorders, fecal microbiota transplantation (FMT) appears to be a promising therapeutic strategy. By means of FMT, the thriving microbiota undergoes reproduction to renew itself and generates bioactive metabolites. This approach is advantageous due to the absence of reported significant side effects, making it deemed safe, even in high-risk patients (Haikal et al. 2019; Wimo et al. 2017). Some research findings suggest that this treatment method is efficient in impairing memory, promoting neurogenesis, and increasing levels of secretory proinflammatory cytokines and Aβ protein. It also reduces the level of chronic inflammation and cognitive impairment in AD conditions. It is also efficiently used for treating PD cases (Ghezzi et al. 2022). In a study involving mouse models of Alzheimer's disease (AD), it was observed that cognitive dysfunction was associated with alterations in the composition of gut microbiota. Consequently, the modification of this microbiota via FMT was found to be effective in mitigating cognitive dysfunction in AD (Reitz 2012).

Differences in both the composition and functionality of the gut microbiota have been observed in both healthy individuals and patients with various neurological disorders. It is recognized that diet plays a role in influencing the composition of the microbiome, consequently impacting the function of the gut-brain axis (GBA). Various therapeutic interventions, such as the utilization of probiotics, have been implemented to address dysbiosis in the gut microbiome. These interventions aim to restore a balance in intestinal microflora, ultimately leading to improved clinical outcomes in neurological disorders (Kerry et al. 2018). Numerous diets have been suggested as helpful in preventing or ameliorating neurodegenerative illnesses because they can affect inflammation within the CNS via the GBA (Wilson et al. 2020). For instance, the Mediterranean diet, which relies mostly on olive oil as its major source of fat, is a well-known healthy diet that is rich in vegetables, whole grains, and is low in dairy. Practice of a Mediterranean diet, according to new research, supports individuals with Parkinson's disease with their motor and cognitive symptoms (Fontana et al. 2021). A diet high in fat and low in carbohydrates known as a "ketogenic diet" is used to treat epilepsy (Ułamek-Kozioł et al. 2019). Probiotics are nonviable food components that promote the host's health and are linked to microbiome regulation (Varesi et al. 2022). Predominantly, probiotics consist of Bifidobacterium and bacteria that produce lactic acid, such as Lactobacillus. Increasing evidence indicates that metabolites produced by probiotics play a crucial role as mediators in host-microbe interactions induced by diet. Additionally, various bacterial species residing in the gut, such as *Bacteroides*, *Clostridium*, *Bifidobacterium*, *Peptostreptococcus*, *Lactobacillus*, and *Ruminococcus*, have been reported to generate numerous tryptophan catabolites, including indole, 3-methylindole, indoleacetic acid (IAA), and tryptamine (Kumar and Sperandio 2019; Russell et al. 2013; Wikoff et al. 2009). Probiotics have been demonstrated in both human and animal models to be beneficial in maintaining intestinal homeostasis by stabilizing the epithelial barrier, increasing SCFA production, modifying the mucosal immune system towards a more immunoregulatory response, and decreasing the generation of proinflammatory cytokines. It is used for treating AD, PD, and ALS in animal models, providing data that enhances the positive aspect of this treatment approach (Di Gioia et al. 2020). Numerous studies employing mouse models have shown that administering probiotics can be advantageous in various neurological disorders, leading to enhancements in cognitive outcomes. In a recent investigation, it was demonstrated that prolonged use of probiotics, comprising six bacterial strains, mitigated motor impairments and exhibited neuroprotective effects on dopaminergic neurons in a genetic mouse model of Parkinson's disease (PD) (Hsieh et al. 2020). Collectively, these discoveries prompted researchers to explore the application of probiotics in diverse neurological dysfunctions.

As per the International Scientific Association for Probiotics and Prebiotics (ISAPP), a prebiotic is defined as nonviable food components that are selectively utilized by host-microbial populations, providing health benefits. Instead of relying on probiotic supplements, prebiotics offer an alternative approach to modulating gut microbial flora. These compounds, recognized for their impact on gastrointestinal health, include nondigestible oligosaccharides (NDOs), human milk oligosaccharides (HMOs), and soluble, fermentable fibers (Hill et al. 2014). A recent study demonstrated that the prebiotic lactulose can enhance cognitive function in mouse models of Alzheimer's disease by influencing autophagy and anti-inflammatory pathways (Lee et al. 2021). As a result, these findings suggest that both probiotics and prebiotics could serve as effective treatments for neurological disorders. Nevertheless, further research is essential to gain a detailed understanding of the underlying mechanisms, as a mere correlation does not necessarily imply causation.

Synbiotics refer to a combination of prebiotics and probiotics, wherein the prebiotics support the growth and metabolism of probiotic microorganisms, enhancing their viability and benefits. This combination positively influences the host by increasing the abundance of beneficial microbes in the gastrointestinal tract (GIT). It is crucial to select an appropriate combination in synbiotics to ensure the survival of probiotic microorganisms in the GIT (Barathikannan et al. 2019; Lorente-Picón and Laguna 2021). Research has indicated that the utilization of synbiotics is more efficacious compared to the isolated use of probiotics or prebiotics alone.

In recent years, the terms "postbiotics" and "paraprobiotics" have been introduced to emphasize that probiotics are not the sole viable compounds capable of positively influencing human health (Sánchez et al. 2017). Postbiotics, alternatively referred to as metabiotics, biogenics, or cell-free supernatants (CFSs), encompass bacterial fermentation metabolites and soluble factors derived from live bacteria or released following bacterial cell lysis (Aguilar-Toalá et al. 2018). Paraprobiotics are characterized as microbial cells that are nonviable or inactivated, with some researchers categorizing them as a subset of postbiotics. These structural components have the potential to initiate biological activity in hosts (Cuevas-González et al. 2020).

2.12 Conclusion

Gut microbiome is complex network which develops during the time of birth, and it grows gradually. Gut microbiome gets involved in the variety of physiological, biological, pathological, and behavioral processes. It acts as positive feedback which helps in the maintenance of various host mechanisms. This homeostasis might be altered by a number of cross talking which occurs between the host gut microbiome with several internal and external factors that lead to the development of different pathogenic conditions. Some of the pathogenic conditions can be reversed by maintaining a healthy gut environment through dietary interventions that promote prebiotics, specific probiotic strains, and potentially FMT offers promising avenues for mitigating the effects of brain aging and neurodegenerative diseases. Brain aging is an inevitable event that occurs during the lifetime of an individual which is a natural and irreversible process. But in case of gut microbiome, cross talk with brain can lead to a variety of neuronal problems including neurodegenerative diseases, epilepsy, stroke, traumatic brain injury, and so on. Among these, neurodegenerative diseases had a wider impact among the people all over the world. Development and progression of neurodegenerative diseases are mainly achieved by brain-gut-microbiota axis. Most of the treatment approaches for GBA-mediated neurodegenerative diseases can be modulated by maintaining a health gut environment. Future research should focus on identifying key gut microbes and metabolites associated with brain health, developing personalized interventions based on individual gut profiles, and exploring the potential of combining gut modulation strategies with other therapies to improve brain health and overall well-being throughout life.

References

Aanerud J, Borghammer P, Chakravarty MM, Vang K, Rodell AB, Jónsdottir KY, Møller A, Ashkanian M, Vafaee MS, Iversen P, Johannsen P, Gjedde A (2012) Brain energy metabolism and blood flow differences in healthy aging. J Cereb Blood Flow Metab 32(7):1177–1187. https://doi.org/10.1038/jcbfm.2012.18

Abdel-Haq R et al (2018) Microbiome–microglia connections via the gut–brain axis. J Exp Med 216(1):41–59. https://doi.org/10.1084/jem.20180794

Afanas'ev I (2010) Signaling and damaging functions of free radicals in aging-free radical theory, hormesis, and TOR. Aging Dis 1(2):75–88

Aguilar-Toalá J, Garcia-Varela R, Garcia H, Mata-Haro V, GonzálezCórdova A, Vallejo-Cordoba B, Hernández-Mendoza A (2018) Postbiotics: an evolving term within the functional foods field. Trends Food Sci Technol 75:105–114. https://doi.org/10.1016/j.tifs.2018.03.009

Azam S, Haque ME, Balakrishnan R, Kim IS, Choi DK (2021) The ageing brain: molecular and cellular basis of neurodegeneration. Front Cell Dev Biol 9:683459. https://doi.org/10.3389/fcell.2021.683459

Aziz Q et al (2012) Gut microbiota and gastrointestinal health: current concepts and future directions. Neurogastroenterol Motil 25(1):4–15. https://doi.org/10.1111/nmo.12046

Barathikannan K, Chelliah R, Rubab M, Daliri EB-M, Elahi F, Kim D-H, Agastian P, Oh S-Y, Oh DH (2019) Gut microbiome modulation based on probiotic application for anti-obesity: a review on efficacy and validation. Microorganisms 7:456. https://doi.org/10.3390/microorganisms7100456

Bayer TA, Wirths O (2010) Intracellular accumulation of amyloid-Beta—a predictor for synaptic dysfunction and neuron loss in Alzheimer's disease. Front Aging Neurosci 2:8. https://doi.org/10.3389/fnagi.2010.00008

Beghi E, Giussani G (2018) Aging and the epidemiology of epilepsy. Neuroepidemiology 51(3–4):216–223. https://doi.org/10.1159/000493484

Berer K, Mues M, Koutrolos M, Rasbi ZA, Boziki M, Johner C, Wekerle H, Krishnamoorthy G (2011) Commensal microbiota and myelin autoantigen cooperate to trigger autoimmune demyelination. Nature 479(7374):538–541. https://doi.org/10.1038/nature10554

Benakis C et al (2020) The microbiome-gut-brain axis in acute and chronic brain diseases. Curr Opin Neurobiol 61:1–9. https://doi.org/10.1016/j.conb.2019.11.009

Berrill JW, Gallacher J, Hood K, Green JT, Matthews SB, Campbell AK, Smith A (2013) An observational study of cognitive function in patients with irritable bowel syndrome and inflammatory bowel disease. Neurogastroenterol Motil 25(11):918–e704. https://doi.org/10.1111/nmo.12219

Biagi E, Nylund L, Candela M, Ostan R, Bucci L, Pini E, Nikkïla J, Monti D, Satokari R, Franceschi C, Brigidi P, De Vos W (2010) Through ageing, and beyond: gut microbiota and inflammatory status in seniors and centenarians. PLoS One 5(5):e10667. https://doi.org/10.1371/journal.pone.0010667

Bjorksten J (1977) Some therapeutic implications of the crosslinkage theory of aging. Adv Exp Med Biol 86B:579–602. https://doi.org/10.1007/978-1-4757-9113-6_33

Burcelin R (2016) Gut microbiota and immune crosstalk in metabolic disease. Mol Metab 5(9):771–781. https://doi.org/10.1016/j.molmet.2016.05.016

Claesson MJ, Cusack S, O'Sullivan O, Greene-Diniz R, de Weerd H, Flannery E, Marchesi JR, Falush D, Dinan T, Fitzgerald G, Stanton C, van Sinderen D, O'Connor M, Harnedy N, O'Connor K, Henry C, O'Mahony D, Fitzgerald AP, Shanahan F, Twomey C, O'Toole PW (2011) Composition, variability, and temporal stability of the intestinal microbiota of the elderly. Proc Natl Acad Sci U S A 108(Suppl 1):4586–4591. https://doi.org/10.1073/pnas.1000097107

Claesson MJ, Jeffery IB, Conde S, Power SE, O'Connor EM, Cusack S, Harris HM, Coakley M, Lakshminarayanan B, O'Sullivan O, Fitzgerald GF, Deane J, O'Connor M, Harnedy N, O'Connor K, O'Mahony D, van Sinderen D, Wallace M, Brennan L, Stanton C, O'Toole PW (2012) Gut microbiota composition correlates with diet and health in the elderly. Nature 488(7410):178–184. https://doi.org/10.1038/nature11319

Cornelius E (1972) Increased incidence of lymphomas in thymectomized mice—evidence for an immunological theory of aging. Experientia 28(4):459. https://doi.org/10.1007/BF02008340

Cuevas-González P, Liceaga A, Aguilar-Toalá J (2020) Postbiotics and paraprobiotics: from concepts to applications. Food Res Int 136:109502. https://doi.org/10.1016/j.foodres.2020.109502

D'Argenio V, Salvatore F (2015) The role of the gut microbiome in the healthy adult status. Clin Chim Acta 451(Pt A):97–102. https://doi.org/10.1016/j.cca.2015.01.003

Davidovic M, Sevo G, Svorcan P, Milosevic DP, Despotovic N, Erceg P (2010) Old age as a privilege of the "selfish ones". Aging Dis 1(2):139–146

Di Gioia D, Bozzi Cionci N, Baffoni L, Amoruso A, Pane M, Mogna L, Gaggìa F, Lucenti MA, Bersano E, Cantello R, De Marchi F, Mazzini L (2020) A prospective longitudinal study on the microbiota composition in amyotrophic lateral sclerosis. BMC Med 18(1):153. https://doi.org/10.1186/s12916-020-01607-9

Diaz Heijtz R, Wang S, Anuar F, Qian Y, Björkholm B, Samuelsson A, Hibberd ML, Forssberg H, Pettersson S (2011) Normal gut microbiota modulates brain development and behavior. Proc Natl Acad Sci U S A 108(7):3047–3052. https://doi.org/10.1073/pnas.1010529108

Dumurgier J, Tzourio C (2020) Epidemiology of neurological diseases in older adults. Rev Neurol 176(9):642–648. https://doi.org/10.1016/j.neurol.2020.01.356

Floyd RA, Hensley K (2002) Oxidative stress in brain aging. Implications for therapeutics of neurodegenerative diseases. Neurobiol Aging 23(5):795–807. https://doi.org/10.1016/s0197-4580(02)00019-2

Fontana L, Ghezzi L, Cross AH, Piccio L (2021) Effects of dietary restriction on neuroinflammation in neurodegenerative diseases. J Exp Med 218(2):e20190086. https://doi.org/10.1084/jem.20190086

Franceschi C, Bonafè M, Valensin S, Olivieri F, De Luca M, Ottaviani E, De Benedictis G (2000) Inflamm-aging. An evolutionary perspective on immunosenescence. Ann N Y Acad Sci 908:244–254. https://doi.org/10.1111/j.1749-6632.2000.tb06651.x

Garaschuk O, Semchyshyn HM, Lushchak VI (2018) Healthy brain aging: interplay between reactive species, inflammation and energy supply. Ageing Res Rev 43:26–45. https://doi.org/10.1016/j.arr.2018.02.003

Ghezzi L, Cantoni C, Rotondo E, Galimberti D (2022) The gut microbiome-brain cross-talk in neurodegenerative diseases. Biomedicines 10(7):1486. https://doi.org/10.3390/biomedicines10071486

Griffin WS (2013) Neuroinflammatory cytokine signaling and Alzheimer's disease. N Engl J Med 368(8):770–771. https://doi.org/10.1056/NEJMcibr1214546

Haikal C, Chen Q-Q, Li J-Y (2019) Microbiome changes: an indicator of Parkinson's disease? Transl Neurodegener 8:1–9. https://doi.org/10.1186/s40035-019-0175-7

Hill C, Guarner F, Reid G, Gibson GR, Merenstein DJ, Pot B, Morelli L, Canani RB, Flint HJ, Salminen S, Calder PC, Sanders ME (2014) The International Scientific Association for Probiotics and Prebiotics consensus statement on the scope and appropriate use of the term probiotic. Nat Rev Gastroenterol Hepatol 11:506–514. https://doi.org/10.1038/nrgastro.2014.66

Hou Y et al (2019) Ageing as a risk factor for neurodegenerative disease. Nat Rev Neurol 15(10):565–581. https://doi.org/10.1038/s41582-019-0244-7

Hsieh T-H, Kuo C-W, Hsieh K-H, Shieh M-J, Peng C-W, Chen Y-C, Chang YL, Huang Y-Z, Chen C-C, Chang P-K, Chen K-Y, Chen H-Y (2020) Probiotics alleviate the progressive deterioration of motor functions in a mouse model of Parkinson's disease. Brain Sci 10:206. https://doi.org/10.3390/brainsci10040206

Joseph JA, Shukitt-Hale B, Casadesus G, Fisher D (2005) Oxidative stress and inflammation in brain aging: nutritional considerations. Neurochem Res 30(6–7):927–935. https://doi.org/10.1007/s11064-005-6967-4

Kandlur A, Satyamoorthy K, Gangadharan G (2020) Oxidative stress in cognitive and epigenetic aging: a retrospective glance. Front Mol Neurosci 13:41. https://doi.org/10.3389/fnmol.2020.00041

Kerry RG, Patra JK, Gouda S, Park Y, Shin H-S, Das G (2018) Benefaction of probiotics for human health: a review. J Food Drug Anal 26:927–939. https://doi.org/10.1016/j.jfda.2018.01.002

Koloski NA, Jones M, Kalantar J, Weltman M, Zaguirre J, Talley NJ (2012) The brain—gut pathway in functional gastrointestinal disorders is bidirectional: a 12-year prospective population-based study. Gut 61(9):1284–1290. https://doi.org/10.1136/gutjnl-2011-300474

Kumar A, Sperandio V (2019) Indole signaling at the host-microbiota-pathogen interface. mBio 10:e01031-19. https://doi.org/10.1128/mBio.01031-19

Lee Y-S, Lai D-M, Huang H-J, Lee-Chen G-J, Chang C-H, Hsieh-Li HM, Lee G-C (2021) Prebiotic lactulose ameliorates the cognitive deficit in Alzheimer's disease mouse model through macro-autophagy and chaperone mediated autophagy pathways. J Agric Food Chem 69:2422–2437. https://doi.org/10.1021/acs.jafc.0c07327

Lorente-Picón M, Laguna A (2021) New avenues for Parkinson's disease therapeutics: disease-modifying strategies based on the gut microbiota. Biomolecules 11:433. https://doi.org/10.3390/biom11030433

Mattson MP, Chan SL, Duan W (2002) Modification of brain aging and neurodegenerative disorders by genes, diet, and behavior. Physiol Rev 82(3):637–672. https://doi.org/10.1152/physrev.00004.2002

Maynard Smith J (1962) Review lectures on senescence. I. The causes of ageing. Proc R Soc Lond Ser B 157(966):115–127

McHugh D, Gil J (2018) Senescence and aging: causes, consequences, and therapeutic avenues. J Cell Biol 217(1):65–77. https://doi.org/10.1083/jcb.201708092

Moore DJ, West AB, Dawson VL, Dawson TM (2005) Molecular pathophysiology of Parkinson's disease. Annu Rev Neurosci 28:57–87. https://doi.org/10.1146/annurev.neuro.28.061604.135718

Morais LH, Schreiber HL 4th, Mazmanian SK (2021) The gut microbiota-brain axis in behaviour and brain disorders. Nat Rev Microbiol 19(4):241–255. https://doi.org/10.1038/s41579-020-00460-0

Müller WE, Eckert A, Kurz C, Eckert GP, Leuner K (2010) Mitochondrial dysfunction: common final pathway in brain aging and Alzheimer's disease—therapeutic aspects. Mol Neurobiol 41(2–3):159–171. https://doi.org/10.1007/s12035-010-8141-5

Pickett EK, Herrmann AG, McQueen J, Abt K, Dando O, Tulloch J, Jain P, Dunnett S, Sohrabi S, Fjeldstad MP, Calkin W, Murison L, Jackson RJ, Tzioras M, Stevenson A, d'Orange M, Hooley

M, Davies C, Colom-Cadena M, Anton-Fernandez A et al (2019) Amyloid Beta and Tau cooperate to cause reversible behavioral and transcriptional deficits in a model of Alzheimer's disease. Cell Rep 29(11):3592–3604.e5. https://doi.org/10.1016/j.celrep.2019.11.044

Priyadarshi A, Khuder SA, Schaub EA, Priyadarshi SS (2001) Environmental risk factors and Parkinson's disease: a metaanalysis. Environ Res 86(2):122–127. https://doi.org/10.1006/enrs.2001.4264

Reitz C (2012) Alzheimer's disease and the amyloid cascade hypothesis: a critical review. Int J Alzheimer's Dis 2012:1–11. https://doi.org/10.1155/2012/369808

Russell WR, Duncan SH, Scobbie L, Duncan G, Cantlay L, Calder AG, Anderson SE, Flint HJ (2013) Major phenylpropanoid-derived metabolites in the human gut can arise from microbial fermentation of protein. Mol Nutr Food Res 57:523–535. https://doi.org/10.1002/mnfr.201200594

Sánchez B, Delgado S, Blanco-Míguez A, Lourenço A, Gueimonde M, Margolles A (2017) Probiotics, gut microbiota, and their influence on host health and disease. Mol Nutr Food Res 61:1600240. https://doi.org/10.1002/mnfr.201600240

Spires-Jones TL, Attems J, Thal DR (2017) Interactions of pathological proteins in neurodegenerative diseases. Acta Neuropathol 134(2):187–205. https://doi.org/10.1007/s00401-017-1709-7

Toricelli M, Pereira AAR, Souza Abrao G, Malerba HN, Maia J, Buck HS, Viel TA (2021) Mechanisms of neuroplasticity and brain degeneration: strategies for protection during the aging process. Neural Regen Res 16(1):58–67. https://doi.org/10.4103/1673-5374.286952

Ułamek-Kozioł M, Czuczwar SJ, Januszewski S, Pluta R (2019) Ketogenic diet and epilepsy. Nutrients 11(10):2510. https://doi.org/10.3390/nu11102510

van Deursen JM (2014) The role of senescent cells in ageing. Nature 509(7501):439–446. https://doi.org/10.1038/nature13193

van Heemst D (2010) Insulin, IGF-1 and longevity. Aging Dis 1(2):147–157

Varesi A, Pierella E, Romeo M, Piccini GB, Alfano C, Bjørklund G, Oppong A, Ricevuti G, Esposito C, Chirumbolo S, Pascale A (2022) The potential role of gut microbiota in Alzheimer's disease: from diagnosis to treatment. Nutrients 14(3):668. https://doi.org/10.3390/nu14030668

Vom Berg J, Prokop S, Miller KR, Obst J, Kälin RE, Lopategui-Cabezas I, Wegner A, Mair F, Schipke CG, Peters O, Winter Y, Becher B, Heppner FL (2012) Inhibition of IL-12/IL-23 signaling reduces Alzheimer's disease-like pathology and cognitive decline. Nat Med 18(12):1812–1819. https://doi.org/10.1038/nm.2965

Wikoff WR, Anfora AT, Liu J, Schultz PG, Lesley SA, Peters EC, Siuzdak G (2009) Metabolomics analysis reveals large effects of gut microflora on mammalian blood metabolites. Proc Natl Acad Sci U S A 106:3698–3703. https://doi.org/10.1073/pnas.0812874106

Wilson AS, Koller KR, Ramaboli MC, Nesengani LT, Ocvirk S, Chen C, Flanagan CA, Sapp FR, Merritt ZT, Bhatti F, Thomas TK, O'Keefe SJD (2020) Diet and the human gut microbiome: an international review. Dig Dis Sci 65(3):723–740. https://doi.org/10.1007/s10620-020-06112-w

Wimo A, Guerchet M, Ali G-C, Wu Y-T, Prina AM, Winblad B, Jönsson L, Liu Z, Prince M (2017) The worldwide costs of dementia 2015 and comparisons with 2010. Alzheimers Dement 13:1–7. https://doi.org/10.1016/j.jalz.2016.07.150

Yin F, Sancheti H, Patil I, Cadenas E (2016) Energy metabolism and inflammation in brain aging and Alzheimer's disease. Free Radic Biol Med 100:108–122. https://doi.org/10.1016/j.freeradbiomed.2016.04.200

Zheng D, Liwinski T, Elinav E (2020) Interaction between microbiota and immunity in health and disease. Cell Res 30(6):492–506. https://doi.org/10.1038/s41422-020-0332-7

The Gut Microbiome and the Central Nervous System (CNS)

3

Amulya Vijay, Faraz Ahmad, Sudesh Ravi, and Anandan Balakrishnan

Abstract

The presence of resident microbial species in the human gastrointestinal tract has been known for a long time. However, only recently have they been recognized as tremendous regulators of overall host pathophysiology. Interestingly, the central nervous system (CNS) features as a key target of the gut microflora. Recent studies indicate that the bidirectional interactions of the microbiota-gut-brain axis play critical roles in influencing a plethora of neuroendocrine and neuroinflammatory mediators, with huge implications for neurotransmitter physiology at the levels of enteric, autonomous, and central nervous systems. This chapter deals with the discussion of the pathophysiological roles of the resident microflora in the developing and adult brains, focusing on the two relevant aspects of blood-brain barrier development and adult hippocampal neurogenesis. Further, dysbiosis of the gut microbiome as a pathogenic factor for the development of CNS disorders (e.g., multiple sclerosis, Alzheimer's and Parkinson's diseases, amyotrophic lateral sclerosis, and stroke) is detailed. We also discuss the possibilities of harnessing the alterations in the gut microbial species for diagnostic, prognostic, and therapeutic purposes for these diseases. Lastly, the implications of the microbiota-gut-brain axis for promoting longevity and healthy ageing are illustrated.

A. Vijay · A. Balakrishnan (✉)
Department of Genetics, Dr ALM PG Institute of Basic Medical Sciences, University of Madras, Chennai, India

F. Ahmad · S. Ravi
School of Bioscience and Technology, Vellore Institute of Technology, Vellore, India

S. Pathak, A. Banerjee (eds.), *Gut Microbiome and Brain Ageing*,
https://doi.org/10.1007/978-981-99-8803-7_3

Keywords

Gut microbiota · Blood-brain barrier · Adult neurogenesis · Neuroendocrine · Neurodegeneration · Neuroinflammation · CNS disorders · Longevity

3.1 Gut Microbiome

In the human digestive system, there exists a complex ecology of microbes known as the gut microbiome. It comprises the collective genomes of microbiota inhabiting humans, including protozoa, archaea, eukaryotes, viruses, and bacteria (Sender et al. 2016). The gut microbiota is now considered an important partner of human cells, interacting with virtually all human cells (Mukherjee et al. 2023).

Even though studies of gut microbes have been ongoing for several years, recent years have seen a significant increase in interest outside the realm of infectious disorders. Human health and disease, including digestion, metabolism, immune responses, and brain functions, are strongly influenced by the gut microbiota. The interplay between the gut microbiota, host variables, and environmental factors has the potential to completely change how diseases are caused and how they are treated in medicine. *Prevotella*, *Ruminococcus*, *Bacteroides*, and *Firmicutes* are only a few of the numerous major bacterial families that have been discovered in human guts (O'Toole and Jeffery 2015). The anaerobic bacteria *Bifidobacterium*, *Lactobacillus*, *Peptostreptococcus*, and *Clostridium* are all found in the colon's low-oxygen environment. The health of the gut is thought to be significantly influenced by these microorganisms (Lozupone et al. 2012).

The equilibrium of the gut microbiome can be disturbed by several factors, including nutrition, antibiotic use, age, exposure to chemicals in the environment, medicine, physical activity, and chronic illnesses (Hasan and Yang 2019). Taking antibiotics can harm the gut microbiota in many ways, including reducing species diversity, altering metabolic activity, and making the body more vulnerable to invasion by pathogenic bacteria (Round and Mazmanian 2009). It has also been demonstrated that exposure to environmental chemicals alters the gut flora. A person's lifestyle choices, including their food, level of activity, and the use of antibiotics, can contribute to dysbiosis (Carding et al. 2015).

Two significant research initiatives to characterize the human microbiome are the Human Microbiome Project (HMP) and MetaHIT. These meta-omics projects have used a range of analyses, including whole-genome shotgun (WGS) metagenomic sequencing of host-associated microbial communities to identify functional genes, 16S ribosomal RNA (rRNA) sequencing to taxonomically characterize the microbiota communities, and meta-transcriptomics to identify active genes and pathways (Proctor 2011).

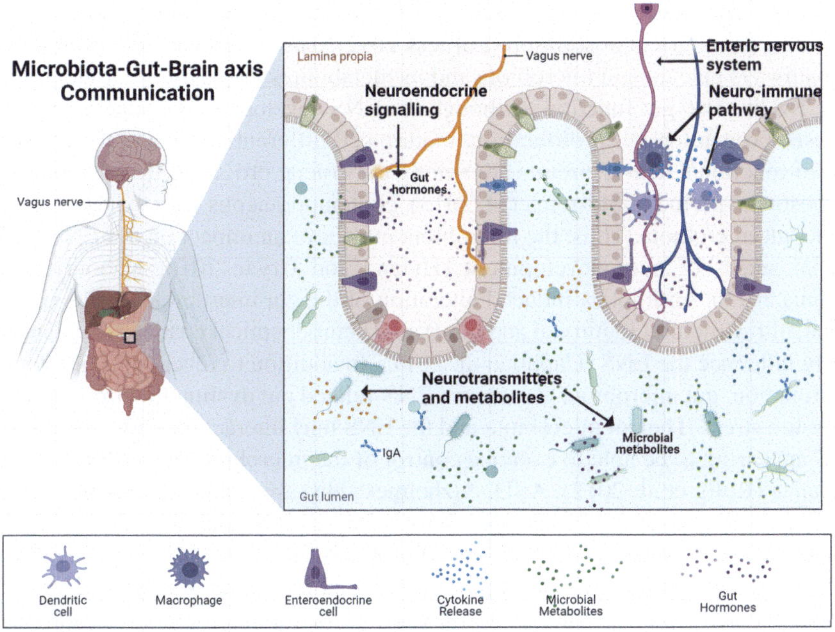

Fig. 3.1 Gut-brain axis communication comprises both direct and indirect signaling including autonomic and enteric nervous systems, neuroimmune pathways, neuroendocrine signaling, neurotransmitters, and metabolites

3.2 Microbiota-Gut-Brain Axis

The gut-brain axis is a crucial network of connections involving several biological systems that enable two-way communication between the gut and the brain which controls intestinal homeostasis and the central nervous system. The resident microbiome is a crucial factor in modulating these gut-brain signaling cascades which has led to the establishment of the theory of a microbiota-gut-brain axis. The communication between the gut and the central nervous system (CNS) comprises both direct and indirect signaling including autonomic and enteric nervous system (ENS), neuroimmune pathways, neuroendocrine signaling, neurotransmitters, and metabolites as shown in Fig. 3.1.

3.2.1 Enteric Nervous System (ENS)

The walls of the gastrointestinal tract consist of a sophisticated network of neurons and glial cells known as the enteric nervous system (ENS). It is frequently referred to as the "second brain" because of its capacity to control different digestive processes while operating independently of the central nervous system (CNS). The

ENS regulates the gastrointestinal tract's movement, the release of digesting enzymes, and nutrient absorption (Furness 2012). These intrinsic and afferent neural pathways give the gut microbiota and its metabolites the chance to affect not just the CNS but also gut function. Although the ENS develops largely during embryogenesis, progenitor cell proliferation, mature cell differentiation, and the construction of functional neural circuits are among the crucial processes that continue into the postnatal period (Goldstein et al. 2013), contemporaneous with the development of the gut microbiota. Thus, the microbiota may have an impact on the ENS during crucial stages of neurodevelopment (Hyland and Cryan 2016; Kabouridis and Pachnis 2015). Antibiotics-induced alteration of the gut microbiota results in functional, ultrastructural (neuronal and glial), and neurochemical changes that can massively influence the ENS (Caputi et al. 2017). In addition to revealing ENS activity and function, gut microbiota also shows ENS-related gut dysfunction brought on by early-life stress. The gut microbiota and the ENS may interact reciprocally, with the ENS appearing to be able to exercise control of the microbiota, according to recent evidence (Rolig et al. 2017). ASD, Alzheimer's disease (AD), Parkinson's disease (PD), and other major CNS disorders have all been associated with ENS (Rao and Gershon 2016) as well as other serious GI problems including Hirschsprung disease and neuropathic chronic intestinal pseudo-obstruction, etc. (Gariepy 2001).

The myenteric plexus (Auerbach's plexus) and the submucosal plexus (Meissner's plexus) are the two primary plexuses that make up the ENS. The myenteric plexus controls gastrointestinal motility and is situated between the longitudinal and circular muscle layers of the gastrointestinal tract. The submucosal plexus is found in the submucosa and controls secretion and absorption (Gershon 1999).

The sympathetic and parasympathetic nervous systems, as well as the vagus nerve, are some of the channels via which the ENS and CNS communicate. Acetylcholine, serotonin, and substance P are only a few of the neurotransmitters and neuropeptides found in the ENS, all of which are crucial for controlling gastrointestinal function (Schemann and Camilleri 2013).

Recent years have seen a significant increase in research focusing on the pathophysiology of the ENS, and there is mounting proof that ENS dysfunction may play a role in several gastrointestinal disorders, including gastroparesis, irritable bowel syndrome, and inflammatory bowel disease. To create new treatments for digestive diseases and gain a better knowledge of the complex relationships between the ENS and other bodily systems, more study is required (Mawe and Hoffman 2013).

The microbiome-gut-brain axis allows for bidirectional communication between the gut microbiome and the ENS (Bienenstock 2012; Gershon and Margolis 2021). The ENS, which controls vital gut processes like motility, nutrition intake, and immune response, is the intestine's intrinsic neural system (Ganz and Ratcliffe 2023). Alterations in the composition of the gut microbiome can promote a condition known as dysbiosis, where the balance between useful and pathogenic bacteria is altered, typically encouraging the latter. The gut microbiota can regulate enteric neurons and glia, altering gut physiology. Numerous neurodevelopmental and neurodegenerative diseases have been associated with dysbiosis (Ojeda et al. 2021). Through endocrine, neurological, and immunological mechanisms, the gut microbiome influences mental function and behaviour. Behaviour changes have been linked to changes in the gut microbiome makeup (Shaik et al. 2020).

The enteric nerve system in the bowel interacts with the gut microbiota to affect intestinal motility and secretion. Early organogenesis sets the stage for the ENS's development, which continues once feeding starts and maintains plasticity until adulthood. The interaction between the intestinal microbiota and the ENS throughout key developmental and disease-pathogenesis stages is being increasingly understood (Ganz and Ratcliffe 2023). In conclusion, the gut microbiota can influence the ENS by controlling enteric neurons and glia, encouraging dysbiosis, encouraging it, impacting behaviour, interfering with intestinal motility and secretion, and interfering with proper growth.

The autonomic nervous system (ANS) controls involuntary bodily functions, such as heartbeat, breathing, and digestion. The ANS, along with the HPA axis, is an intricate network that unconsciously establishes and controls host physiological homeostasis (Jänig 2006). The direct and fastest neuronal connection between the gut and the brain is the tenth cranial nerve called the vagus nerve, which collects information from different visceral organs (Berthoud and Neuhuber 2000). Vagal afferents are assumed to be polymodal because of the wide range of receptors that are expressed on them, meaning that they can respond to a wide range of chemical/mechanical, or hormonal responses (Berthoud et al. 2004). Studies with vagotomies have unequivocally shown how crucial constant bidirectional vagal signaling is for healthy brain function, including host behaviour. Indeed, reports are suggesting that vagus nerve ablation by gastrectomy causes an increase in the occurrence of psychiatric illnesses in humans (Browning and Houseworth 1953; Whitlock 1961). Furthermore, vagotomy induces neurogenic bowel dysfunction, shows symptoms related to psychiatric disease including anxiety-like, fear-related behaviour, sensorimotor gating, learning difficulties, etc. (Klarer et al. 2014, 2017, 2018). Alternatively, vagus nerve stimulation (VNS) also supports its function in mood regulation. VNS has been shown to improve depressive-like behaviours in chronic restraint stress animals by increasing hippocampal BDNF expression and also regulate the release of norepinephrine, 5-HT, and dopamine in anxiety- and depression-associated regions of the brain (Grimonprez et al. 2015; Breit et al. 2018). In addition to the treatment of resistant depression, VNS is being investigated for treating chronic pain, Crohn's disease, and some epilepsies (Breit et al. 2018; Krahl 2012; Penry and Dean 1990; Chakravarthy et al. 2015; Ren et al. 1989; Bonaz et al. 2016).

3.2.2 Neuroendocrine Pathway

The main nonneuronal route of interaction in the microbiota-gut-brain axis is through neuroendocrine systems called the hypothalamic-pituitary-adrenal (HPA) axis. The main purpose of HPA axis activation in response to stress is to prepare the body for the "fight or flight" reaction (Mayer 2000). The hypothalamic paraventricular nucleus (PVN) releases the corticotrophin-releasing factor (CRF) when homeostasis is disrupted during stress, which in turn causes the anterior pituitary to release the adrenocorticotrophic hormone (ACTH). Systemic circulation of ACTH targets the adrenal cortex to release glucocorticoids which interact with lower-affinity glucocorticoid receptors as well as high-affinity mineralocorticoid receptors in the brain (Herman et al. 2016; Smith and Vale 2006; Tsigos and Chrousos 2002). The microbiota affects how the HPA axis functions; studies show that male GF mice exhibit a hyperresponsive HPA axis in response to restraint stress. GF mice have also been found to have altered hippocampus 5-HT1A receptor and NMDA RNA expression which in turn controls CRF release from the hypothalamus. Conversely, the microbiome of stressed animals shows significant modifications with altered HPA axis. There is enough evidence showing irritable bowel syndrome is associated with increased ACTH and cortisol responses to CRF infusion (Dinan et al. 2006) and is linked with microbiota dysbiosis (Eisenstein 2016; Grenham et al. 2011). Apart from the neuroendocrine-HPA axis systems, it also interacts with neuronal pathways via interplay with the vagus nerve and may also influence the neuroimmune-HPA axis.

3.2.3 Neuroimmune Pathway

The major role of the epithelium in the GI tract is to create a physical barrier between the immunological cells in the body and the billions of microbiota by creating a robust luminal-mucosal interface. The primary site of host-microbe interactions takes place at this luminal-mucosal interface, by allowing constant communication via direct physical contact or by the chemicals secreted by them (Fasano and Shea-Donohue 2005). The host immune cells recognize many different types of bacteria by recognizing pathogen-associated molecular patterns once activated; they are capable of attracting inflammatory mediators, cytokine synthesis, and the recruitment of acute inflammatory cells through chemokines (Takeda and Akira 2004). There is plenty of evidence to suggest that the gut microbiota releases cytokines, chemokines, endocrine messengers, neurotransmitters, neuropeptides, and microbial by-products that enter circulation in the blood and lymphatic systems and may regulate the brain and behaviour. It is well-established how a microbial population of the gut can influence both innate and adaptive immune responses at mucosal surfaces by activating particular immune cell populations, influencing their migration, and influencing how they function (Cassel et al. 2008; El Aidy et al. 2015; Kamada et al. 2013; Mazmanian et al. 2005; Powell et al. 2017).

The interactions between the microbiota and microglia, which account for 5–12% of all brain cells, basically drive the microbiota-immune-brain. Microglia are the innate immune cells capable of detecting minute alterations in their environment (Bilbo and Schwarz 2012; Delpech et al. 2015; Hickman et al. 2013; Kim and de Vellis 2005; Nimmerjahn et al. 2005; Tremblay et al. 2011) and mount neuroinflammatory responses by producing a variety of cytokines and chemokines (Kettenmann et al. 2011; Ransohoff and El Khoury 2016; Yamasaki et al. 2014). These chemical messengers in turn recruit other immune cells, to repair affected tissue and restore homeostasis. Microglia have an important role in neuronal growth, remodelling, signalling, and plasticity in addition to their immunological function (Schafer and Stevens 2015). A recent study showed that a diverse GI microbiota is essential for the maintenance and development of healthy microglia (Erny et al. 2015).

3.2.4 Neurotransmitters and Metabolites

Metabolites produced by the microbiota in the gut are thought to be critical for the modulation of both CNS and ENS function including GI motility, mood, and behaviour. By modifying serotoninergic, dopaminergic, glutamatergic, noradrenergic, and GABAergic neurotransmission, the gut bacteria can have an impact on brain function. These neuroactive compounds can be produced by microbiota, and they can also stimulate the synthesis and metabolism of neurotransmitters. The common neurochemical language between host and microbe is one of the core aspects of the idea of microbial endocrinology.

Enteric neurons can synthesize dopamine and norepinephrine but cannot convert norepinephrine into epinephrine (Costa 2000). Research has shown that the microbiota produce an enzyme glucuronidase which is essential for the host to transform physiologically inactive to active forms of norepinephrine and dopamine (Asano et al. 2012). *Bacillus* are known to produce norepinephrine and dopamine (Tsavkelova et al. 2000). *Escherichia*, a common gut microbiome, is known to produce catecholamines, such as norepinephrine (Shishov et al. 2009). *Escherichia* spp. (Richard and Foster 2003) and *Lactobacillus* spp. (Siragusa et al. 2007) have been demonstrated to synthesize GABA, the main inhibitory neurotransmitter of the host nervous system. Also, there is evidence that the amino acid glutamate can be converted by both the host and bacterium into GABA (Smith et al. 1992; Strandwitz et al. 2018). The ability of glutamate decarboxylase, the enzyme that transforms glutamic acid into GABA, exists in the faecal microbiome (Pokusaeva et al. 2017). Host synthesis of GABA can affect the microbiota population which can sense extracellular GABA with their receptors (Guthrie and Nicholson-Guthrie 1989).

3.3 Gut Microbiome and the Brain in the Early Life

3.3.1 Gut Microbiome and Development of the Blood-Brain Barrier (BBB)

The blood-brain barrier (BBB) is a crucial mechanism of CNS protection and homeostasis conferred by the tight junctions of the brain microvascular endothelial cells (BMVECs). Unsurprisingly, an immature BBB is the primary reason for elevated susceptibility of the developing brain to exogenous/systemic toxicants and consequent immune dysfunction. The implications of the resident gut microflora for the development of BBB have been a focus of several research endeavours in recent years. Using mice raised in germ-free conditions, Braniste et al. provided one of the first lines of evidence for gut microbiome-BBB interaction during development. They reported the presence of dysfunctional BBB with increased permeability in neonatal and adult germ-free mice which they attributed to reduced levels of occludin and claudin-5, protein components of BMVEC tight junctions. Moreover, these deficits were attenuated by colonization of pathogen-free butyrate-secreting bacteria such as *Clostridium tyrobutyricum* (Braniste et al. 2014).

Dysbiosis of gut microbiota and disintegrity of BBB is thought to be one of the critical primary mechanisms linking antibiotic exposure and the risk of developing psychiatric and behavioural disorders, like schizophrenia (Kelly et al. 2021) and autism (Eshraghi et al. 2020). Thus, low-dose penicillin treatment during late gestational and early postnatal periods in mice results in permanent deleterious effects on gut microflora composition which translates into sustained immune dysfunction, BBB deficits, and behavioural (anxiety and aggression) and social (preference for social novelty) alterations. Interestingly, colonization with *Lactobacillus rhamnosus* JB-1 may ameliorate these penicillin-induced effects (Leclercq et al. 2017). The molecular mechanisms linking the gut-brain axis and BBB development have however not been discerned completely. It can be speculated that in addition to immune dysregulation, the deregulation of neurotrophin (Bercik et al. 2011) and endocrine (Cani and Knauf 2016) signaling may play critical roles in this respect.

3.4 Gut Microbiome and the Brain in the Adult Life

3.4.1 Gut Microbiome and Adult Neurogenesis

Adult hippocampal neurogenesis is thought to be a crucial pathway of neuroprotection against deficits in cognition, learning, and memory as well as responses to stressful events, chemicals, and insults. Using bromodeoxyuridine (BrdU) staining, Ogbonnaya and coworkers reported amplified hippocampal neurogenesis in adult mice raised under germ-free conditions as opposed to normally raised controls. Interestingly, the induction of colonization of microbiota in the gut of germ-free mice after weaning did not alter the pattern of adult neurogenesis, indicating that only early-life changes in the gut microbiota during a certain critical period may impact hippocampal neurogenic pathways (Yamada 1989). Antibiotics are known to reduce the potential of adult neurogenesis and survival of new neurons. Interestingly, probiotic treatment consisting of bacterial species from the *Lactobacillus* and *Bifidobacterium* genera can attenuate these antibiotic-induced deficits in mice, probably by enhancing the numbers of Ly6C(hi) monocytes (Möhle et al. 2016). Early-life inflammatory bowel diseases result in irreversible dysbiosis of the microflora population and may result in severe neurogenic and behavioural deficits and elevated neuroinflammation into adulthood (Salvo et al. 2020).

Interestingly, faecal microbiota transplantation can confer depressive behaviour from mice subjected to unpredictable chronic mild stress (UCMS) to naïve recipient mice. Further, this is accompanied by a transfer of reduced potential for neurogenesis and repressed proliferation of the hippocampal neural stem cells. The underlying mechanisms of such faecal microbiota transplantation-mediated transfer of depressive symptoms and deficits in adult neurogenesis possibly rely on gut microbiota-driven alterations in fatty acid metabolism and endocannabinoid signalling (Chevalier et al. 2020). Alterations in tryptophan metabolism and serotonergic signalling may constitute another pathway of microflora-driven transfer of decreased neurogenic potential (Siopi et al. 2020). Together, the studies indicate robust participation of the resident gut microbial species in influencing adult neurogenesis (and consequently psychological state) in the host via multiple mechanisms. More recently, Wei et al. have also implicated bacterial species expressing tryptophanase (and producing indole), such as *Escherichia coli* in adult hippocampal neurogenesis. The neurogenic and synaptogenic effects of *Escherichia coli* colonization were proposed to be dependent on aryl hydrocarbon receptor (AhR) and further reported the involvement of β-catenin, neurogenin 2, and vascular endothelial growth factor α (Wei et al. 2021). Bacterial-derived SCFAs such as butyrate have been evidenced to endorse adult neurogenesis in young murine models transplanted with faecal microbiota from older mice, possibly via signalling through the fibroblast growth factor 21 (FGF21) which is known to elicit pro longevity functions (Kundu et al. 2019). These results are supported by the findings of robust stimulation of proliferation, differentiation, and maturation of human neural progenitor cells mediated by SCFAs (Yang et al. 2020).

In a recently published study, transplantation of faecal microbiota from a 5XFAD mouse model of Alzheimer's disease (AD; Sect. 3.5.2.1) into naïve recipient mice resulted in significantly reduced levels of hippocampal neurogenesis, among other deficits including enhanced neuroinflammation and memory impairments (Kim et al. 2021). Finally, the dependence of diet on adult neurogenesis and cognitive functions may also rely on the resident microflora. Thus, the ketogenic diet results in aggravated cognitive impairment induced by hypoxia via alterations in the microbial composition. Further, transplantation of *Bilophila wadsworthia* may lead to a deficient neurogenic potential and cognitive impairment under normal dietary conditions, possibly due to activation of interferon-gamma (INF-γ)-secreting Th1 cells (Olson et al. 2021). In conclusion, there seems to be a robust interaction of the gut-brain axis and cognitive and behavioural functions via the mediation of the machinery for adult hippocampal neurogenesis.

3.5 Gut Microbiota and Central Nervous System (CNS) Disorders

Gut microbiota along with the host regulate the pathophysiology of signalling through the metabolic, immune, and nervous systems through dynamic interactions in a bidirectional manner. Normal functioning of the brain and emotional behaviour depend on a healthy resident microbiota, and in turn the integrity and homeostasis of gut physiology are controlled by the CNS. The tremendous implications of the resident gut microbial species in the CNS pathophysiology are well studied specifically in neuroinflammatory disorders like multiple sclerosis (MS), neurodegenerative disorders including Alzheimer's dementia (AD), Parkinson's disease (PD), amyotrophic lateral sclerosis (ALS), and stroke.

3.5.1 Microbiome in Inflammatory Disease

Microbiota influences neuroinflammation through the "microbiota-gut-brain axis" by both regulating cellular responses by immune cells and via humoral modes through cytokine releases. It plays a critical role in immune surveillance and maintains homeostasis through effectors T cells and mature APCs. In the CNS, glial cells, specifically microglia, are the primary immune effectors and cytokine producers. They can establish homeostasis supports neuronal health and can produce various patterns of cytokine and/or phagocytosis in different pathological conditions.

3.5.1.1 Multiple Sclerosis (MS)

Animal and human studies indicate that dysbiosis of the resident microbial population is crucially involved in the pathogenesis of multiple sclerosis (MS). In this regard, the experimental autoimmune encephalomyelitis (EAE) has proved instrumental in deducing the impact of gut microbiome dyshomeostasis in immune functions, particularly in autoimmune degenerative diseases such as MS. The resident microbial population in the gut elicits massive modulation of the balances between pro- and anti-inflammatory signalling. It is regarded as a prominent environmental factor for autoimmune demyelinating diseases, including MS (Ochoa-Repáraz et al. 2018). This is hardly surprising given that MS is principally an immune dysfunction. Moreover, studies have indicated significant differences between the subjects with MS (especially relapsing-remitting MS) when compared to controls, with an increased presence of the microbial species of the *Pseudomonas*, *Haemophilus*, *Mycoplana*, *Dorea*, and *Blautia* genera in the MS patients, while *Parabacteroides*, *Prevotella*, and *Adlercreutzia* species were abundant (Chen et al. 2016). The phenotype of immune T and B cells are shifted to pro-inflammatory phenotype in MS, a phenomenon which is regulated by the gut microbiota via secretion of its bioactive metabolites. In particular, the lymphoid B-cell populations are known to be susceptible to microbiome alterations in MS (Longbrake and Cross 2016). The major functions of B cells, viz., antigen presentation and proinflammatory cytokine secretion, are aberrantly activated in MS (Ireland et al. 2016). Immune activation of IL-10-secreting CD4+ T cells (Th17 cells) may also represent a notable pathway linking gut microbiome and pathogenesis of MS (Cekanaviciute et al. 2017; Lee et al. 2011). *Clostridium* species, for example, may promote T17 cell activation and are significantly increased in relapsing-remitting MS (RRMS) patients (Miyake et al. 2015). *Bacteroides fragilis*-derived polysaccharide A (PSA) may represent a therapeutic agent against EAE because of its ability to regulate the immune responses of CD39+-CD4+ T cells and FoxP3+ Treg cells in a Toll-like receptor 2 (TLR-2)-dependent manner (Wang et al. 2014). Interestingly, the expression of lipid 654 which is derived from *Bacteroides* spp. in the gut and acts as a TLR-2 ligand is repressed in MS subjects (Farrokhi et al. 2013). *Lactobacillus* spp. is another class of microbial species that are thought to regulate the functions of IL-10-secreting Treg cells in animal models of EAE (Lavasani et al. 2010). Glial activation is another pathogenic mechanism that is thought to be modulated by the gut microbiota in MS. For instance, lipopolysaccharides of *Porphyromonas gingivalis* are recognized by TLR2 and have been shown to activate glial cells and promote neuroinflammation (Shapira et al. 2002).

The blood-brain barrier (BBB; Sect. 3.3.1) is another point of interest linking gut microbiota and MS pathology. SCFAs such as acetate, propionate, and butyrate derived from gut microbial-mediated fermentation of dietary fibre have been shown to potently regulate BBB. Of note, the population of *Clostridium* spp. which produces SCFAs (particularly of the XIVa and IV Clusters) is altered in MS subjects as evaluated by 16S ribosomal RNA (rRNA) sequencing in faecal samples (Miyake et al. 2015). *Faecalibacterium prausnitzii*, a butyrate-producing member of the clostridial cluster IV involved in the promotion of anti-inflammatory signalling, is severely decreased in the colons of MS patients (Swidsinski et al. 2017).

3.5.2 Microbiome in Neurodegenerative Diseases

The gut microbiome is associated with altering susceptibility to and progression of neurodegenerative diseases (NDs) (Kim et al. 2022). When considering the treatment of NDs, the gut microbiota is known to play important roles in human physiology and pathology and in the development of microbiome-based therapies (Fang et al. 2020). Techniques such as germ-free rearing, antibiotic treatment, and faecal microbiota transplantation which are developed by manipulating the microbiome have been shown to have effects that can ultimately modify the progression of NDs (Sampson et al. 2016).

Even though the exact relationship involving the gut microbiota and the NDs remains indistinct, researchers are exploring the intestinal microbiome to decipher the potential pathogenic or therapeutic effects on the disease (Sampson et al. 2016). The gut microbiome is known to affect brain health, and scientists are starting to work out how the gut microbiome can affect NDs (Hill-Burns et al. 2017).

In particular, the gut microbiome has been in the progression of NDs such as AD, PD, multiple sclerosis, amyotrophic lateral sclerosis, and frontotemporal dementia. The gut-brain axis is an interactive yet complex network that involves the linking of the brain to the gut as well as establishing a bidirectional communication between the gastrointestinal tract and the CNS. In many NDs, perturbations of the gut-brain axis have been reported, signifying a possible role in disease pathogenesis and making it a potential therapeutic target. Compositional alterations of the gut microbiome have been linked to dysfunction in the gut-brain axis as well as degeneration and inflammation of the CNS (Ghezzi et al. 2022).

The homeostatic condition of the CNS may be influenced by the gut microbiome by the production of molecules and metabolites that modulates the immune system. The gut microbiome is also known to play a role in ageing and NDs (Kim et al.

2022). Researchers have made some findings stating that microbiome-based therapeutics can be used for the treatment of NDs, and it is necessary to develop a customized treatment plan to maximize their effectiveness (Sampson et al. 2016). Brain health is believed to be impacted by the gut microbiome, and scientists are starting to work out how the gut microbiome can affect NDs (Hill-Burns et al. 2017; Kim et al. 2019).

The microbiome influences brain function and may play a role in NDs by promoting or attenuating inflammation, immunologic activation, and direct effects on protein aggregation (Sampson et al. 2016). Patients with PD and AD are deficient in "good" bacteria, which may help explain the development of these conditions according to studies. A relatively new method of determining the specific species of bacteria linked to neurodegeneration is under limelight to study the origins of these disorders (Hill-Burns et al. 2017). Further research is needed to fully understand the relationship between the microbiome and neurodegenerative diseases and to develop potential therapeutic approaches.

3.5.2.1 Alzheimer's Disease

Alzheimer's disease (AD), a neurodegenerative illness, is the most often identified cause of dementia in elderly people. Recent research has demonstrated that changes in the gut microbiota have a direct impact on cognitive decline and actively contribute to AD (Varesi et al. 2022). The gut microbiome of AD patients differs from control age- and sex-matched persons in composition and has less microbial diversity (Vogt et al. 2017). Any age-related neurological condition, such as AD, and mood problems is influenced by the composition of the gut flora (Varesi et al. 2022).

It is understood that through the microbiota-gut-brain axis (MGBA), the gut microbiota can communicate with the central nervous system (CNS). The gut bacteria, the enteric nervous system, and the central nervous system all participate in the bidirectional MGBA communication system. The gut microbiota affects physiological homeostasis as it should be and plays a role in the aetiology of disorders ranging from obesity to neurodegenerative illnesses like AD. Multiple studies have used 16S rRNA sequencing to identify distinct microbiota alterations in AD patients and transgenic mice, highlighting the important role of *Bacteroides fragilis* in cognitive impairment and brain amyloidosis (Zhang et al. 2017a). The development and course of AD may be influenced by the gut flora. The microbiota-gut-brain axis may be used by probiotics to control cognitive deterioration in AD. However, further research is needed to establish a cause-effect relationship among gut microbiota, diet, and neurodegeneration (Vogt et al. 2017).

Probiotics are living bacteria that give the host health advantages when taken in sufficient quantities. The MGBA may be used to control cognitive deterioration in AD through probiotic intervention. It has been proved that the probiotic strain *Lactobacillus plantarum* PS128 enhances cognitive performance in AD patients. However, there is little clinical evidence between gut microbiota, nutrition, and neurodegeneration to support a cause-and-effect relationship between these factors (Cattaneo et al. 2017). Therefore, more investigations are required to thoroughly comprehend the gut microbiome's role in AD and to create efficient treatments.

The microbiota-gut-brain axis, a two-way communication mechanism that enables the gut to affect the brain via neuronal, endocrine, immunological, and metabolic signalling, influences AD through the gut microbiome (Kowalski and Mulak 2019; Burokas et al. 2015). Through several processes, such as the generation of neurotransmitters, short-chain fatty acids, and immune system modulation, the gut microbiota can influence the CNS (Liu et al. 2020). Additionally, the blood-brain barrier (BBB) and immune system are also impacted by the gut microbiota, which can result in neuroinflammation and the build-up of amyloid-beta (A) (Zhuang et al. 2018). Two pathological markers of AD, the deposition of A and the formation of neurofibrillary tangles, are mediated by chronic neuroinflammation (Shabbir et al. 2021). Because of this, the microbiota-gut-brain axis plays a critical role in the pathophysiology and development of AD.

According to several studies, AD patients had changed gut microbiomes with less microbial diversity and compositional differences from control subjects who were matched for age and sex (Burokas et al. 2015; Zhuang et al. 2018). The gut microbiota may offer new targets for treatments and knowledge of AD pathophysiology (Kowalski and Mulak 2019). Probiotics and dietary changes are two potential treatments. By modifying the host's metabolic system, immunological system, neurological system, and endocrine system, symbiotic gut microbiomes are known to sustain brain health (Shabbir et al. 2021). Therefore, comprehending how the gut microbiota contributes to AD may provide new information about its pathogenesis and potential treatments (Liu et al. 2020).

3.5.2.2 Parkinson's Disease

Parkinson's disease (PD) is a neurodegenerative condition that affects the central nervous system. Recent research has demonstrated that the gut microbiome is critical to the aetiology and management of PD. A growing body of research has connected gut microorganisms to the aetiology and symptoms of PD, highlighting the gut microbiota as an essential regulator of neurodegenerative illnesses. Gastrointestinal symptoms and changes to the enteric nerve system may precede PD (Wallen et al. 2022). The gut-brain axis can be altered by changes in the gut microbiome, which can result in neurodegeneration, oxidative stress, and inflammation (Liang et al. 2021).

A meta-analysis of the PD gut microbiome revealed that although there have been reports of altered gut microbiomes in PD, the results have been inconsistent, most likely because of variations in study design and patient demographics. Another study reported that the pathophysiology and therapy of PD are both affected by the chemicals produced by gut microbes. The research also implies that probiotic-based supplements, dietary changes, and microbial products may be useful in treating PD patients who have gut microbial dysbiosis (Liang et al. 2021).

Studies on the microbiome-gut-brain axis in PD have focused on the gut microbiome's role in the disease's development and treatment. To learn more about the various disease pathways, researchers are examining the alterations in the gut microbiota associated with PD. The possibility of the gut microbiota as a therapeutic target for PD is currently being investigated (Tan et al. 2022).

The pathogenesis and therapy of PD both heavily depend on the gut microbiota. The gut-brain axis can be impacted by changes in the gut microbiome, which can result in neurodegeneration, oxidative stress, and inflammation. Researchers are looking at the altered gut microbiota in PD to learn more about the many disease causes and as a potential therapeutic target. Further research is needed to fully understand the role of the gut microbiome in PD and to develop effective treatments.

Numerous studies have found altered gut microbiomes in PD, although the results are inconsistent, most likely because of variations in the populations and study designs (Petrov et al. 2017). However, PD patients' gut microbiomes differ significantly from those of healthy controls, according to a meta-analysis study (Romano et al. 2021). In particular, butyrate-producing bacteria are less prevalent, and pro-inflammatory microorganisms are more prevalent in PD patients. These alterations in the gut microbiota could be a factor in the neuroinflammation and neurodegeneration seen in Parkinson's disease.

Butyrate is a short-chain fatty acid that is generated by gut bacteria and has anti-inflammatory and neuroprotective properties (Sun and Shen 2018). Patients with PD have lower levels of butyrate-producing bacteria, which could be a factor in the neuroinflammation and neurodegeneration they experience. Proteobacteria, a pro-inflammatory bacterium, are more prevalent in PD patients and may have a role in the gut inflammation that is associated with the disease.

To sum up, PD patients' gut microbiomes shift, with butyrate-producing bacteria declining and pro-inflammatory bacteria rising. These modifications could be a factor in the neuroinflammation and neurodegeneration seen in PD. To better understand the mechanisms underpinning the gut-brain axis in PD and to create microbiome-based treatments for this crippling condition, further study is required.

3.5.2.3 Amyotrophic Lateral Sclerosis (ALS)

Several recent studies have focussed on understanding the implications of resident microbiota dysbiosis in the pathogenesis of amyotrophic lateral sclerosis (ALS), a degenerative condition that primarily affects motor neurons. In addition to serving as a target for environmental triggers for ALS, gut microbiota may also act as modifying (aggravating or inhibiting) factors during the progression of the disease.

Superoxide dismutase 1 transgenic (SOD1-Tg) mice have served as an excellent model for understanding the pathophysiology of ALS, including its links to the resident microbial population. For example, Wu et al. showed a deficiency of butyrate-producing *Butyrivibrio fibrisolvens* in the GI tract of SOD1-Tg mice (Wu et al. 2015); interestingly, they also proposed that exogenous butyrate supplementation could alleviate ALS progression in these mice (Zhang et al. 2017b) (PMID: 28129947). Blacher et al. provided evidence for an altered microbiome in SOD1-Tg mice which showed aggravation of the disease phenotype in germ-free settings and upon treatment with broad-spectrum antibiotics. They proposed *Akkermansia muciniphila* as a beneficial microbial species, and *Ruminococcus torques* and *Parabacteroides distasonis* as harmful ones concerning ALS symptoms. Further, they proposed that bacterial-derived nicotinamide might offer protection against ALS (Blacher et al. 2019). More recently, it has been suggested that alterations in gut microbiome may be linked with inflammatory responses during the pathogenesis of ALS in SOD1-Tg mice. Further, they found evidence for prodromal dysbiosis of the resident microbial populations in the gut, preceding the onset of muscle atrophy and motor dysfunction, the classical symptoms of ALS (Figueroa-Romero et al. 2019). Immune responses may be a relevant target of the gut microbiome in ALS pathogenesis, as supported by a study in SOD1-Tg mice.

Evidence from clinical studies has provided contrasting results for the participation of the gut-brain-microbiome axis in the induction and progression of ALS. Analyses of faecal samples from ALS subjects indicated significantly reduced expression of *Ruminococcus* spp. and altered *Firmicutes/Bacteroidetes* ratio compared to control subjects (Rowin et al. 2017). In concurrence, Di Gioia and coworkers observed substantial changes in the identities of the microbial species in ALS patients vs. controls, with elevated levels of proinflammatory *Cyanobacteria* species and downregulated levels of potentially protective *Bacteroidetes* species. However, the severity of disease symptoms was not found to be associated with microbial dyshomeostasis (Brenner et al. 2018). In agreement with the study by Blacher et al. in SOD1-Tg mice, the risk of developing ALS was proposed to be significantly increased upon repeated use of antibiotics, particularly penicillin and related agents that act on beta-lactamase (Sun et al. 2019).

3.5.2.4 Stroke

Stroke is a severe cerebrovascular condition that is characterized by an interruption of blood flow to the brain. With regards to animal modes, middle cerebral artery occlusion (MCAO) has served as a useful tool to study the pathogenic pathways and factors responsible for stroke. As expected, several environmental factors can affect the pathology of stroke. Hypertension and diabetes are the two major risk factors for stroke. Because of the capability of the resident microbiota to impact the metabolic status of the host, the gut-brain axis is thought to have tremendous implications for the pathogenesis of stroke.

Atherosclerosis is a critical facet of stroke pathogenesis. Gut microbiota are known to influence the formation of atherosclerosis and thrombosis by inducing immune cell activation, possibly via alterations in the levels of trimethylamine *N*-oxide generated by the metabolism of dietary phosphatidylcholine (Haghikia et al. 2018). Indeed, microbial-derived trimethylamine *N*-oxide is associated with cardiovascular risk in human subjects (Tang et al. 2013). Several species of gut microflora are thought to be involved in the development of atherosclerotic plaques, including *Porphyromonas gingivalis (*Hayashi et al. 2011*)* and *Chlamydia pneumonia* (Blessing et al. 2001). Activation of blood platelets and the consequently ensuing thrombosis may also be related to the metabolic action of the resident microbiota. Phenylacetylglutamine and phenylacetylglycine are two major bacterial metabolites produced from the catabolic actions of dietary phenylalanine that can bind to β2 adrenergic receptors on palates to activate them. The consequent induction of thrombosis may significantly elevate the risk for the development of cardiovascular events such as stroke (Nemet et al. 2020). Cholinergic signalling during stroke pathogenesis is also a target of the gut microbial species and metabolites derived from them. Thus, Wu and colleagues demonstrated that microbiota imbalance in favour of species of the *Proteobacteria* and *Deferribacteres* phyla and against *Firmicutes* species may alter cholinergic signalling by modulating the host metabolic profile (Wu et al. 2021).

The gut-brain axis is also thought to influence brain repair mechanisms following stroke via microbial actions on immune responses. Dysregulation of gut microflora populations may be a prognostic mechanism closely linked with the events during and following stroke. Altered levels of Treg and IL-17 secreting γδ T cells induced by microbial dyshomeostasis potentially stimulate stroke development (Benakis et al. 2016). Dysbiosis of the gut microbiota and increased presence of *Enterobacteriaceae* species has been proposed as a pathway for hyperactivation of neuroinflammatory signalling in stroke (Xu et al. 2021). Similar findings of microflora alterations and reduced diversity have been observed in pig guts in an MCAO model and the consequent alterations in immune responses as observed by the elevated plasma levels of TNF-α and IL-6 (Jeon et al. 2020). Gut bacteria-mediated activation of triggering receptor expressed on myeloid cells 1 (TREM1) on peripheral CD11b+CD45+ myeloid cells is thought to allow their migration to its ischaemic brain resulting in significantly elevated levels of neuroinflammation (Liu et al. 2019). Gut-derived activated CD4+ T-cell translocation to meninges has also been shown to exacerbate ischaemic brain injury in a more recent study (Feng et al. 2019).

3.6 Targeting the Microbiome

3.6.1 Diagnosis and Prognosis

The host and its microbiota interact in many ways including as commensals, symbionts, or pathogens and play diverse roles in human pathophysiology by modulating direct and indirect pathways as mentioned earlier. Dysbiosis of these gut microbiota homeostasis might contribute to many life-threatening diseases. Microbial dysbiosis can be a potential biomarker diagnosis, prognosis, and treatment response for multiple diseases. Moreover, metabolites and small molecules derived from microbiota can trigger host inflammatory activation in the CNS contributing to the pathogens of multiple neuropathologies. These metabolites or small molecules along with host-derived cytokines act as potentially efficient diagnostics and prognostic markers. Studies in drug-naïve patients diagnosed with early-stage Parkinson's disease (PD) suggest that plasma levels of gut microbiota-derived trimethylamine *N*-oxide can be used as a biomarker in early PD. Dysbiosis of the gut microbiota, specifically enrichment of *Enterococcus* and depletion of *Prevotella*, may be related to enhanced risk of development of hypertensive intracerebral haemorrhage and stroke-associated pneumonia (SAP). These gut microbiota compositions along with the serum cytokine profiles are potential biomarkers for the ICH. The gut presence of *Prevotellaceae* is tremendously decreased in PD, and this may serve as a sensitive and specific biomarker for the disease.

3.6.2 Therapeutic Targets

The gut-brain axis serves as a conduit for information transfer between the CNS and the gut microbiota. The creation of short-chain fatty acids, modifications to neurotransmitters and gut peptides, and microbial effects on chronic inflammation and immunological function are just a few of the ways that the gut microbiota can have an impact on the CNS. Numerous neurological illnesses, including as obesity, cardiometabolic disease, spinal cord injury, traumatic brain injury, stroke, and PD, have the gut microbiota linked to their pathophysiology (Grochowska et al. 2022).

According to studies, microbial modification could be able to help treat or prevent certain neurological conditions (Chan et al. 2022). For instance, a healthy gut microbiome seems to enhance the effectiveness of brain tumour treatment. Microbiome dysbiosis can be treated in several ways, including faecal microbiota transplantation, oral probiotics, bacteriophages, genetic manipulation of gut microbiota, and vagus nerve stimulation (Willman et al. 2022).

It has been discovered that CNS infections reduce gut microbiota alpha diversity (Grochowska et al. 2022). One of the contributing aspects to the CNS's ability to maintain homeostasis is the taxonomical composition of the gut. Scientific evidence has put a lot of emphasis on the effectiveness of diet as a therapeutic target and how different dietary interventions may modify gut microbiota in different ways. Due to its hypothesized effects on gut microbiota modification, neuronal stability, and epileptogenesis prevention, the ketogenic diet (KD) in particular has a well-established role in other disorders including intractable epilepsy (Kaviyarasan et al. 2022).

Faecal microbiota transplant, oral probiotics, bacteriophages, genetic alteration of gut microbiota, and vagus nerve stimulation are a few of the main ways microbiome dysbiosis may be improved (Grochowska et al. 2022). Additionally, it has been demonstrated that the ketogenic diet modifies the gut flora and may one day be used to treat neurological illnesses. Research is still primarily focused on the effectiveness of diet as a target of treatment and how various dietary treatments may affect gut microbiota in different ways.

Additionally, the gut microbiome of patients with neuroinfections has been examined. It was discovered that, when compared to controls, the prevalence of bacterial taxa such as *Clostridium*, *Anaerostipes*, *Lachnobacterium*, *Lachnospira*, and *Roseburia* was lower in patients with neuroinfection. Although there were no differences in beta diversity, patients with neuroinfections had reduced within-sample diversity as measured by alpha diversity measurements. Furthermore, despite alpha diversity measurements coming close to statistical significance, short-term antibiotic therapy had no discernible impact on the gut microbiota.

There is rising interest in the possibility of manipulating the microbiome as a therapeutic target for a variety of neurological illnesses due to the complicated interaction between the gut microbiota and the central nervous system. To find the most efficient treatments for gut-brain axis illnesses, however, and to establish direct causal links between altered gut microbiota and a variety of disorders, more studies are required.

3.7 Microbiota as Regulators of Longevity and Their Anti-aging Functions

The ageing and lifespan processes have been revealed to be greatly influenced by the gut flora (Sonowal et al. 2017). According to studies, the gut microbiota changes significantly as we age, resulting in increased disease vulnerability and immune system dysregulation. It has been discovered that among older persons who are ageing normally and successfully, alpha diversity increases with age. Extremely long-lived people's gut microbiota makeup has also been investigated, and some studies have shown certain microbial signatures linked to longevity (Zhang et al. 2017a). By activating the innate immune system, age-related gut dysbiosis may contribute to unhealthy ageing and decreased longevity (Biagi et al. 2010). Studies on aged rats fed a low-fat diet enhanced with *Lactobacillus* and subjected to calorie restriction have revealed improved organ function (O'Toole 2017). It is becoming increasingly clear that the microbes in the human gut play crucial roles in health and disease and may influence the ageing process (Kong et al. 2016).

According to recent studies, the gut flora can affect ageing and longevity. For instance, research has revealed that older persons' gut microbiota is less varied than that of younger ones and that this reduction in diversity is linked to a higher risk of developing age-related disorders. Probiotics and faecal microbiota transplantation are two further therapies that have been found to modify the gut microbiota and increase health and lifespan in animal models. These results imply that a possible anti-aging target may be the gut microbiota (Backhed et al. 2007).

The gut microbiota may regulate innate immunity, sarcopenia, and cognitive function alterations that are brought on by ageing and are all components of frailty (O'Toole and Jeffery 2015). The human gut microbiome has an impact on host health either directly or through gut bacteria linked to autoimmunity. Through microbial, viral, and nutritional therapies, the gut microbiota may be revived, which may enhance immunological homeostasis, guard against oxidative stress, and halt the ageing process (Singh et al. 2023). By preserving intestinal epithelial integrity, promoting digestion, developing the intestinal immune system, and suppressing the growth of harmful bacteria, the gut microbiota affects the long-term homeostasis of metazoans. Age-related chronic diseases are linked to abnormal changes in the gut microbiota (Kim et al. 2022).

3.8 Conclusion

The central nervous system (CNS), including brain circuitry, neurophysiology, and behaviour, has been shown to critically depend on the gut microbiota for proper functioning. The gut-brain axis is an essential component involved in a variety of CNS disorders, where the gut-brain axis is implicated, such as Alzheimer's disease, Parkinson's disease, Autism spectrum disorder, Anxiety and depression. Over the last few years, research has demonstrated that the gut microbiome plays a role in basic neurodegenerative processes from the creation of the blood-brain barrier to neurogenesis. The vagus nerve connects the gut to the brain directly, and bacteria can trigger the enteric nervous system's afferent neurons. Microbial species dysbiosis can cause aberrant immunological signalling, and host homeostasis imbalance, leading to the progression of illness related to CNS. The gut-brain axis describes the relationship between bacteria and their interaction with the brain, which results in changes in CNS status. The function of the gut microbiome and the gut-brain interplay in specific CNS illnesses has been established, and it has been proposed that a healthy gut microbiota is significant to the prevention and treatment of CNS diseases.

Furthermore, mechanistic research into the gut-brain axis is required to better understand the underlying causes of CNS illnesses and to develop new therapeutic targets. The creation of accurate biomarkers for CNS illnesses may aid in the identification of persons at risk of developing these diseases as well as the monitoring of disease progression. The incorporation of gut-brain axis research into clinical practice may result in new CNS diagnostic and therapeutic strategies such as faecal microbiota transplantation (FMT) and probiotics, targeting microbiome metabolites may open up new avenues for the treatment of CNS illnesses. Personalized medicine may be used to design and target specific gut microbiome and its metabolite patterns linked to CNS illnesses.

To summarize, effective treatment options for CNS disorders remain an active subject due to the intricacy of the aetiologies and the scarcity of appropriate biomarkers in people. The gut microbiome and gut-brain axis are a potential field of research to understand the mechanics of this complicated system and to design effective biomarkers and treatments that could lead to new CNS disease treatment strategies.

References

Asano Y, Hiramoto T, Nishino R, Aiba Y, Kimura T, Yoshihara K et al (2012) Critical role of gut microbiota in the production of biologically active, free catecholamines in the gut lumen of mice. Am J Physiol Liver Physiol [Internet] 303:G1288–G1295. https://www.physiology.org/doi/10.1152/ajpgi.00341.2012

Backhed F, Ley RE, Sonnenburg JL, Peterson DA, Gordon JI (2007) Host-microbe interactions in the gut microbiota. Science 318(5857):1380–1384

Benakis C, Brea D, Caballero S, Faraco G, Moore J, Murphy M et al (2016) Commensal microbiota affects ischemic stroke outcome by regulating intestinal γδ T cells. Nat Med [Internet] 22:516–523. http://www.ncbi.nlm.nih.gov/pubmed/27019327

Bercik P, Denou E, Collins J, Jackson W, Lu J, Jury J et al (2011) The intestinal microbiota affect central levels of brain-derived neurotropic factor and behavior in mice. Gastroenterol [Internet] 141:599–609, 609.e1–3. http://www.ncbi.nlm.nih.gov/pubmed/21683077

Berthoud H-R, Neuhuber WL (2000) Functional and chemical anatomy of the afferent vagal system. Auton Neurosci [Internet] 85:1–17. https://linkinghub.elsevier.com/retrieve/pii/S1566070200002150

Berthoud HR, Blackshaw LA, Brookes SJH, Grundy D (2004) Neuroanatomy of extrinsic afferents supplying the gastrointestinal tract. Neurogastroenterol Motil [Internet] 16:28–33. https://onlinelibrary.wiley.com/doi/10.1111/j.1743-3150.2004.00471.x

Biagi E, Nylund L, Candela M, Ostan R, Bucci L, Pini E et al (2010) Through ageing, and beyond: gut microbiota and inflammatory status in seniors and centenarians. Ahmed N, editor. PLoS One [Internet] 5:e10667. https://dx.plos.org/10.1371/journal.pone.0010667

Bienenstock J (2012) Commensal communication to the brain: pathways and behavioral consequences. Microb Ecol Heal Dis [Internet] 23. http://www.microbecolhealthdis.net/index.php/mehd/article/view/19007

Bilbo SD, Schwarz JM (2012) The immune system and developmental programming of brain and behavior. Front Neuroendocrinol [Internet] 33:267–286. https://linkinghub.elsevier.com/retrieve/pii/S0091302212000386

Blacher E, Bashiardes S, Shapiro H, Rothschild D, Mor U, Dori-Bachash M et al (2019) Potential roles of gut microbiome and metabolites in modulating ALS in mice. Nat [Internet] 572:474–480. http://www.ncbi.nlm.nih.gov/pubmed/31330533

Blessing E, Campbell LA, Rosenfeld ME, Chough N, Kuo CC (2001) Chlamydia pneumoniae infection accelerates hyperlipidemia induced atherosclerotic lesion development in C57BL/6J mice. Atherosclerosis [Internet] 158:13–17. http://www.ncbi.nlm.nih.gov/pubmed/11500169

Bonaz B, Sinniger V, Hoffmann D, Clarençon D, Mathieu N, Dantzer C et al (2016) Chronic vagus nerve stimulation in Crohn's disease: a 6-month follow-up pilot study. Neurogastroenterol Motil [Internet] 28:948–953. https://onlinelibrary.wiley.com/doi/10.1111/nmo.12792

Braniste V, Al-Asmakh M, Kowal C, Anuar F, Abbaspour A, Tóth M et al (2014) The gut microbiota influences blood-brain barrier permeability in mice. Sci Transl Med [Internet] 6:263ra158. http://www.ncbi.nlm.nih.gov/pubmed/25411471

Breit S, Kupferberg A, Rogler G, Hasler G (2018) Vagus nerve as modulator of the brain–gut axis in psychiatric and inflammatory disorders. Front Psychiatry [Internet] 9:44. http://journal.frontiersin.org/article/10.3389/fpsyt.2018.00044/full

Brenner D, Hiergeist A, Adis C, Mayer B, Gessner A, Ludolph AC et al (2018) The fecal microbiome of ALS patients. Neurobiol Aging [Internet] 61:132–137. http://www.ncbi.nlm.nih.gov/pubmed/29065369

Browning JS, Houseworth JH (1953) Development of new symptoms following medical and surgical treatment for duodenal ulcer. Psychosom Med [Internet] 15:328–336. http://journals.lww.com/00006842-195307000-00006

Burokas A, Moloney RD, Dinan TG, Cryan JF (2015) Microbiota regulation of the mammalian gut–brain axis. Adv Appl Microbiol 91:1–62. https://linkinghub.elsevier.com/retrieve/pii/S0065216415000027

Cani PD, Knauf C (2016) How gut microbes talk to organs: the role of endocrine and nervous routes. Mol Metab [Internet] 5:743–752. http://www.ncbi.nlm.nih.gov/pubmed/27617197

Caputi V, Marsilio I, Filpa V, Cerantola S, Orso G, Bistoletti M et al (2017) Antibiotic-induced dysbiosis of the microbiota impairs gut neuromuscular function in juvenile mice. Br J Pharmacol [Internet] 174:3623–3639. https://onlinelibrary.wiley.com/doi/10.1111/bph.13965

Carding S, Verbeke K, Vipond DT, Corfe BM, Owen LJ (2015) Dysbiosis of the gut microbiota in disease. Microb Ecol Health Dis [Internet] 26:26191. http://www.ncbi.nlm.nih.gov/pubmed/4315779

Cassel SL, Sutterwala FS, Flavell RA (2008) The tiny conductor: immune regulation via commensal organisms. Cell Host Microbe [Internet] 3:340–341. https://linkinghub.elsevier.com/retrieve/pii/S193131280800156X

Cattaneo A, Cattane N, Galluzzi S, Provasi S, Lopizzo N, Festari C et al (2017) Association of brain amyloidosis with pro-inflammatory gut bacterial taxa and peripheral inflammation markers in cognitively impaired elderly. Neurobiol Aging [Internet] 49:60–68. https://linkinghub.elsevier.com/retrieve/pii/S019745801630197X

Cekanaviciute E, Yoo BB, Runia TF, Debelius JW, Singh S, Nelson CA et al (2017) Gut bacteria from multiple sclerosis patients modulate human T cells and exacerbate symptoms in mouse models. Proc Natl Acad Sci U S A [Internet] 114:10713–10718. http://www.ncbi.nlm.nih.gov/pubmed/28893978

Chakravarthy K, Chaudhry H, Williams K, Christo PJ (2015) Review of the uses of vagal nerve stimulation in chronic pain management. Curr Pain Headache Rep [Internet] 19:54. http://link.springer.com/10.1007/s11916-015-0528-6

Chan DG, Ventura K, Villeneuve A, Du Bois P, Holahan MR (2022) Exploring the connection between the gut microbiome and Parkinson's disease symptom progression and pathology: implications for supplementary treatment options. J Parkinsons Dis [Internet] 12:2339–2352. https://www.medra.org/servlet/aliasResolver?alias=iospress&doi=10.3233/JPD-223461

Chen J, Chia N, Kalari KR, Yao JZ, Novotna M, Paz Soldan MM et al (2016) Multiple sclerosis patients have a distinct gut microbiota compared to healthy controls. Sci Rep [Internet] 6:28484. http://www.ncbi.nlm.nih.gov/pubmed/27346372

Chevalier G, Siopi E, Guenin-Macé L, Pascal M, Laval T, Rifflet A et al (2020) Effect of gut microbiota on depressive-like behaviors in mice is mediated by the endocannabinoid system. Nat Commun [Internet]. 11:6363. http://www.ncbi.nlm.nih.gov/pubmed/33311466

Costa M (2000) Anatomy and physiology of the enteric nervous system. Gut [Internet] 47:15iv–19iv. https://gut.bmj.com/lookup/doi/10.1136/gut.47.suppl_4.iv15

Delpech J-C, Madore C, Nadjar A, Joffre C, Wohleb ES, Layé S (2015) Microglia in neuronal plasticity: influence of stress. Neuropharmacology [Internet] 96:19–28. https://linkinghub.elsevier.com/retrieve/pii/S0028390815000052

Dinan TG, Quigley EMM, Ahmed SMM, Scully P, O'Brien S, O'Mahony L et al (2006) Hypothalamic-pituitary-gut axis dysregulation in irritable bowel syndrome: plasma cytokines as a potential biomarker? Gastroenterology [Internet] 130:304–311. https://linkinghub.elsevier.com/retrieve/pii/S0016508505023942

Eisenstein M (2016) Microbiome: bacterial broadband. Nature [Internet] 533:S104–S106. https://www.nature.com/articles/533S104a

El Aidy S, Dinan TG, Cryan JF (2015) Gut microbiota: the conductor in the orchestra of immune–neuroendocrine communication. Clin Ther [Internet] 37:954–967. https://linkinghub.elsevier.com/retrieve/pii/S0149291815001393

Erny D, Hrabě de Angelis AL, Jaitin D, Wieghofer P, Staszewski O, David E et al (2015) Host microbiota constantly control maturation and function of microglia in the CNS. Nat Neurosci [Internet] 18:965–977. https://www.nature.com/articles/nn.4030

Eshraghi RS, Davies C, Iyengar R, Perez L, Mittal R, Eshraghi AA (2020) Gut-induced inflammation during development may compromise the blood-brain barrier and predispose to autism spectrum disorder. J Clin Med [Internet] 10:27. http://www.ncbi.nlm.nih.gov/pubmed/33374296

Fang P, Kazmi SA, Jameson KG, Hsiao EY (2020) The microbiome as a modifier of neurodegenerative disease risk. Cell Host Microbe [Internet] 28:201–222. https://linkinghub.elsevier.com/retrieve/pii/S1931312820303498

Farrokhi V, Nemati R, Nichols FC, Yao X, Anstadt E, Fujiwara M et al (2013) Bacterial lipodipeptide, Lipid 654, is a microbiome-associated biomarker for multiple sclerosis. Clin Transl Immunol [Internet] 2:e8. http://www.ncbi.nlm.nih.gov/pubmed/25505950

Fasano A, Shea-Donohue T (2005) Mechanisms of disease: the role of intestinal barrier function in the pathogenesis of gastrointestinal autoimmune diseases. Nat Clin Pract Gastroenterol Hepatol [Internet] 2:416–422. https://www.nature.com/articles/ncpgasthep0259

Feng Y, He X, Luo S, Chen X, Long S, Liang F et al (2019) Chronic colitis induces meninges traffic of gut-derived T cells, unbalances M1 and M2 microglia/macrophage and increases ischemic brain injury in mice. Brain Res [Internet] 1707:8–17. http://www.ncbi.nlm.nih.gov/pubmed/30445026

Figueroa-Romero C, Guo K, Murdock BJ, Paez-Colasante X, Bassis CM, Mikhail KA et al (2019) Temporal evolution of the microbiome, immune system and epigenome with disease progression in ALS mice. Dis Model Mech [Internet] 13:dmm041947. http://www.ncbi.nlm.nih.gov/pubmed/31597644

Furness JB (2012) The enteric nervous system and neurogastroenterology. Nat Rev Gastroenterol Hepatol [Internet] 9:286–294. https://www.nature.com/articles/nrgastro.2012.32

Ganz J, Ratcliffe EM (2023) Who's talking to whom: microbiome-enteric nervous system interactions in early life. Am J Physiol Liver Physiol [Internet] 324:G196–G206. https://journals.physiology.org/doi/10.1152/ajpgi.00166.2022

Gariepy CE (2001) Intestinal motility disorders and development of the enteric nervous system. Pediatr Res [Internet] 49:605–613. https://www.nature.com/doifinder/10.1203/00006450-200105000-00001

Gershon MD (1999) The enteric nervous system: a second brain. Hosp Pract [Internet] 34:31–52. http://www.tandfonline.com/doi/full/10.3810/hp.1999.07.153

Gershon MD, Margolis KG (2021) The gut, its microbiome, and the brain: connections and communications. J Clin Invest [Internet] 131:e143768. https://www.jci.org/articles/view/143768

Ghezzi L, Cantoni C, Rotondo E, Galimberti D (2022) The gut microbiome–brain crosstalk in neurodegenerative diseases. Biomedicines [Internet] 10:1486. https://www.mdpi.com/2227-9059/10/7/1486

Goldstein A, Hofstra R, Burns A (2013) Building a brain in the gut: development of the enteric nervous system. Clin Genet [Internet] 83:307–316. https://onlinelibrary.wiley.com/doi/10.1111/cge.12054

Grenham S, Clarke G, Cryan JF, Dinan TG (2011) Brain? Gut? Microbe communication in health and disease. Front Physiol [Internet] 2:94. http://journal.frontiersin.org/article/10.3389/fphys.2011.00094/abstract

Grimonprez A, Raedt R, Baeken C, Boon P, Vonck K (2015) The antidepressant mechanism of action of vagus nerve stimulation: evidence from preclinical studies. Neurosci Biobehav Rev [Internet] 56:26–34. https://linkinghub.elsevier.com/retrieve/pii/S0149763415001736

Grochowska M, Laskus T, Paciorek M, Pollak A, Lechowicz U, Makowiecki M et al (2022) Patients with infections of the central nervous system have lowered gut microbiota alpha diversity. Curr Issues Mol Biol [Internet] 44:2903–2914. https://www.mdpi.com/1467-3045/44/7/200

Guthrie GD, Nicholson-Guthrie CS (1989) gamma-Aminobutyric acid uptake by a bacterial system with neurotransmitter binding characteristics. Proc Natl Acad Sci [Internet] 86:7378–7381. https://pnas.org/doi/full/10.1073/pnas.86.19.7378

Haghikia A, Li XS, Liman TG, Bledau N, Schmidt D, Zimmermann F et al (2018) Gut microbiota-dependent trimethylamine N-oxide predicts risk of cardiovascular events in patients with stroke and is related to proinflammatory monocytes. Arterioscler Thromb Vasc Biol [Internet] 38:2225–2235. http://www.ncbi.nlm.nih.gov/pubmed/29976769

Hasan N, Yang H (2019) Factors affecting the composition of the gut microbiota, and its modulation. PeerJ [Internet] 7:e7502. https://peerj.com/articles/7502

Hayashi C, Viereck J, Hua N, Phinikaridou A, Madrigal AG, Gibson FC et al (2011) Porphyromonas gingivalis accelerates inflammatory atherosclerosis in the innominate artery of ApoE deficient mice. Atherosclerosis [Internet] 215:52–59. http://www.ncbi.nlm.nih.gov/pubmed/21251656

Herman JP, McKlveen JM, Ghosal S, Kopp B, Wulsin A, Makinson R et al (2016) Regulation of the hypothalamic-pituitary-adrenocortical stress response. Compr Physiol [Internet] Wiley 6:603–621. https://onlinelibrary.wiley.com/doi/10.1002/cphy.c150015

Hickman SE, Kingery ND, Ohsumi TK, Borowsky ML, Wang L, Means TK et al (2013) The microglial sensome revealed by direct RNA sequencing. Nat Neurosci [Internet] 16:1896–1905. https://www.nature.com/articles/nn.3554

Hill-Burns EM, Debelius JW, Morton JT, Wissemann WT, Lewis MR, Wallen ZD et al (2017) Parkinson's disease and Parkinson's disease medications have distinct signatures of the gut microbiome. Mov Disord [Internet] 32:739–749. https://onlinelibrary.wiley.com/doi/10.1002/mds.26942

Hyland NP, Cryan JF (2016) Microbe-host interactions: influence of the gut microbiota on the enteric nervous system. Dev Biol [Internet] 417:182–187. https://linkinghub.elsevier.com/retrieve/pii/S0012160616301518

Ireland SJ, Guzman AA, Frohman EM, Monson NL (2016) B cells from relapsing remitting multiple sclerosis patients support neuro-antigen-specific Th17 responses. J Neuroimmunol [Internet] 291:46–53. http://www.ncbi.nlm.nih.gov/pubmed/26857494

Jänig W (2006) The integrative action of the autonomic nervous system: neurobiology of homeostasis. Integr Action Auton Nerv Syst Neurobiol Homeost 1–610

Jeon J, Lourenco J, Kaiser EE, Waters ES, Scheulin KM, Fang X et al (2020) Dynamic changes in the gut microbiome at the acute stage of ischemic stroke in a pig model. Front Neurosci [Internet] 14:587986. http://www.ncbi.nlm.nih.gov/pubmed/33343283

Kabouridis PS, Pachnis V (2015) Emerging roles of gut microbiota and the immune system in the development of the enteric nervous system. J Clin Invest [Internet] 125:956–964. http://www.jci.org/articles/view/76308

Kamada N, Chen GY, Inohara N, Núñez G (2013) Control of pathogens and pathobionts by the gut microbiota. Nat Immunol [Internet] 14:685–690. https://www.nature.com/articles/ni.2608

Kaviyarasan S, Chung Sia EL, Retinasamy T, Arulsamy A, Shaikh MF (2022) Regulation of gut microbiome by ketogenic diet in neurodegenerative diseases: a molecular crosstalk. Front Aging Neurosci [Internet] 14:1015837. https://www.frontiersin.org/articles/10.3389/fnagi.2022.1015837/full

Kelly JR, Minuto C, Cryan JF, Clarke G, Dinan TG (2021) The role of the gut microbiome in the development of schizophrenia. Schizophr Res [Internet] 234:4–23. http://www.ncbi.nlm.nih.gov/pubmed/32336581

Kettenmann H, Hanisch U-K, Noda M, Verkhratsky A (2011) Physiology of microglia. Physiol Rev [Internet] 91:461–553. https://www.physiology.org/doi/10.1152/physrev.00011.2010

Kim SU, de Vellis J (2005) Microglia in health and disease. J Neurosci Res [Internet] 81:302–313. https://onlinelibrary.wiley.com/doi/10.1002/jnr.20562

Kim S, Kwon S-H, Kam T-I, Panicker N, Karuppagounder SS, Lee S et al (2019) Transneuronal propagation of pathologic α-synuclein from the gut to the brain models Parkinson's disease. Neuron [Internet] 103:627–641.e7. https://linkinghub.elsevier.com/retrieve/pii/S089662731930488X

Kim N, Jeon SH, Ju IG, Gee MS, Do J, Oh MS et al (2021) Transplantation of gut microbiota derived from Alzheimer's disease mouse model impairs memory function and neurogenesis in C57BL/6 mice. Brain Behav Immun [Internet] 98:357–365. http://www.ncbi.nlm.nih.gov/pubmed/34500036

Kim KH, Chung Y, Huh J-W, Park DJ, Cho Y, Oh Y et al (2022) Gut microbiota of the young ameliorates physical fitness of the aged in mice. Microbiome [Internet] 10:238. https://microbiomejournal.biomedcentral.com/articles/10.1186/s40168-022-01386-w

Klarer M, Arnold M, Günther L, Winter C, Langhans W, Meyer U (2014) Gut vagal afferents differentially modulate innate anxiety and learned fear. J Neurosci [Internet] 34:7067–7076. https://www.jneurosci.org/lookup/doi/10.1523/JNEUROSCI.0252-14.2014

Klarer M, Weber-Stadlbauer U, Arnold M, Langhans W, Meyer U (2017) Cognitive effects of subdiaphragmatic vagal deafferentation in rats. Neurobiol Learn Mem [Internet] 142:190–199. https://linkinghub.elsevier.com/retrieve/pii/S1074742717300692

Klarer M, Krieger J-P, Richetto J, Weber-Stadlbauer U, Günther L, Winter C et al (2018) Abdominal vagal afferents modulate the brain transcriptome and behaviors relevant to schizophrenia. J Neurosci [Internet] 38:1634–1647. https://www.jneurosci.org/lookup/doi/10.1523/JNEUROSCI.0813-17.2017

Kong F, Hua Y, Zeng B, Ning R, Li Y, Zhao J (2016) Gut microbiota signatures of longevity. Curr Biol [Internet] 26:R832–R833. https://linkinghub.elsevier.com/retrieve/pii/S0960982216309277

Kowalski K, Mulak A (2019) Brain-gut-microbiota axis in Alzheimer's disease. J Neurogastroenterol Motil [Internet] 25:48–60. http://www.jnmjournal.org/journal/view.html?doi=10.5056/jnm18087

Krahl S (2012) Vagus nerve stimulation for epilepsy: a review of the peripheral mechanisms. Surg Neurol Int [Internet] 3:47. http://surgicalneurologyint.com/surgicalint-articles/vagus-nerve-stimulation-for-epilepsy-a-review-of-the-peripheral-mechanisms/

Kundu P, Lee HU, Garcia-Perez I, Tay EXY, Kim H, Faylon LE et al (2019) Neurogenesis and prolongevity signaling in young germ-free mice transplanted with the gut microbiota of old mice. Sci Transl Med [Internet] 11:eaau4760. http://www.ncbi.nlm.nih.gov/pubmed/31723038

Lavasani S, Dzhambazov B, Nouri M, Fåk F, Buske S, Molin G et al (2010) A novel probiotic mixture exerts a therapeutic effect on experimental autoimmune encephalomyelitis mediated by IL-10 producing regulatory T cells. PLoS One [Internet]. 5:e9009. http://www.ncbi.nlm.nih.gov/pubmed/20126401

Leclercq S, Mian FM, Stanisz AM, Bindels LB, Cambier E, Ben-Amram H et al (2017) Low-dose penicillin in early life induces long-term changes in murine gut microbiota, brain cytokines and behavior. Nat Commun [Internet]. 8:15062. http://www.ncbi.nlm.nih.gov/pubmed/28375200

Lee YK, Menezes JS, Umesaki Y, Mazmanian SK (2011) Proinflammatory T-cell responses to gut microbiota promote experimental autoimmune encephalomyelitis. Proc Natl Acad Sci U S A [Internet] 108(Suppl):4615–4622. http://www.ncbi.nlm.nih.gov/pubmed/20660719

Liang Y, Cui L, Gao J, Zhu M, Zhang Y, Zhang H-L (2021) Gut microbial metabolites in Parkinson's disease: implications of mitochondrial dysfunction in the pathogenesis and treatment. Mol Neurobiol [Internet] 58:3745–3758. https://link.springer.com/10.1007/s12035-021-02375-0

Liu Q, Johnson EM, Lam RK, Wang Q, Bo Ye H, Wilson EN et al (2019) Peripheral TREM1 responses to brain and intestinal immunogens amplify stroke severity. Nat Immunol [Internet] 20:1023–1034. http://www.ncbi.nlm.nih.gov/pubmed/31263278

Liu S, Gao J, Zhu M, Liu K, Zhang H-L (2020) Gut microbiota and dysbiosis in Alzheimer's disease: implications for pathogenesis and treatment. Mol Neurobiol [Internet] 57:5026–5043. https://link.springer.com/10.1007/s12035-020-02073-3

Longbrake EE, Cross AH (2016) Effect of multiple sclerosis disease-modifying therapies on B cells and humoral immunity. JAMA Neurol [Internet] 73:219–225. http://www.ncbi.nlm.nih.gov/pubmed/26720195

Lozupone CA, Stombaugh JI, Gordon JI, Jansson JK, Knight R (2012) Diversity, stability and resilience of the human gut microbiota. Nature [Internet] 489:220–230. https://www.nature.com/articles/nature11550

Mawe GM, Hoffman JM (2013) Serotonin signalling in the gut—functions, dysfunctions and therapeutic targets. Nat Rev Gastroenterol Hepatol [Internet] 10:473–486. https://www.nature.com/articles/nrgastro.2013.105

Mayer EA (2000) The neurobiology of stress and gastrointestinal disease. Gut [Internet]. 47:861–869. https://gut.bmj.com/lookup/doi/10.1136/gut.47.6.861

Mazmanian SK, Liu CH, Tzianabos AO, Kasper DL (2005) An immunomodulatory molecule of symbiotic bacteria directs maturation of the host immune system. Cell [Internet] 122:107–118. https://linkinghub.elsevier.com/retrieve/pii/S0092867405004514

Miyake S, Kim S, Suda W, Oshima K, Nakamura M, Matsuoka T et al (2015) Dysbiosis in the gut microbiota of patients with multiple sclerosis, with a striking depletion of species belonging to clostridia XIVa and IV clusters. PLoS One [Internet] 10:e0137429. http://www.ncbi.nlm.nih.gov/pubmed/26367776

Möhle L, Mattei D, Heimesaat MM, Bereswill S, Fischer A, Alutis M et al (2016) Ly6C(hi) monocytes provide a link between antibiotic-induced changes in gut microbiota and adult hippocampal neurogenesis. Cell Rep [Internet] 15:1945–1956. http://www.ncbi.nlm.nih.gov/pubmed/27210745

Mukherjee S, Ovchinnikova G, Stamatis D, Li CT, Chen I-MA, Kyrpides NC et al (2023) Standardized naming of microbiome samples in Genomes Online Database. Database [Internet] 2023. https://academic.oup.com/database/article/doi/10.1093/database/baad001/7042581

Nemet I, Saha PP, Gupta N, Zhu W, Romano KA, Skye SM et al (2020) A cardiovascular disease-linked gut microbial metabolite acts via adrenergic receptors. Cell [Internet] 180:862–877.e22. http://www.ncbi.nlm.nih.gov/pubmed/32142679

Nimmerjahn A, Kirchhoff F, Helmchen F (2005) Resting microglial cells are highly dynamic surveillants of brain parenchyma in vivo. Science (80) [Internet] 308:1314–1318. https://www.science.org/doi/10.1126/science.1110647

O'Toole PW (2017) The gut microbiome: driving our health into old age? Cell Host Microbe 22(6):728–730

O'Toole PW, Jeffery IB (2015) Gut microbiota and aging. Science (80) [Internet] 350:1214–1215. https://www.science.org/doi/10.1126/science.aac8469

Ochoa-Repáraz J, Kirby TO, Kasper LH (2018) The gut microbiome and multiple sclerosis. Cold Spring Harb Perspect Med [Internet] 8:a029017. http://www.ncbi.nlm.nih.gov/pubmed/29311123

Ojeda J, Ávila A, Vidal PM (2021) Gut microbiota interaction with the central nervous system throughout life. J Clin Med [Internet] 10:1299. https://www.mdpi.com/2077-0383/10/6/1299

Olson CA, Iñiguez AJ, Yang GE, Fang P, Pronovost GN, Jameson KG et al (2021) Alterations in the gut microbiota contribute to cognitive impairment induced by the ketogenic diet and hypoxia. Cell Host Microbe [Internet] 29:1378–1392.e6. http://www.ncbi.nlm.nih.gov/pubmed/34358434

Penry JK, Dean JC (1990) Prevention of intractable partial seizures by intermittent vagal stimulation in humans: preliminary results. Epilepsia [Internet] 31:S40–S43. https://onlinelibrary.wiley.com/doi/10.1111/j.1528-1157.1990.tb05848.x

Petrov VA, Saltykova IV, Zhukova IA, Alifirova VM, Zhukova NG, Dorofeeva YB et al (2017) Analysis of gut microbiota in patients with Parkinson's disease. Bull Exp Biol Med [Internet] 162:734–737. http://link.springer.com/10.1007/s10517-017-3700-7

Pokusaeva K, Johnson C, Luk B, Uribe G, Fu Y, Oezguen N et al (2017) GABA-producing Bifidobacterium dentium modulates visceral sensitivity in the intestine. Neurogastroenterol Motil [Internet] 29:e12904. https://onlinelibrary.wiley.com/doi/10.1111/nmo.12904

Powell N, Walker MM, Talley NJ (2017) The mucosal immune system: master regulator of bidirectional gut–brain communications. Nat Rev Gastroenterol Hepatol [Internet] 14:143–159. https://www.nature.com/articles/nrgastro.2016.191

Proctor LM (2011) The human microbiome project in 2011 and beyond. Cell Host Microbe [Internet]. 10:287–291. https://linkinghub.elsevier.com/retrieve/pii/S1931312811002940

Ransohoff RM, El Khoury J (2016) Microglia in health and disease. Cold Spring Harb Perspect Biol [Internet] 8:a020560. http://cshperspectives.cshlp.org/lookup/doi/10.1101/cshperspect.a020560

Rao M, Gershon MD (2016) The bowel and beyond: the enteric nervous system in neurological disorders. Nat Rev Gastroenterol Hepatol [Internet] 13:517–528. https://www.nature.com/articles/nrgastro.2016.107

Ren K, Randich A, Gebhart GF (1989) Vagal afferent modulation of spinal nociceptive transmission in the rat. J Neurophysiol [Internet] 62:401–415. https://www.physiology.org/doi/10.1152/jn.1989.62.2.401

Richard HT, Foster JW (2003) Acid resistance in Escherichia coli. Adv Appl Microbiol 52:167–186. https://linkinghub.elsevier.com/retrieve/pii/S0065216403010074

Rolig AS, Mittge EK, Ganz J, Troll JV, Melancon E, Wiles TJ et al (2017) The enteric nervous system promotes intestinal health by constraining microbiota composition. Gore J, editor. PLoS Biol [Internet] 15:e2000689. https://dx.plos.org/10.1371/journal.pbio.2000689

Romano S, Savva GM, Bedarf JR, Charles IG, Hildebrand F, Narbad A (2021) Meta-analysis of the Parkinson's disease gut microbiome suggests alterations linked to intestinal inflammation. NPJ Park Dis [Internet] 7:27. https://www.nature.com/articles/s41531-021-00156-z

Round JL, Mazmanian SK (2009) The gut microbiota shapes intestinal immune responses during health and disease. Nat Rev Immunol [Internet] 9:313–323. https://www.nature.com/articles/nri2515

Rowin J, Xia Y, Jung B, Sun J (2017) Gut inflammation and dysbiosis in human motor neuron disease. Physiol Rep [Internet] 5:e13443. http://www.ncbi.nlm.nih.gov/pubmed/28947596

Salvo E, Stokes P, Keogh CE, Brust-Mascher I, Hennessey C, Knotts TA et al (2020) A murine model of pediatric inflammatory bowel disease causes microbiota-gut-brain axis deficits in adulthood. Am J Physiol Gastrointest Liver Physiol [Internet] 319:G361–G374. http://www.ncbi.nlm.nih.gov/pubmed/32726162

Sampson TR, Debelius JW, Thron T, Janssen S, Shastri GG, Ilhan ZE et al (2016) Gut microbiota regulate motor deficits and neuroinflammation in a model of Parkinson's disease. Cell [Internet] 167:1469–1480.e12. http://www.ncbi.nlm.nih.gov/pubmed/27912057

Schafer DP, Stevens B (2015) Microglia function in central nervous system development and plasticity. Cold Spring Harb Perspect Biol [Internet] 7:a020545. http://cshperspectives.cshlp.org/lookup/doi/10.1101/cshperspect.a020545

Schemann M, Camilleri M (2013) Functions and disorders of the enteric nervous system. Nat Rev Gastroenterol Hepatol 10(6):338–351. https://doi.org/10.1038/nrgastro.2013.35

Sender R, Fuchs S, Milo R (2016) Revised estimates for the number of human and bacteria cells in the body. PLoS Biol [Internet] 14:e1002533. https://dx.plos.org/10.1371/journal.pbio.1002533

Shabbir U, Arshad MS, Sameen A, Oh D-H (2021) Crosstalk between gut and brain in Alzheimer's disease: the role of gut microbiota modulation strategies. Nutrients [Internet] 13:690. https://www.mdpi.com/2072-6643/13/2/690

Shaik L, Kashyap R, Thotamgari SR, Singh R, Khanna S (2020) Gut-brain axis and its neuro-psychiatric effects: a narrative review. Cureus [Internet] 12:e11131. https://www.cureus.com/articles/39481-gut-brain-axis-and-its-neuro-psychiatric-effects-a-narrative-review

Shapira L, Ayalon S, Brenner T (2002) Effects of Porphyromonas gingivalis on the central nervous system: activation of glial cells and exacerbation of experimental autoimmune encephalomyelitis. J Periodontol [Internet] 73:511–516. http://www.ncbi.nlm.nih.gov/pubmed/12027253

Shishov VA, Kirovskaya TA, Kudrin VS, Oleskin AV (2009) Amine neuromediators, their precursors, and oxidation products in the culture of Escherichia coli K-12. Appl Biochem Microbiol [Internet] 45:494–497. http://link.springer.com/10.1134/S0003683809050068

Singh B, Mal G, Sharma D, Al Nakshabandi A, Marotta F (2023) Gut microbiome and virome: anti-aging interventions. In: Gut microbiota in aging and chronic diseases. Springer International Publishing, Cham, pp 65–82

Siopi E, Chevalier G, Katsimpardi L, Saha S, Bigot M, Moigneu C et al (2020) Changes in gut microbiota by chronic stress impair the efficacy of fluoxetine. Cell Rep [Internet] 30:3682–3690.e6. http://www.ncbi.nlm.nih.gov/pubmed/32187541

Siragusa S, De Angelis M, Di Cagno R, Rizzello CG, Coda R, Gobbetti M (2007) Synthesis of γ-aminobutyric acid by lactic acid bacteria isolated from a variety of Italian cheeses. Appl Environ Microbiol [Internet] 73:7283–7290. https://journals.asm.org/doi/10.1128/AEM.01064-07

Smith SM, Vale WW (2006) The role of the hypothalamic-pituitary-adrenal axis in neuroendocrine responses to stress. Dialogues Clin Neurosci [Internet] 8:383–395. https://www.tandfonline.com/doi/full/10.31887/DCNS.2006.8.4/ssmith

Smith DK, Kassam T, Singh B, Elliott JF (1992) Escherichia coli has two homologous glutamate decarboxylase genes that map to distinct loci. J Bacteriol [Internet] 174:5820–5826. https://journals.asm.org/doi/10.1128/jb.174.18.5820-5826.1992

Sonowal R, Swimm A, Sahoo A, Luo L, Matsunaga Y, Wu Z et al (2017) Indoles from commensal bacteria extend healthspan. Proc Natl Acad Sci [Internet] 114:E7506–E7515. https://pnas.org/doi/full/10.1073/pnas.1706464114

Strandwitz P, Kim KH, Terekhova D, Liu JK, Sharma A, Levering J et al (2018) GABA-modulating bacteria of the human gut microbiota. Nat Microbiol [Internet] 4:396–403. https://www.nature.com/articles/s41564-018-0307-3

Sun M-F, Shen Y-Q (2018) Dysbiosis of gut microbiota and microbial metabolites in Parkinson's Disease. Ageing Res Rev [Internet] 45:53–61. https://linkinghub.elsevier.com/retrieve/pii/S1568163718300680

Sun J, Zhan Y, Mariosa D, Larsson H, Almqvist C, Ingre C et al (2019) Antibiotics use and risk of amyotrophic lateral sclerosis in Sweden. Eur J Neurol [Internet] 26:1355–1361. http://www.ncbi.nlm.nih.gov/pubmed/31087715

Swidsinski A, Dörffel Y, Loening-Baucke V, Gille C, Göktas Ö, Reißhauer A et al (2017) Reduced mass and diversity of the colonic microbiome in patients with multiple sclerosis and their improvement with ketogenic diet. Front Microbiol [Internet] 8:1141. http://www.ncbi.nlm.nih.gov/pubmed/28702003

Takeda K, Akira S (2004) Microbial recognition by Toll-like receptors. J Dermatol Sci [Internet] 34:73–82. https://linkinghub.elsevier.com/retrieve/pii/S0923181103002226

Tan AH, Lim SY, Lang AE (2022) The microbiome–gut–brain axis in Parkinson disease—from basic research to the clinic. Nat Rev Neurol [Internet] 18:476–495. https://www.nature.com/articles/s41582-022-00681-2

Tang WHW, Wang Z, Levison BS, Koeth RA, Britt EB, Fu X et al (2013) Intestinal microbial metabolism of phosphatidylcholine and cardiovascular risk. N Engl J Med [Internet] 368:1575–1584. http://www.ncbi.nlm.nih.gov/pubmed/23614584

Tremblay M-È, Stevens B, Sierra A, Wake H, Bessis A, Nimmerjahn A (2011) The role of microglia in the healthy brain: figure 1. J Neurosci [Internet] 31:16064–16069. https://www.jneurosci.org/lookup/doi/10.1523/JNEUROSCI.4158-11.2011

Tsavkelova EA, Botvinko IV, Kudrin VS, Oleskin AV (2000) Detection of neurotransmitter amines in microorganisms with the use of high-performance liquid chromatography. Dokl Biochem Proc Acad Sci USSR, Biochem Sect [Internet] 372:115–117. http://www.ncbi.nlm.nih.gov/pubmed/10935181

Tsigos C, Chrousos GP (2002) Hypothalamic–pituitary–adrenal axis, neuroendocrine factors and stress. J Psychosom Res [Internet] 53:865–871. https://linkinghub.elsevier.com/retrieve/pii/S0022399902004294

Varesi A, Pierella E, Romeo M, Piccini GB, Alfano C, Bjørklund G et al (2022) The potential role of gut microbiota in Alzheimer's disease: from diagnosis to treatment. Nutrients [Internet] 14:668. https://www.mdpi.com/2072-6643/14/3/668

Vogt NM, Kerby RL, Dill-McFarland KA, Harding SJ, Merluzzi AP, Johnson SC et al (2017) Gut microbiome alterations in Alzheimer's disease. Sci Rep [Internet] 7:13537. https://www.nature.com/articles/s41598-017-13601-y

Wallen ZD, Demirkan A, Twa G, Cohen G, Dean MN, Standaert DG et al (2022) Metagenomics of Parkinson's disease implicates the gut microbiome in multiple disease mechanisms. Nat Commun [Internet] 13:6958. https://www.nature.com/articles/s41467-022-34667-x

Wang Y, Telesford KM, Ochoa-Repáraz J, Haque-Begum S, Christy M, Kasper EJ et al (2014) An intestinal commensal symbiosis factor controls neuroinflammation via TLR2-mediated CD39 signalling. Nat Commun [Internet] 5:4432. http://www.ncbi.nlm.nih.gov/pubmed/25043484

Wei GZ, Martin KA, Xing PY, Agrawal R, Whiley L, Wood TK et al (2021) Tryptophan-metabolizing gut microbes regulate adult neurogenesis via the aryl hydrocarbon receptor. Proc Natl Acad Sci U S A [Internet] 118:e2021091118. http://www.ncbi.nlm.nih.gov/pubmed/34210797

Whitlock FA (1961) Some psychiatric consequences of gastrectomy. BMJ [Internet] 1:1560–1564. https://www.bmj.com/lookup/doi/10.1136/bmj.1.5239.1560

Willman J, Willman M, Reddy R, Fusco A, Sriram S, Mehkri Y et al (2022) Gut microbiome and neurosurgery: implications for treatment. Clin Transl Discov [Internet] 2:e139. https://onlinelibrary.wiley.com/doi/10.1002/ctd2.139

Wu S, Yi J, Zhang Y-G, Zhou J, Sun J (2015) Leaky intestine and impaired microbiome in an amyotrophic lateral sclerosis mouse model. Physiol Rep [Internet] 3:e12356. http://www.ncbi.nlm.nih.gov/pubmed/25847918

Wu W, Sun Y, Luo N, Cheng C, Jiang C, Yu Q et al (2021) Integrated 16S rRNA gene sequencing and LC-MS analysis revealed the interplay between gut microbiota and plasma metabolites in rats with ischemic stroke. J Mol Neurosci [Internet] 71:2095–2106. http://www.ncbi.nlm.nih.gov/pubmed/33954858

Xu K, Gao X, Xia G, Chen M, Zeng N, Wang S et al (2021) Rapid gut dysbiosis induced by stroke exacerbates brain infarction in turn. Gut [Internet]. http://www.ncbi.nlm.nih.gov/pubmed/33558272

Yamada E (1989) Further observation on the intraepithelial nerve fibers of rabbit ocular ciliary epithelium. Arch Histol Cytol [Internet] 52:191–195. http://www.ncbi.nlm.nih.gov/pubmed/2570599

Yamasaki R, Lu H, Butovsky O, Ohno N, Rietsch AM, Cialic R et al (2014) Differential roles of microglia and monocytes in the inflamed central nervous system. J Exp Med [Internet] 211:1533–1549. https://rupress.org/jem/article/211/8/1533/41731/Differential-roles-of-microglia-and-monocytes-in

Yang LL, Millischer V, Rodin S, MacFabe DF, Villaescusa JC, Lavebratt C (2020) Enteric short-chain fatty acids promote proliferation of human neural progenitor cells. J Neurochem [Internet] 154:635–646. http://www.ncbi.nlm.nih.gov/pubmed/31784978

Zhang L, Wang Y, Xiayu X, Shi C, Chen W, Song N et al (2017a) Altered gut microbiota in a mouse model of Alzheimer's disease. J Alzheimer's Dis [Internet] 60:1241–1257. https://www.medra.org/servlet/aliasResolver?alias=iospress&doi=10.3233/JAD-170020

Zhang Y-G, Wu S, Yi J, Xia Y, Jin D, Zhou J et al (2017b) Target intestinal microbiota to alleviate disease progression in amyotrophic lateral sclerosis. Clin Ther [Internet] 39:322–336. http://www.ncbi.nlm.nih.gov/pubmed/28129947

Zhuang Z-Q, Shen L-L, Li W-W, Fu X, Zeng F, Gui L et al (2018) Gut microbiota is altered in patients with Alzheimer's disease. Yu J-T, editor. J Alzheimer's Dis [Internet] 63:1337–1346. https://www.medra.org/servlet/aliasResolver?alias=iospress&doi=10.3233/JAD-180176

Gut-Brain Interplay and Cognitive Degeneration

4

Kalaiselvi Krishnamoorthy, Ponnulakshmi Rajagopal, and Selvaraj Jayaraman

Abstract

The dysregulation of the gut microbiota has been associated with the development of numerous gastrointestinal and extraintestinal disorders, and it has been found to be a critical regulator of many disease conditions. This correlation between the gut microbiota-brain axis and neurodegeneration has been established by recent research, which suggests that changes in the gut microbiome may be a contributing factor. This has made it possible to investigate novel microbiota-based treatments. Recent advances in research have greatly increased our understanding of the importance of the gut microbiota; this gut ecosystem includes viruses, archaea, bacteria, fungi, yeast, and eukaryote. Increasing evidence demonstrates how dynamic gut microbiota changes can affect brain physiology and behavior. Cognition was generally assumed to be controlled only by the central nervous system. The process of neurodegeneration, cerebrovascular illnesses, and cognitive dysfunction is now understood to be regulated and influenced by several non-nervous system variables, including the gut-resident bacteria of the gastrointestinal tract. The gut microbiome not only produces metabolites but also interacts with the immune system, potentially impacting the brain. Dysbiosis in the gut microbiome can lead to chronic inflammation, which has been associated with various neuropsychiatric conditions such as anxiety and depression. Overall, the gut microbiome is increasingly recognized as an impor-

K. Krishnamoorthy · S. Jayaraman (✉)
Department of Biochemistry, Centre of Molecular Medicine and Diagnostics (COMManD), Saveetha Dental College and Hospital, Saveetha Institute of Medical and Technical Sciences, Saveetha University, Chennai, India
e-mail: selvarajj.sdc@saveetha.com

P. Rajagopal
Department of Central Research Laboratory, Meenakhsi Ammal Dental College and Hospitals, Meenakhsi Academy of Higher Education and Research, Chennai, India

S. Pathak, A. Banerjee (eds.), *Gut Microbiome and Brain Ageing*,
https://doi.org/10.1007/978-981-99-8803-7_4

tant factor in brain function and mental health. Elucidating the intricate interactions within the gut-brain axis holds promise for developing novel therapeutic strategies to treat and manage neuropsychiatric disorders. Based on these observations, this chapter discusses how the gut microbiome is intricate in brain function.

Keywords

Microbiota · Stress · Food habit · Lifestyle · Brain function

4.1 Introduction

The Human Microbiome Project and the Metagenomics of the Human Digestive Tract Project were launched in 2007 to study the diversity of microbial communities within the human body. However, it is inaccurate to state that the colonization of the human gut by microorganisms begins in utero and is initiated by various microbial groups found in the amniotic fluid and placenta (Mariat et al. 2009). The interaction between the gut and the brain is termed the gut-brain axis. Microorganisms in the gut flora can be infectious, neutral, or beneficial to the host. The microbiota ecosystem includes all bacteria, fungi, viruses, and archaea (Swidsinski et al. 2005). Microorganisms have been an integral part of human history and play a critical role in maintaining human physiology. The human body is home to a vast array of microorganisms, most of which are bacteria, viruses, and fungi that reside in the gastrointestinal tract. These microorganisms, especially bacteria, are involved in various physiological functions within the body, such as host metabolism and immunity. In recent years, there has been a surge of research on the role of the gut microbiome (GM) in influencing the brain through the microbiome-gut-brain axis. The gut and central nervous system (CNS) are connected via a two-way communication pathway known as the gut-brain axis. This axis is believed to be influenced by multiple mechanisms, such as the immune system, endocrine system, and vagus nerve, to facilitate this communication. Studies have shown that alterations in the GM may be associated with a range of neurological and psychiatric conditions, including depression, anxiety, and autism spectrum disorders. Researchers are investigating the potential of using probiotics, prebiotics, and other interventions to modulate GM to prevent and treat these conditions. Overall, the growing body of research on the microbiome-gut-brain axis is shedding light on the intricate relationship between microorganisms and human health, providing new avenues for understanding and treating neurological and psychiatric disorders (Cryan et al. 2019; Vaher et al. 2022).

The trillions of microbes that reside in and on us are referred to as the "human microbiome" (Ursell et al. 2012). The study of the microbiome has advanced tremendously over the past two decades. The group of microorganisms is called the GM, and the majority of our microbial population lives in the gut. This GM plays a major role in human health and the proper function of body metabolism. For the digestion of food particles, GM will be used, and they can produce SCFAs that protect the host from carcinogens and help the intestinal mucosa develop an immune system. Many biological systems, including immune function, food nutrient digestion, host behavior, and cognitive function, have a substantial impact on the components of GM (Guo et al. 2019).

4.2 Influencing Factors on Gut Microbiota

The composition of the GM can be influenced by a variety of factors, including genetics, diet, age, lifestyle, medications, environmental exposures, mode of birth, diet, stress, physical exercise, medications, etc. (Fig. 4.1). The GM is recognized to have a critical role in several physiological processes, such as digestion, metabolism, immune function, and even behavior. In order to promote gut health and prevent or treat dysbiosis-related diseases, it is essential to comprehend the factors that influence GM and their impact on health (Cresci and Bawden 2015; Kumbhare et al. 2019). In this context, exploring the influencing factors on GM can help to promote a better understanding of the complex interplay between GM and human health. The significant factors that will be discussed are listed below.

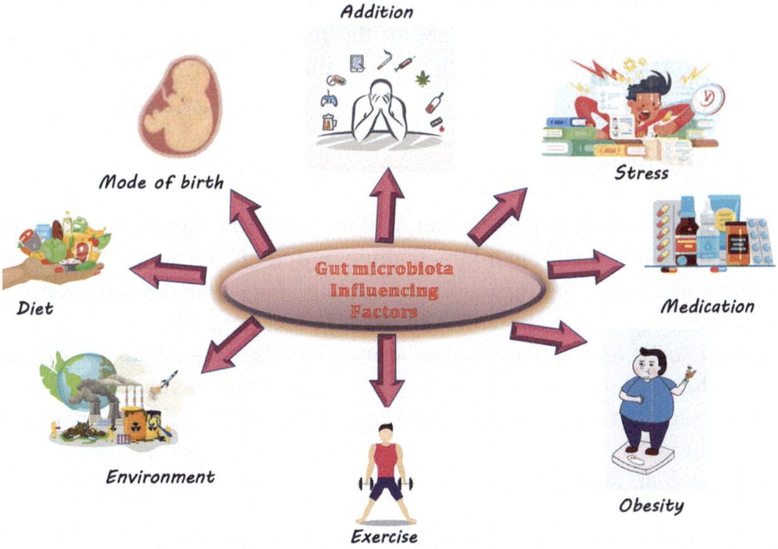

Fig. 4.1 Represents the influencing factors involved in the GM. The composition and diversity of the gut microbiota are influenced by a variety of factors including diet, environment, etc.

4.3 Mode of Birth

Birth serves as the first opportunity for large-scale bacterial colonization. When a baby is born naturally through vaginal delivery, the mother's vaginal microbiota is first exposed to the child when they move through the birth canal. Dominguez researched Cesarean section and normal delivery mother and newborn baby's microbial development. They find that after 20mins of birth, the microbiota of vaginally born children is similar to that of their mother's vagina, whereas infants delivered through Cesarean section have microbial communities similar to those found on human skin (Dominguez-Bello et al. 2010).

4.4 Diet

This is another factor that commonly affects GM composition and brain behavior. In general diet, habits relate to the composition of the microbiota in the gut (Turnbaugh et al. 2008). This Western diet habit leads to the obese and many metabolic-related disorders, because this food has high levels of fat, salts, and sugars. The animal model (Rat) study involved administering a high-fat diet (HFD) to the animals, leading to changes in GM levels as a consequence of this dietary intake (Hildebrandt et al. 2009; Devkota et al. 2012). Studies involving human volunteers who were provided with diets that controlled their dietary intake have been conducted to examine the effects of different nondigestible carbohydrates on the composition of GM over short periods. By switching the major nondigestible carbohydrate in the diet from wheat bran to resistant starch (or vice versa), while maintaining a constant intake of protein, fat, and total carbohydrate, researchers observed rapid and reversible changes in the representation of specific bacterial groups in the GM of overweight participants within a few days (Walker et al. 2011).

4.5 Environment

The environment is another general factor; at the same time, a polluted environment can disturb the GM. Some environmentally exposed chemicals have already been proven that could create an impact on the GA, e.g., *beta-glucuronidases*, *azoreductase*, *sulfatases*, *beta-lyase*, and *nitroreductase* (Claus et al. 2017).

Bisphenol A (BPA) is one of the endocrine-disrupting chemicals, and for plastic product production, it's generally used. In an animal model study exposure to the BPA 120 (mg/mL) in male CD-1 mice (21 days old) for 10 weeks of treatment, the results conclude BPA caused a decreased level in the species diversity of gut microbes (Lai et al. 2016).

4.6 Exercise

Regular moderate level of exercise is required for healthy brain activity, and it leads to a reduced level of stress and an increased level of immunity. At the same time, excess level of physical exercise affects brain plasticity, influencing cognition and health (Weinberg and Gould 1999; Bermon et al. 2015). Overall, it is important to engage in regular moderate exercise to promote healthy brain activity, reduce stress, and boost immunity. However, excessive levels of physical exercise may have negative effects on brain plasticity, cognition, and overall health, so it is important to find a balance that works for each individual.

4.7 Obesity

In obese conditions, GM plays an immense role: modulation of the human GM could lead to weight gain for underweight children, and also it can lose weight for obese condition individuals.

In the last 5 years, there has been a significant focus on investigating the relationship between GM and its role in regulating body weight and metabolism. Research has shown that the GM plays a critical role in the control of energy balance, appetite, and metabolism and that changes in its composition may lead to the development of metabolic disorders and obesity (Turnbaugh and Gordon 2009). Studies investigating the potential role of GM in obesity have primarily focused on its peripheral control of food intake. However, it remains unclear whether GM can also influence central control of food intake, and therefore, further research is necessary to explore this potential connection. Currently, investigations are underway to examine whether GM plays a role in regulating the side effects of centrally acting psychotropic medications, including atypical antipsychotics, which may cause obesity as a possible side effect. These studies are derived from observations made on rats that were administered olanzapine, a medication known to impact the GM composition (Manco 2012; Davey et al. 2012).

4.8 Medication

Many therapeutic medications are used for the diagnosis process. Among these drugs, antibiotic drugs play a vital role in GA alterations. Commercially used drugs like metformin, proton pump inhibitors, laxatives, and statins were also shown to have robust effects on GM (Antunes et al. 2011). Metatranscriptomic techniques were also used to examine GM samples exposed to multiple medications. The effect of short-term exposure of human feces to nonantibiotic medications such as stomach acid suppressor (nizatidine), cardiac glycosides, an analgesic (phenacetin), an anthelmintic (levamisole), and sulfasalazine on the expression of microbial genes linked with drug import and metabolism (Maurice et al. 2013).

4.9 Stress-Induced Alterations in Gut Microbiota

Numerous preclinical studies have demonstrated that stress can impact the GM in various animals, including pigs, resulting in alterations to its composition and ecology (Mudd et al. 2016); rodents (Golubeva et al. 2015); and horses (Mach and Fuster-Botella 2017). Pregnancy period time stress leads to the alteration of the offspring's GM. Depression, anxiety, and stress are highly comorbid conditions, which mean that they often occur together in individuals. This may be because they share overlapping biological mechanisms and manifestations. As a result, researchers studying the GM-brain axis often investigate these conditions together. Depression and anxiety are two of the most prevalent psychiatric disorders worldwide, affecting millions of people. These disorders fall under the category of mood disorders and may arise due to the disruption of the body's ability to regulate its internal processes in response to stress and return to a state of balance or homeostasis. Chronic stress can lead to dysregulation of the hypothalamic-pituitary-adrenal (HPA) axis, a key component of the body's stress response system. This can result in alterations in neurotransmitter function, inflammation, and other physiological changes that can lead to depression and anxiety condition. Research indicates that GM may have a role in modifiable mood and behavior, potentially through its communications with the CNS via the gut-brain axis. As a result, researchers are investigating the use of probiotics and other interventions targeting GM as potential treatments for these conditions (McEwen and Wingfield 2003; Cryan et al. 2019; Morais et al. 2021).

4.10 Addiction

Addiction to the drug could lead to the modification of GM. Most of the recent research focus on the impact of drug addiction modified the GM level. Chronic inhalation of alcohol vapor induced significant alterations in the GM of mice. Specifically, the abundance of certain bacterial groups was altered, and there was an overall decrease in bacterial diversity. These changes in the GM were associated with increased levels of systemic inflammation and altered behavior in the mice (Peterson et al. 2017). Other studies have also found that drug addiction, including addiction to opioids, cocaine, and methamphetamine, can lead to alterations in the GM (Galley et al. 2014). These changes in GM may contribute to the development and maintenance of addiction, as well as other adverse health outcomes associated with drug use. Overall, these studies suggest that addiction can lead to significant changes in the gut microbiome, which may contribute to the development and maintenance of addiction, as well as other adverse health outcomes associated with addiction. More studies are required to gain a deeper understanding of the intricate relationships between GM and addiction.

4.11 GM Involved in Brain Function and Behavior

American scientist Gershon in the nineteenth century elucidated the relationship between the gut and the brain (Grenham et al. 2011). Neurological networks, immunologic, and endocrine systems all place an immense role in the gut-brain communication processes. The changes of microbiota in the term are defined as dysbiosis. This dysbiosis is linked with many metabolic-related disorders such as obesity, diabetes, colorectal cancer, and rheumatoid arthritis, CVD (Shreiner et al. 2015; Young 2017). Interactions with the CNS take place according to the structure known as the "microbiota-gut-brain axis." Interaction between the gut, intestinal microbiota, and the brain involves the release of different metabolites such as SCFAs, structural components of bacteria, and signaling molecules.

4.12 The Gut Microbiome and Diseases

The human body's relationship with its microbes is complex and interconnected. Disrupting this delicate balance can have serious consequences for the body and its microbial inhabitants (Gomaa 2020). GM has been linked to a growing list of diseases and health conditions, ranging from gastrointestinal disorders to metabolic diseases and even neurological disorders. Some of the disorders are discussed below (Fig. 4.2).

4.13 Multiple Sclerosis (MS)

MS is an autoimmune disorder that affects the CNS. It is characterized by the gradual onset of symptoms such as motor deficits, blurred vision, and changes in sensation that often appear spontaneously and without any warning signs. These symptoms may worsen over time, leading to significant disability and impairment in daily life. Unlike some other neurological disorders, MS often presents without a clear prodrome or warning signs prior to the onset of symptoms (Dendrou et al. 2015). The etiology of MS appears to be complex and unclear, involving interactions of environmental and genetic factors. Among these impacts, GM stands out as one of the most significant environmental contributors. The "hygiene hypothesis" states that reducing exposure to infections in childhood may raise the risk of autoimmune and allergy illnesses (Ascherio 2013; Oksenberg 2013; Fleming and Fabry 2007).

Currently, there is no therapy designed to intentionally modify the GM in individuals with MS for beneficial effects (Jangi et al. 2016; Tremlett et al. 2016). However, identifying the links between GM and MS can create opportunities for developing innovative therapeutic interventions. One such intervention is dietary modification, which can lead to rapid changes in GM. Plant-based diets, SCFA supplementation, and probiotics are among the dietary modifications that could potentially improve GM in MS patients, but they require further evaluation. Furthermore,

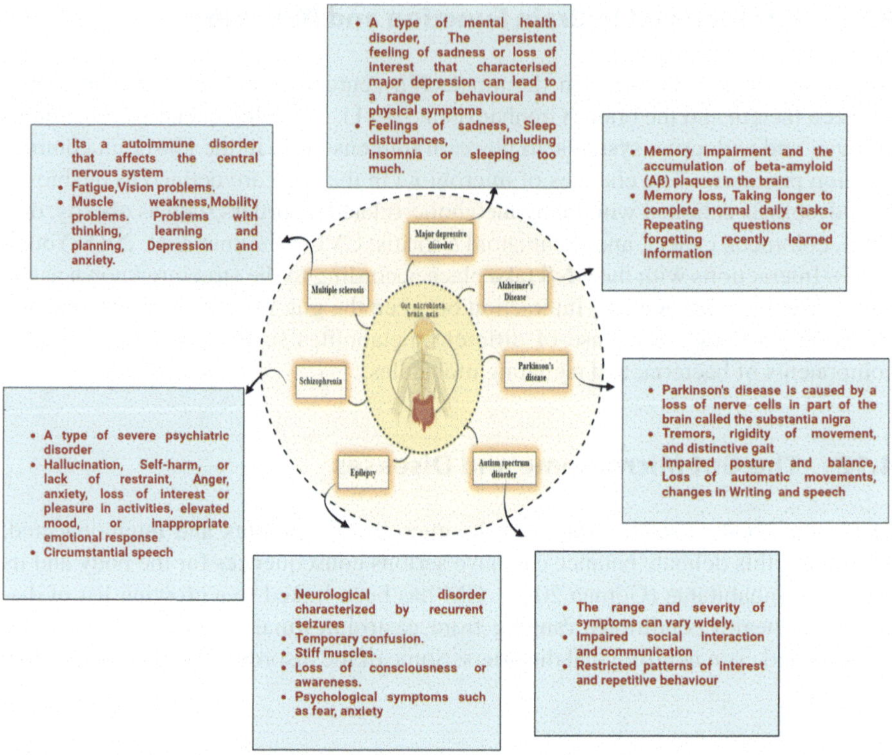

Fig. 4.2 Illustration represents the GM alteration causing major brain-related disorders with their symptoms

the diet should be considered as a potential confounding factor when studying GM in MS (David et al. 2014; Budhram et al. 2017). In a study by Borody and colleagues, fecal microbiota transplantation (FMT) was administered to three MS patients who were experiencing severe constipation. The results showed that FMT reduced their neurological symptoms and improved their walking ability. Nevertheless, it is important to acknowledge that the study had certain limitations, including a small sample size and a lack of a control group, which restricts the generalizability and reliability of the findings (Borody et al. 2013).

4.14 Major Depressive Disorder (MDD)

MDD is a type of mental illness (Kessler et al. 2003). Growing evidence suggests that GM is crucial for controlling human behavior and brain activity (Mayer et al. 2014; McGuinness et al. 2022). Comparing patients with MDD to healthy controls, observational studies revealed apparent changes in the composition of the GM;

however, it is unclear which particular taxa of bacteria are responsible for differences between groups.

Several observational studies examined the GM of patients with MDD. Additional research has shown that probiotics and symbiotics can help patients with their symptoms of anxiety or depression. It has been demonstrated that a plant-based diet can modify gut function and helps reduce depressive symptoms. These findings suggest that MDD may have impacted the composition of the GM (Chang et al. 2022; Horn et al. 2022). A recent study on MDD condition reported a higher relative abundance of *Firmicutes*, *Actinobacteria*, and *Bacteroidetes* in individuals with MDD. Furthermore, the study demonstrated that fecal transplantation of microbiota from MDD patients to germ-free mice resulted in depressive-like behavioral changes in the mice (Zheng et al. 2016). The above findings indicate that MDD can disturb the GM, at the same time targeting the GM may provide a new option for its treatment and management.

4.15 Alzheimer's Disease (AD)

AD is the most common neurodegenerative disorder worldwide, characterized by memory impairment and the accumulation of beta-amyloid (Aβ) plaques in the brain (Kelley and Petersen 2007; Malampati et al. 2020). Multiple factors link GM to ASD symptoms, with early life events playing a significant role in shaping the composition of microbial communities. Delivery mode, for example, can have a major impact on the development of GM, with infants delivered vaginally having different microbiota than those delivered via Cesarean section. Other early life factors, such as diet and antibiotic use, can also shape the composition of GM and potentially impact the development of ASD symptoms. The causes behind this AD pathogenesis are difficult to understand. AD is defined by a severe deficit in memory, cognitive, and physical performance, resulting in a decrease in mental, behavioral, and functional activities that disrupt day-to-day life activities. Aging with poor nutrition intake leads to AD condition; in this condition, modified rich nutrient composition of diet could manage a good GM (Tiwari et al. 2019; Kesika et al. 2021). In AD conditions, the gut-brain axis, GM arrangement, and probiotic intervention act as a major role. Probiotics are organisms that are alive that are beneficial to humans, particularly the digestive system. Probiotics are sometimes called "good bacteria," since they aid in maintaining a healthy gut (Joint FAO 2002). A multifactorial etiology appears to be likely with brain inflammation changes; notably modified microglia activity plays a major role (Wyss-Coray 2006). Experimental data denoted that changes in the GM could modify intestinal activity and systemic-related inflammation which leads to the distribution in the blood-brain barrier function (Jiang et al. 2017; Askarova et al. 2020; Le Page et al. 2018). At the same time, GM dysfunctions are related to the early phases of AD pathogenesis; it can promote immunosenescence, neuroinflammation, oxidative stress, and cytokine production (Leblhuber et al. 2020).

4.16 Parkinson's Disease (PD)

It is one of the major neurodegenerative diseases. Above 65 years of age groups of people generally affected by PD. In this PD condition, α-synuclein plays a vital role; it's a type of protein involved in the communication between the brain and gut.

In addition to affecting the brain, PD can also impact other organs in the body. Notably, PD patients commonly experience gastrointestinal problems, with constipation affecting around 80% of individuals with PD. This suggests that PD may involve disruptions to the enteric nervous system (ENS), a complex network of neurons that controls digestive functions in the gut. Accumulation of the α-synuclein protein in the ENS, similar to what occurs in the brain in PD, may contribute to these gastrointestinal symptoms. Hence, it is essential to consider the gut-brain axis in the pathology of PD and explore potential therapeutic strategies that target ENS (Ueki and Otsuka 2004; Parashar and Udayabanu 2017). Although PD is typically thought of as a movement illness, it has long been understood that the symptoms extend beyond motor dysfunction, since PD patients frequently experience non-motor symptoms, such as cognitive impairment (Reeve et al. 2014; Aarsland et al. 2017; Perez-Pardo et al. 2017). Symptoms of PA include tremors, rigid movements, and a unique gait, but these motor impairments and brain pathology may not become evident until the disease has progressed significantly. At this stage, many of the dopaminergic neurons in the substantia nigra, which control motor function, have degenerated or lost their axons. This loss of dopamine is thought to be the primary cause of Parkinson's motor symptoms. This loss of function can lead to the motor symptoms of PD, such as tremors and difficulty with movement (Cheng et al. 2010). New evidence suggests that certain lifestyle factors may play a role in the development of PD. Specifically, recent research has shown that smoking and coffee consumption are linked to a lower risk of developing PD. It is believed that these protective effects may be due to changes in the GM (Scheperjans et al. 2015).

4.17 Autism Spectrum Disorder (ASD)

ASD is a neurodevelopmental disorder that is defined by challenges in communication, social interaction, and repetitive or restricted behaviors. These symptoms can appear as early as 1 year of age in some infants. While ASD is believed to have a strong genetic component, environmental factors can also play a role in the development of co-occurring medical conditions, including anxiety. It is not uncommon for up to 40% of children with ASD to experience significant anxiety. Therefore, early intervention and support can help improve outcomes for individuals with ASD and their families (Risch et al. 2014; Li and Zhou 2016).

While the symptoms of ASD can vary greatly from person to person, they are typically characterized by difficulties in three main behavioral domains: social communication, social interaction, and repetitive behaviors or interests. The social communication difficulties that are commonly associated with ASD include challenges in using and interpreting nonverbal communication, such as facial expressions and gestures, as well as difficulties with verbal communication, including challenges with language comprehension and expression. Social interaction difficulties can involve challenges in forming and maintaining social relationships, interpreting social cues, and engaging in reciprocal social interactions. People with ASD may struggle to understand the perspectives of others and may have difficulty with imaginative or creative play. Repetitive behaviors and interests may include a wide range of behaviors, such as repeating specific phrases or movements, becoming intensely focused on a particular topic or activity, and exhibiting rigid adherence to routines or schedules. Overall, while the presentation of symptoms can vary significantly among individuals with ASD, these three domains of behavior are commonly affected and are used as primary criteria for diagnosis (Lenroot and Yeung 2013).

4.18 Epilepsy

Epilepsy is a neurological disorder characterized by recurrent seizures, which affect millions of people worldwide. Recent studies have suggested a potential link between GM and the development and progression of epilepsy. More than 70 million individuals worldwide suffer from epilepsy, a chronic neurological condition that has a significant social and financial impact. It is defined by relapses and unprovoked sudden seizures. An elevated risk of mental illness is present in patients with epilepsy, which can result in higher levels of disability and mortality rates. This underscores the importance of addressing the mental health needs of individuals with epilepsy to enhance their overall health outcomes and reduce disability and mortality. There have been limited studies investigating the potential impact of changes in the GM on epilepsy (Dahlin and Prast-Nielsen 2019; Şafak et al. 2020). These changes in GM may contribute to the development and progression of epilepsy through several mechanisms. For example, alterations in the GM can lead to increased intestinal permeability and systemic inflammation, which may trigger seizures. In addition, the GM plays a crucial role in modulating the immune system, and dysbiosis can lead to immune dysfunction, which may lead to epilepsy. Additionally, the gut-brain axis has been implicated in the pathogenesis of epilepsy, with evidence suggesting that the GM may influence brain function and behavior through various pathways, such as the production of neurotransmitters and SCFAs. A ketogenic diet (KD), which is high in fat and low in carbohydrates, has been shown to be an effective treatment for drug-resistant epilepsy (Kwan and Brodie 2000). Another study investigated the impact of a ketogenic diet on the GM and seizure activity in mice with epilepsy induced by pilocarpine. The study found that the ketogenic diet improved the GM composition, including an increase in beneficial bacteria such as *Lactobacillus* and *Bifidobacterium*, and reduced seizure activity in the mice (Olson et al. 2018).

Overall, these findings suggest a potential link between the GM and epilepsy. Further research is needed to elucidate the precise mechanisms underlying this relationship and to explore the potential therapeutic strategies targeting the GM in the management of epilepsy.

4.19 Schizophrenia

Schizophrenia is a severe psychiatric disorder characterized by a range of symptoms, including delusions, hallucinations, and disorganized thinking. Recent studies have suggested a potential link between GM and the development and progression of schizophrenia (Rantala et al. 2022). The exact cause of schizophrenia is yet unclear. Several investigations have demonstrated that people with schizophrenia have gastrointestinal inflammation and gut bacterial infections (Fond et al. 2015; Yolken et al. 2015). Shen conducted a cross-sectional study that used the 16S sequencing method to compare the GM of individuals with schizophrenia to that of healthy controls. The predominant bacterial phyla in the GM of healthy individuals are *Firmicutes*, *Actinobacteria*, *Bacteroidetes*, *Proteobacteria*. However, in individuals with schizophrenia, the abundance of *Proteobacteria* was higher than in healthy controls (Shen et al. 2018). Castro-Nallar and their colleague conducted a study on the GM of individuals with schizophrenia and found that the proportion of *Firmicutes* was higher in individuals with schizophrenia compared to healthy controls. However, their study did not find a significant difference in the abundance of *Proteobacteria* between the two groups. This suggests that there may be a variety of disorders related to *Firmicutes* in the digestive tract of individuals with schizophrenia, while the imbalance of *Proteobacteria* may be more noticeable in the gut (Castro-Nallar et al. 2015). A study done by Nguyen and colleagues investigated the GM composition in individuals with schizophrenia compared to healthy controls. The study found significant differences in the GM of individuals with schizophrenia, including a decrease in bacterial diversity and an increase in harmful bacteria such as *Clostridium*.

These changes in the GM may contribute to the development and progression of schizophrenia through several mechanisms. For example, alterations in the GM can lead to increased intestinal permeability and systemic inflammation, which may trigger or exacerbate symptoms of schizophrenia. The immune system is influenced by the GM, and dysbiosis can cause immune dysfunction that may contribute to the development of schizophrenia (Nguyen et al. 2018). Additionally, the gut-brain axis may play a role in the pathogenesis of schizophrenia, as GM can impact brain function and behavior via various pathways, including the production of neurotransmitters and SCFAs. Taken together, these findings suggest a potential link between GM and schizophrenia. Further research is needed to elucidate the precise mechanisms underlying this relationship and to explore the potential therapeutic strategies targeting GM in the management of schizophrenia.

4.20 GM Effect on Behavior and Cognition

The two-way communication between the GM and the brain can be attributed to various mechanisms, including neural, immune, metabolic, and endocrine pathways. The immune system is essential for the two-way exchange of information between the stomach and the brain. Pro-inflammatory cytokines including IL-1,

IL-6, and TNF indicate systemic inflammation and are upregulated in the serum. These cytokines are linked to patient symptoms like sadness and illness (Kelley et al. 2003; Myers 2008).

According to the GM-brain axis, there are two-way interactions between the bacteria in the gut and several aspects of brain function, such as emotional behavior and cognitive function (Rhee et al. 2009; Cryan and Dinan 2012). The human GM is referred to as the "second brain" of the human since it generates most neurotransmitters found in the brain and is known to have a key role in neural development, cognition, and behavior (Dinan and Cryan 2017).

A chronic illness called cognitive impairment is the main factor in the increasing loss of functioning and disability. Notably, a few approaches have been suggested to address the dysbiosis of GM, including the use of probiotics, dietary changes, and fecal microbiota transplants (Zmora et al. 2019). Changes in the GM caused by inflammation, infection, or drugs, especially antibiotic therapy, can have extraintestinal effects, including changes in the brain. Behavioral changes, such as depression, anxiety, and cognitive impairments, have recently been identified as additional microbial targets (Collins et al. 2012). Gareau and colleagues conducted a study on germ-free mice modal; they stated that the microbiota is essential for normal cognitive development (Gareau et al. 2011).

4.21 GM Influences the Brain Axis

The brain-gut axis is a two-way communication pathway that facilitates the transmission of information between the brain and the gut in mammals. This communication occurs through multiple pathways, which include the nervous system, the hypothalamic-pituitary-adrenal (HPA) axis, and the immune system. These pathways enable the brain and gut to communicate and interact with each other in various ways (Forsythe et al. 2010).

The GM hypothesis proposes that the complex community of microorganisms inhabiting the gastrointestinal tract can have a profound impact on brain function and behavior through the gut-brain or microbiota-gut-brain axis. This communication pathway is characterized by a bidirectional exchange of signals between the GM, the enteric nervous system, the immune system, and the CNS. Recent research has highlighted the important role of GM in the development of mental disorders, including depression, anxiety, and schizophrenia. Alterations in the GM have been observed in individuals with these disorders, and there is evidence that changes in the GM can contribute to the development of these conditions (Kelly et al. 2016). Regulating the GM has the potential to improve dysfunction in the gut-brain axis as well as alleviate abnormalities in the immune system, HPA axis, and brain. This is consistent with the GM hypothesis, which suggests that the GM can influence brain function and behavior. Therefore, targeting the GM may be a promising direction for the therapy and prevention of mental disorders. By understanding the complex interactions between the GM and the brain, we can develop effective treatments for these conditions (Rieder et al. 2017; Liang et al. 2018).

4.22 Mechanisms and Pathways for the Microbiota Affecting Cognition

Apart from this, several mechanisms and pathways through which the GM can affect cognition. These include the following:

Production of neurotransmitters: The GM can produce neurotransmitters such as serotonin, dopamine, and gamma-aminobutyric acid (GABA) that are involved in regulating mood, behavior, and cognitive function. Dysregulation of these neurotransmitters has been linked to cognitive impairment and mental disorders such as depression and anxiety.

Regulation of the immune system: The GM plays a critical role in regulating the immune system, and alterations in the GM can lead to immune dysfunction and inflammation, which have been linked to cognitive impairment and neurodegenerative diseases such as Alzheimer's and Parkinson's disease.

Production of metabolites: The GM produces metabolites such as SCFAs, which can cross the blood-brain barrier and affect cognitive function. SCFAs have been shown to improve memory and learning in animal models.

Modulation of the HPA axis: The GM can regulate the HPA axis, which is involved in the stress response and has been linked to cognitive impairment and mental disorders. Alterations in the GM can lead to dysregulation of the HPA axis and contribute to cognitive dysfunction.

Direct neural communication: The GM can communicate with the brain through the vagus nerve, which connects the gut and the brain. This direct neural communication may play a role in the regulation of cognitive function. Overall, GM can affect cognition through a variety of mechanisms and pathways. Understanding these complex interactions between the GM and the brain is essential for developing effective treatments for cognitive dysfunction and mental disorders.

4.23 GM Modification for Therapeutic Changes

GM, which is a complex ecosystem of microorganisms that inhabit the gastrointestinal tract, has been increasingly recognized as a key factor in maintaining human health (Fig. 4.3). Modifications to GM have been shown to have therapeutic benefits for a variety of health conditions. One of the ways to modify the GM is through dietary interventions. Consuming a diet that is high in fiber, fruits, vegetables, and fermented foods can promote the growth of beneficial bacteria in the gut, leading to a more diverse and healthier GM. This approach is effective in improving insulin sensitivity in obese individuals, reducing inflammation in patients with inflammatory bowel disease, and alleviating symptoms of irritable bowel syndrome. Another way to modify the GM is with probiotics, which are live microorganisms that are intended to confer health benefits when consumed. Probiotics are effective in preventing and treating antibiotic-associated diarrhea, as well as improving symptoms of lactose intolerance and reducing inflammation in patients with ulcerative colitis (Thursby and Juge 2017). Fecal microbiota transplantation (FMT), which involves

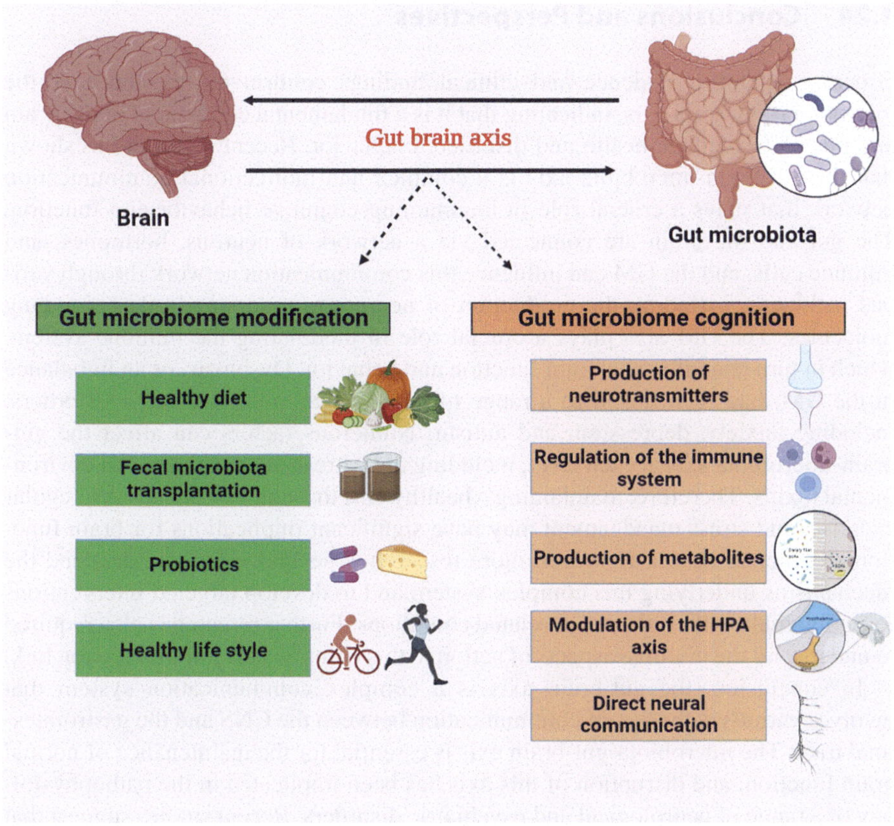

Fig. 4.3 Illustration represents the GM modification as a beneficial therapeutic tool and pathways through which the GM can affect the cognition process

transferring fecal material from a healthy donor to a recipient, is another method for modifying the GM. FMT is effective in treating recurrent *Clostridium difficile* infection, a serious gastrointestinal infection that is difficult to treat with antibiotics. FMT is also being investigated as a potential treatment for other conditions, such as inflammatory bowel disease and multiple sclerosis (Gupta et al. 2016). In addition to these approaches, researchers are also exploring the use of prebiotics, which are nondigestible food ingredients that promote the growth of beneficial bacteria in the gut, and postbiotics, which are metabolic by-products of probiotic bacteria that have therapeutic effects. Taken together, modifying the GM through dietary interventions, probiotics, FMT, and other methods has the potential to provide therapeutic benefits for a variety of health conditions. Further research is needed to better understand the mechanisms behind these effects and to develop effective interventions (Fong et al. 2020).

4.24 Conclusions and Perspectives

Growing scientific evidence and clinical findings confirm the existence of the microbiota-gut-brain axis, indicating that it is a fundamental component of the brain and poised to regulate health and disordered behavior. Recent research has shown that the gut-brain-microbiota axis is a complex and bidirectional communication network that plays a crucial role in maintaining cognitive behavior and function. The gut and the brain are connected via a network of neurons, hormones, and immune cells, and the GM can influence this communication network through various pathways, including the production of neurotransmitters and other signaling molecules. The GM also plays a crucial role in modulating the immune system, which in turn can influence brain function and behavior. Dysbiosis, or an imbalance in the GM, has been linked to a range of neurological and psychiatric disorders, including anxiety, depression, and autism. Numerous factors can affect the gut-brain-microbiota axis at each level, including diet, stress, medications, and environmental toxins. Therefore, maintaining a healthy GM through a balanced diet, regular exercise, and stress management may have significant implications for brain function and mental health. However, more research is needed to fully understand the mechanisms underlying this complex system and to develop targeted interventions for individuals with dysbiosis or related conditions. Further research is also required to understand the multiple aspects of action in this complicated communication link.

In conclusion, the gut-brain axis is a complex communication system that involves various pathways of communication between the CNS and the gastrointestinal tract. The microbiota-gut-brain axis is essential for the maintenance of normal brain function, and disruption of this axis has been implicated in the pathophysiology of a range of neurological and psychiatric disorders. Recent studies suggest that GM plays a crucial role in the gut-brain axis and has the potential to modulate brain function and behavior through various mechanisms, including neurotransmitter production, immune modulation, and the production of microbial metabolites. Alterations in GM composition and diversity have been linked to various cognitive disorders. GM is a promising therapeutic target for these cognitive disorders, and studies have shown that interventions such as probiotics, prebiotics, and fecal microbiota transplantation can improve cognitive function in animal models and humans. The gut-brain axis is a critical area of research that holds great promise for the development of novel therapies for cognitive disorders. Understanding the underlying mechanisms of the gut-brain axis could lead to new insights into the pathophysiology of cognitive degeneration and the development of more effective interventions that target GM. Emerging evidence suggests that GM plays an important role in brain function and behavior and that alterations in GM composition and diversity may contribute to the development of cognitive disorders such as Alzheimer's disease and other forms of dementia. Therefore, targeting GM through interventions such as probiotics, prebiotics, and FMT has been proposed as a potential therapeutic strategy for the prevention and treatment of cognitive disorders. However, much more research is needed to fully understand the mechanisms underlying the gut-brain axis and to develop safe and effective interventions that target

GM for the treatment of cognitive disorders. Despite the promising results from preclinical studies, human clinical trials are needed to evaluate the efficacy and safety of these interventions.

References

Aarsland D, Creese B, Politis M, Chaudhuri KR, Ffytche DH, Weintraub D, Ballard C (2017) Cognitive decline in Parkinson disease. Nat Rev Neurol 13(4):217–231. https://doi.org/10.1038/nrneurol.2017.27

Antunes LC, Han J, Ferreira RB, Lolić P, Borchers CH, Finlay BB (2011) Effect of antibiotic treatment on the intestinal metabolome. Antimicrob Agents Chemother 55:1494–1503. https://doi.org/10.1128/AAC.01664-10

Ascherio A (2013) Environmental factors in multiple sclerosis. Expert Rev Neurother 13(12 Suppl):3–9. https://doi.org/10.1586/14737175.2013.865866

Askarova S, Umbayev B, Masoud AR, Kaiyrlykyzy A, Safarova Y, Tsoy A et al (2020) The links between the gut microbiome, aging, modern lifestyle and Alzheimer's disease. Front Cell Infect Microbiol 10:104. https://doi.org/10.3389/fcimb.2020.00104

Bermon S, Petriz B, Kajėnienė A, Prestes J, Castell L, Franco OL (2015) The microbiota: an exercise immunology perspective. Exerc Immunol Rev 21:70–79. PMID: 25825908

Borody TJ, Paramsothy S, Agrawal G (2013) Fecal microbiota transplantation: indications, methods, evidence, and future directions. Curr Gastroenterol Rep 15:1–7

Budhram A, Parvathy S, Kremenchutzky M, Silverman M (2017) Breaking down the gut microbiome composition in multiple sclerosis. Mult Scler (Houndmills, Basingstoke, England) 23(5):628–636. https://doi.org/10.1177/1352458516682105

Castro-Nallar E, Bendall ML, Pérez-Losada M, Sabuncyan S, Severance EG, Dickerson FB, Schroeder JR, Yolken RH, Crandall KA (2015) Composition, taxonomy and functional diversity of the oropharynx microbiome in individuals with schizophrenia and controls. PeerJ 3:e1140. https://doi.org/10.7717/peerj.1140

Chang L, Wei Y, Hashimoto K (2022) Brain–gut–microbiota axis in depression: a historical overview and future directions. Brain Res Bull 182:44–56. https://doi.org/10.1016/j.brainresbull.2022.02.004

Cheng HC, Ulane CM, Burke RE (2010) Clinical progression in Parkinson disease and the neurobiology of axons. Ann Neurol 67(6):715–725. https://doi.org/10.1002/ana.21995

Claus SP, Guillou H, Ellero-Simatos S (2017) The gut microbiota: a major player in the toxicity of environmental pollutants? NPJ Biofilms Microbiomes 3:17001

Collins SM, Surette M, Bercik P (2012) The interplay between the intestinal microbiota and the brain. Nat Rev Microbiol 10(11):735–742. https://doi.org/10.1038/nrmicro2876

Cresci GA, Bawden E (2015) Gut microbiome: what we do and don't know. Nutr Clin Pract 30(6):734–746. https://doi.org/10.1177/0884533615609899

Cryan JF, Dinan TG (2012) Mind-altering microorganisms: the impact of the gut microbiota on brain and behaviour. Nat Rev Neurosci 13:701–712

Cryan JF, O'Riordan KJ, Cowan CSM, Sandhu KV, Bastiaanssen TFS, Boehme M, Codagnone MG, Cussotto S, Fulling C, Golubeva AV, Guzzetta KE, Jaggar M, Long-Smith CM, Lyte JM, Martin JA, Molinero-Perez A, Moloney G, Morelli E, Morillas E, O'Connor R et al (2019) The microbiota-gut-brain axis. Physiol Rev 99(4):1877–2013. https://doi.org/10.1152/physrev.00018.2018

Dahlin M, Prast-Nielsen S (2019) The gut microbiome and epilepsy. EBioMedicine 44:741–746. https://doi.org/10.1016/j.ebiom.2019.05.024

Davey KJ, O'Mahony SM, Schellekens H, O'Sullivan O, Bienenstock J, Cotter PD, Dinan TG, Cryan JF (2012) Gender-dependent consequences of chronic olanzapine in the rat: effects on body weight, inflammatory, metabolic and microbiota parameters. Psychopharmacology 221(1):155–169. https://doi.org/10.1007/s00213-011-2555-2

David LA, Maurice CF, Carmody RN, Gootenberg DB, Button JE, Wolfe BE, Ling AV, Devlin AS, Varma Y, Fischbach MA, Biddinger SB, Dutton RJ, Turnbaugh PJ (2014) Diet rapidly and reproducibly alters the human gut microbiome. Nature 505(7484):559–563. https://doi.org/10.1038/nature12820

Dendrou CA, Fugger L, Friese MA (2015) Immunopathology of multiple sclerosis. Nat Rev Immunol 15(9):545–558. https://doi.org/10.1038/nri3871

Devkota S, Wang Y, Musch MW, Leone V, Fehlner-Peach H, Nadimpalli A, Antonopoulos DA, Jabri B, Chang EB (2012) Dietary-fat-induced taurocholic acid promotes pathobiont expansion and colitis in Il10-/- mice. Nature 487(7405):104–108. https://doi.org/10.1038/nature11225. PMID: 22722865; PMCID: PMC3393783

Dinan TG, Cryan JF (2017) Microbes, immunity, and behavior: psychoneuroimmunology meets the microbiome. Neuropsychopharmacology 42(1):178–192. https://doi.org/10.1038/npp.2016.103

Dominguez-Bello MG, Costello EK, Contreras M, Magris M, Hidalgo G, Fierer N, Knight R (2010) Delivery mode shapes the acquisition and structure of the initial microbiota across multiple body habitats in newborns. Proc Natl Acad Sci U S A 107(26):11971–11975. https://doi.org/10.1073/pnas.1002601107. Epub 2010 Jun 21. PMID: 20566857; PMCID: PMC2900693

Fleming J, Fabry Z (2007) The hygiene hypothesis and multiple sclerosis. Ann Neurol 61(2):85–89. https://doi.org/10.1002/ana.21092

Fond G, Boukouaci W, Chevalier G, Regnault A, Eberl G, Hamdani N, Dickerson F, Macgregor A, Boyer L, Dargel A, Oliveira J, Tamouza R, Leboyer M (2015) The "psychomicrobiotic": targeting microbiota in major psychiatric disorders: a systematic review. Pathol Biol 63(1):35–42. https://doi.org/10.1016/j.patbio.2014.10.003

Fong W, Li Q, Yu J (2020) Gut microbiota modulation: a novel strategy for prevention and treatment of colorectal cancer. Oncogene 39(26):4925–4943. https://doi.org/10.1038/s41388-020-1341-1

Forsythe P, Sudo N, Dinan T, Taylor VH, Bienenstock J (2010) Mood and gut feelings. Brain Behav Immun 24(1):9–16. https://doi.org/10.1016/j.bbi.2009.05.058

Galley JD, Nelson MC, Yu Z, Dowd SE, Walter J, Kumar PS, Lyte M, Bailey MT (2014) Exposure to a social stressor disrupts the community structure of the colonic mucosa-associated microbiota. BMC Microbiol 14:189. https://doi.org/10.1186/1471-2180-14-189

Gareau MG, Wine E, Rodrigues DM, Cho JH, Whary MT, Philpott DJ et al (2011) Bacterial infection causes stress-induced memory dysfunction in mice. Gut 60(3):307–317

Golubeva AV, Crampton S, Desbonnet L, Edge D, O'Sullivan O, Lomasney KW, Zhdanov AV, Crispie F, Moloney RD, Borre YE, Cotter PD, Hyland NP, O'Halloran KD, Dinan TG, O'Keeffe GW, Cryan JF (2015) Prenatal stress-induced alterations in major physiological systems correlate with gut microbiota composition in adulthood. Psychoneuroendocrinology 60:58–74. https://doi.org/10.1016/j.psyneuen.2015.06.002. Epub 2015 Jun 17 PMID: 26135201

Gomaa EZ (2020) Human gut microbiota/microbiome in health and diseases: a review. Antonie Van Leeuwenhoek 113(12):2019–2040. https://doi.org/10.1007/s10482-020-01474-7

Grenham S, Clarke G, Cryan JF, Dinan TG (2011) Brain-gut-microbe communication in health and disease. Front Physiol 2:94. https://doi.org/10.3389/fphys.2011.00094. PMID: 22162969; PMCID: PMC3232439

Guo TT, Song D, Ho CT, Zhang X, Zhang CD, Cao JX, Wu ZF (2019) J Agric Food Chem 67:8847

Gupta S, Allen-Vercoe E, Petrof EO (2016) Fecal microbiota transplantation: in perspective. Ther Adv Gastroenterol 9(2):229–239. https://doi.org/10.1177/1756283X15607414

Hildebrandt MA, Hoffmann C, Sherrill-Mix SA, Keilbaugh SA, Hamady M, Chen YY, Knight R, Ahima RS, Bushman F, Wu GD (2009) High-fat diet determines the composition of the murine gut microbiome independently of obesity. Gastroenterology. 137(5):1716–24.e1–2. https://doi.org/10.1053/j.gastro.2009.08.042. Epub 2009 Aug 23. PMID: 19706296; PMCID: PMC2770164

Horn J, Mayer DE, Chen S, Mayer EA (2022) Role of diet and its effects on the gut microbiome in the pathophysiology of mental disorders. Transl Psychiatry 12(1):164. https://doi.org/10.1038/s41398-022-01922-0

Jangi S, Gandhi R, Cox LM, Li N, von Glehn F, Yan R, Patel B, Mazzola MA, Liu S, Glanz BL, Cook S, Tankou S, Stuart F, Melo K, Nejad P, Smith K, Topçuolu BD, Holden J, Kivisäkk P, Chitnis T et al (2016) Alterations of the human gut microbiome in multiple sclerosis. Nat Commun 7:12015. https://doi.org/10.1038/ncomms12015

Jiang C, Li G, Huang P, Liu Z, Zhao B (2017) The gut microbiota and Alzheimer's disease. J Alzheimers Dis 58:1–15. https://doi.org/10.3233/JAD-161141

Joint FAO (2002) WHO working group report on drafting guidelines for the evaluation of probiotics in food. London, ON, Canada, p 30

Kelley BJ, Petersen RC (2007) Alzheimer's disease and mild cognitive impairment. Neurol Clin 25(3):577–v. https://doi.org/10.1016/j.ncl.2007.03.008

Kelley KW, Bluthé RM, Dantzer R, Zhou JH, Shen WH, Johnson RW, Broussard SR (2003) Cytokine-induced sickness behavior. Brain Behav Immun 17(Suppl 1):S112–S118. https://doi.org/10.1016/s0889-1591(02)00077-6

Kelly JR, Clarke G, Cryan JF, Dinan TG (2016) Brain-gut-microbiota axis: challenges for translation in psychiatry. Ann Epidemiol 26(5):366–372. https://doi.org/10.1016/j.annepidem.2016.02.008

Kesika P, Suganthy N, Sivamaruthi BS, Chaiyasut C (2021) Role of gut-brain axis, gut microbial composition, and probiotic intervention in Alzheimer's disease. Life Sci 264:118627. https://doi.org/10.1016/j.lfs.2020.118627. Epub 2020 Oct 22. PMID: 33169684

Kessler RC, Berglund P, Demler O, Jin R, Koretz D, Merikangas KR, Rush AJ, Walters EE, Wang PS, National Comorbidity Survey Replication (2003) The epidemiology of major depressive disorder: results from the National Comorbidity Survey Replication (NCS-R). JAMA 289(23):3095–3105. https://doi.org/10.1001/jama.289.23.3095

Kumbhare SV, Patangia DVV, Patil RH, Shouche YS, Patil NP (2019) Factors influencing the gut microbiome in children: from infancy to childhood. J Biosci 44(2):49

Kwan P, Brodie MJ (2000) Early identification of refractory epilepsy. N Engl J Med 342(5):314–319. https://doi.org/10.1056/NEJM200002033420503

Lai KP, Chung YT, Li R, Wan HT, Wong CK (2016) Bisphenol A alters gut microbiome: comparative metagenomics analysis. Environ Pollut 218:923–930. https://doi.org/10.1016/j.envpol.2016.08.039. Epub 2016 Aug 21 PMID: 27554980

Le Page A, Dupuis G, Frost EH, Larbi A, Pawelec G, Witkowski JM et al (2018) Role of the peripheral innate immune system in the development of Alzheimer's disease. Exp Gerontol 107:59–66. https://doi.org/10.1016/j.exger.2017.12.019

Leblhuber F, Steiner K, Geisler S, Fuchs D, Gostner JM (2020) On the possible relevance of bottom-up pathways in the pathogenesis of Alzheimer's disease. Curr Top Med Chem 20:1415–1421

Lenroot RK, Yeung PK (2013) Heterogeneity within autism spectrum disorders: what have we learned from neuroimaging studies? Front Hum Neurosci 7:733. https://doi.org/10.3389/fnhum.2013.00733

Li Q, Zhou JM (2016) The microbiota-gut-brain axis and its potential therapeutic role in autism spectrum disorder. Neuroscience 324:131–139. https://doi.org/10.1016/j.neuroscience.2016.03.013

Liang S, Wu X, Hu X, Wang T, Jin F (2018) Recognizing depression from the microbiota-gut-brain axis. Int J Mol Sci 19(6):1592. https://doi.org/10.3390/ijms19061592

Mach N, Fuster-Botella D (2017) Endurance exercise and gut microbiota: a review. J Sport Health Sci. 6(2):179–197. https://doi.org/10.1016/j.jshs.2016.05.001. Epub 2016 May 10. PMID: 30356594; PMCID: PMC6188999

Malampati S, Song JX, Chun-Kit Tong B, Nalluri A, Yang CB, Wang Z, Gopalkrishnashetty Sreenivasmurthy S, Zhu Z, Liu J, Su C, Krishnamoorthi S, Iyaswamy A, Cheung KH, Lu JH, Li AM (2020) Targeting aggrephagy for the treatment of Alzheimer's disease. Cells 9(2):311. https://doi.org/10.3390/cells9020311

Manco M (2012) Gut microbiota and developmental programming of the brain: from evidence in behavioral endophenotypes to novel perspective in obesity. Front Cell Infect Microbiol 2:109. https://doi.org/10.3389/fcimb.2012.00109

Mariat D, Firmesse O, Levenez F, Guimarães V, Sokol H, Doré J, Corthier G, Furet JP (2009) The Firmicutes/Bacteroidetes ratio of the human microbiota changes with age. BMC Microbiol 9:123. https://doi.org/10.1186/1471-2180-9-123. PMID: 19508720; PMCID: PMC2702274

Maurice CF, Haiser HJ, Turnbaugh PJ (2013) Xenobiotics shape the physiology and gene expression of the active human gut microbiome. Cell 152(1–2):39–50. https://doi.org/10.1016/j.cell.2012.10.052

Mayer EA, Knight R, Mazmanian SK, Cryan JF, Tillisch K (2014) Gut microbes and the brain: paradigm shift in neuroscience. J Neurosci 34(46):15490–15496. https://doi.org/10.1523/JNEUROSCI.3299-14.2014

McEwen BS, Wingfield JC (2003) The concept of allostasis in biology and biomedicine. Horm Behav 43(1):2–15. https://doi.org/10.1016/s0018-506x(02)00024-7

McGuinness AJ, Davis JA, Dawson SL, Loughman A, Collier F, O'Hely M, Simpson CA, Green J, Marx W, Hair C, Guest G, Mohebbi M, Berk M, Stupart D, Watters D, Jacka FN (2022) A systematic review of gut microbiota composition in observational studies of major depressive disorder, bipolar disorder and schizophrenia. Mol Psychiatry 27(4):1920–1935. https://doi.org/10.1038/s41380-022-01456-3

Morais LH, Schreiber HL 4th, Mazmanian SK (2021) The gut microbiota-brain axis in behaviour and brain disorders. Nat Rev Microbiol 19(4):241–255. https://doi.org/10.1038/s41579-020-00460-0

Mudd AT, Alexander LS, Berding K, Waworuntu RV, Berg BM, Donovan SM, Dilger RN (2016) Dietary prebiotics, milk fat globule membrane, and lactoferrin affects structural neurodevelopment in the young piglet. Front Pediatr 4:4. https://doi.org/10.3389/fped.2016.00004. PMID: 26870719; PMCID: PMC4740374

Myers JS (2008) Proinflammatory cytokines and sickness behavior: implications for depression and cancer-related symptoms. Oncol Nurs Forum 35(5):802–807. https://doi.org/10.1188/08.ONF.802-807

Nguyen TT, Kosciolek T, Eyler LT, Knight R, Jeste DV (2018) Overview and systematic review of studies of microbiome in schizophrenia and bipolar disorder. J Psychiatr Res 99:50–61. https://doi.org/10.1016/j.jpsychires.2018.01.013

Oksenberg JR (2013) Decoding multiple sclerosis: an update on genomics and future directions. Expert Rev Neurother 13(12 Suppl):11–19. https://doi.org/10.1586/14737175.2013.865867

Olson CA, Vuong HE, Yano JM, Liang QY, Nusbaum DJ, Hsiao EY (2018) The gut microbiota mediates the anti-seizure effects of the ketogenic diet. Cell 174(2):497. https://doi.org/10.1016/j.cell.2018.06.051

Parashar A, Udayabanu M (2017) Gut microbiota: implications in Parkinson's disease. Parkinsonism Relat Disord 38:1–7. https://doi.org/10.1016/j.parkreldis.2017.02.002

Perez-Pardo P, Kliest T, Dodiya HB, Broersen LM, Garssen J, Keshavarzian A, Kraneveld AD (2017) The gut-brain axis in Parkinson's disease: possibilities for food-based therapies. Eur J Pharmacol 817:86–95. https://doi.org/10.1016/j.ejphar.2017.05.042

Peterson VL, Jury NJ, Cabrera-Rubio R, Draper LA, Crispie F, Cotter PD, Dinan TG, Holmes A, Cryan JF (2017) Drunk bugs: chronic vapour alcohol exposure induces marked changes in the gut microbiome in mice. Behav Brain Res 323:172–176. https://doi.org/10.1016/j.bbr.2017.01.049. Epub 2017 Feb 1. PMID: 28161446; PMCID: PMC5518606

Rantala MJ, Luoto S, Borráz-León JI, Krams I (2022) Schizophrenia: the new etiological synthesis. Neurosci Biobehav Rev 142:104894. https://doi.org/10.1016/j.neubiorev.2022.104894

Reeve A, Simcox E, Turnbull D (2014) Ageing and Parkinson's disease: why is advancing age the biggest risk factor? Ageing Res Rev 14(100):19–30. https://doi.org/10.1016/j.arr.2014.01.004

Rhee SH, Pothoulakis C, Mayer EA (2009) Principles and clinical implications of the brain-gut-enteric microbiota axis. Nat Rev Gastroenterol Hepatol 6:306–314

Rieder R, Wisniewski PJ, Alderman BL, Campbell SC (2017) Microbes and mental health: a review. Brain Behav Immun 66:9–17. https://doi.org/10.1016/j.bbi.2017.01.016

Risch N, Hoffmann TJ, Anderson M, Croen LA, Grether JK, Windham GC (2014) Familial recurrence of autism spectrum disorder: evaluating genetic and environmental contributions. Am J Psychiatry 171(11):1206–1213. https://doi.org/10.1176/appi.ajp.2014.13101359

Şafak B, Altunan B, Topçu B, Eren Topkaya A (2020) The gut microbiome in epilepsy. Microb Pathog 139:103853. https://doi.org/10.1016/j.micpath.2019.103853

Scheperjans F, Pekkonen E, Kaakkola S, Auvinen P (2015) Linking smoking, coffee, urate, and Parkinson's disease—a role for gut microbiota? J Parkinsons Dis 5(2):255–262. https://doi.org/10.3233/JPD-150557

Shen Y, Xu J, Li Z, Huang Y, Yuan Y, Wang J, Zhang M, Hu S, Liang Y (2018) Analysis of gut microbiota diversity and auxiliary diagnosis as a biomarker in patients with schizophrenia: a cross-sectional study. Schizophr Res 197:470–477. https://doi.org/10.1016/j.schres.2018.01.002

Shreiner AB, Kao JY, Young VB (2015) The gut microbiome in health and in disease. Curr Opin Gastroenterol 31(1):69–75

Swidsinski A, Loening-Baucke V, Lochs H, Hale LP (2005) Spatial organization of bacterial flora in normal and inflamed intestine: a fluorescence in situ hybridization study in mice. World J Gastroenterol 11(8):1131–1140. https://doi.org/10.3748/wjg.v11.i8.1131. PMID: 15754393; PMCID: PMC4250702

Thursby E, Juge N (2017) Introduction to the human gut microbiota. Biochem J 474(11):1823–1836. https://doi.org/10.1042/BCJ20160510

Tiwari S, Atluri V, Kaushik A, Yndart A, Nair M (2019) Alzheimer's disease: pathogenesis, diagnostics, and therapeutics. Int J Nanomedicine 14:5541–5554. https://doi.org/10.2147/IJN.S200490. PMID: 31410002; PMCID: PMC6650620

Tremlett H, Fadrosh DW, Faruqi AA, Hart J, Roalstad S, Graves J, Spencer CM, Lynch SV, Zamvil SS, Waubant E, US Network of Pediatric MS Centers (2016) Associations between the gut microbiota and host immune markers in pediatric multiple sclerosis and controls. BMC Neurol 16(1):182. https://doi.org/10.1186/s12883-016-0703-3

Turnbaugh PJ, Gordon JI (2009) The core gut microbiome, energy balance and obesity. J Physiol 587(Pt 17):4153–4158. https://doi.org/10.1113/jphysiol.2009.174136

Turnbaugh PJ, Bäckhed F, Fulton L, Gordon JI (2008) Diet-induced obesity is linked to marked but reversible alterations in the mouse distal gut microbiome. Cell Host Microbe 3(4):213–223. https://doi.org/10.1016/j.chom.2008.02.015. PMID: 18407065; PMCID: PMC3687783

Ueki A, Otsuka M (2004) Life style risks of Parkinson's disease: association between decreased water intake and constipation. J Neurol 251(Suppl 7):vII18–vII23. https://doi.org/10.1007/s00415-004-1706-3

Ursell LK, Metcalf JL, Parfrey LW, Knight R (2012) Defining the human microbiome. Nutr Rev 70(Suppl 1):S38–S44. https://doi.org/10.1111/j.1753-4887.2012.00493.x. PMID: 22861806; PMCID: PMC3426293

Vaher K, Bogaert D, Richardson H, Boardman JP (2022) Microbiome-gut-brain axis in brain development, cognition and behavior during infancy and early childhood. Dev Rev 66:101038

Walker AW, Ince J, Duncan SH, Webster LM, Holtrop G, Ze X, Brown D, Stares MD, Scott P, Bergerat A, Louis P, McIntosh F, Johnstone AM, Lobley GE, Parkhill J, Flint HJ (2011) Dominant and diet-responsive groups of bacteria within the human colonic microbiota. ISME J 5(2):220–230. https://doi.org/10.1038/ismej.2010.118

Weinberg R, Gould D (1999) Foundations of sport and exercise psychology (No. Ed. 2). Human Kinetics Publishers (UK) Ltd.

Wyss-Coray T (2006) Inflammation in Alzheimer disease: driving force, bystander or beneficial response? Nat Med 12(9):1005–1015. https://doi.org/10.1038/nm1484

Yolken RH, Severance EG, Sabunciyan S, Gressitt KL, Chen O, Stallings C, Origoni A, Katsafanas E, Schweinfurth LA, Savage CL, Banis M, Khushalani S, Dickerson FB (2015) Metagenomic sequencing indicates that the oropharyngeal phageome of individuals with schizophrenia differs from that of controls. Schizophr Bull 41(5):1153–1161. https://doi.org/10.1093/schbul/sbu197

Young VB (2017) The role of the microbiome in human health and disease: an introduction for clinicians. BMJ 356:j831. https://doi.org/10.1136/bmj.j831

Zheng P, Zeng B, Zhou C, Liu M, Fang Z, Xu X, Zeng L, Chen J, Fan S, Du X, Zhang X, Yang D, Yang Y, Meng H, Li W, Melgiri ND, Licinio J, Wei H, Xie P (2016) Gut microbiome remodeling induces depressive-like behaviors through a pathway mediated by the host's metabolism. Mol Psychiatry 21(6):786–796. https://doi.org/10.1038/mp.2016.44

Zmora N, Suez J, Elinav E (2019) You are what you eat: diet, health and the gut microbiota. Nat Rev Gastroenterol Hepatol 16(1):35–56. https://doi.org/10.1038/s41575-018-0061-2

Neurological Disorders: The Impact of Alteration in Gut Microbiome and Inflammation on Disease Development

5

Jeane Rebecca Roy, Coimbatore Sadagopan Janaki, Ponnulakshmi Rajagopal, and Selvaraj Jayaraman

Abstract

The physiology and pathology of humans are significantly influenced by the human microbiota. In addition to helping to preserve the homeostasis of the gastrointestinal system, the gut microbiota, and microbial metabolites also send signals to other bodily systems, such as the brain. The immune and endocrine systems both have an impact on the functioning of the central nervous system. Notable immune cells called microglia, keep the central nervous system's functions and homeostasis in check. The gut microbiota's significance in controlling microglial growth and function has been addressed. The gut and brain regulate each other bidirectionally via multiple mechanisms and pathways, including neural, neuroendocrine, and immunological signals. Recently, research has focused more on the relationship between immune-related neurological diseases such as multiple sclerosis, alzheimer's disease, and neuroinflammation and the microbiota-gut-brain axis. The involvement of the microbiota-gut-brain axis in immune-related neurological diseases has been studied using a variety of tech-

J. R. Roy · C. S. Janaki
Department of Anatomy, Bhaarath Medical College and Hospital, Bharath Institute of Higher Education and Research (BIHER), Chennai, India

P. Rajagopal
Central Research Laboratory, Meenakshi Ammal Dental College and Hospitals, Meenakshi Academy of Higher Education and Research, Chennai, India

S. Jayaraman (✉)
Department of Biochemistry, Centre of Molecular Medicine and Diagnostics (COMManD), Saveetha Dental College and Hospital, Saveetha Institute of Medical and Technical Sciences, Saveetha University, Chennai, India
e-mail: selvarajj.sdc@saveetha.com

S. Pathak, A. Banerjee (eds.), *Gut Microbiome and Brain Ageing*,
https://doi.org/10.1007/978-981-99-8803-7_5

niques, including GF studies, infection investigations, probiotic research, antibiotic studies, and fecal transplantation studies. The purpose of this chapter is to increase understanding of the significance of the gut microbiome, the possible mechanisms underlying changed gut microbiota and inflammation, and the pathogenic changes related to neurological disorders.

Keywords

Altered gut microbiome · Neuroinflammation · Developmental disorders · Neurodegenerative disorders · Gut microbiota brain signaling

5.1 Introduction

The physiology and pathology of humans are significantly influenced by the human microbiota. Bacteria, bacteriophages, yeasts, fungi, and viruses make up the mammalian gut microbiota. In humans, in the initial 3 years, it continues to grow and stabilize but gets affected throughout a lifetime by many environmental and lifestyle factors. The metabolism of xenobiotics, dietary fiber fermentation, the production of vitamins, the defense against pathogens, metabolic equilibrium, and the control of the immune system are some of the biological processes carried out by the gut microbiota (Lavelle and Sokol 2020). Birth brings about the development of the gut microbiome (GM), which changes in many ways over the course of a person's lifetime. Infants get their mothers' gut microbiomes at birth. The distinctiveness of the GM is low and erratic throughout the first 3 years of life. Healthy individuals, who primarily have Bacteroidetes and Firmicutes, have a steady microbial composition starting at the age of three and it largely stays the same throughout adulthood. Many variables, such as lifestyle choices, genetics, age, geography, medications, illnesses, and nutrition, can have an impact on the distribution and species diversity of the various bacteria in the gut. The GM again experiences significant alterations in elderly adults (>65 years old) (Tao et al. 2020).

Moreover, an extensive part of how illness outcomes are regulated is by the gut microbiome. A diet deficient in fiber diminishes the variety of the colon microbiota and disturbs the microbiota composition, which is defined by the loss of distal colon microbiota and a decline in the neighboring mucosal community. Long-term high-fat diets have been made known to increase the number of harmful bacteria like Faecalibacterium spp. while decreasing or eliminating the number of beneficial bacteria such as Alistipes and Bacteroides spp. Dietary fat may disrupt the pathways that produce arachidonic acid and lipopolysaccharide (LPS), and it may also cause inflammation to flare on a frequent basis. Consequently, the symbiotic interaction between the host and the gut microbiota is crucial for preserving human health.

This chapter intends to raise awareness of the importance of the gut microbiome, the potential processes underlying altered gut microbiome and inflammation, and the concerned pathological alterations in neurological disorders.

5.2 Influence of Gut Microbiota on Brain Function and Development

The body receives benefits from the gut microbiota when dietary nutrients are transformed into microbial compounds that interact with one another as well as with host cells, subsequently affecting one's health and illness state. In addition to helping to preserve the homeostasis of the gastrointestinal (GI) system, the gut microbiota and microbial metabolites also send signals to other bodily systems, such as the brain (Emge et al. 2016).

The complex process of brain development typically starts in the third prenatal week and lasts until late adolescence. The involvement of GI microflora has lately been recognized as one of the many elements that can influence brain development. Early on in life, the gut microbiota is varied and rich, and its alteration during this pivotal stage can impact brain function and development (Wang et al. 2018). Better cognitive outcomes were seen in newborns with greater Bacteroides levels, whereas those with higher levels of gut microbiota's alpha diversity (the variety of species found inside each person) had poorer values on the total composite score, expressive language scales, and visual reception (Carlson et al. 2018). Furthermore, given that our GIT is home to more than a hundred trillion bacteria, multiple studies have looked at the link between a healthy microbiome and the appropriate development of brain systems and neural circuits (Borre et al. 2014).

The immune and endocrine systems both have an impact on the functioning of the central nervous system (CNS) by either activating the enteric nervous system of the entire digestive system inadvertently or directly via input transmissions from the vagus nerve during early life colonization of the gut microbiota (Ding and Liu 2019). Studies have shown that the lack of comparable results in vagotomized animal models lends weight to the idea that the vagal nerve plays a role in the microbiota-gut-brain axis (MGBA), which can be impacted by specific probiotics (Bravo et al. 2011).

Research on germ-free (GF) animals or animals given broad-spectrum antibiotic treatment is extensively used to examine the effects of the gut microbiota's total lack of development and behavior (Wang et al. 2018). The learning, recognition, and behavioral abilities of GF mice were compromised. Furthermore, the levels of important neurotransmitters, including 5-hydroxytryptamine (5-HT), serotonin, and brain-derived neurotrophic factor (BDNF), are changed in these mice as compared to control mice. Early childhood may be a vulnerable era for the microbiota in the gut to control brain growth and behavioral roles because GF mice subjected to gut microbiota exhibit behaviors that are similar to those of particular pathogen-free mice (Bercik et al. 2011). Recent research indicates that the diversity of the gut microbiota is strongly correlated with the functional connectivity between the sensorimotor and parietal lobes, the anterior cingulate and insula, and the amygdala with the midbrain in resting-state functional MRI of 1-year-old infants (Gao et al. 2019).

Notable immune cells called microglia keep the central nervous system's functions and homeostasis in check. The gut microbiota's significance in controlling microglial growth and function has been addressed in research findings. The immature phenotypes and Tight junction (TJ) protein expression decreased, and the inflammatory cytokine profiles that affect the basal surveillance state were altered in GF mice and mice treated with antibiotics, both of which had severe microglial abnormalities, and resulted in increased BBB permeability, which may permit hazardous chemicals to invade the brain and induce neuroinflammation and injury (Erny et al. 2015). These investigations highlight the part played by the gut microbiota in controlling brain function and neuroimmunity.

5.3 Gut Microbiota-Brain Signaling via the Immune System

The body defends itself from microbial invasion through the gut by gut-associated lymphoid tissues (GALT). The blood-brain barrier (BBB) can be crossed by T cells, dendritic cells (DCs), and macrophages from the gut and GALT, which can then influence neurons and glia. In the choroid plexus, the brain harbors resident immune cells including macrophages and DCs, parenchymal cells like microglia, and CSF cells like leukocytes (Prinz and Priller 2017). So, the gut microbiota, which influences the host immune system, may additionally regulate these immune cells as well. Previous works displayed germ-free mice had microglia with global cellular proportion and maturation defects, reducing innate immune responses, thus pointing out that gut microbiota regulates microglial activation and homeostasis, which impact the CNS immune system (Wu and Wu 2012).

It is known that the brain also releases molecules linked to innate immunity, namely cytokines and Toll-like receptors (TLRs), as well as molecules associated with adaptive immunity that play important modulatory roles in brain development, like antibody receptors and the major histocompatibility complex (MHC). There may be a connection between autoimmunity and the CNS peripheral immune system, according to the existence of meningeal lymphatic vessels. Additionally, microglia and lymphocytes are necessary for the proper connectivity of brain circuits and are capable of regulating cognition (Morimoto and Nakajima 2019). Immune cells such as T cells, neutrophils, NK cells, and macrophages can enter the brain and interact with the CNS similarly to how microglia do, which can affect the function of the brain. Microglia can regenerate in the CNS from embryonic progenitor cells and undergo phagocytosis, antigen presentation, and brain physiological functions and further to this, severely influence behavior and neurodegenerative illnesses. Most parts of the brain's vasculature have tissue-specific selective BBB features that permit the movement of necessary molecules while preventing the entrance of potentially harmful materials or cells (Wu et al. 2023).

Several works of literature show that the GI microbiome develops brain immunity as well as influences the maturation of microglia cells. GI microbiome triggers the microglial cell maturation in the initial years of embryonic development and paves the way for the proper functioning of microglia in adulthood. SCFAs from the gut microbiota may influence microglia maturation (Silva et al. 2020). Immune factors, cytokines, and chemokines circulate through the vagus nerve and circumventricular organs to affect the brain. Pro-inflammatory cytokines can increase BBB permeability in the brain by causing neuroinflammation. Inflammatory responses, immune cell infiltration, and reactive gliosis can result from BBB leakage, causing neurodegeneration (Jeon et al. 2021). The neurotransmitters like serotonin, dopamine, and glutamate in the brain are affected by cytokines while intestinal microbes regulate brain function via non-inflammatory cytokines. In neonatal mice, antibiotic exposure decreased plasma granulocyte colony-stimulating factor (G-CSF). G-CSF crosses the BBB to stimulate brain neurogenesis and protect against Alzheimer's disease, ischemic injury, and Parkinson's disease (Prakash et al. 2013; Meuer et al. 2006). Studies with adult mice treated with antibiotics end up losing gut microbial diversity and hippocampus neurogenesis and altered gut microbiome can be restored with the help of probiotics and had more Ly6C high monocytes and better neurogenesis (Möhle et al. 2016). Thus, the gut microbiota may promote G-CSF production to treat neurodegenerative diseases and normal neurodevelopment. The immunomodulatory effects of GM are shown in Fig. 5.1.

Normal immune responses are triggered by the immune system and microbiota. However, immune system failure causes autoimmune and inflammatory diseases. It was noted that anxiety-like behavior increased in GF and control rodents without an intestinal microbiome, so targeting the restoration of gut microbiota in early life could cure this behavior. Dysbiosis can disrupt microbial metabolism, which can directly affect anxiety-related disorders by affecting the vagus nerve, immune system, and inflammatory responses. LPS and amyloids from the GI microbiome contribute to proinflammatory cytokines and signaling pathways associated with neurodegenerative inflammation. Under different immunological conditions, gut microbial communities recruit different immune cells to influence gut immune responses. *Proteobacteria, Fusobacteria, Firmicutes, Cyanobacteria, Bacteroidetes, Verrucomicrobia* are the predominant microbial communities in the mammalian gut microbiome, according to NGS and metagenome analysis. The host's homeostasis depends on the GIT's heterogeneity but can negatively affect the host. The GI microbiome is crucial to hosting immunity, as GF mice have fewer immature gut-associated lymphoid tissues (GALT), antimicrobial peptides (AMPs), and immunoglobins A (IgAs) than wild-type mice due to imbalanced immune system development. Additionally, the enteric immune system differentiates commensals from pathogens and determines immunological tolerance to them (Kabat et al. 2014).

SCFAs inhibit NF-κB and histone deacetylase (HDAC) activity, promoting gut-barrier function and anti-inflammatory effects as well as increasing regulatory T lymphocyte function and number in the enteric system (Silva et al. 2020). Intestinal immune function is impacted by microbial gut flora tryptophan compounds that bind to the aryl hydrocarbon receptor (AhR). IL-22 is activated by a commensal gut bacterium called Lactobacillus reuteri, which also generates indole-3-aldehyde and catabolizes tryptophan (Schiering et al. 2017). Spermine, Diamine, spermidine, and polyamines are examples of arginine derivatives generated by the intestinal microflora that affect immune responses by improving the homeostasis of the intestinal mucosa and resident immune cells (Bekebrede et al. 2020). Thus, intestinal flora shapes gut microbial communities and regulates pathogens on the intestinal mucosa,

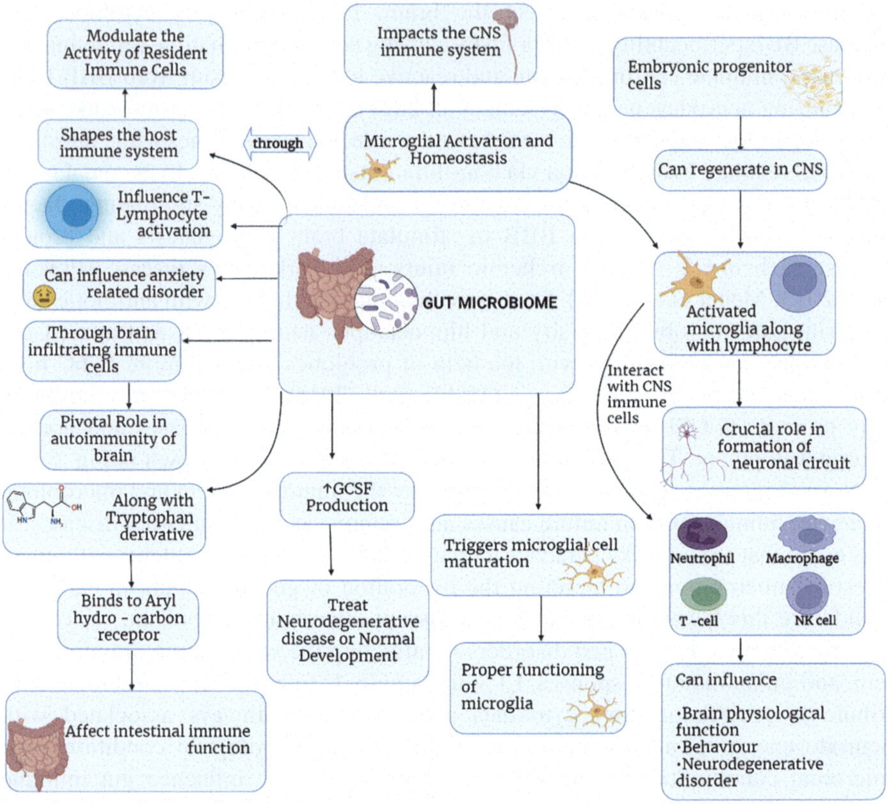

Fig. 5.1 The immunomodulatory effects of GM. GM by activating the microglia shapes the host immune system and modulates the activity of resident immune cells. It also impacts the CNS immune system. Activated microglia along with lymphocyte plays a key task in the formation of the neuronal circuit. These microglial cells are derived from embryonic progenitor cells and they regenerate in CNS. The activated microglial cells can interact with CNS immune cells and influence the physiological function of the brain, behavior, and neurodegenerative disorder. GM can also trigger microglial cell maturation, combine with tryptophan derivative of microbes to affect intestinal immune function, influence anxiety-related disorders, influence T lymphocyte activation, play a pivotal role in autoimmunity of the brain. Gut Microbiome can increase GCSF production thereby treating neurodegenerative diseases

modulating immune system development and function, and previous reports confirmed that GF mice have fewer lamina propria associated CD4+ lymphocytes due to fewer T helper 1 (Th1) and Th17 cells, IL-17, and IL-22. The vital function of immune cells in immunology are formed by specific bacterial species. Th17 cells can defend against pathogens and are implicated in autoimmune disease pathogenesis. These cells decrease when segmented filamentous bacteria (SFB) in the GI microflora decrease or disappear, indicating their potential to differentiate Th17 cells and induce IL-22 and IL-17 (Yang et al. 2021).

5.4 Neuroinflammation

Gut microbiota modulates brain-resident immune cells as well. Neuroinflammation activates the gut and brain immune systems, contributing to neurological disorders. Toll-like receptors (TLRs) on immune cells recognize pathogen-associated molecular patterns (PAMPs) and microbe-associated molecular patterns (MAMPs) and activate and as a result, TNF-α, IL-6, IL-1β and IL-17A, produced by activated immune cells, get into the circulation of the brain via the BBB and may cause or worsen several neurological disorders (Hug et al. 2018). In the gut and spinal cord of the experimental autoimmune encephalomyelitis (EAE) model, GF mice generate less interferon (IFNγ) and IL-17A. Segmented filamentous bacteria (SFB) colonization induces Th17 and Th1 responses in the intestine and brain and promotes EAE symptoms in GF mice, demonstrating that gut microbiota impacts neuroinflammation via immunological responses. While Bacteroides fragilis and Prevotella histicola colonization promote Treg function to suppress EAE (Mangalam et al. 2017). Th17 cells can be activated by SFB colonization in the gut and result in symptoms resembling ASD; however, neutralizing antibodies can lessen ASD-related behavioral impairments. Neuroinflammation is brought on by inflammasome activation, which results in caspase-1 development and the production of IL-1 and IL-18. Pro-inflammatory cytokine production and inflammasome pathways involved in various neurological illnesses can be activated by specific MAMPs. Studies show mice caspase-1 deficient mice have diminished anxiety and depressive-like behaviors following chronic stress (Yang et al. 2019).

CNS microglia affect brain development, homeostasis, and pathology and involve in cytokine release, complement activation, and phagocytosis. Previous studies have shown that depleted gut microbiota is associated with immature microglia with altered inflammatory gene expression (Salter and Stevens 2017). Similar to microglia, astrocytes are immune cells that govern the blood-brain barrier (BBB), manage the growth and repair of the central nervous system (CNS), and provide antigens. Through type 1 IFN signaling, bacterial tryptophan metabolites and AHR activation in astrocytes control inflammation of CNS (Rothhammer et al. 2016). The homeostasis of immune cells may play a role in regulating gut microbiota-brain communication.

5.4.1 Altered Gut Microbiome

A surge in the population of bacterial pathogens and a drop in the prevalence of beneficial microorganisms is known as dysbiosis, which is an alteration in the gut microflora, often characterized by commensal loss, a bloom of pathobionts, and loss of diversity. The gut microbiome undergoes structural and functional alterations because of dysbiosis. As a result, the metagenomic capabilities of microbial communities as well as their taxonomic composition are altered. "Gain of function dysbiosis" causes chronic inflammation by promoting an excess of pathogenic flora. "Loss of function dysbiosis," distinguished by the lack of beneficial bacteria, favors the emergence of several disorders such as irritable bowel syndrome (IBS), obesity, colorectal cancer, chronic kidney disease, cardiovascular disease, metabolic syndrome, and many others (Wilkins et al. 2019). When beneficial flora decreases and pathogenic flora increases, some diseases can develop. The interruption of microbial homeostasis due to the conversion of SCFA-producing microbiota into toxic metabolite-producing microbiota interferes with gut-brain communication resulting in the emergence of neuro-disorders such as Alzheimer's disease, Parkinson's disease, anxiety, and depression. Commensal bacteria might impact brain function by directly altering immunological and neuronal receptors in the ENS and CNS (Jalandra et al. 2022).

The GI tract, which is inhabited by the majority of beneficial microbes, links up with the vast number of immune cells. Healthy microbiota boosts barrier immunity and helps protect their ecological habitat by strengthening the immune system. Healthy microbiota boosts barrier immunity and helps protect their ecological habitat by strengthening the immune system. Bacteroides fragilis, segmented filamentous bacteria (SFB), Clostridia, and Mucispirillum directly influence immune system maturation and specialization. To preserve its homeostatic interaction with the microbiota, the host reduces tissue inflammation, microbial translocation, and bacteria contact with the epithelial cell surface. This segregation is accomplished via the interaction of epithelial cells, mucus, immune cells, antimicrobial peptides, and immunoglobulin A (IgA) in the GI system, which has the maximum density of commensals (Belkaid and Hand 2014). Pattern recognition receptors (PRR) control the colonization of microbes. A tight junction protein complex is disengaged by dysbiosis, which promotes the entry of bacteria or bacteria-related components into the lamina propria and lowers the levels of the zonula occludens toxin. Pathogens, harmful external, and internal signals like oxidative stress and inflammatory cytokines produced by tissues or cells, can all activate TLRs. In steady-state environments, TLR4 exists at a low level in the intestinal mucosa and may be activated in response to lipopolysaccharides (LPSs) made by gram-negative bacteria. Intestinal inflammation and intestinal barrier disruption were reduced in TLR4 knockout mice. Activation of the TLR4-mediated inflammatory signaling pathway can result from microbiota dysbiosis (Ding and Liu 2019).

Although a person's microbiota tends to remain relatively stable over time, there are still daily variations within a person, and dramatic changes can happen after significant disruptions. Antibiotics, infection or colonization by foreign commensal microorganisms, significant dietary changes, or non-infectious diseases that affect GI function are common causes of disruption. The microbiota may recover and revert to its natural composition after the selection pressure is removed, depending on the extent and duration of the disturbance. As a result, numerous risk factors have been related to the pathogenesis of gut dysbiosis. Obesity and high-fat and high-sugar diets have been linked to reproducible gut microbiota changes. The constitution of the microbiota can be modulated and maintained by environmental factors such as food, exercise, maternal surroundings, delivery method, and interaction with sources of new microbiota like probiotics. Gut dysbiosis is also linked to social stressors and exposure to xenobiotics like heavy metals and pesticides (Sommer and

Bäckhed 2013). The physiology of the host may be negatively impacted by repeated disruptions, which might hinder healing.

Dysbiosis thus causes persistent, low-grade irritation making gut permeability and inflammation as the risk factors for several illnesses, including IBD, advanced liver disease, chronic kidney disease, diabetes, and so on. TLR in innate immune cells, such as CD4+ T helper (Th) cell subsets, myeloid cells, and mast cells, recognize LPS and peptidoglycan found in intestinal microbes as pathogen-related or damage-associated molecular patterns, respectively. This results in the onset of chronic systemic low-grade inflammation. In addition to this, dysbiotic microbiota hijacks the host's immune system by altering inflammasome signaling, modulating TLR signaling, and degrading secretory IgA through bacteria-derived compounds (sIgA) (Levy et al. 2017). By directly adhering to the intestinal epithelium, SFB increases IgA synthesis that constrains the pro-inflammatory property of commensal bacteria and activates Th17 cells and serum amyloid A protein. In contrast, most other bacteria use SCFAs to control the homeostasis of regulatory T lymphocytes (Treg). Gut microbiota is closely associated with the balance of Th17 and Treg via Th17/Treg axis (Omenetti and Pizarro 2015). Specific components of GM cause the production of pro-inflammatory cytokines such as IL-6, IL-23, IL-1β as well as the production and expansion of Th17 cells. Commensal bacteria and their metabolites promote the generation of intestinal Tregs that actively induce mucosal tolerance. In dysbiosis, a decrease in the number of SCFA-producing bacteria results in an unbalanced production of Tregs and irritated mucosa (Levy et al. 2017). Polysaccharide A, which is produced by B. fragilis and TGF-β from intestinal epithelial cells, controls Treg cells. TGF-β is mainly expressed and activated upon short-chain fatty acid (SCFA) stimulation or by dendritic cells (DCs) after bacterial contact. TGF-β levels in the gut are modulated by the gut microbiota which impacts the development and function of immune cells to regulate microbiota sequestration within the mammalian lumen (Bauché and Marie 2017). In a study, more than 30 different human Clostridia strains were injected into GF mice, and this caused a threefold rise in Tregs than the uncolonized controls. However, adding of just one strain from the same Clostridia collection resulted in an even more subdued Treg response (Atarashi et al. 2013). This implies that higher microbial diversity enhances host immune responses, whilst lower microbial diversity may promote inflammatory processes.

In addition to TGF-β signaling, the gut microbiota and its metabolites affect the integrity of the intestinal epithelium by activating signaling pathways such as AhR, STAT3, and mTOR. Excessive energy intake can result in the excessive activation of mTORC1 signaling, which can alter homeostasis and cause metabolic diseases (Bauché and Marie 2017). Numerous mTOR inhibitors, including rapamycin, metformin, and resveratrol, are used to treat metabolic disorders, specifically obesity and type 2 diabetes (T2DM). Strangely, all these medications change the gut microbiome (Levy et al. 2017). Metformin and resveratrol inhibit mTORC1 by activating AMP-activated protein kinase, also known as AMP-activated protein kinase. Rapamycin is the traditional mTORC1 inhibitor, but over time, it will also inhibit mTORC2 and Akt, impairing insulin action and upsetting glucose homeostasis. Strong evidence that the host's gut microbiota can control the activation of AMPK and its downstream effectors, like mTORC1, was found in germ-free mice, which were resistant to HFD-induced obesity (Noureldein and Eid 2018).

Inactivation of tuberin (TSC2) will increase mTORC1 pathway activity, which will decrease the development of intestinal goblet and Paneth cells. Due to this suppression, mucus and antimicrobial peptide production will decrease, which will result in dysbiosis of the gut microbiota. However, mTORC2 signaling has a role in the control of the immunological state and leads to the inactivation of Akt, leading to the impairment of phosphorylation of FoxO1 (Noureldein and Eid 2018). FoxO1 regulates the formation of antimicrobial peptides by binding to the gene promoter's regulatory region, which affects gut microbiota composition. When immune cells are activated by stress, they release more pro-inflammatory cytokines (IL-6, TNF-α, and IL-1β), as well as a number of chemokines (CXCL-1, MIP 1α and monocyte chemoattractant protein-1) (Zeng et al. 2013).

Alterations in the GM composition have a direct impact on intestinal barrier integrity, GI motility, and mucus secretion. Disruption in gut epithelial structure may allow the translocation of commensal bacteria and toxic signals into the host which cause both inflammation and infection. Glial cell, synaptic pruning, and myelination function as well as immunological, neurotransmitter, and stress hormone production are all supported by balanced GM and its derived metabolic profiles. Thus, via controlling neuronal circuits, the microbiota gut brain (MGB) axis affects higher psychological processes such as mood, emotion, cognition, and memory. Through the MGB axis, gut dysbiosis alters the integrity of the intestinal membrane and negatively influences the endocrine, neurological, and immunological systems (Anand et al. 2022). Further, a variety of bacteria, including pathogenic, commensal, and probiotic microbes, can activate neural pathways and the central nervous system's signaling process, which can lead to the onset of mental illnesses such as depression and anxiety. In addition, Lactobacillus rhamnosus, Helicobacter pylori, L. paracasei, Pseudomonas, E. coli, and B. longum have been mentioned while exploring directed interactions between the central nervous system and the GIT (Anand et al. 2022). In addition to affecting the gut-brain axis, the gut microbiota also affects immunity, metabolism, and brain function. Additionally, it has been demonstrated that factors such as diet, sleep habits, and antimicrobial exposure alter brain function by altering the gut microbiome (Arneth 2018).

Abnormalities in the metabolic profile of GM and GM-derived organisms lead to immune-driven inflammation, insufficient neurotransmitter release, elevated stress response, premature and malfunctioning glial cells, defective synaptic pruning, and impaired myelination. Therefore, dysregulation of the MGB axis affects higher psychological functions such as cognition, emotion, mood, and memory by altering neuronal circuitry. Because of the altered microbial variety, composition, and metabolites, there may be fewer goblet cells and less mucus produced, resulting in a thin mucosal layer. Leaky gut (leaky gut syndrome) is a condition in which the changed microbiota reduces the expression of tight junction proteins (such as claudin-5 and occludin) between intestinal epithelial cells, increasing permeability (Anand et al. 2022). This leads to the increased translocation of pathobionts and their toxic components such as LPS and peptidoglycan into the systemic circulation leading to increased secretion of pro-inflammatory cytokines (like interleukin (IL)-18, IL-1, IL-6, and TNF-α), and alter the levels and synthesis of neurotransmitters (noradrenaline, glutamate, dopamine, and 5-HT), lead to faulty MGB signaling. A leaky gut causes disruptions in the tight and anchoring junction proteins in the frontal cortex, hippocampus, and striatal regions of the brain, results in chronic systemic inflammation and eventually alters the function of the brain (Erny et al. 2015). In the brain parenchyma, increased BBB permeability leads to an increase in the release of chemokines, endocrine (stress) messengers, and cytokines as well as excessive immune cell trafficking and toxic microbial metabolites. Neurotrophin synthesis, myelination, and changes in neurotransmitters and their related receptors are all signs of a changed neuroimmune status (Arneth 2018; Erny et al. 2015). Taking these findings together, it can be shown that GD inhibits MGB signaling and sets off the pro-inflammatory cascade, both of which are directly associated with the beginning of NPD symptoms.

5.5 Microbiota Influenced the Blood-Brain Barrier and the Structure of the Brain

Specialized endothelial cells in the microvasculature make up a semipermeable barrier BBB, which divides the brain and peripheral blood. Several conditions are linked to microbially induced BBB disruption, including Parkinson's disease, Alzheimer's disease, schizophrenia, and autism spectrum disorders (ASDs) (Heijtz et al. 2011; Fiorentino et al. 2016). The modulation of the BBB by numerous bacterial metabolites and neurotransmitters is produced in the gut. Several rodent models have focused that the loss of intestinal microbiota can bring about the permeability of BBB, and its functionality can be restored when the gut microbiota is free of pathogens (Braniste et al. 2014). The enhanced BBB permeability and its subsequent development into Alzheimer's disease with the deposition of amyloid-peptide are also possible effects of metabolic illnesses like diabetes (Spadoni et al. 2017).

Microbiota dysbiosis affects the protective properties of the BBB, such as tight junction expression and permeability regulation that can bring about behavioral changes. The microbiota dysbiosis can affect the protective properties of the BBB, through tight junction expressions such as permeability modulation, and can lead to behavioral alterations. Previous works have shown that microbiota is necessary for the typical development of the hippocampus and microglial morphology, displaying the interconnection of brain morphology and GM (Zhu et al. 2020). Earlier MRI studies have displayed that there is a correlation between the relative abundance of actinobacteria and better microstructural organization of the amygdala, hypothalamus, and thalamus. Structural alteration can influence motor coordination, focus, and cognitive test (Fernandez-Real et al. 2015).

5.6 Neurological Disorders

The GI microbiota is sensitive to progressive functional loss due to neurological disorders. Pathogenic processes like the weakening of the gut epithelial barrier, the loss of enteric neurons, and altered mucosal immune function can result in an excessive generation of pro-inflammatory cytokines. Along with aging, many other factors as previously mentioned in this chapter are linked to considerable changes in the composition of the gut microbiota, like decreased microbial richness and diversity and a drop in bacteria with anti-inflammatory capabilities, along with alterations in GI physiology. Brain functions, such as cognition and memory, can be regulated by the gut microbiota through this pathway of communication called the microbiota-gut-brain axis (Holmes et al. 2020).

The biochemical signals cause the brain to react, leading to mood, behavior, and cognition changes. Clinical trials have examined a variety of methods for altering the microbiome, including fecal microbiota transplantation (FMT), the administration of probiotics or antibiotics, or dietary therapies that are non-invasive yet may alter the host immune systems (Anand et al. 2022). Ailments of the central and peripheral nervous systems, known as neurological disorders, may involve the autonomous nervous system, spinal cord, brain, neuromuscular plaque, cranial and peripheral nerves and their roots (Holmes et al. 2020). Here, we highlight the neurological conditions that may be impacted by GM-immune interaction, including developmental abnormalities and neurodegeneration.

5.6.1 Development Disorders

5.6.1.1 Autism Spectrum Disorder

Autism spectrum disorder (ASD) is a neurodevelopmental disease that is distinguished by poor social interaction and the appearance of repetitive or stereotyped behaviors. Over the past few decades, ASD incidence has been steadily rising. ASD may result from an amalgamation of factors associated with genes and the environment. Due to immunological dysregulation and digestive issues, anomalies in the peripheral, enteric, and neuroimmune systems have been linked to ASD. Gastrointestinal disruptions are more common in ASD patients and a rise in intestinal permeability, which results in higher serum levels of toxins and bacterial metabolites, might cause immunological reactions that affect social behavior and brain function. The cerebellum and cerebral cortex of postmortem ASD patients' brains exhibit greater microglia and astroglia activation, and the cerebrospinal fluid and cortical regions of the brain have higher amounts of proinflammatory cytokines (Vuong and Hsiao 2017).

Earlier research work into the changes in the gut microbiota composition of ASD patients observed that the proportion of Bacteroidetes/Firmicutes spp., whereas the proportion of strains of Sutterella, Odoribacter, and Butyricimonas spp., were considerably raised while the relative proportion of Veillonella and Streptococcus spp., were all substantially lowered (Liu et al. 2022). In ASD patients, behavioral and emotional symptoms, such as temper tantrums, sleep issues, irritability, and aggression have a strong association with the intensity of their GI symptoms. Prolonged GI disturbances are thought to be a risk factor for the development of ASD and gastric symptoms are often recorded in ASD patients, which may be attributed to a neurologic reason instead of a gastroenteric cause (Vuong and Hsiao 2017). Children with ASD and GI issues had lower amounts of the genera Coprococcus, Enterococcus, Prevotella, Bacteroides, Veillonellaceae, Bifidobacterium, and Escherichia coli while having greater levels of the genera Lactobacillus, Faecalibacterium, and Ruminococcus (Xu et al. 2019). Clostridium histolyticum, which is known for producing toxins, was found in higher concentrations in the feces of children with ASD. The Vancomycin medicament has been reported to ameliorate ASD symptoms by lowering the levels of Clostridia, which indicates that C. histolyticum-group bacteria may be a factor in ASD-like symptoms (Xu et al. 2019; Vuong and Hsiao 2017).

Several studies have shown that immune response regulation has a role in the influence of altered gut microbiota during pregnancy on the development of ASD in children. The effector cytokine IL-17A and Th17 cells were thought to be the main contributors in mothers for maternal immune activation (MIA)-induced ASD in offspring in a murine model of maternal immune activation. IL-17 receptors were elevated in both the mother and the fetus, and maternal IL-17A promoted aberrant brain growth and ASD-like behavioral symptoms in children (Choi et al. 2016). These imply that altering the composition of the gut microbiota in the mother and inhibiting IL-17A signaling may be effective preventative measures for ASD. The increased gut permeability, ASD-related abnormalities, and altered microbial composition in MIA children were all alleviated by the administration of the human commensal Bacteroides fragilis (Xu et al. 2019). IL 6, IL 1β, TNF-α, and monocyte chemotactic protein 1 are more prevalent in cerebral fluid in ASD individuals, which is indicative of mucosal inflammation. As a result, the influx of immune cells such as natural killer cells, monocytes, and CD3+ TNFα+ or CD3+ IFNγ+ cells, produce pro-inflammatory cytokines, that lead to immunological dysfunction in the GI tract (Chaidez et al. 2014). Higher serum LPS levels associated with the leaky gut syndrome have been linked to social skill deficiencies in autistic patients, indicating persistent inflammatory responses that disrupt brain circuitry signaling. ASD individuals had decreased GABA concentrations and greater isopropanol and SCFAs concentrations in their feces. Children with ASD had increased quantities of SCFAs in their feces, including isobutyric acid, caproic acid, valeric acid, isovaleric acid, butyric acid, propionic acid, and acetic acid (Kang et al. 2018). Rats had aberrant movements, cognitive impairments, and decreased social relations after intraventricular infusion of propionic acid. Additionally, rats treated with propionic acid had more neuroinflammation and oxidative stress in their brain tissue and were also found to have glial cell differentiation, gliosis, and the release of pro-inflammatory cytokines in an in vitro investigation utilizing human neural stem cells (Xu et al. 2019).

ASD may be impacted by several gut microbiota-related issues from the prenatal maternal aspects such as maternal health, antibiotic use, the birth method, and feeding habits. In a rodent experiment, giving a mother a high-fat diet resulted in gut microbial dysbiosis and deficiencies in social behavior in the offspring, highlighting maternal obesity is strongly correlated with ASD offspring. Furthermore, anomalies in fetal neurodevelopment and ASD have been linked to females taking various antibiotics during pregnancy and could be brought about by antibiotic-induced immunological activation and changes in the mother's GM. C-section deliveries resulted in social and behavioral impairments in the offspring in a mouse model of ASD, meanwhile, oxytocin therapy restored the mice born by c-section's low sociability. Children with longer feeding intervals with continued breastfeeding as well as longer for more than 6 months were found to have a lower chance of developing ASD and ASD-related GI symptoms (Connolly et al. 2016). Thus, immunological responses and gut microbiota during pregnancy and the first few years of life may contribute to the development of ASD-like behaviors in children.

5.6.1.2 Attention Deficit / Hyperactivity Disorder (ADHD)

Several works have reported that the gut microbiota may be involved in the pathophysiological processes that result in attention deficit hyperactivity disorder (ADHD) symptoms. The gut microbiome affects the catecholaminergic neurotransmission system's metabolic pathways or the expression of the genes that code for these neurotransmitter transporters. ADHD has been linked to higher levels of oxidative and nitrosative (NO) stress indicators and lower antioxidant concentrations, as well as altered dopaminergic neurons' mitochondrial quantity and function (Avcil et al. 2021). When the mitochondrial activity is dysregulated, the byproducts of the oxidative activities that result in the production of adenosine triphosphate (ATP), reactive oxygen species (ROS), and reactive oxygen-nitrogen species (RONS), formed rapidly. Excessive ROS/RONS oxidize the polyunsaturated fatty acids (PUFAs) that constitute the neuronal membranes, altering the process of apoptosis and increasing the synthesis of SCFAs as a result of alterations in the microbiome. ROS/RONS cause a vicious cycle by inducing the generation of inflammatory cytokines, microglial activation, and NOD-like receptor family, pyrin domain-containing 3 (NLRP3) inflammasome activation (Checa-Ros et al. 2021; Solleiro-Villavicencio and Rivas-Arancibia 2018).

Several literatures have reported that ADHD patients have elevated levels of pro-inflammatory cytokines such as IFNγ, IL-1, IL-13, and IL-6 in their serum. The bacterium Dialister spp., which has been demonstrated to positively correlate with an altered temperament and impulsiveness in toddlers, has an antagonistic relationship with the pro-inflammatory cytokine IL-6. In contrast to healthy controls (HC), ADHD patients showed considerably lower amounts of Dialister spp. and higher levels of pro-inflammatory cytokines (Martínez et al. 2013). Limiting the expression of these cytokines may be an essential preventative approach in the treatment of ADHD as excessive levels of pro-inflammatory interleukins are known to be associated with neuroinflammation that can induce decreased cortical volume and negative behavior (Checa-Ros et al. 2021). The presence of higher inflammatory markers in ADHD patients might lend backing to the idea that low levels of Feacalbacteria spp. generate inflammation, which interferes with brain development and ultimately leads to the pathophysiology of ADHD (Bull-Larsen and Mohajeri 2019).

With the aid of next-generation 16S rDNA sequencing, Aarts et al. assessed the variations in the microbial diversity in fecal samples from ADHD patients. Bifidobacterium, a member of the phylum Actinobacteria, however, significantly increased in the ADHD cohort. Fascinatingly, the Bifidobacterium genus had significantly higher levels of the enzyme cyclohexadienyl dehydratase (CDT), which is involved in the synthesis of phenylalanine, a dopaminergic precursor and even altered reward anticipation responses, a neurological feature of ADHD, may be explained by the microbiome diversity between patients with ADHD and controls in their study. Parental reports of ADHD symptoms were negatively correlated with the prevalence of the Faecalibacterium species, and it was discovered that the concentration of this bacterium was significantly lower in the ADHD group. By sharing sequences with microorganisms that can metabolize the GABA neurotransmitter, the genus Ruminococcaceae_UGC_004 was used in a different study to support the idea that the gut microbiome plays a role in the neurotransmitter systems that are connected to the pathophysiology of ADHD (Bull-Larsen and Mohajeri 2019). The ADHD group showed decreased beta diversity and substantial changes in alpha diversity. In comparison to controls, ADHD samples were found to have a considerably greater abundance of Bacteroidaceae at the family level. It was noted that the decrease in alpha diversity was not seen in ADHD patients' fathers and controls, but only in their mothers. This study raises the possibility that changes in the microbiome makeup may be transferred from mothers to offspring (Prehn-Kristensen et al. 2018). Wan et al. discovered that among ADHD patients, Odoribacter (order Bacteroidales) and Enterococcus concentrations were significantly higher than those of Ruminococcaceae and Faecalibacterium. The species of Paraprevotella xylaniphila, Veillonella parvula, Bacteroides caccae and Odoribacter splanchnicus significantly increased in the ADHD group. The genes that encode the enzymes that are a part of the dopaminergic synaptic pathways were altered in the ADHD group. Pieces of literature support that Faecalibacterium may have anti-inflammatory effects, and that aberrant levels may increase the development of inflammatory markers that may play a role in the pathophysiology of ADHD. Levodopa may be excessively converted into dopamine in the digestive tract by Enterococcus, a bacterium that has been associated with neurotransmitter release. Additionally, it has been observed that mice deprived of the 5-HT transporter have much higher levels of the bacteria Enterococcus, which can lead to lower levels of 5-HT, a neurotransmitter that is also crucial in the pathophysiology of ADHD. Contrarily, Bacteroides coprocola and Lactobacillus were much less prevalent in the ADHD group than Sutterella stercoricannis, Bacteroides uniformis, Bacteroides ovatus, the genus Fusobacterium, and Ruminococcaceae_UGC_004. The phylum Actinobacterium significantly declined because of micronutrient supplementation, with the order Bifidobacteriales declining by 25% along with higher concentrations of the species Collinsella. As a result, there was an improvement in aggression, emotional control, and general function (Cheng et al. 2020). The findings show that adding extra micronutrients may have an impact on how many putative probiotic bacterial species are present.

Due to the diversity and composition of the gut microbiota, which oscillates throughout the 24-h cycle of light and darkness, it is probable that the gut, like the other tissues, possesses its own peripheral circadian clocks that eventually rely on the central clock. In animal tests, the variety and abundance of various species, namely Bacteroidetes and Clostridia, changed throughout the cycle of light and shade. Bacteroidetes were more prevalent and there were more bacteria during the mice's active phase, but Firmicutes predominated during the rest phase and there were fewer bacteria present. Melatonin also appears to influence the Firmicutes: Bacteroidetes ratio and the richness and diversity of the gut microbiota in mice (Leone et al. 2015). The microbial constitution and integrity of the colonic mucosa in sleep-deprived rats, whose plasma melatonin levels were lower than in controls, were examined by Gao et al. to determine the changes brought on by melatonin treatment. The sleep-deprived mice had significantly lower microbial diversity and richness and significantly higher Firmicutes: Bacteroidetes ratios, while a significant drop in Faecalibacterium, Bacteroides, and Akkermansia, was seen, along with an increase in Aeromonas. Sleep deprivation also drastically decreased the expression of TJ proteins in goblet cells. On the other hand, following melatonin administration, the Firmicutes: Bacteroidetes ratio, and the richness-diversity indices were reverted to levels like those of the control group. Aeromonas levels significantly decreased after melatonin therapy, while the concentrations of Akkermansia, Bacteroides, and Faecalibacterium increased. Melatonin also markedly boosted the expression of TJ proteins and goblet cell numbers. When the disruptive effects of lack of sleep were neutralized, the ratio of pro-inflammatory to anti-inflammatory cytokines and the redox status also changed (Acuña-Castroviejo et al. 2017).

5.6.2 Neurodegeneration

5.6.2.1 Parkinson's Disease

Parkinson's disease (PD) is a multicentric, progressive neurodegenerative disorder that disrupts dopaminergic nerve cells in the substantia nigra due to the accumulation of α-synuclein (α-syn). This encourages Lewy bodies, which are circular lamellated eosinophilic cytoplasmic inclusions, to gradually assemble. Immune dysregulation and diminished cellular and humoral immune responses in PD display the relationship between autoimmune disorders and PD. PD risk is associated with genes that control leukocyte/lymphocyte activity and cytokine-mediated signaling. There is growing evidence that dysbiosis-related intestinal inflammation and intestinal inflammatory responses have a pathogenic function in PD. Increased intestinal permeability, which is prevalent among PD patients and could possibly be an indication of gut microbiota problems, has been linked to motor impairments, and microglial activation. One of the clinical premotor indicators of PD may be α-syn neurodegeneration in the ENS associated with GI wall physiological changes and chronic constipation. The gut microbiota likely impacts enteric neurons involved in α-syn secretion are likely impacted by the gut microbiota (Sampson et al. 2016).

There is a positive correlation between the relative richness of the gut Enterobacteriaceae community and the severity of gait problems and postural instability in PD. There were found to be reduced levels of SCFA and lower levels of butyrate-producing bacteria in PD patients, which may potentially cause enhanced mucosal permeability and exposure to coliform bacteria's systemic endotoxins (Gerhardt and Mohajeri 2018). The unregulated expression and misfolding of the α-syn colon may be brought on by the increase in gut permeability and systemic exposure to bacterial endotoxins caused by the decrease in Prevotellaceae population. Although the abundances of Butyricicoccus, Clostridium, and Roseburia were positively connected with SCFA levels, those of Phascolarctobacterium, Intestinimonas, Akkermansia, Sporobacter, Flavonifractor, and Shigella/Escherichia were adversely correlated with SCFA levels (Sampson et al. 2016). In the fecal samples of people with PD, there is a substantial drop in ghrelin and Prevotellaceae species rather than in the relative counts of Enterobacteriaceae that biosynthesize thiamine, folate, and neuroactive SCFAs and thus incorporating these vitamins and SCFAs to a diet could assist with tackling Parkinson's disease by eventually causing an upsurge Lactobacillaceae and Prevotellaceae (Tan et al. 2020a, b).

The organization of the intestinal barrier depends on occludin and other tight-junction proteins. Intestinal permeability is enhanced by dysbiosis of the gut, which destroys the occludins. In PD patients, dysbiosis impairs intestinal barrier function, activates the immune system, and causes a systemic inflammatory response and can result in constipation and colonic inflammation due to the long-term existence of PD. Therefore, LPS and other bacterial neurotoxins pass the intestinal wall and enter the bloodstream to compromise the epithelial barrier of the gut. NF-κB and TLR4 are responsible for the generation of TNF-α, IL-1β and IL-6 that cause systemic inflammation when bacterial LPS is present in the bloodstream and progressively breakdown BBB leading to α-syn buildup (Sampson et al. 2016). Increased levels of TLR4 or bacterial endotoxin-specific ligand expression, CD3+ T cells, and other cytokines open the door to BBB breakdown, which may cause dopaminergic neuronal death in the substantia nigra. This shows that one of the major causes triggering PD neurodegeneration may be TLR4-interceded inflammation in the brain or intestinal inflammation (Perez-Pardo et al. 2019).

The WHO defines probiotics as live microorganisms that, when given to the host in the proper dosages, promote their health. It has been demonstrated that probiotics, such as lactobacilli and bifidobacteria, can treat PD-like disorders (Tan et al. 2020a, b). L-DOPA, an important dopamine precursor molecule, can be converted by Bacillus spp., a probiotic bacterium, into dopamine by using DOPA decarboxylase. People with Parkinson's disease (PD) are said to benefit from regular consumption of fermented milk products containing Lactobacillus casei shirota because it reduces the levels of fecal staphylococci (Cassani et al. 2011).

Researchers found that specific bacteria can create beneficial phenolic acids during bacterial fermentation in vitro experiments. The synthesis of metabolites obtained from flavanol-rich preparation (FRP) that affect inflammation or α-synuclein misfolding varies depending on the heterogenic humanized gnotobiotic mice used in the research. Microbial communities in the gut interfere with the misfolding and toxicity of α-synuclein, which is a primary pathogenic mechanism of PD and related α-synucleinopathies. In the early stages of PD, digestive problems might arise; this may help in the early detection of the condition before motor symptoms like tremors and rigidity appear. A vegetarian diet that includes SCFAs has also been proven to increase UDPRS III levels while lowering the daily effective dose of levodopa-equivalent. Thus, research on the gastric flora can reveal information regarding the toxicity of PD medications (Brown and Goldman 2020).

5.6.2.2 Alzheimer's Disease

Neuroinflammation, beta-amyloid (Aβ) plaque accumulation, and neurofibrillary tau tangles are some of the pathological characteristics of AD. In the CNS, the Aβ polymerizes into fibrils through self-aggregation, resulting in further neurotoxicity and inflammation. The peripheral immune system is significantly influenced by gut microbes, making this a promising possible link to AD pathogenesis. The peripheral innate and adaptive immune system can be modulated by cell components and metabolites from a changed gut microbial ecology, which can then affect CNS neuroinflammatory activity (Seo and Holtzman 2020). The compounds linked with pathogens in the GI tract may activate mast cells, which then migrate to other tissues and release inflammatory mediators including cytokines and chemokines as well as reactive oxygen species, and these mechanisms serve as a defense against infections, which also have an impact on the microglia, astrocytes, and blood vessels in the brain. Chronic mast cell activation may result in severe neuroinflammation and accelerate the degeneration of neurons. The proinflammatory mediators and innate immune cells including mast cells and microglia might affect BBB permeability. Furthermore, elevated BBB permeability may promote immune cell or mediator infiltration into the brain, accelerating neuroinflammation (Zhu et al. 2020).

By drawing in hemocytes to the brain and producing neuroinflammation, dysbiosis accelerates the advancement of Alzheimer's disease. A decrease in microbial diversity and a disproportion in the gut microbiota are present in AD patients. It has been shown that pathogenic gut microbiota such as Escherichia coli, Bacteroides, and Ruminococcus are more prevalent in AD patients, whereas beneficial gut microbiota such as Dialister, Eubacterium rectale, and Bifidobacterium are less prevalent (Seo and Holtzman 2020). In an AD rodent model, it has been demonstrated that neuroinflammation, which in turn affects amyloidosis, is modulated by antibiotic-induced microbiota modification. Aggregates of Aβ- and tau-related pathologies in the brains of AD mice may be lessened by the introduction of healthy gut flora (Kim et al. 2020). L. plantarum administration could prevent cognitive dysfunction by reducing tau hyperphosphorylation and Aβ plaque development in AD rodents (Song et al. 2022).

A critical molecule needed for intestinal homeostasis, P-glycoprotein, was found to be expressed at low levels in a study of AD elders, demonstrating a direct link between intestinal inflammation and microbiota imbalance. In fact, variations in the gut microbiota may affect the body's levels of tryptophan and serotonin as well as the synthesis of important brain-supporting chemicals including dopamine, norepinephrine, and brain-derived neurotrophic factor (BDNF). As previously indicated, the gut microbiota's generation of SCFAs, such as butyrate, propionate, and acetate, plays a positive function in energy production, gut epithelia homeostasis, and immunological modulation (Morris et al. 2017). Dysbiosis can affect their synthesis, which favors the deposition of Aβ plaques, metabolic dysfunctions, and microglia dysregulation, all of which accelerate cognitive loss. Furthermore, a decline in butyrate-producing bacteria, as seen in AD, has been associated with T cell dysregulation, epithelial barrier leakage, and elevated bacterial translocation (Doifode et al. 2021). Therefore, circulating LPS from Gram-negative endobacteria, also known as metabolic endotoxemia, stimulates BBB disruption and systemic inflammation via TLR4 to promote neuroinflammation. Trimethylamine *N*-oxide (TMAO) and amyloid are two toxic chemicals that might arise due to intestinal dysbiosis. Recent studies have found associations between the microbial metabolite TMAO and elevated beta-amyloid production, peripheral immune response activation, elevated oxidative stress, intestinal mucosal barrier dysfunction, platelet hyperactivity, and BBB permeability (Morris et al. 2017). Thus, bile acids from bacteria and cholesterol are more likely to enter the brain and some endobacteria have the capacity to create gasotransmitter molecules, including nitric oxide (NO), hydrogen (H2), ammonia (NH3), methane (CH4), and hydrogen sulfide (H2S), which appears to be essential for the correct neuronal function and whose modification contributes to AD pathogenesis (Oleskin and Shenderov 2016).

The idea of fecal microbial transplantation (FMT) has also been researched to slow cognitive decline, improve gut immune responses, and trigger anti-inflammatory cascades in a rat model of AD. Studies have shown inflammation and microbial dysbiosis in the gut of transgenic mice using the $ADLP^{APT}$ (AD-like pathology with amyloid and neurofibrillary tangles) when analyzed with wild-type (WT) mice. $ADLP^{APT}$ mice that received a fecal transplant from WT donor mice displayed an interesting combination of improved cognition, decreased gut and systemic inflammation, and reduced amyloid and Tau pathology in the brain (Kim et al. 2020).

The use of probiotics has been advocated as beneficial for AD. In a mouse model of β-amyloid injected animal model, lactobacilli and bifidobacteria were proven to be helpful in lowering memory and learning impairment. Remarkably, mini-mental status examination (MMSE) scores increased considerably after 12 weeks of consuming Lactobacillus and Bifidobacterium species in fermented milk in a randomized clinical trial of 60 AD patients (Akbari et al. 2016). The measures of cognition in patients with AD improve when Helicobacter pylori are eradicated with triple eradication therapy. The possibility that probiotics or antibiotics could be used as future AD treatment medicines is encouraging. In a recent study, it was discovered that probiotics and exercise may improve cognitive function, reduce the quantity of Aβ plaques in the hippocampus, and subsequently slow down the advancement of Alzheimer's disease in mice (Doifode et al. 2021).

5.6.2.3 Multiple Sclerosis

The CNS is affected by the chronic inflammatory disease known as multiple sclerosis (MS), which causes demyelination and neurodegeneration. According to recent studies, the gut commensal microbial communities can be viewed as a unique environmental risk factor for MS as well as other immune-mediated illnesses.

In experimental autoimmune encephalomyelitis (EAE) rats, the most widely used inflammatory demyelinating disease model, as well as MS patients, changes in the microbiome and the prevalence of "leaky gut" have been discovered. Faulty Treg function and raised proinflammatory cell infiltration are the immunological alterations associated with EAE (Buscarinu et al. 2017). The beneficial gut microbiota can restrict astrocyte pathogenicity, maintain BBB permeability, and activate microglia. Bifidobacterium and lactic acid-producing bacteria like Lactobacillus have been shown to lessen the severity of EAE symptoms in animal models (Zhu et al. 2020). Archaea are widely distributed in MS, but Firmicutes, Butyricimonas, and Bacteroidetes phyla are scarce or absent. The induction of FoxP3+ Tregs, which control inflammation, is facilitated by species such as Bacteroides and Clostridia species. In the gut microbiota, TLR2 identifies polysaccharide A of Bacteroides fragilis to protect against EAE-related CNS demyelination and inflammation. Haemophilus, Blautia, Psuedomonas, and Dorea genera were increased in MS patients, however, Prevotella, Parabacteroides, and Adlercreutzia were drastically decreased (Buscarinu et al. 2017). A recent study found that GABA produced by Lactobacillus brevis can directly affect adaptive immune cells and antigen-presenting cells (APCs) in response to myelin proteins, which may lessen inflammation and relieve EAE symptoms (Zhu et al. 2020). Serotonin levels were shown to be lower in MS patients and EAE animal models. By inhibiting the growth of T cells, IL-17, and IFNγ-release, and stimulating the synthesis of IL-10, elevated serotonin levels may lessen the severity of sickness. Serotonin may also affect the pathogenic processes in MS by polarizing macrophages into M2 macrophages (Sacramento et al. 2018). Although dopamine levels are lesser in MS patients than in healthy people, there is a higher percentage of IL-17 and IFNγ-producing cells in MS patients, suggesting that dopamine may have a suppressive effect. Dopamine has the potential to be therapeutic for the management of relapsing-remitting MS by preventing peripheral blood mononuclear cells (PBMCs) from producing IFN- and IL-17 (Melnikov et al. 2016). There will be a considerable decrease in SCFA concentrations and a loss of bacteria that produce SCFAs, in individuals suffering from MS and thus have altered gut microbiomes. Propionic acid-treated rodents exhibited decreased lymphocyte infiltration, demyelination, and activation of Tregs in the small intestine in the EAE animal model. It was also shown that 2 weeks of propionic acid treatment for MS patients reduced Th1 and Th17 while increasing Treg and genes that promote Treg. Propionic acid treatment for 3 years also decreased brain atrophy and the annual relapse rate of MS (Duscha et al. 2020). Additionally, in EAE animals, the stimulation of GLP-1/GLP-1R signaling in microglia reduced spinal cord damage and alleviated clinical symptoms. Probiotic therapy and fecal microbial transplantation have been shown to reduce MS symptoms in other studies as well (Zhu et al. 2020).

5.6.2.4 Amyotrophic Lateral Sclerosis

Lower motor neurons in lower cranial nerve motor nuclei and upper motor neurons in the corticobulbar tract both gradually degenerate because of ALS. It causes the weakness of the muscles responsible for mastication, mouth-cheek movement, and the shielding of the airway during swallowing occurs from dysfunction of either of these neural groups. Therefore, those who have ALS typically consume food and beverages more slowly, generally in smaller quantities. Moreover, patients with substantial corticobulbar dysfunction frequently lament that they have developed a severe intolerance to spicy food, as this can frequently result in laryngospasm. The bulbar dysfunction in ALS is likely to have an impact on the oral microbiome causing difficulty in oral hygiene that progressively impacts the intestinal microbiome. In ALS, the weakness of airway muscles can result in pneumonia due to the combination of frequent aspiration, difficulty in the expulsion of secretions, and insufficient lung expansion. Eventually, immobility can cause pressure sores and skin infections, so also gastrostomy tubes too get infected. These lead to the frequent usage of antibiotics which may influence the microbiota in ALS patients (Calvo et al. 2022).

Several pieces of work demonstrated that there was decreased junctional protein expression and increased permeability and leakiness with different expressions and less diversity of the genus and class levels in the GM of ALS patients (Gotkine et al. 2020). Altered gut microbiota have an impact on the progression of ALS, and was found to have a lower Firmicutes-to-Bacteriodetes ratio and fewer beneficial bacteria such as Anaerostipes, Lachnospiraceae, and Oscillobacter than healthy individuals. Another study displayed that ALS patients had altered Firmicutes to Bacteriodetes ratios together with elevated fecal inflammatory markers, indicating that the imbalance of the gut microbiota may be related to the pathogenesis of the illness (Calvo et al. 2022). Insufficient quantities of bacterial products such as butyrate and SCFA cause intestinal epithelium permeability to increase in dysbiosis, which is linked to ALS and causes systemic dissemination. This causes dendritic cells and macrophages to generate proinflammatory cytokines, which alter immune homeostasis (Casani-Cubel et al. 2021). It was also noted that there will be a decrease in the protein expression of tight and adherent junctions, which leads to increased intestinal epithelial permeability. In the early stages of ALS, a marked bacterial decline in Butyrivibrio fibrisolvens and Firmicutes. Anaerostipes and the bacterial genus Dorea, which is known for its toxic microorganisms, have both seen significant increases; on the other hand, the Faecalibacterium, Bacteroides, Prevotella genus and some genera of the Lachnospiraceae family, which are known for their beneficial microorganisms for ALS patients, have seen significant decreases (Zhai et al. 2019). Presymptomatic SOD1 transgenic (SOD1-Tg) mice that are susceptible to ALS have already shown dysbiosis that has been connected to the presence of neurodegenerative Ruminococcus torques bacterium. Whereas, the presence of Akkermansia muciniphila is linked in alleviating ALS symptoms, potentially increasing motor neurons, and raising GABA/glutamate ratios in the hippocampus. A similar increase was found in a study related to nicotinamide rise that improved the motor and functional abilities of ALS (Zhu et al. 2020). Tryptophan levels in plasma are raised by the bacteria Bifidobacterium infantis, which has an impact on serotonin transmission in the brain. They even boost the production of SCFAs, which are produced through bacterial fermentation and have similar neuroactive features. The GABA system is one of the primary methods by which intestinal bacteria regulate brain chemistry and it has been demonstrated that some species of Lactobacillus and Bifidobacterium segregate and elevate GABA, by the Lactobacillus rhamnosus, Lactobacillus brevis, and Bifidobacterium dentium (Calvo et al. 2022; Zhu et al. 2020). No matter what part of the brain or type of neuron is afflicted, all these evidence show a consistent link between bacterial flora and the commencement of neurodegeneration.

5.6.2.5 Huntington's Disease

A progressive neurodegenerative condition with motor, cognitive, and psychiatric abnormalities, Huntington's disease (HD) is monogenic, completely penetrant, and fully symptomatic. The huntingtin (HTT) gene on chromosome 4 is responsible for HD, which is broadly expressed in peripheral organs such as the gut and skeletal muscles. HD is brought about by the increase of CAG trinucleotide repeats in exon 1 of the huntingtin (HTT) gene. Mutant huntingtin (mHTT) protein expression in the GI tract results in GI dysfunction, including reduced gut motility, diarrhea, and malabsorptive absorption of nutrients. Both in HD patients and in several transgenic mouse models of HD, malabsorption is correlated with the amount of weight loss that is a hallmark of HD (Du et al. 2021). However, several pieces of evidence point to gut dysbiosis as a likely cause of HD. Furthermore, weight loss, a typical HD symptom, may be hampered by gastrointestinal dysfunction. In relation to body weight loss and motor impairment, there was an increase in the presence of Bacteroidetes and a decrease in the number of Firmicutes. There are correlations between the presence of a systemic chronic inflammatory condition in HD patients and altered gut microbiota, such as Intestinimonas sp. and Bilophila sp., and plasma levels of cytokines (IL-4 and IL-6, respectively) (Patil et al. 2021).

In an R6/1 mice model of HD, 16S RNA sequencing was carried out to analyze the GM, and was observed that the microbiome's composition varied significantly. The amount of water in the feces of HD mice at 12 weeks of age was also considerably altered. There was a proportionate decrease in Firmicutes and an increase in Bacteroidetes, whereas HD male mice showed higher microbiota diversity than WT control mice, but the HD females did not show any such differences (Kong et al. 2020). Studies on R6/2 mice indicated changes in the GM's microbial composition, including a higher percentage of Proteobacteria and Bacteroidetes and lower concentrations of Firmicutes in comparison to levels maintained in the control group (Stan et al. 2020). Gubert et al. discovered changes in the microbiome composition at the level of orders like Lachnospirales, Bacteroidales, and Oscillospirales in a highly intriguing work employing the R6/1 mouse strain. This investigation revealed that all HD animals exhibit more alpha diversity than WT mice. HD patients were clinically evaluated by means of 16S RNA sequencing on stool samples and cognitive tests and had considerably lower species richness (alpha diversity) and different microbial community composition (beta diversity) than the control group, according to the microbiome evaluation. The findings of the sequencing study reveal statistically significant differences in Euryarchaeota, Firmicutes, and Verrucomicrobia at the phylum level in HD males. Similar changes were seen in families like Erysipelotrichaceae, Akkermansiaceae, Acidaminococcaceae, Enterobacteriaceae, Bifidobacteriacea, Lachnospiraceae, Bacteroidaceae, Christensenellaceae, Clostridiaceae, Flavobacteriaceae, Coriobacteriaceae, Rikenellaceae, Eggerthellaceae, Methanobacteriaceae, Peptococcaceae, and Peptostreptococcaceae. The analysis revealed a positive correlation between the elevated Intestinimonas bacterial count and anti-inflammatory cytokine IL-4. The genus Bilophila is negatively correlated with pro-inflammatory IL-6 levels (Patil et al. 2021). Moreover, it was discovered that there were significant connections between Porphyromonas and IL-4, IL-10, and IL-13 as well as positive correlations between Clostridium XVIII, TNF-α, and IL-8 (Du et al. 2021). The outcomes of these studies unambiguously show a connection between HD patients' immunological responses and the composition of their gut microbiomes.

5.6.2.6 Epilepsy

A persistent propensity to have seizures, together with emotional and cognitive problems, are the hallmarks of the brain condition epilepsy. The hippocampal formation is important for the development and control of a number of brain processes, including memory and learning, and is susceptible to neurological conditions like epilepsy. The most prevalent form of epilepsy, temporal lobe epilepsy (TLE), alters the excitability and circuitry of the hippocampal region under the influence of both aberrant neurogenesis and neuronal degeneration.

Despite a paucity of clear evidence, the gut microbiome may have a role in the formation and activity of seizures by altering microbial signals to the hippocampus. According to Sewal et al., intraperitoneal administration of LPS increased the propensity for seizures in rats and increased BBB permeability while also raising levels of proinflammatory cytokines in the brain. The mechanisms underlying epileptogenesis in the temporal lobe have also been linked to impaired synaptosomal transport of GABA and glutamate, which is essential to quickly end neurotransmission by eliminating extracellular transmitters. Rodents treated chronically with Lactobacillus rhamnosus (JB-1) had higher GABA(A2) and lower GABA(B1b) receptor expression in the hippocampus (Ogbonnaya et al. 2015). A similar study by Janik et al., demonstrated Lactobacillus rhamnosus (JB-1) treatment in mice enhanced hippocampus GABA and glutamate levels. These studies clearly indicate that the gut microbiome affects the hippocampus's GABAergic and glutamatergic transmission. Numerous strains of the Lactobacillus, Lactococcus, and Streptococcus genera, as well as Bifidobacterium have been reported to release GABA/glutamate. Uncontrolled epilepsy increased the Firmicutes/Bacteroidetes ratio. Firmicutes-phylum bacteria have the ability to control neurotransmitter levels. In patients with temporal lobe epilepsy (TLE), excessive and uncontrolled neuronal activity is one of the most obvious signs of an epileptic brain. The enteric nervous system (ENS) has shown a potential involvement of the gut microbiome in the control of neuronal excitability. It is interesting to note that a subset of ENS neurons in the colon of

Sprague-Dawley rats responded to the probiotic Lactobacillus reuteri by becoming significantly more excitable, increasing the number of action potentials per depolarizing pulse, and opening calcium-dependent potassium channels. These findings imply that substantial changes in the gut microbiome, or the entire lack of gut microbes in GF animals, are related to modifications in neuronal excitability in the ENS and CNS. Significantly, higher levels of α-diversity were connected to an unexpectedly higher number of uncommon gut bacterial species. Studies showed that more Lactobacillus and Bifidobacterium were associated with fewer annual seizures. The metabolism of the antiepileptic medication zonisamide can be influenced by the intestinal gut microbiota. Vagus nerve stimulation and the ketogenic diet (KD) are alternative therapy methods for uncontrolled epilepsy. A KD has been connected to altered intestinal microbiota composition and function, as well as a decrease in the frequency of seizures in epileptic patients. Meanwhile, a number of studies have shown that probiotic supplementation has a beneficial impact on epilepsy (Geng et al. 2022).

5.6.2.7 Stroke

Recent investigations have demonstrated that an altered gut microbiome influences post-stroke outcomes through a variety of pathways, including metabolites, endotoxemia, local and systemic inflammation, immune and neurological systems, and gut leakiness. Stroke results in gut dysmotility, raises gut-barrier permeability, and promotes the transfer of microorganisms and microbial byproducts into the bloodstream, including trimethyl amine-*N*-oxide (TMAO) and LPS. These modifications hasten systemic inflammatory response and symptom aggravation, frequently resulting in poor prognosis. The ischemic brain disrupts gut-brain axis signaling via neuronal or HPA axis mechanisms in stroke triggered gut dysbiosis. The activation of innate immune cells by metabolic endotoxemia leads to chronic systemic inflammation and enhanced immune cell migration into the brain, triggering the BBB to break down and generating neuroinflammation. The presence of gut dysbiosis is supported by the fact that 50% of stroke patients experience GI symptoms. Ischemic stroke causes intestinal barrier integrity to be compromised, mucus secretion to be diminished, and intestinal microorganisms to be transported into the bloodstream and the extraintestinal organs (Singh et al. 2016). The intestine and the brain's local immune cells are impacted by gut dysbiosis. Upon microglial activation in the early stages of a stroke, peripheral immune cells, particularly monocytes and T- and B-lymphocytes, infiltrate the area.

Recent studies have found a link between TMAO and an increased risk of cardiovascular and cerebrovascular disorders, indicating that this metabolite may have the ability to treat these conditions by modifying gut flora. The raised plasma levels of TMAO were dose-dependently associated with a higher risk of stroke and cardiovascular events in a longitudinal study involving more than 4000 cases. The levels of TMAO decreased after antibiotic therapy, emphasizing the role of certain bacterial species in the gut for the formation of this molecule. Preclinical studies show that phosphatidylcholine metabolites such as choline and TMAO can boost the expression of macrophage scavenger receptors that contribute in atherosclerosis, which is most likely brought on by the presence of bacterial species that live in the gut (Yin et al. 2015).

Lowered diversity of the gut microbiota after stroke has been clearly demonstrated in experimental stroke models in mice. The GI microflora was altered by antibiotics in a way that also reduced the production of an IL-17-associated chemokine and slowed the migration of proinflammatory IL-17γδ T cells. So, by controlling intestinal T-cell infiltration to the brain, intestinal bacterial species arise to govern neuroinflammation following a stroke. When compared to healthy subjects, clinical research on stroke patients revealed altered Firmicutes-to-Bacteroidetes ratios, higher abundances of opportunistic pathogens (Megasphaera, Enterobacter, and Desulfovibrio), lower abundances of beneficial SCFA-producing bacteria like that of Faecalibacterium, Anaerostipes, Blautia, Lachnospiraceae, Roseburia, and Bacteroides (Benakis et al. 2016). Additionally, after-stroke FMT containing SCFA-producers such as Lactobacillus fermentum, Bifidobacterium longum, Faecalibacterium prausnitzii, and Clostridium symbiosum reduced post-stroke cognitive impairments and inflammation while also increasing plasma, gut, and brain SCFA concentrations, aiding in post-stroke recovery in elderly models. The number and diversity of Bacteroidetes have also been observed to have decreased after the stroke. According to a different study, individuals who have had a transient ischemic attack or stroke have higher concentrations of opportunistic pathogens such as Desulfovibrio, Enterobacter, Megasphaera, and Osicillibacter and lower concentrations of commensal or helpful genera such as Bacteroides, Faecalibacterium, and Prevotella. Additionally, an association between the severity of a stroke and the higher prevalence of Prevotellaceae and Peptococcaceae is seen (Zhu et al. 2020). In a mouse model of ischemia/reperfusion, therapy with a specific bacterial strain, specifically C. butyricum, improved cognitive performance and reduced neuronal harm. After traumatic brain injury, adding psychobiotics to one's diet has been proposed to lessen comorbidity and psychiatric outcomes. To fully understand the potential of such microbial therapeutic approaches, more clinical research is required.

5.7 Conclusion and Future Perspective

In the upcoming years, it is anticipated that there will be an increase in the prevalence of neurological illnesses, necessitating greater research. More and more evidence point to the possibility that changes in the gut microbiota or problems with the MGB axis may have an impact on brain function, either directly or indirectly. The gut microbiota appears to play a role in controlling neurophysiological function and cognition via bacterial metabolites, neurotransmitters, IECs, and the immune system. Research has recently focused on the relationship between immune-related neurological illnesses and the microbiota-gut-brain axis.The involvement of the MGB axis in immune-related neurological illnesses has been investigated using a variety of techniques, including GF studies, fecal transplantation studies, probiotic research, and antibiotic studies. Animal models provide useful tools for investigating how gut microbiota contributes to the development of neurological disease. The clear biological differences between humans and other animals along with the complexity, uniqueness, and vitality of the human gut microbiota make it more challenging to put these findings to the diagnosis and treatment of neurological diseases in people. The complexity of the BBB must be considered in studies on the dual interaction between the GM and the brain. Modern technologies are being developed to determine and verify biological mechanisms of action and to develop treatments for neurological diseases. As a result, more pertinent clinical intervention studies should be conducted to clarify the mechanisms that govern how the microbiota interfaces with the human brain. Researchers can find more thorough reasons and the impacts of underlying pathways with the help of interventional techniques like probiotics, prebiotics, and fecal transplantation therapies. Adopting microbial therapies, such as probiotic/prebiotic medications, FMT, etc., as a standard therapeutic approach for the early detection and management of neurological disorders comes with a number of challenges. The acceptability and safety of microbial therapeutic interventions in high-risk populations like children, the elderly, and people with compromised immune systems is another significant concern. Any therapeutic intervention involving microbes in populations with elevated risk has a higher risk, hence extra care should be taken to minimize any possible side effects. Additional research on the impact of microbial therapeutic interventions, potential combinations with other therapies, the right sample size, and longer follow-up studies should thus be taken into consideration. It is necessary to conduct extensive studies to determine the most advantageous single or microbial formulation for each specific neurologic condition as the therapeutic effects on neurologic disorders differ and depend on the therapeutic bacterial strain. The effects of herbal medicine on the gut-brain axis require further focused research. Additionally, there is a need to ascertain the long-term safety of such therapeutic intervention because the usage of probiotics is expanding tremendously. While animal models provide valuable tools for studying the gut-brain connection, translating these findings to humans requires careful consideration of the complexity and unique characteristics of the human gut microbiome. Future research should bridge the gap between preclinical findings and human applications. The role of the BBB in mediating communication between the gut and the brain needs further exploration. Understanding how gut-derived molecules interact with the BBB is essential for developing effective therapeutic strategies. Well-designed clinical trials are necessary to evaluate the efficacy and safety of potential interventions targeting the gut microbiome for preventing or managing neurological disorders. This includes studying the effects of probiotics, prebiotics, and fecal microbiota transplantation (FMT) in human populations. Tailoring interventions based on individual gut microbiome profiles holds promise for optimizing treatment strategies and minimizing side effects. Addressing challenges associated with microbial therapies, such as ensuring safety and acceptability in high-risk populations (children, elderly, immunocompromised), is crucial. Further research on potential side effects, optimal dosing schedules, and long-term safety of these interventions is necessary. Identifying the most beneficial microbial formulations for specific neurological disorders is critical. Different bacterial strains may have varying therapeutic effects, requiring targeted approaches for each condition. Investigating the potential of herbal medicine to modulate the gut-brain axis and its impact on neurological health warrants further exploration. Assessing the long-term safety and efficacy of various gut microbiome-based interventions remains essential.

References

Acuña-Castroviejo D, Rahim I, Acuña-Fernández C, Fernández-Ortiz M, Solera-Marín J, Sayed RKA, Díaz-Casado ME, Rusanova I, López LC, Escames G (2017) Melatonin, clock genes and mitochondria in sepsis. Cell Mol Life Sci 74(21):3965–3987

Akbari E, Asemi Z, Daneshvar Kakhaki R, Bahmani F, Kouchaki E, Tamtaji OR, Hamidi GA, Salami M (2016) Effect of probiotic supplementation on cognitive function and metabolic status in Alzheimer's disease: a randomized, double-blind and controlled trial. Front Aging Neurosci 8:256

Anand N, Gorantla VR, Chidambaram SB (2022) The role of gut dysbiosis in the pathophysiology of neuropsychiatric disorders. Cells 12(1):54

Arneth BM (2018) Gut-brain axis biochemical signalling from the gastrointestinal tract to the central nervous system: gut dysbiosis and altered brain function. Postgrad Med J 94(1114):446–452

Atarashi K, Tanoue T, Oshima K, Suda W, Nagano Y, Nishikawa H, Fukuda S, Saito T, Narushima S, Hase K, Kim S, Fritz JV, Wilmes P, Ueha S, Matsushima K, Ohno H, Olle B, Sakaguchi S, Taniguchi T, Morita H et al (2013) Treg induction by a rationally selected mixture of Clostridia strains from the human microbiota. Nature 500(7461):232–236

Avcil S, Uysal P, Yenisey Ç, Abas BI (2021) Elevated melatonin levels in children with attention deficit hyperactivity disorder: relationship to oxidative and nitrosative stress. J Atten Disord 25(5):693–703

Bauché D, Marie JC (2017) Transforming growth factor β: a master regulator of the gut microbiota and immune cell interactions. Clin Transl Immunol 6(4):e136

Bekebrede AF, Keijer J, Gerrits WJ, Boer VCD (2020) The molecular and physiological effects of protein-derived polyamines in the intestine. Nutrients 12(1):197

Belkaid Y, Hand TW (2014) Role of the microbiota in immunity and inflammation. Cell 157(1):121–141

Benakis C, Brea D, Caballero S, Faraco G, Moore J, Murphy M, Sita G, Racchumi G, Ling L, Pamer EG, Iadecola C, Anrather J (2016) Commensal microbiota affects ischemic stroke outcome by regulating intestinal γδ T cells. Nat Med 22(5):516–523

Bercik P, Denou E, Collins J et al (2011) The intestinal microbiota affect central levels of brain-derived neurotropic factor and behavior in mice. Gastroenterology 141(2):599–609.e6093. https://doi.org/10.1053/j.gastro.2011.04.052

Borre YE, O'Keeffe GW, Clarke G, Stanton C, Dinan TG, Cryan JF (2014) Microbiota and neurodevelopmental windows: implications for brain disorders. Trends Mol Med 20(9):509–518. https://doi.org/10.1016/j.molmed.2014.05.002

Braniste V, Al-Asmakh M, Kowal C, et al. (2014) The gut microbiota influences blood-brain barrier permeability in mice [published correction appears in Sci Transl Med. 2014 Dec 10;6(266):266er7. Guan, Ng Lai [corrected to Ng, Lai Guan]]. Sci Transl Med. 2014;6(263):263ra158. doi:https://doi.org/10.1126/scitranslmed.3009759

Bravo JA, Forsythe P, Chew MV et al (2011) Ingestion of lactobacillus strain regulates emotional behavior and central GABA receptor expression in a mouse via the vagus nerve. Proc Natl Acad Sci USA 108(38):16050–16055. https://doi.org/10.1073/pnas.1102999108

Brown EG, Goldman SM (2020) Modulation of the microbiome in Parkinson's disease: diet, drug, stool transplant, and beyond. Neurotherapeutics 17(4):1406–1417

Bull-Larsen S, Mohajeri MH (2019) The potential influence of the bacterial microbiome on the development and progression of ADHD. Nutrients 11(11):2805

Buscarinu MC, Cerasoli B, Annibali V, Policano C, Lionetto L, Capi M, Mechelli R, Romano S, Fornasiero A, Mattei G, Piras E, Angelini DF, Battistini L, Simmaco M, Umeton R, Salvetti M, Ristori G (2017) Altered intestinal permeability in patients with relapsing-remitting multiple sclerosis: a pilot study. Mult Scler (Houndmills, Basingstoke, England) 23(3):442–446

Carlson AL, Xia K, Azcarate-Peril MA et al (2018) Infant gut microbiome associated with cognitive development. Biol Psychiatry 83(2):148–159. https://doi.org/10.1016/j.biopsych.2017.06.021

Calvo AC, Valledor-Martín I, Moreno-Martínez L, Toivonen JM, Osta R (2022) Lessons to learn from the gut microbiota: a focus on amyotrophic lateral sclerosis. Genes 13(5):865

Casani-Cubel J, Benlloch M, Sanchis-Sanchis CE, Marin R, Lajara-Romance JM, de la Rubia Orti JE (2021) The impact of microbiota on the pathogenesis of amyotrophic lateral sclerosis and the possible benefits of polyphenols. An overview. Metabolites 11(2):120

Cassani E, Privitera G, Pezzoli G, Pusani C, Madio C, Iorio L, Barichella M (2011) Use of probiotics for the treatment of constipation in Parkinson's disease patients. Minerva Gastroenterol Dietol 57(2):117–121

Chaidez V, Hansen RL, Hertz-Picciotto I (2014) Gastrointestinal problems in children with autism, developmental delays or typical development. J Autism Dev Disord 44(5):1117–1127

Checa-Ros A, Jeréz-Calero A, Molina-Carballo A, Campoy C, Muñoz-Hoyos A (2021) Current evidence on the role of the gut microbiome in ADHD pathophysiology and therapeutic implications. Nutrients 13(1):249

Cheng S, Han B, Ding M, Wen Y, Ma M, Zhang L, Qi X, Cheng B, Li P, Kafle OP, Liang X, Liu L, Du Y, Zhao Y, Zhang F (2020) Identifying psychiatric disorder-associated gut microbiota using microbiota-related gene set enrichment analysis. Brief Bioinform 21(3):1016–1022

Choi GB, Yim YS, Wong H, Kim S, Kim H, Kim SV, Hoeffer CA, Littman DR, Huh JR (2016) The maternal interleukin-17a pathway in mice promotes autism-like phenotypes in offspring. Science (New York, N.Y.) 351(6276):933–939

Connolly N, Anixt J, Manning P, Ping-I Lin D, Marsolo KA, Bowers K (2016) Maternal metabolic risk factors for autism spectrum disorder-An analysis of electronic medical records and linked birth data. Autism Res 9(8):829–837

Diaz Heijtz R, Wang S, Anuar F et al (2011) Normal gut microbiota modulates brain development and behavior. Proc Natl Acad Sci USA 108(7):3047–3052. https://doi.org/10.1073/pnas.1010529108

Ding J, Liu Q (2019) Toll-like receptor 4: a promising therapeutic target for pneumonia caused by Gram-negative bacteria. J Cell Mol Med 23(9):5868–5875

Doifode T, Giridharan VV, Generoso JS, Bhatti G, Collodel A, Schulz PE, Forlenza OV, Barichello T (2021) The impact of the microbiota-gut-brain axis on Alzheimer's disease pathophysiology. Pharmacol Res 164:105314

Du G, Dong W, Yang Q, Yu X, Ma J, Gu W, Huang Y (2021) Altered gut microbiota related to inflammatory responses in patients with Huntington's disease. Front Immunol 11:603594

Duscha A, Gisevius B, Hirschberg S, Yissachar N, Stangl GI, Eilers E et al (2020) Propionic acid shapes the multiple sclerosis disease course by an immunomodulatory mechanism. Cell 180(6):1067–1080

Emge JR, Huynh K, Miller EN et al (2016) Modulation of the microbiota-gut-brain axis by probiotics in a murine model of inflammatory bowel disease. Am J Physiol Gastrointest Liver Physiol 310(11):G989–G998. https://doi.org/10.1152/ajpgi.00086.2016

Erny D, Hrabě de Angelis AL, Jaitin D, Wieghofer P, Staszewski O, David E, Keren-Shaul H, Mahlakoiv T, Jakobshagen K, Buch T, Schwierzeck V, Utermöhlen O, Chun E, Garrett WS, McCoy KD, Diefenbach A, Staeheli P, Stecher B, Amit I, Prinz M (2015) Host microbiota constantly control maturation and function of microglia in the CNS. Nat Neurosci 18(7):965–977

Fernandez-Real JM, Serino M, Blasco G et al (2015) Gut microbiota interacts with brain microstructure and function. J Clin Endocrinol Metab 100(12):4505–4513. https://doi.org/10.1210/jc.2015-3076

Fiorentino M, Sapone A, Senger S et al (2016) Blood-brain barrier and intestinal epithelial barrier alterations in autism spectrum disorders. Mol Autism 7:49. Published 2016 Nov 29. https://doi.org/10.1186/s13229-016-0110-z

Gao W, Salzwedel AP, Carlson AL et al (2019) Gut microbiome and brain functional connectivity in infants-a preliminary study focusing on the amygdala. Psychopharmacology 236(5):1641–1651. https://doi.org/10.1007/s00213-018-5161-8

Geng ZH, Zhu Y, Li QL, Zhao C, Zhou PH (2022) Enteric nervous system: the bridge between the gut microbiota and neurological disorders. Front Aging Neurosci 14:810483

Gerhardt S, Mohajeri MH (2018) Changes of colonic bacterial composition in Parkinson's disease and other neurodegenerative diseases. Nutrients 10(6):708

Gotkine M, Kviatcovsky D, Elinav E (2020) Amyotrophic lateral sclerosis and intestinal microbiota—toward establishing cause and effect. Gut Microbes 11(6):1833–1841

Holmes A, Finger C, Morales-Scheihing D, Lee J, McCullough LD (2020) Gut dysbiosis and age-related neurological diseases; an innovative approach for therapeutic interventions. Transl Res 226:39–56

Hug H, Mohajeri MH, La Fata G (2018) Toll-like receptors: regulators of the immune response in the human gut. Nutrients 10(2):203

Jalandra R, Dhar R, Pethusamy K, Sharma M, Karmakar S (2022) Dysbiosis: gut feeling. F1000Research 11(911):911

Jeon MT, Kim KS, Kim ES, Lee S, Kim J, Hoe HS, Kim DG (2021) Emerging pathogenic role of peripheral blood factors following BBB disruption in neurodegenerative disease. Ageing Res Rev 68:101333

Kabat AM, Srinivasan N, Maloy KJ (2014) Modulation of immune development and function by intestinal microbiota. Trends Immunol 35(11):507–517

Kang DW, Ilhan ZE, Isern NG, Hoyt DW, Howsmon DP, Shaffer M, Lozupone CA, Hahn J, Adams JB, Krajmalnik-Brown R (2018) Differences in fecal microbial metabolites and microbiota of children with autism spectrum disorders. Anaerobe 49:121–131

Kim MS, Kim Y, Choi H, Kim W, Park S, Lee D, Kim DK, Kim HJ, Choi H, Hyun DW, Lee JY, Choi EY, Lee DS, Bae JW, Mook-Jung I (2020) Transfer of a healthy microbiota reduces amyloid and tau pathology in an Alzheimer's disease animal model. Gut 69(2):283–294

Kong G, Cao KL, Judd LM, Li S, Renoir T, Hannan AJ (2020) Microbiome profiling reveals gut dysbiosis in a transgenic mouse model of Huntington's disease. Neurobiol Dis 135:104268

Lavelle A, Sokol H (2020) Gut microbiota-derived metabolites as key actors in inflammatory bowel disease. Nat Rev Gastroenterol Hepatol 17(4):223–237

Leone V, Gibbons SM, Martinez K, Hutchison AL, Huang EY, Cham CM, Pierre JF, Heneghan AF, Nadimpalli A, Hubert N, Zale E, Wang Y, Huang Y, Theriault B, Dinner AR, Musch MW, Kudsk KA, Prendergast BJ, Gilbert JA, Chang EB (2015) Effects of diurnal variation of gut microbes and high-fat feeding on host circadian clock function and metabolism. Cell Host Microbe 17(5):681–689

Levy M, Kolodziejczyk AA, Thaiss CA, Elinav E (2017) Dysbiosis and the immune system. Nat Rev Immunol 17(4):219–232

Liu L, Wang H, Chen X, Xie P (2022) Gut microbiota: a new insight into neurological diseases. Chin Med J 136:1261. https://doi.org/10.1097/CM9.0000000000002212

Mangalam A, Shahi SK, Luckey D, Karau M, Marietta E, Luo N et al (2017) Human gut-derived commensal bacteria suppress CNS inflammatory and demyelinating disease. Cell Rep 20(6):1269–1277

Martínez I, Lattimer JM, Hubach KL, Case JA, Yang J, Weber CG, Louk JA, Rose DJ, Kyureghian G, Peterson DA, Haub MD, Walter J (2013) Gut microbiome composition is linked to whole grain-induced immunological improvements. ISME J 7(2):269–280

Melnikov M, Belousova O, Murugin V, Pashenkov M, Boyko A (2016) The role of dopamine in modulation of Th-17 immune response in multiple sclerosis. J Neuroimmunol 292:97–101

Meuer K, Pitzer C, Teismann P, Krüger C, Göricke B, Laage R et al (2006) Granulocyte-colony stimulating factor is neuroprotective in a model of Parkinson's disease. J Neurochem 97(3):675–686

Möhle L, Mattei D, Heimesaat MM et al (2016) Ly6C(hi) monocytes provide a link between antibiotic-induced changes in gut microbiota and adult hippocampal neurogenesis. Cell Rep 15(9):1945–1956

Morimoto K, Nakajima K (2019) Role of the immune system in the development of the central nervous system. Front Neurosci 13:916
Morris G, Berk M, Carvalho A, Caso JR, Sanz Y, Walder K, Maes M (2017) The role of the microbial metabolites including tryptophan catabolites and short chain fatty acids in the pathophysiology of immune-inflammatory and neuroimmune disease. Mol Neurobiol 54(6):4432–4451
Noureldein MH, Eid AA (2018) Gut microbiota and mTOR signaling: insight on a new pathophysiological interaction. Microb Pathog 118:98–104
Ogbonnaya ES, Clarke G, Shanahan F, Dinan TG, Cryan JF, O'Leary OF (2015) Adult hippocampal neurogenesis is regulated by the microbiome. Biol Psychiatry 78(4):e7–e9
Oleskin AV, Shenderov BA (2016) Neuromodulatory effects and targets of the SCFAs and gasotransmitters produced by the human symbiotic microbiota. Microb Ecol Health Dis 27:30971
Omenetti S, Pizarro TT (2015) The Treg/Th17 Axis: a dynamic balance regulated by the gut microbiome. Front Immunol 6:639
Patil RS, Vyas SG, Quazi WT, Tembhurnikar HJ, Milmile PS, Umekar MJ (2021) The gut microbiome in Huntington disease: a review. GSC Biol Pharm Sci 15(3):317–326
Perez-Pardo P, Dodiya HB, Engen PA, Forsyth CB, Huschens AM, Shaikh M, Voigt RM, Naqib A, Green SJ, Kordower JH, Shannon KM, Garssen J, Kraneveld AD, Keshavarzian A (2019) Role of TLR4 in the gut-brain axis in Parkinson's disease: a translational study from men to mice. Gut 68(5):829–843
Prakash A, Medhi B, Chopra K (2013) Granulocyte colony-stimulating factor (GCSF) improves memory and neurobehavior in an amyloid-β induced experimental model of Alzheimer's disease. Pharmacol Biochem Behav 110:46–57
Prehn-Kristensen A, Zimmermann A, Tittmann L, Lieb W, Schreiber S, Baving L, Fischer A (2018) Reduced microbiome alpha diversity in young patients with ADHD. PLoS One 13(7):e0200728
Prinz M, Priller J (2017) The role of peripheral immune cells in the CNS in steady state and disease. Nat Neurosci 20(2):136–144
Rothhammer V, Mascanfroni ID, Bunse L, Takenaka MC, Kenison JE, Mayo L et al (2016) Type I interferons and microbial metabolites of tryptophan modulate astrocyte activity and central nervous system inflammation via the aryl hydrocarbon receptor. Nat Med 22(6):586–597
Sacramento PM, Monteiro C, Dias ASO, Kasahara TM, Ferreira TB, Hygino J, Wing AC, Andrade RM, Rueda F, Sales MC, Vasconcelos CC, Bento CAM (2018) Serotonin decreases the production of Th1/Th17 cytokines and elevates the frequency of regulatory $CD4^+$ T-cell subsets in multiple sclerosis patients. Eur J Immunol 48(8):1376–1388
Salter MW, Stevens B (2017) Microglia emerge as central players in brain disease. Nat Med 23(9):1018–1027
Sampson TR, Debelius JW, Thron T, Janssen S, Shastri GG, Ilhan ZE, Challis C, Schretter CE, Rocha S, Gradinaru V, Chesselet MF, Keshavarzian A, Shannon KM, Krajmalnik-Brown R, Wittung-Stafshede P, Knight R, Mazmanian SK (2016) Gut microbiota regulate motor deficits and neuroinflammation in a model of Parkinson's disease. Cell 167(6):1469–1480.e12
Schiering C, Wincent E, Metidji A, Iseppon A, Li Y, Potocnik AJ, Omenetti S, Henderson CJ, Wolf CR, Nebert DW, Stockinger B (2017) Feedback control of AHR signalling regulates intestinal immunity. Nature 542(7640):242–245
Seo DO, Holtzman DM (2020) Gut microbiota: from the forgotten organ to a potential key player in the pathology of Alzheimer's disease. J Gerontol A Biol Sci Med Sci 75(7):1232–1241
Silva YP, Bernardi A, Frozza RL (2020) The role of short-chain fatty acids from gut microbiota in gut-brain communication. Front Endocrinol 11:25
Singh V, Roth S, Llovera G, Sadler R, Garzetti D, Stecher B, Dichgans M, Liesz A (2016) Microbiota dysbiosis controls the neuroinflammatory response after stroke. J Neurosci 36(28):7428–7440
Solleiro-Villavicencio H, Rivas-Arancibia S (2018) Effect of chronic oxidative stress on neuroinflammatory response mediated by $CD4^+$T cells in neurodegenerative diseases. Front Cell Neurosci 12:114
Sommer F, Bäckhed F (2013) The gut microbiota—masters of host development and physiology. Nat Rev Microbiol 11(4):227–238
Song X, Zhao Z, Zhao Y, Wang Z, Wang C, Yang G, Li S (2022) *Lactobacillus plantarum* DP189 prevents cognitive dysfunction in D-galactose/$AlCl_3$ induced mouse model of Alzheimer's dis-

ease via modulating gut microbiota and PI3K/Akt/GSK-3β signaling pathway. Nutr Neurosci 25(12):2588–2600

Spadoni I, Fornasa G, Rescigno M (2017) Organ-specific protection mediated by cooperation between vascular and epithelial barriers. Nat Rev Immunol 17(12):761–773. https://doi.org/10.1038/nri.2017.100

Stan TL, Soylu-Kucharz R, Burleigh S, Prykhodko O, Cao L, Franke N et al (2020) Increased intestinal permeability and gut dysbiosis in the R6/2 mouse model of Huntington's disease. Sci Rep 10(1):18270

Tan EK, Chao YX, West A, Chan LL, Poewe W, Jankovic J (2020a) Parkinson disease and the immune system—associations, mechanisms and therapeutics. Nat Rev Neurol 16(6):303–318

Tan AH, Hor JW, Chong CW, Lim SY (2020b) Probiotics for Parkinson's disease: current evidence and future directions. JGH Open 5(4):414–419

Tao J, Li S, Gan RY, Zhao CN, Meng X, Li HB (2020) Targeting gut microbiota with dietary components on cancer: effects and potential mechanisms of action. Crit Rev Food Sci Nutr 60(6):1025–1037

Vuong HE, Hsiao EY (2017) Emerging roles for the gut microbiome in autism spectrum disorder. Biol Psychiatry 81(5):411–423

Wang S, Harvey L, Martin R et al (2018) Targeting the gut microbiota to influence brain development and function in early life. Neurosci Biobehav Rev 95:191–201. https://doi.org/10.1016/j.neubiorev.2018.09.002

Wilkins LJ, Monga M, Miller AW (2019) Defining dysbiosis for a cluster of chronic diseases. Sci Rep 9(1):1–10

Wu D, Chen Q, Chen X, Han F, Chen Z, Wang Y (2023) The blood-brain barrier: structure, regulation, and drug delivery. Signal Transduct Target Ther 8(1):217. Published 2023 May 25. https://doi.org/10.1038/s41392-023-01481-w

Wu HJ, Wu E (2012) The role of gut microbiota in immune homeostasis and autoimmunity. Gut Microbes 3(1):4–14

Xu M, Xu X, Li J, Li F (2019) Association between gut microbiota and autism spectrum disorder: a systematic review and meta-analysis. Front Psych 10:473

Yang Y, Wang H, Kouadir M, Song H, Shi F (2019) Recent advances in the mechanisms of NLRP3 inflammasome activation and its inhibitors. Cell Death Dis 10(2):128

Yang Q, Wang Y, Jia A, Wang Y, Bi Y, Liu G (2021) The crosstalk between gut bacteria and host immunity in intestinal inflammation. J Cell Physiol 236(4):2239–2254

Yin J, Liao SX, He Y, Wang S, Xia GH, Liu FT et al (2015) Dysbiosis of gut microbiota with reduced trimethylamine-N-oxide level in patients with large-artery atherosclerotic stroke or transient ischemic attack. J Am Heart Assoc 4(11):e002699

Zeng H, Yang K, Cloer C, Neale G, Vogel P, Chi H (2013) mTORC1 couples immune signals and metabolic programming to establish T(reg)-cell function. Nature 499(7459):485–490

Zhai CD, Zheng JJ, An BC, Huang HF, Tan ZC (2019) Intestinal microbiota composition in patients with amyotrophic lateral sclerosis: establishment of bacterial and archaeal communities analyses. Chin Med J 132(15):1815–1822

Zhu S, Jiang Y, Xu K, Cui M, Ye W, Zhao G, Jin L, Chen X (2020) The progress of gut microbiome research related to brain disorders. J Neuroinflammation 17(1):25

Influence of Altered Gut Microbiota in Cellular Senescence

6

Bhuvaneswari Ponnusamy and Selvaraj Jayaraman

Abstract

Senescence is a physiological process a cell undergoes. This process is often considered as a double-edged sword as it as both beneficial and non-beneficial effects in the living system. Various cell types (e.g., immune cells, intestinal epithelial cells) are susceptible to undergo senescence reversibly or irreversibly. In this chapter, we elaborately discuss gut microbes and their influence on cellular senescence. Ageing is an inevitable action. As we age, there is a gradual depreciation in the body's functioning and immune system and are thus prone to various disease conditions. Gut microbes are essential symbiotic organisms present in the intestinal tract of the host orchestrating a wide range of physiological actions and helping in the orderly functioning of the tissues or organs. Normally as we age there is a significant alteration in the microbial communities in the gut. This dysbiosis of microbial population breaches the functioning and homeostasis of gut microbes. This leads to the formation of senescence associated secretory phenotype and leads to formation of inflammatory microenvironment. Inflammation is a basic pathological feature for various age-related diseases including cancer, diabetes, etc. Practicing good diet practices, diet rich in probiotics, prebiotics, and naturally occurring polyphenols (e.g., fruits, vegetables, green tea) are found to enhance growth of beneficial gut microbes and may offer potential strategies to mitigate the negative effects of age-related gut dysbiosis and cellular senescence.

B. Ponnusamy · S. Jayaraman (✉)
Department of Biochemistry, Centre of Molecular Medicine and Diagnostics (COMManD), Saveetha Dental College and Hospital, Saveetha Institute of Medical and Technical Sciences, Saveetha University, Chennai, India
e-mail: selvarajj.sdc@saveetha.com

S. Pathak, A. Banerjee (eds.), *Gut Microbiome and Brain Ageing*,
https://doi.org/10.1007/978-981-99-8803-7_6

Keywords

Aging · Cellular senescence · Inflammation · Microbiota · SASP · Diet · Natural polyphenols

6.1 Introduction

Cellular senescence is a state of permanent cell cycle arrest characterized by a loss of replicative capacity and a distinct set of phenotypic changes (Kumari and Jat 2021). However, they are metabolically active. There are a number of extracellular and intracellular triggers that can impose senescence to a cell, including DNA damage, telomere shortening, oxidative stress, oncogene activation, epigenomic alterations, disarranged chromatins, loss of protein homeostasis, lack of mitochondrial function, inflammation, xenobiotics, and malnourishment (Surova and Zhivotovsky 2013; Passos et al. 2010). Those cells undergoing senescence show distinct characteristic traits namely cell growth arrest, lysosomal activity, damaged macromolecules, and formation of senescence-Associated Secretory Phenotype (SASP) (Gorgoulis et al. 2019; Lee et al. 2006; Saleh et al. 2018). And the morphological features show an increased cell size, flat appearance, and polynucleated with big vacuoles (Campisi and D'Adda Di Fagagna 2007). Senescence is currently recognized as a dynamic and multifaceted process, wherein the characteristics of senescent cells (SC) undergo continuous evolution and diversification in a manner that depends on the specific context involved (Boisvert et al. 2018). These cells typically maintain their viability despite changes in metabolic role and are often resistant to cell death (Zhu et al. 2015).

Senescence process is a protective mechanism that forms a barrier against the division of defective or potentially harmful cells, including those at risk of becoming cancerous. Senescence serves important physiological functions such as normal development, tissue homeostasis, tissue remodeling, repair, wound healing, insulin secretion by pancreatic beta cells, and limiting tumor progression. Its primary role is to prevent the propagation of dysfunctional, damaged, or transformed cells to subsequent generations, thereby safeguarding the integrity of the organism (Collado et al. 2007; Muñoz-Espín and Serrano 2014). However, the accumulation of SC over time can have negative consequences. SC can contribute to tissue dysfunction and age-related diseases by promoting chronic inflammation, destroying tissue regeneration and repair mechanisms, and altering the surrounding microenvironment. The chronic presence of SC is noted in many conditions, such as carcinogenesis, neurodegenerative diseases, cardiovascular disease, and frailty (Muñoz-Espín and Serrano 2014). SCs have been documented in a range of age-related diseases, including but not limited to atherosclerosis, diabetes, lung disease (Chandrasekaran et al. 2017). Thus, senescence is a program that exhibits dual effects, with both advantageous and destructive impacts on a cell (Schosserer et al. 2017).

The human microbiome consists of the combined family of various bacteria, viruses, archaea, protozoa, and eukaryotes residing within us. These microorganisms coexist symbiotically in various locations throughout the human body, such as the oral region, reproductive organs, trachea, skin, and digestive system (Lloyd-Price et al. 2016). It is approximately said that ~10^{13}–10^{14} microbes are noted in human gut that makes up around to a ratio of 1:1 (microbial cells to human cells) (Sender et al. 2016). This residing micro biota majorly forms bacterial population, namely Firmicutes, Bacteroidetes, and Actinobacteria (Tap et al. 2009). The relationship between the gut microbiota and the living system is a symbiotic one that is maintained and balanced through an intricate web of crosstalks between molecules in metabolism, immunity, and neuroendocrine region. This communication is likely facilitated by metabolites produced by the microbes, which have diverse effects and can act as signaling molecules to regulate the interplay between the host's neurological, immune, and inflammatory systems. This interconnectedness potentially establishes physiological connections between the gut and other organ systems in the body (Kho and Lal 2018). Various external factors, such as intake of antibiotics, dietary components, as well as psychological and physical stress, along with host-related factors, can disrupt the balance of the gut microbiome. This imbalance is referred as microbiome dysbiosis that can negatively impact the normal functioning of the microorganisms, compromising the overall well-being of the host. It can also lead to an overgrowth of specific members of the microbiota, including potentially harmful pathogenic organisms, resulting in an imbalance in the production of microbial-derived products or metabolites that may be detrimental to the host. This dysregulation can contribute to a wide range of diseases affecting local, systemic, or remote organs. Numerous investigations are ongoing to explore the connection between microbial dysbiosis and cellular senescence. This chapter enlightens the importance of gut microbiome alterations and its role in cellular senescence.

6.2 Stress and Senescence

Cellular senescence, simply termed as cell cycle arrest, is orchestrated upon activation of p53/p21, p16Ink4a/pRb, DNA damage, chromatin remodeling, senescence-associated β-galactosidase (SA-β-gal) activity, alterations in metabolic pathways (mTOR/AMPK), diminished production of mitochondrial reactive oxygen species (ROS), cellular hypertrophy, and the emergence of SASP (Herranz and Gil 2018). Some argue that the evolution of cellular senescence serves as a mechanism to safeguard cells exposed to chronic stress-induced damage, thereby providing protection against tumorigenesis (Sager 1991). This is because SASP has the potential to recruit immune cells, namely macrophages, natural killer cells, or T cells, for the efficient removal of senescent cells (Song et al. 2020).

Various intrinsic and extrinsic stress plays an upper hand in the cellular senescence (Fig. 6.1). They are responsible for the changes in mitochondrial functioning, ATP generation, and initiation of Nf-KB cascade (Bulua et al. 2011). The imbalances in gut bacteria and reduced diversity can initiate a detrimental cycle. In this cycle, mitochondrial damage can increase the release free radicals, which, in turn,

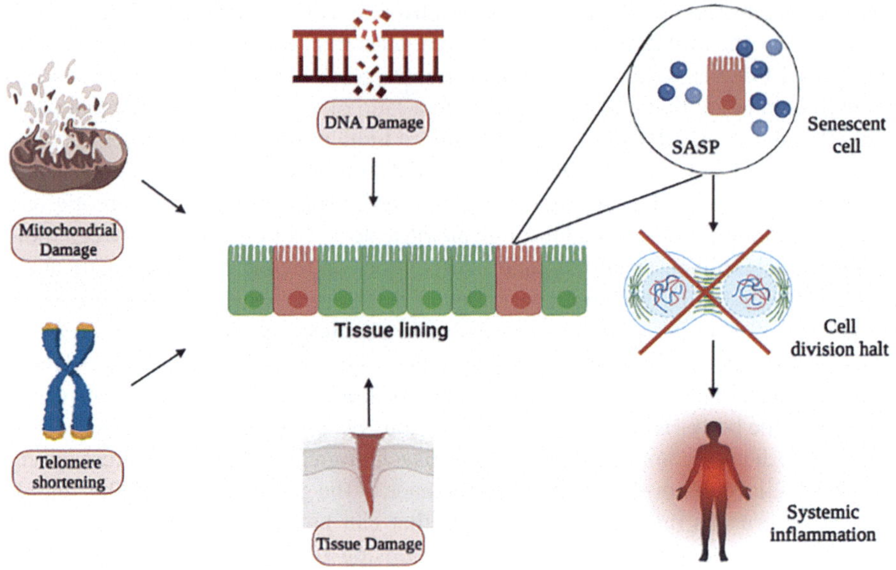

Fig. 6.1 Representation of various stress factors in cellular senescence

exert sustained oxidative stress and cause further damage to macromolecules. This leads to damage in the genetic material (DNA), thereby pushing the cells into irreversible halt cell division and develop SASP characteristics (Childs et al. 2014). Thus, ultimately having an impact in the tissue and organ functioning, and exhibiting characteristics commonly associated with aging. Further the replicative stress also pave way to the formation of SCs. In general, in cellular division, chromosomes experience a loss of the protective telomere cap, which is essential for their stability. With each subsequent division, telomeres progressively shorten, eventually reaching a point where they can no longer support further replication. This triggers the activation of the cellular senescence program, leading to a state of growth arrest and functional decline in the affected cells (Campisi 1997). Condensing of telomere is also contributed by the ROS stress (Epel et al. 2004) thereby resulting in SC generation. Recent studies also reveal that immune cell undergoing senescent program is associated with loss of immune surveillance and cytotoxic effect that can promote the accumulation of SCs in the host system (Song et al. 2020).

There is a time to time communication between brain and microbiota residing in the gut. This interaction between the brain and the gut, known as the gut-brain axis, is essential for the host homeostasis. It is also revealed that anxiety, stress, and mood have a negative projection on physiological functioning of gut microbes. However, the miscommunication of gut and brain is not only responsible for these temporary disturbances but to more persistent conditions. For instance, disorders of mood of an individual are often connected with digestive disorders such as irritable bowel syndrome (Fond et al. 2014). This may be due to the disruption of the balance in the microbiome composition in the gut and can result in chronic inflammation. Moving towards the urban culture including the frequent usage of antibiotics, intake of high fat diet, and pressurized lifestyles, contribute to gut dysbiosis. Moreover, a reduced diversity, characterized by little quantity and irregular bacterial colony distribution in the gut floor is also noticed. Dysbiosis and low diversity can impact various aspects such as food cravings, metabolism, stress response, and mood, thereby compromising immune function and overall health (Madison and Kiecolt-Glaser 2019).

6.3 Gut Microbial Dysbiosis and Associated Pathology

A versatile range of microbiota is found to be anchoring the digestive tract of a living system. They are found to survive in the host biological system through a symbiotic dealing, thereby helping in the digestion and metabolism of dietary intakes (Goldsmith and Sartor 2014). These microbes are highly concentrated in the colonic region by forming colonies. These colonies comprise bacteria, archaea, bacteriophages, viruses, and meiofauna (Norman et al. 2014). The intestines of the human body harbor approximately 100 trillion bacteria, a quantity that surpasses the total count of human cells in the body by tenfold (Andoh 2016). Comprising over 1000 distinct bacterial species, this group is dominated by approximately 100 species, accounting for nearly 99% of the total population (Guarner 2005). The type of diet we consume plays an important role in the prevalence of a specific microbial species. For instance, a dietary intake of plant-based carbohydrate-rich food results in an increase of the phylum *Bacteroidetes*, specifically the *Prevotella* species. This is due to the abundance of enzymes in the metabolism of carbohydrate, namely glycosyltransferases, glycoside hydrolases, and polysaccharide lyases. On the other hand, a diet rich in fats and proteins is found to increase the prevalence of *Bacteroides* species (Wu et al. 2011).

An important physiological role of colon is to produce short-chain fatty acids (SCFAs), namely acetate, propionate, and butyrate, from the indigestible fiber content in the food. This action is facilitated by the anaerobic microbes, mainly *Bifidobacteria*, *Firmicutes*, and *Bacteroidetes* residing in the colonic tract by the process of fermentation (Andoh 2016). SCFA plays an essential role in maintaining the physiology of colon, i.e., serves as an important source maintaining anionic concentration, serving as energy for colonic epithelial cells (~10% of the total energy from food intake), and they also play a vital role in obstructing the growth of pathogenic gram negative bacteria (Simpson and Campbell 2015).

Maintaining a harmonious bacterial culture in intestines is advantageous for any organism. Deviations in the morphology of the microbial community that lead to an imbalance between beneficial commensal, habitat in the gut, and potentially harmful pathogens can compromise the integrity of the gut and disrupt intestinal homeostasis. The altered state of the gut microbiota can arise from the depletion of good microbes, the multiplication of detrimental organisms or a general reduction in microorganisms (DeGruttola et al. 2016). Various investigations have been conducted to prove the interconnection between age and the microbial dysbiosis in the gut (Vaiserman et al. 2020). They represent that, upon aging an organism gradually show an enhancement in the number of facultative anaerobes, with a significant decrease in the presence of beneficial probiotic bacterial species and there has been a noticeable change in the ratio of firmicutes to Bacteroidetes. Furthermore, the gut microbiota is known to exert a crucial influence on the lifespan of an individual. A current report displays the relation between gut microbiome composition and human survival. The report shows that an increase in the amount of Bacteroides or reduced gut microbiome is linked with reduction of lifespan and promotes morbidity (Wilmanski et al. 2021). Studies have also revealed that gut microbiome is associated with development of chronic pathological conditions, namely type II diabetes via altering the glucose breakdown, energy usage, inflammation, and gut dysbiosis (Gurung et al. 2020). Other chronic conditions include cancer. There are a number of studies investigating the association of gut dysbiosis in the promotion of various oncologies, namely, cancer in colon, rectal, stomach, respiratory units, pancreas, prostate, breast, and brain region (Sharma 2022). Furthermore, gut microbiota plays a crucial role in the development of tumor in a distant area from the gut by controlling the release of various metabolic byproducts and toxins in to circulation (Parida and Sharma 2021). Similarly, studies are ongoing in the identifying the connecting point between gut microbes and cardiovascular disease (CVD) precipitation (Wang et al. 2011). There is an increased permeability in the intestinal cells upon aging, resulting in enhanced diffusion of harmful bacterial metabolites into the blood stream leading to persistent inflammation. Increase in trimethylamine-*N*-oxide (TMAO), a byproduct released by microbes, was noted in patients with heart failure stating that TMAO increases the probability of acquiring CVD (Tang et al. 2014). Gut dysbiosis is also related to the development of immunosenescence and inflammaging (Conway and Duggal 2021). The immunological-aging theory proposes that immunosenescence, is a fundamental factor that result in increased circulating ROS species and cytokines, also known as oxi-inflamm aging (De la Fuente and Miquel 2009). Figure 6.2 represents the pathological conditions associated with gut dysbiosis.

6.4 Immune Senescence and Gut Microbiome

Gut microbiome has the tendency and potency to maintain and regulate surveillance of the gastrointestinal tract by coordinating the immune system. However, the functionality and physiology of the gut microbiome is maintained by the mucosal immune system (Kamada and Núñez 2014). It is well known that the regulatory T cells (Treg) play an important role in maintaining the homeostasis of the immune system by suppressing the action or proliferation of the cytotoxic T cells. Under normal conditions, microbial population of *Clostridium* genus shows an enhanced production of these Tregs through the utilization of butyrate released by the microbial fermentation (Furusawa et al. 2013). This was evidenced in a study by Furusawa et al. where germ-free mice lacking microbiota in their gut exhibited an obstruction in the production of SCFA, which had a negative impact on the generation of Treg cells. However, when the mice were supplemented with SCFAs or introduced with

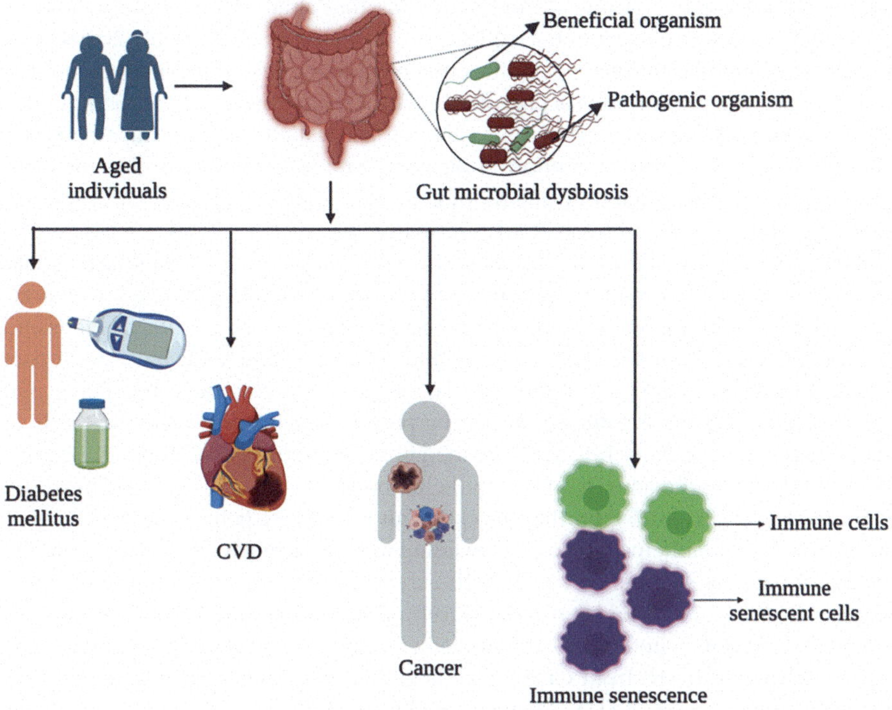

Fig. 6.2 Pathology of gut dysbiosis

commensal microbes there was an improvement in the formation of Treg cells (Furusawa et al. 2013). This clears that the metabolic byproducts released by the gut microbiome play an essential role in the homeostasis of Treg cells.

In the same aspect, studies were conducted to identify the role of gut microbes in production and maintenance of the effector T cells. An important class of T helper cells is Th17 distinguished by the release of interleukins (IL)—17A, 17F, 21, and 22. These IL act as a powerful proinflammatory marker and eliciting potent action upon inflammation. They actively involved in the production of tumor necrosis factor (TNF)-α and IL-1β. A study involving gut microbes-free germ-free mice exhibited a deficit of Th1 and Th17 cell, which upon reconstitution of the microbes got reversed showing the importance of the gut microbiome in the maintenance of the immune cells (Ivanov et al. 2008).

Immune system alteration is also evident with age related gut microbe changes. One important immune system surveillance found in intestinal region is mediated by Peyer's patches. This unique type of cells are found throughout the small intestines and sometimes also in large intestines. They are small lymphatic nodes found in the intestine made of microfold cells (M cells) (Chan et al. 2020). The elderly individuals show drop in the count of Mcells in the Peyer's patch that resulted in reduced expression of the expression of ZO-1 (zonula occludens-1), JAMs (junctional adhesion molecules), and occludins thus gradually reading to loss of tight junction intestinal epithelial cells and gut leakage (Chan et al. 2020). Gut microbiome secretes various metabolites that play vital role in boosting the immune system. Fermentation of polysaccharides by Bacteroides leads to the production of SCFAs such as acetate and propionate, while thick-walled fungi produce butyrate. These SCFAs is essential in supporting integrity and functionality of the intestinal epithelium. The role of butyrate and propionate is to regulate and maintain a harmonious physiology and immune reactions within the intestinal tract, whereas the acetate serves as a substrate for lipogenesis and gluconeogenesis (Macfarlane and Macfarlane 2011). Gut bacteria produce and release polyamines, derived from arginine, which can inhibit LPS (lipopolysaccharides) induced release of cytokines and macrophages. Additionally, both primary and secondary bile acids have an impact on the pro-inflammatory cytokines levels, phagocytic cells, dendritic cells, and Kupffer cells. Taurine a naturally occurring protein found in fish and meat aids in the breakdown of primary bile acids by the gut microbiota, is essential in maintaining and enhancing the activity of the epithelial barrier by promoting the production of IL-18 (Kaplanski 2018). Furthermore, the gut microbiota helps to limit the population of T-follicular helper (Tfh) cells residing in Peyer's patch, in turn reducing the release of bacteria-specific IgA by B-cells across the epithelial cells lining intestine (Faas et al. 2017). Bacteroides fragilis secretes a substance called Polysaccharide A (PSA) that exhibits a potent anti-inflammatory impact. It achieves this by stimulating the release of IL-10 from CD4+ T cells and altering the balance between TH1 and TH2 cells in favor of TH1 (Postler and Ghosh 2017).

Alterations in the gut microbiota can contribute to an upregulation of pro-inflammatory factors, leading to chronic inflammation associated with aging. One example is the enhanced permeability of the intestinal mucosa due to insufficient production of short-chain fatty acids (SCFAs). This increased permeability allows intestinal bacteria to translocate into the bloodstream. Another study says that use of antibiotics prior to the cancer treatment with immune checkpoint inhibiters (ICI) showed a decrease in the efficacy of ICI during cancer treatment (Pinato et al. 2019). A loss in the equilibrium of gut microbes is the main underlying reason affecting various treatment strategies in cancer. Gut microbiota is found to influence the maturation of lymphoid cells and myeloid cells (Thaiss et al. 2016; Zhang et al. 2015).

6.5 Interlink Between Age-Related Alterations in Gut and Cellular Senescence

Intestines are majorly made of two types of cells namely epithelial and subepithelial fibroblasts which act as a shield to protect the underlying tissues from various detrimental factors. However, as people age, their intestinal tissues gradually develop a chronic SASP and SCs. IECs' ability of IECs to serve as barriers may be compromised by upsetting their regular operations. Consequently, it may result in greater gut permeability and a higher risk of infection and inflammation (Sharma 2022). As we age, our gut flora constantly changes and evolves. There are a number of findings suggesting that changes in gut microbiome are strongly associated with aging. There are a number of factors that affect the microbial growth namely vaginal infections that could alter the mother's uterine area, transfer of mother's microbes to the growing fetus, type of delivery undergone by the pregnant ladies also has an impact on the initial inoculum of fetus, breast milk fed or supplementary fed fetus show changes in the microbe composition, and others including hereditary, diet, environmental exposure and use of antibiotic can affect the gut microbiome formation (Tamburini et al. 2016). Moreover, variations in the composition of microbes in the gastrointestinal tract are observed upon aging. For example, one study found that aged people had increased concentrations of Bacteroidetes and Firmicutes (Galkin et al. 2020).

The consumption of alcohol, antibiotics, probiotics, and dietary habits has all been linked to aging of the gut microbiota. In aged mice, the gene expression of oxidative stress and senescence markers were elevated (Moorefield et al. 2017) in intestinal stem cells. Furthermore, exposure to radiation led to premature cellular senescence and the development of a SASP in intestinal stem cells in live animals (Kumar et al. 2019). A recent study revealed that the presence of SCs expressing p16Inka4a and p21 increases with age in different organs of the human body, including the tissues of the colon. This finding indicates that docking of SCs in host intestines is directly associated with the aging (Idda et al. 2020). A study found that the epithelial cells of intestine displayed significant signs of senescence, as indicated by the rise in the levels of p16Ink4a, p21Cip1, and increased activity of SA-β-gal, in both normal mice and a mouse model with accelerated aging (Ercc1−/Δ) (Yousefzadeh et al. 2020). In another study, increased damage to the genetic material consequently increased the levels of senescence markers (p53/p21WAF1), stimulating regulatory molecules associated with SASP (NFκB, p38MAPK, Cox-2), and stress markers were noted in the intestines of aged mice (Sharma et al. 2022). These findings indicate that the intestinal tissue becomes more susceptible to spontaneous genotoxic stress as a result of aging. Moreover, data suggest that a consistent increase in senescence markers, including SA-βgal activity and p21, in intestinal epithelial organoids derived from aged mice on comparison with young individuals. This indicates that the age-related upregulation of senescence markers is consistently present in the organoids derived from the intestinal epithelium of older mice (Uchida et al. 2018). The evidence clearly indicates that both intestinal epithelial cells and stem cells undergo cellular senescence as a result of aging, leading to various functional changes and disturbances in gastrointestinal homeostasis. Additionally, the continuous secretion of a SASP increases a condition favorable for inflammation and potentially promotes carcinogenesis. These effects negatively impact gut permeability, immune activation, and the composition of the gut microbes (Uchida et al. 2018). A study results showed that imbalanced gut microbiota can influence the formation of SASP. Also, the study explains that alterations in the microbial composition were due to obesity and release of byproduct bacterial colonies called DCA (deoxycholic acid). This metabolite triggers formation of SASP in stellate hepatic cells, thus creating an inflammatory microenvironment and ultimately resulting in liver cancer in experimental mice model (Yoshimoto et al. 2013).

Surprisingly, upon inhibition of DCA production or reducing gut bacteria through antibiotic treatment effectively prevented the development of hepatocellular carcinoma (HCC) in obese mice. This finding highlights the potential of targeting DCA production or modulating the gut microbiota as potential strategies for preventing HCC in individuals with obesity (Yoshimoto et al. 2013). A different investigation on contribution of SASP in the colon was studied to find its correlation with formation colorectal cancer. The SASP activity was due to the presence of stromal-senescent cells in the colon thus enhancing cell division, metastasis, and invasion in colorectal cancer cell lines, by activating the signaling pathways (MAPK and PI3K) (Guo et al. 2019). It is important to acknowledge that while cellular senescence is generally associated with aging-related changes is also important in maintaining the cellular homeostasis by preventing the proliferation of cancer cells. Despite the negative aspects of cellular senescence in aging, promoting senescence in cancer cells is often seen as a favorable approach as certain metabolic byproducts, of microbes in the intestines, butyrate is said to have the capacity to restrict growth of cancer cells by increasing p21, p27, and SASP regulator gene expression. Additionally, these metabolites have the ability to suppress genotoxic factors linked with oncogenesis (Ebert et al. 2003).

6.6 Role of Gut Microbiome in Amelioration of Cellular Senescence

There is a growing body of evidence indicating that the gut microbiome exerts a significant influence on our health with the diet we intake. It is conceivable that we can enhance our well-being by manipulating our dietary choices. Through microbial fermentation, the gut microbiome utilizes nutrients from our diet as fuel to generate unique set of bioactive metabolic products, including short-chain fatty acids (SCFAs), phenols, neurotransmitters, hormones, endotoxins, and ammonia (Oliphant and Allen-Vercoe 2019) which then get access into bloodstream and influence various organs systems throughout the body. As it is already well known that, the formation of SCs can be primarily by the generation of oxidative stress. And upon the redox or antioxidant activity the accumulation of these SCs can be reduced. In this context, upon aging there is an increase in the generation of oxidative stress inside the host, upon providing proper diet or antioxidant supplements can help in improving the host form oxidative damage. In this regard, gut microbiome is found to be engaged in generation of various metabolites that are potent antioxidant and anti-inflammatory activity (Chang et al. 2014). These metabolites thus have the potential in managing the generation of SCs (Rossi et al. 2020). Moreover, colon microbiota ferment complex phytomolecules, particularly polyphenols, that are not well absorbed in the small intestine (Parkar et al. 2013). This fermentation process results in the production of various bioactive byproducts, that exhibit versatile and more potent action compared to the actual compound (Wang et al. 2018).

Diet has a vital part in homeostasis of microbes residing in the gut. Diet has become a prominent factor in predicting the composition of gut bacteria, surpassing one's genetic makeup. The food we consume determines the flourishing of specific bacteria in the gut, and reciprocally, these gut bacteria assist in the process of digestion (Carmody et al. 2015). In broad term, plant protein, unsaturated fats, and fiber contribute to the promotion of a beneficial gut microbiota, whereas intake of animal protein, saturated fats, and simple or artificial sugars in abundance have opposite effect (Madison and Kiecolt-Glaser 2019). Current evidence indicates that there are interconnected relationships between stress/mood, diet, and microbes that can result in either detrimental or beneficial cycle. These complex connections between the mind, body, humans, and bacteria shed light on both the capacity to adapt and the development of chronic diseases.

Nutraceuticals play a major role in the removal of SCs form the host. Widely investigated nutraceuticals in the attenuation of senescence are quercetin, oleuropein aglycone (OLE), epigallocatechin gallate (EGCG), and fisetin. Quercetin, flavonoid compound, is a naturally occurring phytochemical in various plants and vegetables. It is well recognized compound for its potential antioxidant, anti-inflammatory and immune system regulating properties. This flavonoid is actively involved in removal of SC mediated through phosphoinositide 3-kinase pathway. PI3K pathway is widely involved in cellular growth and survival process, thus the regulatory properties of quercetin inhibit the PI3K pathway mediated SC growth and induces apoptosis (Soto-Gamez and Demaria 2017). Moreover, combination therapy of quercetin with Dasatinib, an anti-cancer drug, also showed a potent eradication of SC (Chondrogianni et al. 2010). The mRNA levels of markers associated with SCs, namely p16 and p21, as well as inflammatory markers related to SASP (such as CXCL1, IL-1β, IL-6, MCP-1, and TNF-α), exhibited marked reduced expression in all sections of the intestines in aged mice treated with combinational therapy (Quercetin + Dasatinib), in comparison to aged mice treated with a placebo. Similarly phenolic compounds OLE and EGCG have shown remarkable contribution in the deletion of SCs (Kaur et al. 2020).

Intake of probiotics and prebiotics play a major role in nourishment of gut microbiota (Mack 2005). Various researches are being directing probiotics as an anti-senescent agent. A study by Sharma and Padwad, showed that probiotics influences the microbes in gut and directly alters cellular senescence (Sharma and Padwad 2020). Recent research conducted in 2020 explains the advantages of probiotic bacteria Lactobacillus fermentum in amelioration of SCs of a senescence induced mice via influencing the mTOR signaling thereby attenuating SASP factors present in cells (Kumar et al. 2020). Another innovative approach is that the incorporation of Probiotic bacterial with naturally occurring polyphenolic compounds resulted in a new approach in attenuation of the SCs (Sharma and Padwad 2020). Various studies were conducted testing the combinational administration of probiotics and bioactive phenolic compounds. Specifically, a symbiotic formulation comprised of Triphala, an Ayurvedic herb abundant in polyphenols, and a combination of probiotics including Lactobacillus plantarum, Bifidobacteria longum spp. infantis, and L. fermentum, exhibited remarkable effects by extending the lifespan of drosophila, with a reduction in ROS stress, and low-level inflammation (Westfall et al. 2018).

Furthermore, a diet abundant in dietary fibers that promote the production of short-chain fatty acids (SCFAs) could provide additional assistance in addressing age-related microbial dysbiosis and, consequently, restraining the senescent phenotype (Matt et al. 2018). An illustrative example is the use of butyrate supplementation, which has shown the capacity to counteract age-related dysbiosis of the gut microbiota.

6.7 Conclusion

In conclusion, it is evident that a number of external and internal factors play an important part in the modification of microbial colonies in the gut. The host and the microbes residing within it is a complex symbiotic system that is involves a variety of beneficial applications. However, when the symbiosis is affected, they can result in various deleterious pathological conditions. Upon dysregulation, the gut microbiome is involved in release of various metabolites that initiate cellular damage and diseases. Cellular senescence is one such outcome of altered microbiota. The increased inflammatory microenvironment in the gut region underlay a pathway for cellular senescence. Senescent cells are double edged sword that is highly regulated. They are crucial for maintaining cellular homeostasis, inhibiting excess cell proliferation and induction of apoptosis. However, upon disregulation these cells promote an inflammatory condition affecting the neighboring cells. Restoration of the dysbiosis of gut microbiome plays an important role in the senolytic or senostatic effect. Diet, probiotic, plant poly phenols play a vital role in restoration process thereby deleting the senescent cells from the body. Research is ongoing to fully elucidate the specific mechanisms by which gut microbiome dysbiosis leads to cellular senescence and how various factors, such as diet and plant polyphenols, influence this process. This knowledge will pave the way for the development of targeted interventions aimed at restoring gut microbiome balance to prevent or alleviate cellular senescence-related diseases. Future research may explore personalized approaches tailored to individual gut microbiome profiles for optimal efficacy and minimal side effects. Additionally, investigating the long-term safety and efficacy of these interventions remains crucial. By continuing to explore this intricate relationship between the gut microbiome and cellular senescence, researchers can potentially unlock novel therapeutic strategies for various age-related and chronic diseases.

References

Andoh A (2016) Physiological role of gut microbiota for maintaining human health. Digestion 93(3):176–181. https://doi.org/10.1159/000444066

Boisvert MM, Erikson GA, Shokhirev MN, Allen NJ (2018) The aging astrocyte transcriptome from multiple regions of the mouse brain. Cell Rep 22:269–285. https://doi.org/10.1016/j.celrep.2017.12.039

Bulua AC, Simon A, Maddipati R, Pelletier M, Park H, Kim KY, Sack MN, Kastner DL, Siegel RM (2011) Mitochondrial reactive oxygen species promote production of proinflammatory cytokines and are elevated in TNFR1-associated periodic syndrome (TRAPS). J Exp Med 208(3):519–533. https://doi.org/10.1084/jem.20102049

Campisi J (1997) The biology of replicative senescence. Eur J Cancer (Oxford, England: 1990) 33(5):703–709. https://doi.org/10.1016/S0959-8049(96)00058-5

Campisi J, D'Adda Di Fagagna F (2007) Cellular senescence: when bad things happen to good cells. Nat Rev Mol Cell Biol 8:729–740. https://doi.org/10.1038/nrm2233

Carmody RN, Gerber GK, Luevano JM Jr, Gatti DM, Somes L, Svenson KL, Turnbaugh PJ (2015) Diet dominates host genotype in shaping the murine gut microbiota. Cell Host Microbe 17(1):72–84. https://doi.org/10.1016/j.chom.2014.11.010

Chan JF, Yuan S, Kok KH, To KK, Chu H, Yang J, Xing F, Liu J, Yip CC, Poon RW, Tsoi HW, Lo SK, Chan KH, Poon VK, Chan WM, Ip JD, Cai JP, Cheng VC, Chen H, Hui CK et al (2020) A familial cluster of pneumonia associated with the 2019 novel coronavirus indicating person-to-person transmission: a study of a family cluster. Lancet (London, England) 395(10223):514–523. https://doi.org/10.1016/S0140-6736(20)30154-9

Chandrasekaran A, Idelchik MDPS, Melendez JA (2017) Redox control of senescence and age-related disease. Redox Biol 11:91–102. https://doi.org/10.1016/j.redox.2016.11.005

Chang PV, Hao L, Offermanns S, Medzhitov R (2014) The microbial metabolite butyrate regulates intestinal macrophage function via histone deacetylase inhibition. Proc Natl Acad Sci 111:2247. https://doi.org/10.1073/pnas.1322269111

Childs BG, Baker DJ, Kirkland JL, Campisi J, van Deursen JM (2014) Senescence and apoptosis: dueling or complementary cell fates? EMBO Rep 15(11):1139–1153. https://doi.org/10.15252/embr.201439245

Chondrogianni N, Kapeta S, Chinou I, Vassilatou K, Papassideri I, Gonos ES (2010) Anti-ageing and rejuvenating effects of quercetin. Exp Gerontol 45(10):763–771. https://doi.org/10.1016/j.exger.2010.07.001

Collado M, Blasco MA, Serrano M (2007) Cellular senescence in cancer and aging. Cell 130:223–233. https://doi.org/10.1016/j.cell.2007.07.003

Conway J, Duggal NA (2021) Ageing of the gut microbiome: potential influences on immune senescence and inflammageing. Ageing Res Rev 68:101323. https://doi.org/10.1016/j.arr.2021.101323

De la Fuente M, Miquel J (2009) An update of the oxidation-inflammation theory of aging: the involvement of the immune system in oxi-inflamm-aging. Curr Pharm Des 15(26):3003–3026. https://doi.org/10.2174/138161209789058110

DeGruttola AK, Low D, Mizoguchi A, Mizoguchi E (2016) Current understanding of dysbiosis in disease in human and animal models. Inflamm Bowel Dis 22(5):1137–1150. https://doi.org/10.1097/MIB.0000000000000750

Ebert MN, Klinder A, Peters WH, Schäferhenrich A, Sendt W, Scheele J, Pool-Zobel BL (2003) Expression of glutathione S-transferases (GSTs) in human colon cells and inducibility of GSTM2 by butyrate. Carcinogenesis 24(10):1637–1644. https://doi.org/10.1093/carcin/bgg122

Epel ES, Blackburn EH, Lin J, Dhabhar FS, Adler NE, Morrow JD, Cawthon RM (2004) Accelerated telomere shortening in response to life stress. Proc Natl Acad Sci U S A 101(49):17312–17315. https://doi.org/10.1073/pnas.0407162101

Faas MM, Sáez T, de Vos P (2017) Extracellular ATP and adenosine: the Yin and Yang in immune responses? Mol Asp Med 55:9–19. https://doi.org/10.1016/j.mam.2017.01.002

Fond G, Loundou A, Hamdani N, Boukouaci W, Dargel A, Oliveira J, Roger M, Tamouza R, Leboyer M, Boyer L (2014) Anxiety and depression comorbidities in irritable bowel syndrome (IBS): a systematic review and meta-analysis. Eur Arch Psychiatry Clin Neurosci 264:651–660

Furusawa Y, Obata Y, Fukuda S, Endo TA, Nakato G, Takahashi D, Nakanishi Y, Uetake C, Kato K, Kato T, Takahashi M, Fukuda NN, Murakami S, Miyauchi E, Hino S, Atarashi K, Onawa S, Fujimura Y, Lockett T, Clarke JM et al (2013) Commensal microbe-derived butyrate induces the differentiation of colonic regulatory T cells. Nature 504(7480):446–450. https://doi.org/10.1038/nature12721

Galkin F, Mamoshina P, Aliper A, Putin E, Moskalev V, Gladyshev VN, Zhavoronkov A (2020) Human gut microbiome aging clock based on taxonomic profiling and deep learning. iScience 23(6):101199. https://doi.org/10.1016/j.isci.2020.101199

Goldsmith JR, Sartor RB (2014) The role of diet on intestinal microbiota metabolism: downstream impacts on host immune function and health, and therapeutic implications. J Gastroenterol 49(5):785–798. https://doi.org/10.1007/s00535-014-0953-z

Gorgoulis V, Adams PD, Alimonti A, Bennett DC, Bischof O, Bishop C, Campisi J, Collado M, Evangelou K, Ferbeyre G, Gil J, Hara E, Krizhanovsky V, Jurk D, Maier AB, Narita M, Niedernhofer L, Passos JF, Robbins PD, Schmitt CA, Sedivy J, Vougas K, von Zglinicki T, Zhou D, Serrano M, Demaria M (2019) Cellular senescence: defining a path forward. Cell 179(4):813–827. https://doi.org/10.1016/j.cell.2019.10.005

Guarner F (2005) The intestinal flora in inflammatory bowel disease: normal or abnormal? Curr Opin Gastroenterol 21(4):414–418

Guo Y, Ayers JL, Carter KT, Wang T, Maden SK, Edmond D, Newcomb P P, Li C, Ulrich C, Yu M, Grady WM (2019) Senescence-associated tissue microenvironment promotes colon cancer formation through the secretory factor GDF15. Aging Cell 18(6):e13013. https://doi.org/10.1111/acel.13013

Gurung M, Li Z, You H, Rodrigues R, Jump DB, Morgun A, Shulzhenko N (2020) Role of gut microbiota in type 2 diabetes pathophysiology. EBioMedicine 51:102590. https://doi.org/10.1016/j.ebiom.2019.11.051

Herranz N, Gil J (2018) Mechanisms and functions of cellular senescence. J Clin Invest 128(4):1238–1246. https://doi.org/10.1172/JCI95148

Idda ML, McClusky WG, Lodde V, Munk R, Abdelmohsen K, Rossi M, Gorospe M (2020) Survey of senescent cell markers with age in human tissues. Aging 12:4052–4066. https://doi.org/10.18632/aging.102903

Ivanov II, Frutos RL, Manel N, Yoshinaga K, Rifkin DB, Sartor RB, Finlay BB, Littman DR (2008) Specific microbiota direct the differentiation of IL-17-producing T-helper cells in the mucosa of the small intestine. Cell Host Microbe 4(4):337–349. https://doi.org/10.1016/j.chom.2008.09.009

Kamada N, Núñez G (2014) Regulation of the immune system by the resident intestinal bacteria. Gastroenterology 146(6):1477–1488. https://doi.org/10.1053/j.gastro.2014.01.060

Kaplanski G (2018) Interleukin-18: biological properties and role in disease pathogenesis. Immunol Rev 281(1):138–153. https://doi.org/10.1111/imr.12616

Kaur A, Macip S, Stover CM (2020) An appraisal on the value of using nutraceutical based senolytics and senostatics in aging. Front Cell Dev Biol 8:218. https://doi.org/10.3389/fcell.2020.00218

Kho ZY, Lal SK (2018) The human gut microbiome—a potential controller of wellness and disease. Front Microbiol 9:1835. https://doi.org/10.3389/fmicb.2018.01835

Kumar S, Suman S, Fornace AJ, Datta K (2019) Intestinal stem cells acquire premature senescence and senescence associated secretory phenotype concurrent with persistent DNA damage after heavy ion radiation in mice. Aging 11(12):4145–4158. https://doi.org/10.18632/aging.102043

Kumar R, Sharma A, Gupta M, Padwad Y, Sharma R (2020) Cell-free culture supernatant of probiotic lactobacillus fermentum protects against H2O2-induced premature senescence by suppressing ROS-Akt-mTOR axis in murine preadipocytes. Probiotics Antimicrob Proteins 12(2):563–576. https://doi.org/10.1007/s12602-019-09576-z

Kumari R, Jat P (2021) Mechanisms of cellular senescence: cell cycle arrest and senescence associated secretory phenotype. Front Cell Dev Biol 9:645593. https://doi.org/10.3389/fcell.2021.645593

Lee BY, Han JA, Im JS, Morrone A, Johung K, Goodwin EC, Kleijer WJ, DiMaio D, Hwang ES (2006) Senescence-associated beta-galactosidase is lysosomal beta-galactosidase. Aging Cell 5(2):187–195. https://doi.org/10.1111/j.1474-9726.2006.00199.x

Lloyd-Price J, Abu-Ali G, Huttenhower C (2016) The healthy human microbiome. Genome Med 8:51. https://doi.org/10.1186/s13073-016-0307-y

Macfarlane GT, Macfarlane S (2011) Fermentation in the human large intestine: its physiologic consequences and the potential contribution of prebiotics. J Clin Gastroenterol 45 Suppl:S120–S127. https://doi.org/10.1097/MCG.0b013e31822fecfe

Mack DR (2005) Probiotics-mixed messages. Can Fam Physician 51(11):1455–1464

Madison A, Kiecolt-Glaser JK (2019) Stress, depression, diet, and the gut microbiota: human-bacteria interactions at the core of psychoneuroimmunology and nutrition. Curr Opin Behav Sci 28:105–110. https://doi.org/10.1016/j.cobeha.2019.01.011

Matt SM, Allen JM, Lawson MA, Mailing LJ, Woods JA, Johnson RW (2018) Butyrate & dietary soluble fiber improve neuroinflammation associated with aging in mice. Front Immunol 9:1832

Moorefield EC, Andres SF, Blue RE, Van Landeghem L, Mah AT, Santoro MA, Ding S (2017) Aging effects on intestinal homeostasis associated with expansion and dysfunction of intestinal epithelial stem cells. Aging 9(8):1898–1915. https://doi.org/10.18632/aging.101279

Muñoz-Espín D, Serrano M (2014) Cellular senescence: from physiology to pathology. Nat Rev Mol Cell Biol 15:482–496. https://doi.org/10.1038/nrm3823

Norman JM, Handley SA, Virgin HW (2014) Kingdom-agnostic metagenomics and the importance of complete characterization of enteric microbial communities. Gastroenterology 146(6):1459–1469. https://doi.org/10.1053/j.gastro.2014.02.001

Oliphant K, Allen-Vercoe E (2019) Macronutrient metabolism by the human gut microbiome: major fermentation by-products and their impact on host health. Microbiome 7:91. https://doi.org/10.1186/s40168-019-0704-8

Parida S, Sharma D (2021) The microbiome and cancer: creating friendly neighborhoods and removing the foes within. Cancer Res 81(4):790–800. https://doi.org/10.1158/0008-5472.CAN-20-2629

Parkar SG, Trower TM, Stevenson DE (2013) Fecal microbial metabolism of polyphenols and its effects on human gut microbiota. Anaerobe 23:12–19. https://doi.org/10.1016/j.anaerobe.2013.07.009

Passos JF, Nelson G, Wang C, Richter T, Simillion C, Proctor CJ, Miwa S, Olijslagers S, Hallinan J, Wipat A, Saretzki G, Rudolph KL, Kirkwood TB, von Zglinicki T (2010) Feedback between p21 and reactive oxygen production is necessary for cell senescence. Mol Syst Biol 6:347. https://doi.org/10.1038/msb.2010.5

Pinato DJ, Howlett S, Ottaviani D, Urus H, Patel A, Mineo T, Brock C, Power D, Hatcher O, Falconer A, Ingle M, Brown A, Gujral D, Partridge S, Sarwar N, Gonzalez M, Bendle M, Lewanski C, Newsom-Davis T, Allara E et al (2019) Association of prior antibiotic treatment with survival and response to immune checkpoint inhibitor therapy in patients with cancer. JAMA Oncol 5(12):1774–1778. https://doi.org/10.1001/jamaoncol.2019.2785

Postler TS, Ghosh S (2017) Understanding the holobiont: how microbial metabolites affect human health and shape the immune system. Cell Metab 26(1):110–130. https://doi.org/10.1016/j.cmet.2017.05.008

Rossi T, Vergara D, Fanini F, Maffia M, Bravaccini S, Pirini F (2020) Microbiota-derived metabolites in tumor progression and metastasis. Int J Mol Sci 21:5786. https://doi.org/10.3390/ijms21165786

Sager R (1991) Senescence as a mode of tumor suppression. Environ Health Perspect 93:59–62. https://doi.org/10.1289/ehp.919359

Saleh T, Tyutynuk-Massey L, Cudjoe EK, Idowu MO, Landry JW, Gewirtz DA (2018) Non-cell autonomous effects of the senescence-associated secretory phenotype in cancer therapy. Front Oncol 8:164. https://doi.org/10.3389/fonc.2018.00164

Schosserer M, Grillari J, Breitenbach M (2017) The dual role of cellular senescence in developing tumors and their response to cancer therapy. Front Oncol 7:278. https://doi.org/10.3389/fonc.2017.00278

Sender R, Fuchs S, Milo R (2016) Revised estimates for the number of human and bacteria cells in the body. PLoS Biol 14:e1002533. https://doi.org/10.1371/journal.pbio.1002533

Sharma R (2022) Emerging interrelationship between the gut microbiome and cellular senescence in the context of aging and disease: perspectives and therapeutic opportunities. Probiotics Antimicrob Proteins 14(4):648–663. https://doi.org/10.1007/s12602-021-09903-3

Sharma R, Padwad Y (2020) Probiotic bacteria as modulators of cellular senescence: emerging concepts and opportunities. Gut Microbes 11(3):335–349. https://doi.org/10.1080/19490976.2019.1697148

Sharma R, Kumar R, Sharma A, Goel A, Padwad Y (2022) Long-term consumption of green tea EGCG enhances murine health span by mitigating multiple aspects of cellular senescence in mitotic and post-mitotic tissues, gut dysbiosis, and immunosenescence. J Nutr Biochem 107:109068. https://doi.org/10.1016/j.jnutbio.2022.109068

Simpson HL, Campbell BJ (2015) Review article: dietary fibre-microbiota interactions. Aliment Pharmacol Ther 42(2):158–179. https://doi.org/10.1111/apt.13248

Song P, An J, Zou MH (2020) Immune clearance of senescent cells to combat ageing and chronic diseases. Cells 9(3):671. https://doi.org/10.3390/cells9030671

Soto-Gamez A, Demaria M (2017) Therapeutic interventions for aging: the case of cellular senescence. Drug Discov Today 22(5):786–795. https://doi.org/10.1016/j.drudis.2017.01.004

Surova O, Zhivotovsky B (2013) Various modes of cell death induced by DNA damage. Oncogene 32:3789–3797. https://doi.org/10.1038/onc.2012.556

Tamburini S, Shen N, Wu HC, Clemente JC (2016) The microbiome in early life: implications for health outcomes. Nat Med 22(7):713–722. https://doi.org/10.1038/nm.4142

Tang WH, Wang Z, Fan Y, Levison B, Hazen JE, Donahue LM, Wu Y, Hazen SL (2014) Prognostic value of elevated levels of intestinal microbe-generated metabolite trimethylamine-N-oxide in patients with heart failure: refining the gut hypothesis. J Am Coll Cardiol 64(18):1908–1914. https://doi.org/10.1016/j.jacc.2014.02.617

Tap J, Mondot S, Levenez F, Pelletier E, Caron C, Furet JP, Ugarte E, Muñoz-Tamayo R, Paslier DL, Nalin R, Dore J, Leclerc M (2009) Towards the human intestinal microbiota phylogenetic core. Environ Microbiol 11(10):2574–2584. https://doi.org/10.1111/j.1462-2920.2009.01982.x

Thaiss CA, Zmora N, Levy M, Elinav E (2016) The microbiome and innate immunity. Nature 535(7610):65–74. https://doi.org/10.1038/nature18847

Uchida R, Saito Y, Nogami K, Kajiyama Y, Suzuki Y, Kawase Y, Nakaoka T, Muramatsu T, Kimura M, Saito H (2018) Epigenetic silencing of Lgr5 induces senescence of intestinal epithelial organoids during the process of aging. NPJ Aging Mech Dis 5:1. https://doi.org/10.1038/s41514-018-0031-5

Vaiserman A, Romanenko M, Piven L, Moseiko V, Lushchak O, Kryzhanovska N, Guryanov V, Koliada A (2020) Differences in the gut Firmicutes to Bacteroidetes ratio across age groups in healthy Ukrainian population. BMC Microbiol 20(1):221. https://doi.org/10.1186/s12866-020-01903-7

Wang Z, Klipfell E, Bennett BJ, Koeth R, Levison BS, Dugar B, Feldstein AE, Britt EB, Fu X, Chung YM, Wu Y, Schauer P, Smith JD, Allayee H, Tang WH, DiDonato JA, Lusis AJ, Hazen SL (2011) Gut flora metabolism of phosphatidylcholine promotes cardiovascular disease. Nature 472(7341):57–63. https://doi.org/10.1038/nature09922

Wang LC, Pan TM, Tsai TY (2018) Lactic acid bacteria-fermented product of green tea and Houttuynia cordata leaves exerts antiadipogenic and anti-obesity effects. J Food Drug Anal 26:973–984. https://doi.org/10.1016/j.jfda.2017.11.009

Westfall S, Lomis N, Prakash S (2018) Longevity extension in Drosophila through gut-brain communication. Sci Rep 8(1):8362. https://doi.org/10.1038/s41598-018-25382-z

Wilmanski T, Diener C, Rappaport N, Patwardhan S, Wiedrick J, Lapidus J, Earls JC, Zimmer A, Glusman G, Robinson M, Yurkovich JT, Kado DM, Cauley JA, Zmuda J, Lane NE, Magis AT, Lovejoy JC, Hood L, Gibbons SM, Orwoll ES, Price ND (2021) Gut microbiome pattern reflects healthy ageing and predicts survival in humans. Nat Metab 3(2):274–286. https://doi.org/10.1038/s42255-021-00348-0

Wu GD, Chen J, Hoffmann C, Bittinger K, Chen YY, Keilbaugh SA, Bewtra M, Knights D, Walters WA, Knight R, Sinha R, Gilroy E, Gupta K, Baldassano R, Nessel L, Li H, Bushman FD, Lewis JD (2011) Linking long-term dietary patterns with gut microbial enterotypes. Science 334(6052):105–108. https://doi.org/10.1126/science.1208344

Yoshimoto S, Loo TM, Atarashi K, Kanda H, Sato S, Oyadomari S, Iwakura Y, Oshima K, Morita H, Hattori M, Honda K, Ishikawa Y, Hara E, Ohtani N (2013) Obesity-induced gut microbial metabolite promotes liver cancer through senescence secretome. Nature 499(7456):97–101. https://doi.org/10.1038/nature12347

Yousefzadeh MJ, Zhao J, Bukata C, Wade EA, McGowan SJ, Angelini LA, Bank MP, Gurkar AU, McGuckian CA, Calubag MF, Kato JI, Burd CE, Robbins PD, Niedernhofer LJ (2020) Tissue specificity of senescent cell accumulation during physiologic and accelerated aging of mice. Aging Cell 19:e13094. https://doi.org/10.1111/acel.13094

Zhang D, Chen G, Manwani D, Mortha A, Xu C, Faith JJ, Burk RD, Kunisaki Y, Jang JE, Scheiermann C, Merad M, Frenette PS (2015) Neutrophil ageing is regulated by the microbiome. Nature 525(7570):528–532. https://doi.org/10.1038/nature15367

Zhu Y, Tchkonia T, Pirtskhalava T, Gower AC, Ding H, Giorgadze N, Palmer AK, Ikeno Y, Hubbard GB, Lenburg M, O'Hara SP, LaRusso NF, Miller JD, Roos CM, Verzosa GC, LeBrasseur NK, Wren JD, Farr JN, Khosla S, Stout MB, McGowan SJ, Fuhrmann-Stroissnigg H, Gurkar AU, Zhao J, Colangelo D, Dorronsoro A, Ling YY, Barghouthy AS, Navarro DC, Sano T, Robbins PD, Niedernhofer LJ, Kirkland JL (2015) The Achilles' heel of senescent cells: from transcriptome to senolytic drugs. Aging Cell 14(4):644–658. https://doi.org/10.1111/acel.12344

Molecular Interplay of Oxidative Stress and Gut Microbiome in Aging

7

Ahamed Basha Abdul Bari, Madhan Krishnan, and Shyamaladevi Babu

Abstract

Oxidative stress results from an imbalance between reactive oxygen species production and the capacity of the organism to counteract them. These species are reactive substances that may deteriorate lipids, proteins, and DNA, which causes cellular malfunction and acts as etiological factors for various age-related disorders, including aging. The gut microbiota contains trillions of microorganisms, including bacteria, protozoa, viruses, and fungi, as well as their combined genetic material, that are present in the gastrointestinal tract. These microbiomes can also influence oxidative stress through their metabolic activities. As we age, the gut microbiome undergoes changes in composition and diversity, which can alter its metabolic activities and contribute to oxidative stress through mechanisms like production of reactive oxygen species, and modulation of antioxidant defences. For example, certain gut bacteria can produce reactive oxygen species as a byproduct of their metabolism, while others can generate antioxidants that neutralise them. The balance between reactive oxygen species-producing and antioxidant-producing bacteria can shift as we age, leading to increased oxidative stress. In addition, the gut microbiome can influence the host's antioxidant defence system by producing short-chain fatty acids and other metabolites that can modulate gene expression and signalling pathways. One of the short-chain fatty acids, butyrate, can result in an increase in the production of antioxidant enzymes such as catalase and superoxide dismutase and reduce oxidative stress in the gut and other organs. Additionally, available literature also indicates that

A. B. A. Bari (✉)
Department of Physiology, Chettinad Hospital and Research Institute, Chettinad Academy of Research and Education, Kelambakkam, Tamilnadu, India

M. Krishnan · S. Babu
Research, Chettinad Hospital and Research Institute, Chettinad Academy of Research and Education, Kelambakkam, Tamilnadu, India

S. Pathak, A. Banerjee (eds.), *Gut Microbiome and Brain Ageing*,
https://doi.org/10.1007/978-981-99-8803-7_7

oxidative stress can affect the gut microbiome by changing its structure and functionality through the immune response, which in turn influences the aging process. Oxidative byproducts can damage bacterial membranes, DNA, and proteins, leading to microbial cell death and dysbiosis. Dysbiosis, on the other hand, might increase oxidative stress by lowering the production of short-chain fatty acids and other antioxidant metabolites. Overall, the molecular interaction between oxidative stress and the gut microbiome in aging is intricate and complicated. Further research is required to understand the mechanisms underlying these interactions and identify strategies to promote a healthy gut microbiome and reduce oxidative stress in the aging process.

Keywords

Oxidative stress · Antioxidants · Gut microbiome · Aging

7.1 Introduction

Oxidative stress (OS) is a biological process that occurs when the production of reactive oxygen species (ROS) outweighs the ability of cells to detoxify these reactive species (Hajam et al. 2022). ROS can harm cellular components such as DNA, proteins, and lipids, resulting in cellular dysfunction and even cell death. Numerous biochemical processes produce large amounts of ROS as an unavoidable by-product, and in some cases, such as in activated neutrophils, these free radicals are produced to assist in the clearance of host tissue (Parker et al. 2021). They can also be produced in the body as a direct result of electromagnetic radiation from the environment (Schuermann and Mevissen 2021). A variety of tissues may sustain damage if antioxidant defences are insufficient. It is generally accepted that OS plays a crucial role in the aging process (Hajam et al. 2022). As we age, our bodies become less efficient at neutralising ROS, which are byproducts of the regular metabolic processes that take place inside cells. When all body systems are considered, the scope of oxidative damage is enormous, and it leads to systemic damage (cardiovascular, neural, endocrine, etc.) and becomes an etiological factor for various diseases, including age-related disorders (Luo et al. 2020).

The gut microbiota (GMA) is a collection of microorganisms such as viruses, protozoa, and others in the gastro intestinal tract GIT, and the gut microbiome (GME) is the whole ecosystem, including microorganisms, their genetic material, and environmental factors. The GMA controls several physiological processes, including digestion, nutrition absorption, and immune system function. Thus, the GME is vital to the host's health and proper functioning (Heintz-Buschart and Wilmes 2018). The microbiome of the gut is critically vital to the body's ability to control levels of OS. These microorganisms create metabolites, the effects of which in the body can either be pro-oxidant or antioxidant (Shandilya et al. 2022).

Additionally, there is rising evidence to suggest that the GME may have a role in the aging process and the development of age-related disorders (Coman and Vodnar 2020), including OS. The composition and variety of the GME shift as we get older, with a reduction in the number of bacteria that are helpful to our health and an increase in the number of bacteria that could be detrimental. This shift in the GMA is known as 'dysbiosis', and it has been linked to the development of age-related disorders, particularly those connected to OS, such as cardiovascular disease, respiratory disease, and neurodegenerative diseases (Cryan et al. 2019; Saint-Criq et al. 2021; Zhang et al. 2022). This is supported by the fact that there has been an increase in the number of studies that have investigated this hypothesis. Further, studies have indicated that therapies that target the GME, such as probiotics and prebiotics, can help lower OS and inflammation throughout the body and exhibit a protective effect in the aging process. Considering the association between OS and GME as age advances, this chapter discusses the molecular interplay of the OS and the GME in aging process.

7.2 Physiology and Biochemistry of Oxidative Stress

Free radicals are reactive oxygen and nitrogen species (RONS) that are primarily produced endogenously; they can be found in the cytoplasm, endoplasmic reticulum/microsomes, mitochondria, lysosomes, and peroxisomes (Di Meo et al. 2016). The mitochondrion, is the predominant source of endogenous free radicals and is responsible for the production of the vast majority of free radicals that are generated inside the body. Free radicals are produced whenever electrons are moved across the mitochondrial membrane in order to generate ATP. This process is necessary for life but may be harmful if not controlled. The electrons in the electron transport chain's complex I or complex III, where the superoxide is generated, bind diatomic oxygen molecules to create superoxide $O2^{\bullet -}$, which is subsequently released either within the mitochondrial matrix or in the intermembrane space (Ashok et al. 2022). NADH is reduced by the NADH dehydrogenase at complex I, and electron leakage from this complex caused by flavin mononucleotide's partial reduction can result in ROS. Cytochrome c is reduced, and ubiquinone is oxidised at complex III. The semiquinone anion (Q^-), which can transfer electrons to O2 to produce $O2^{*-}$, is produced as a byproduct of coenzyme Q renewal. As an alternative, H_2O_2 could act as a precursor in the Fenton reaction or in the Haber-Weiss reaction, which produces the hydroxyl radical (OH^*). Additionally, oxidative cell damage has been linked more and more to metabolites. Among them, hydroxyalkenals and advanced glycation products (AGEs) are noteworthy (Moldogazieva et al. 2019). AGEs are created when reducing sugars, such as glucose, interact non-enzymatically over a period of weeks with amino groups on proteins, occasionally lipids, and DNA. The proteins complement, fibrinogen, tubulin, myelin, collagen, and tubulin have all been discovered to be targets. A number of pathologies, including the emergence of amyloid plaques in the progression of Alzheimer's disease and other age-related disorders (Banerjee et al. 2021), have been linked to AGEs (Twarda-Clapa et al. 2022).

Within a cell, these radicals can be produced by either accepting or rejecting the donation of one electron, causing them to function as either reductants or oxidants. The reactive radical and non-radical derivatives of oxygen and nitrogen are referred to as reactive oxygen species (ROS) and reactive nitrogen species (RNS), respectively. Reactive oxygen and nitrogen species, often known as RONS, are produced by all aerobic cells. RONS are crucial contributors in both the natural aging process as well as disorders that are related with aging (Speer and McKune 2021). In addition to its usage in the detection of harmful effects, the production of RONS is important for a number of other processes, including signalling, the immunological defence system, and the extraction of energy from organic molecules (Wu et al. 2022).

7.3 Aging and Oxidative Stress

Aging is a complex process influenced by numerous factors. It is generally accepted that the basic hallmarks of aging (Dodig et al. 2019), which include genomic instability, epigenetic changes, and loss of proteostasis, are the fundamental causes of damage at the cellular level, which gradually accumulates over time. Because of this, dysfunctions in nutrition sensing and mitochondrial function, as well as cellular senescence, occur. Further, when the tissue homeostatic systems are unable to compensate for the damage, age-related functional deficits take place, which include depleted stem cell populations and disruptions in intercellular communication.

According to aging theory, OS plays a significant role in the aging process, and our cells become less efficient at repairing oxidative damage as we age, resulting in an accumulation of damage over time (López-Otín et al. 2023). The several hallmarks of aging as mentioned above are greatly impacted by OS. Further, it is characterised by an increasing susceptibility to disease, the functional preservation of tissue homeostasis, and a progressive decline in the efficiency of biochemical and physiological processes. It is a multifaceted process that is both genetically and environmentally determined in terms of epigenetics in aging-related progressive loss of organ and tissue function (Morris et al. 2019). The ‘free radical theory of aging’, which was renamed the ‘oxidative stress theory of aging’ in subsequent years, is based on the hypothesis of structural damage (Hajam et al. 2022). This hypothesis states that age-related functional losses are caused by the buildup of oxidative damage caused by RONS to macromolecules (lipids, DNA, and proteins). Increased amounts of ROS are most likely to be the cause of cellular senescence, which is a physiological process that stops cellular reproduction as a reaction to replication-related damage. The exact mechanism that causes OS-induced aging is still unclear. In addition to the irreversible senescence-associated secretory phenotypes (SASP), which include the secretion of insoluble proteins and components of the extracellular matrix (ECM), senescent cells can secrete soluble molecules (such as interleukins, chemokines, and growth factors). RONS cause cellular senescence by acting on various SASP components (Roger et al. 2021). These components include regulation of the functions of mammalian target of rapamycin complexes, production of interleukin-1, upregulation of matrix metalloproteinases, inhibition of the FOXO (Forkhead box) protein, and controlling the p53/p21 and p16INK4a/pRB senescence pathways.

7.4 List of Age-Related Diseases in Oxidative Stress Conditions

Cardiovascular disease, chronic renal disease, cancers at various sites and organs, neurodegenerative diseases, and biliary diseases are among the numerous acute and chronic pathological processes and are all impacted by OS, cellular senescence, and consequently, SASP factors (Roger et al. 2021). Age-related increases in the prevalence of cardiovascular diseases are directly correlated with mortality rates; cardiovascular disease is the leading cause of death in older adults (North and Sinclair 2012). As byproducts of mitochondrial respiration, ROS are primarily produced; as the main site of oxidative damage and a key contributor to aging, mitochondria are a key target. Recent investigation has connected mitochondrial dysfunction to a range of age-related illnesses, including cancer (Das et al. 2021) and neurodegenerative diseases (Marquez-Exposito et al. 2022).

Additionally, senescent smooth muscle cells undergo osteoblastic transdifferentiation under the control of SASP, which is related to vascular calcification. Brain tissue biopsies demonstrate elevated p16, matrix metalloproteinases, and interleukin-6 levels in a variety of neurodegenerative conditions, including Alzheimer's disease (AD) (Gonzales et al. 2021). Cholangitis, biliary cirrhosis, osteoarthritis, and chronic obstructive pulmonary disease all have damaging SASP profiles that include interleukin-6, 8, and matrix metalloproteinases (Tabibian et al. 2014). In view of the tight relationship that exists between OS, inflammation, and the aging process, the oxidation-inflammatory hypothesis of aging has been proposed. This theory is also known as the oxi-inflamm-aging theory (Martínez de Toda et al. 2021). According to this hypothesis, the aging process is characterised by a decline in homeostasis that is triggered by chronic exposure to OS. This kind of stress has a significant influence on regulatory systems, such as the neurological, endocrine, and immunological systems. As a result of the immune system's subsequent activation, there is an inflammatory state, which feeds on chronic OS and inflammation to increase age-related morbidity and mortality. Thus, it is evident from the above that OS is an important influencer in the physiology of aging (Hajam et al. 2022).

7.5 Gut Microbiota

The GMA refers to the diverse collection of bacteria, virus, fungi, etc. that colonise the gastrointestinal system. It is believed that there are a very large number of microorganisms in the gastrointestinal tract (GIT), which comprises around ten times as many bacterial cells as human cells and more than one hundred times as much genetic material (the microbiome) as the human genome. According to a revised estimate (Thursby and Juge 2017), the proportion of bacterial cells to human cells is really much closer to being equal to one to one. Because the human body contains an extremely high number of bacterial cells, it is commonly referred to as a 'superorganism', which includes both the host and the microorganisms that live inside the host. Human GME research has expanded in the last decade. New technologies have raised concerns about the GMA's effects on human health and illness. The GMA affects metabolism, immunity, and behaviour. Age, environment, stress, nutrition, health conditions, and drug exposure modify our gut flora throughout the entire life cycle. A number of benefits are provided to the host by the microbiota. These benefits are achieved by the microbiota through a variety of physiological processes, such as improving gut integrity or reshaping the intestinal epithelium (Sidhu and van der Poorten 2017; Schmidt et al. 2018), obtaining energy (Barko et al. 2018), defending against pathogens, and controlling host immunity (Barko et al. 2018; Ducarmon et al. 2019). However, the altered microbial composition known as dysbiosis has the potential to cause these mechanisms to become dysfunctional. As increasingly sophisticated approaches to profile and characterise complex ecosystems are developed and the GMA moves towards clinical medicine (Gebrayel et al. 2022), evidence of a function for the microbiota in a variety of disorders affecting the intestines and other parts of the body has progressively accumulated.

7.6 Major Functions of Gut Microbiota

7.6.1 Metabolic Effects

Microbes that live in the digestive tract of mammals have the ability to degrade polysaccharides and starches, which are inaccessible to the host owing to the host's restricted ability to use them as sources of nutrition. A study (Flint et al. 2012) reviewed how changes in the metabolism of vitamins, short-chain fatty acids (SCFAs), lipids, atherogenic compounds, cholesterol, and gases by gut microbes can affect changes in the host's susceptibility to various age related metabolic diseases. The most common metabolites and microbial enzymes that affect the host have been identified, despite their scope and complexity, thanks to analytical advances. Important metabolites include bile acids, SCFAs, dopamine, tryptamine, amino acids, histamine, vitamins, serotonin, paracresol, and phenylacetylglutamine. Additionally, a variety of microbial metabolites can affect the activity of the hepatic and hormonal system in addition to the way they affect the host's metabolic pathways (Bull and Plummer 2014).

7.6.2 Host Protection and Anti-inflammatory or Immunomodulatory Effects

Numerous intestinal bacteria generate antimicrobial compounds, compete with each other for nutrients, and attach to the gut lining, inhibiting the colonisation of pathogens. This behaviour is known as the barrier or competitive-exclusion effect. Pathogenic bacteria can attach to the host cells of the gut wall and then enter the epithelial cells. In studies carried out in the lab, it was discovered that nonpathogenic bacteria compete with bacteria that are pathogenic for these attachment sites at the border of intestinal epithelial cells, avoiding the adherence and subsequent entry of infectious, invasive bacteria. The enteric microbiota may surpass pathogenic bacteria in a resource-contesting situation because bacteria are competing for nutrients in their immediate environment and sustain their collective habitat by sharing and consuming all resources (Ratajczak et al. 2019; Chi et al. 2021). Bacteriocins, which are abundant among gastrointestinal bacteria and have the ability to synthesise antimicrobial substances, are another method by which bacteria can stop their competitors from expanding (Ghosh et al. 2022a).

Exopolysaccharides, as well as other microbially derived macromolecules, also lessen gut inflammation. A number of microbial compounds, including SCFAs, have anti-inflammatory properties. The absence of taxa with anti-inflammatory activity, like *Faecalibacterium prausnitzii*, was a key finding in the initial findings of altered microbiomes in the gut in people with inflammatory bowel disease. Its anti-inflammatory properties have been associated with the peptides, bacterial quorum-sensing molecule *N*-acyl homoserine lactone, and salicylic acid, produced by this anaerobe (Ghosh et al. 2022b).

7.6.3 Neurological Effects

Gut bacteria may affect behaviour, cognition, and mood in accordance with the proposed pathways: the enteric nervous system, the absorption of metabolites produced by bacteria into the circulation of the blood, and the regulation of immunoinflammatory reactions in the central nervous system (Dinan and Cryan 2017; Barrio et al. 2022). As a result of the gut microflora's modulatory effect on the central nervous system, specific probiotics, also known as psychobiotics, have been developed to reduce stress (Dinan and Cryan 2017; Barrio et al. 2022). The potential benefits of modifying the microbiome in humans for neurological and mental health issues are still up for debate. However, cognitive impairment is very common even in the absence of serious underlying neurological disorders such as Alzheimer's (Jiang et al. 2017) or Parkinson's disease, so this research is pertinent to aging (Jiang et al. 2017).

7.6.4 The Gut Microbiome During Biological Aging

Aging is characterised by heterogeneity. As people age, they diverge from one another more and more. Even within the same body, different tissues deteriorate at varying rates. Aging can be studied in a variety of ways, including 'normal' aging (also called average or typical aging) (Tower 2015), pathological aging (also called accelerated aging brought on by particular diseases) (Khaltourina et al. 2020), and successful aging (Martin et al. 2015). Centenarians and other extremely long-lived individuals serve as prime examples of successful aging. Despite avoiding or surviving the majority of the illnesses that cause morbidity and mortality in the majority of other older adults, they may still exhibit some telltale signs of aging. From these old adults, we can learn the best aging processes. Inflammatory aging still occurs in nonagenarians and centenarians, but less so than in 'normal' older adults. According to a study, concurrent anti-inflammatory responses counteract their pro-inflammatory status (Franceschi et al. 2007). Thus, aging affects almost all the physiological functions in the body, including the GME. The GME exhibits variations over the course of a person's lifetime, and its composition has an impact on their level of health and susceptibility to disease.

The GME is highly dynamic immediately after birth, but it gradually stabilises as the immune system develops, liquids are weaned from the diet, and lifestyle factors are normalised in early adulthood (Badal et al. 2020). Even though the composition of the GME remains, if unaffected, relatively stable during early to mid-adulthood, as a person ages, it enters a period of distinct shifts occur in the diversity and functional ability. This is the case despite the fact that this composition remains relatively stable during early to mid-adulthood (Renson et al. 2020). It has been shown that the diversity as well as the composition of the GME change as a result of aging in many different species. These species include worms (such as Caenorhabditis elegans), flies (such as Drosophila), rats, dogs, primates, elephants, zebras, and humans, amongst others. This could be caused by a variety of biological and environmental factors, including immunosenescence, altered gastrointestinal tract physiology, increased exposure to medications, the onset of age-related diseases, and dietary changes associated with long-term care facilities (Renson et al. 2020). Mid-to-late adulthood is when age-related changes in the GME start to manifest in humans, though the precise timing of this change may vary depending on an individual's genetic, environmental, and lifestyle characteristics (Badal et al. 2020). In general, aging is linked to changes in the GMA's functional capacity, altered microbial diversity, increased bacterial community uniqueness, and a decline in core microbial genera. Importantly, there is a clear difference between studies that look at the GME in different older populations.

Variations in geography, diet, environment, and culture, of which can have an impact on the GME, as well as variations in the microbiome analysis methodology, may be to blame for this variation (Wilmanski et al. 2021). Due to these discrepancies in study results, it appears that it is currently impossible to determine the precise time when a person's GME transitions into an 'elderly' state. However, an aging-related drift in the GME was less common in individuals who were in poorer health, indicating that the change in the microbiome that takes place as we age may be advantageous for and indicative of host health. A 4-year follow-up study also discovered a connection between a higher risk of mortality and continuing to have a high prevalence of bacteroides or a low microbial distinctive quality (Wilmanski et al. 2021). This study suggests that certain bacterial genera may play a significant role in the microbiota's ability to function as a novel lifespan marker in elderly individuals.

The GME influences longevity in all species and is the cause of many age-related changes, according to mounting evidence. The host and the microbiome are both physiologically affected by aging, and host-microbiota interactions may have an impact on aging as a whole (Bana and Cabreiro 2019). The microbiome has a significant impact on how the immune system reacts, and dysregulation of this response can prolong inflammatory states (Honda and Littman 2016). Pro-inflammatory and anti-inflammatory activities are imbalanced as a result of age-related immune system decline (Rea et al. 2018). Age-related changes in pro-inflammatory status (also known as 'inflammaging') increase the likelihood of developing chronic diseases and disabilities, such as cardiovascular disease, metabolic disease, cancer, cognitive decline, frailty, and mortality (Das et al. 2021). Additionally, according to another study, the 'gut-brain axis' enables communication between the gut and the brain (Basak et al. 2022) as well as the modulation of behaviour, including higher-order cognitive functions (Socała et al. 2021; Agirman et al. 2021; Mayer et al. 2022). The microbiome offers an intriguing viewpoint for comprehending the physical and cognitive aspects of aging. Given the general decline in health, it is unclear what 'healthy aging' entails. The distinction between healthy and unhealthy aging is not well understood.

7.7 Extreme Aging vs Microbial Traces

Healthy aging, which increases longevity, and biological aging, which causes physiologic decline. While the GME signatures associated with increased mortality may be detrimental, it is possible that advantageous microbes particular to or more prevalent in healthy, long-living people could support the health of their host. Further, certain microbial signatures evolve as a person ages and are particularly obvious in centenarians (people who have lived for 100 years or more). Higher concentrations of several taxa, including Christensenellaceae and Akkermansia, which have previously been linked to health, were found in the microbiomes of the gut of centenarians in a well-studied Italian cohort, suggesting they could promote health maintenance as people age (Biagi et al. 2016; Palmas et al. 2022). Other studies conducted at the time discovered that the Lachnospiraceae, Bacteroidaceae, and Ruminococcaceae families suffered declines due to aging (Badal et al. 2020). Another South Korean study found that the faecal microbiome of centenarians had higher concentrations of Collinsella, Akkermansia, Christensenellaceae, and Clostridium and lower concentrations of Prevotella and Faecalibacterium (Juárez-Fernández et al. 2020).

This was connected to the GME's increased predicted capacity to contribute to the biosynthesis of different *N*-glycans (Koropatkin et al. 2012), glycosphingolipids, and the phosphatidylinositol signalling pathway, according to a study (Kim et al. 2019). Clinical data may highlight potential health-fostering or health-damaging characteristics of particular bacteria, which may inspire research into whether they may also support cognitive health during extreme aging (Badal et al. 2020; Ghosh et al. 2022b), despite the fact that many studies have reported conflicting results regarding the nature of the GME and aging. It is enticing to speculate that longevity-associated bacterial taxa may contribute to the aging host's environment by fostering longevity and, possibly, by shielding the host from environmental stresses. This would encourage the extreme limits of aging in human life. Early research on the African turquoise killifish (Nothobranchius furzeri) demonstrated that the GMA does indeed play a causal role in longevity. According to a few research studies (Smith et al. 2017; Hu and Brunet 2018), transferring the GMA of young fish to middle-aged fish resulted in an increase in both the fish's lifespan and their motor behaviour. This was accompanied by specific transcriptional alterations associated with the fish's intestinal immunity. In a mouse model of progeria, an accelerated form of aging in which the GME is similarly changed, progeroid animals who received the faecal microbiota from wild-type mice also displayed a noticeable increase in longevity and health span (Bárcena et al. 2019). Although whole faecal microbiota transplantation has successfully prolonged life in these animal models, it is not yet known whether it can do the same for humans. Faecal microbiota transplantation is an example of a clinical technique that requires a lot of work.

Further, a more successful approach to dealing with the aging microbiome may involve using specific probiotic microbes that have been shown to have health benefits as we age. According to a study (Bárcena et al. 2019), administration of the bacteria *Akkermansia muciniphila* alone was sufficient to have similar effects on extending lifespan in progeroid mice. In fact, a single bacterial strain might be the key to extending life (Grajeda-Iglesias et al. 2021; Cheng and Xie 2021). However, other studies have demonstrated that elevated *Akkermansia* has a protective effect in models of neurologic diseases (Cheng and Xie 2021) such as Parkinson's disease, multiple sclerosis, epilepsy, amyotrophic lateral sclerosis, and Alzheimer's disease. This suggests that elevated *Akkermansia* may have a beneficial effect on aging and other health disorders (Dao et al. 2016; Rodrigues et al. 2022). These studies did not evaluate whether those microbiota-targeted interventions might have an effect on brain health in advanced age; or, some bacteria might be more harmful than helpful. It is possible that unintended interactions among bacteria and their hosts speed up aging because studies on germ-free mice show that they live about 17% longer than mice that are not infected with a specific pathogen (Tazume et al. 1991). Eliminating these age-related potentially pathogenic microbes is thus another intervention strategy that may be used, perhaps through specific antibiotic use or CRISPR knockout of particular genes or pathogens. However, the effectiveness of these tactics is currently constrained by our current understanding of what causes the bacteria to behave pathologically, and thus further research studies are necessary.

7.8 Molecular Interplay Between Oxidative Stress and Gut Microbiome in Aging

Growing evidence to suggest that there is a molecular interplay between OS and the GME in the aging process (Guo et al. 2022). Several studies have shown that changes in the GME can lead to an increase in OS, while OS can also influence the composition and activity of the GME and ultimately influence the physiological aging process (Brunt et al. 2019; Mossad et al. 2022; Chen et al. 2022).

One of the most important ways in which the GME can cause OS is through the production of bacterial metabolites (Shandilya et al. 2022). These metabolites can have either beneficial or detrimental effects on the host, depending on the composition and concentration of the metabolite and it influences both normal healthy as well as unhealthy aging (aging due to various pathological disorders). One example of a bacterial metabolite that has been linked to OS is trimethylamine *N*-oxide (TMAO) (Ke et al. 2018; Brunt et al. 2020). TMAO is formed by bacteria in the stomach during the metabolic processes of choline and carnitine, both of which are found in high quantities in red meat and other animal products. It has been shown that those who have high levels of TMAO have an increased chance of developing cardiovascular disease as well as other age-related illnesses (Ke et al. 2018; Brunt et al. 2020). It has been demonstrated that TMAO increases OS through a variety of different pathways (Ke et al. 2018; Shanmugham et al. 2023). To begin, TMAO has the ability to activate NADPH oxidases, which are enzymes that are responsible for the production of ROS in cells. Second, TMAO has the potential to block the antioxidant enzymes SOD and catalase from performing their normal functions. Finally, TMAO has been shown to encourage the production of advanced glycation end products, often known as AGEs. AGEs are proteins and lipids that have been altered by OS and are linked to a number of disorders that are connected with aging (Chaudhuri et al. 2018; Rungratanawanich et al. 2021; Taguchi et al. 2021).

Another way in which the GME can influence OS and aging is through its effects on the immune system (Tan et al. 2018; Ghosh et al. 2022b). It has been demonstrated that the GME may affect immune function, and it has also been demonstrated that changes in the GME that occur with aging can lead to an immunological response in a dysregulated manner. Immune responses that are not well regulated might be a contributing factor in the development of chronic inflammation, which is a primary cause of OS. For instance, there is evidence to suggest that age-related changes in the GME (Mariat et al. 2009) can lead to an increase in bacteria that are pro-inflammatory, such as Firmicutes and Proteobacteria, and a reduction in bacteria that are anti-inflammatory, such as Bacteroidetes. This decrease in anti-inflammatory bacteria may contribute to the development of conditions such as inflammatory bowel disease (Mariat et al. 2009). These alterations in the GMA have been linked to an increase in circulating pro-inflammatory cytokines, such as interleukin-6 and tumour necrosis factor-alpha, both of which can contribute to OS and the aging process (Schirmer et al. 2016; Guo et al. 2021). It is also possible for the bacteria in the gut to engage with immune cells located in the gut-associated lymphoid tissue, which can lead to the activation of immunological responses that can contribute to inflammation and OS and impact age-related changes (Schirmer et al. 2016; Guo et al. 2021).

In addition to the impacts that the GME has on OS, there is evidence to suggest that OS can alter the composition and activity of the GME (Dumitrescu et al. 2018; Hu et al. 2020; Ni et al. 2022; Shandilya et al. 2022). These effects are in addition to the effects that the GME has on OS. For instance, it has been demonstrated that ROS have antibacterial properties, and it has also been demonstrated that high levels of ROS can contribute to dysbiosis in the GME (Ballard and Towarnicki 2020), which can lead to an increase in pathogenic bacteria and a reduction in helpful bacteria.

In addition, there is evidence to suggest that OS can boost the development of bacteria that are known to cause inflammation, such as proteobacteria (Rizzatti et al. 2017), while simultaneously limiting the growth of bacteria that are known to cause anti-inflammatory effects, such as bifidobacteria (Ballard and Towarnicki 2020). This imbalance in the GMA can contribute to chronic inflammation and further aggravate OS-induced aging.

Further, according to the findings of another study, a number of the bacteria that dwell in the colon are capable of converting sulphide into hydrogen sulphide (H2S). These bacteria include Fusobacterium, Escherichia, Salmonella, Klebsiella, and Streptococcus. Intestinal epithelial cells transform H2S into thiosulfate (S2O32−) as a means of ensuring their own continued existence. When there is inflammation in the intestinal tract, thiosulfate has the ability to undergo oxidation, which would result in the production of tetrathionate ions (S4O62−) and influence the hallmarks of aging like genomic stability to a greater extent and thus link GME, oxidation, and aging (Wilkie et al. 2021; Buret et al. 2022). Thus, to summarise, there are several mechanisms that exist between the GME (Gut dysbiosis) and OS that are interlinked and influence the process of healthy as well as unhealthy aging (Fig. 7.1).

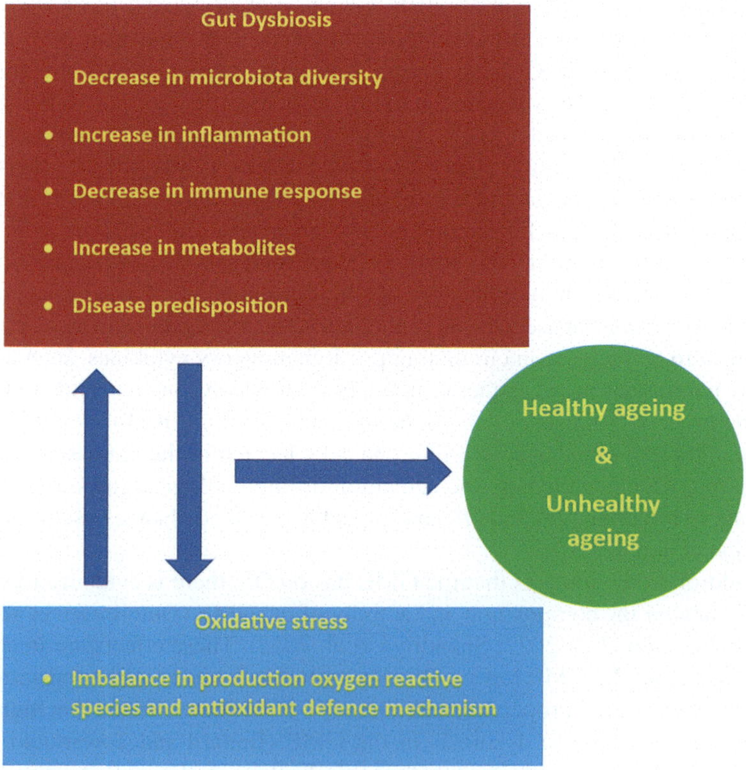

Fig. 7.1 Gut dysbiosis vs OS in the aging process

7.9 Conclusion

In conclusion, there is a complicated molecular interaction between the effects of OS and the microbiota in the gut that occurs during the aging process. Alterations that occur in the composition and activity of the GME as a result of aging can cause an increase in OS, while OS itself can influence the GME's composition and activity. It is likely that this interaction will go in both directions, with one aspect contributing to the other.

To further explain the processes that underlie this interaction, additional study is required, as is the identification of possible therapies that might control the GMA as well as OS in order to promote healthy aging. Dietary therapies, such as the use of prebiotics (e.g., dietary fibers) and probiotics strains that have shown promise in reducing oxidative stress or promoting the growth of beneficial gut bacteria, as well as pharmaceutical interventions, such as antioxidant therapy, may be included as a category of interventions. Overall, gaining a better understanding of the molecular interplay that occurs between OS and the GME in the aging process has the potential to lead to the development of new approaches that promote healthy aging and prevent diseases associated with advanced age. Specific molecular pathways by which OS and the GME interact are crucial. By investigating the potential of prebiotics (dietary fibres) and specific probiotic strains that have demonstrated effectiveness in reducing OS or enhancing the growth of beneficial gut bacteria, experts can create precise interventions to regulate both the GME and OS. Investigating the efficacy of antioxidant therapies or drugs that target specific pathways involved in the OS-GME interaction could hold promise for preventing or treating age-related diseases. Tailoring interventions based on individual gut microbiome profiles for optimal effectiveness and minimal side effects has the potential to usher in a new era of personalized medicine. Assessing the long-term safety and efficacy of these interventions remains vital. This improved understanding has the potential to significantly improve health and lifespan in the future.

References

Agirman G, Yu KB, Hsiao EY (2021) Signaling inflammation across the gut-brain axis. Science 374:1087–1092. https://doi.org/10.1126/science.abi6087

Ashok A, Andrabi SS, Mansoor S et al (2022) Antioxidant therapy in oxidative stress-induced neurodegenerative diseases: role of nanoparticle-based drug delivery systems in clinical translation. Antioxidants (Basel, Switzerland) 11:408. https://doi.org/10.3390/antiox11020408

Badal VD, Vaccariello ED, Murray ER et al (2020) The gut microbiome, aging, and longevity: a systematic review. Nutrients 12:3759. https://doi.org/10.3390/nu12123759

Ballard JWO, Towarnicki SG (2020) Mitochondria, the gut microbiome and ROS. Cell Signal 75:109737. https://doi.org/10.1016/j.cellsig.2020.109737

Bana B, Cabreiro F (2019) The microbiome and aging. Annu Rev Genet 53:239–261. https://doi.org/10.1146/annurev-genet-112618-043650

Banerjee A, Marotta F, Sriramulu S et al (2021) Beyond physical exercise: the role of nutrition, gut microbiota and nutraceutical supplementation in reducing age-related sarcopenia. Curr Aging Sci 14:94–104. https://doi.org/10.2174/1874609814666210203090458

Bárcena C, Valdés-Mas R, Mayoral P et al (2019) Healthspan and lifespan extension by fecal microbiota transplantation into progeroid mice. Nat Med 25:1234–1242. https://doi.org/10.1038/s41591-019-0504-5

Barko PC, McMichael MA, Swanson KS, Williams DA (2018) The gastrointestinal microbiome: a review. J Vet Intern Med 32:9–25. https://doi.org/10.1111/jvim.14875

Barrio C, Arias-Sánchez S, Martín-Monzón I (2022) The gut microbiota-brain axis, psychobiotics and its influence on brain and behaviour: a systematic review. Psychoneuroendocrinology 137:105640. https://doi.org/10.1016/j.psyneuen.2021.105640

Basak S, Das RK, Banerjee A et al (2022) Maternal obesity and gut microbiota are associated with fetal brain development. Nutrients 14:4515. https://doi.org/10.3390/nu14214515

Biagi E, Franceschi C, Rampelli S et al (2016) Gut microbiota and extreme longevity. Curr Biol 26:1480–1485. https://doi.org/10.1016/j.cub.2016.04.016

Brunt VE, Gioscia-Ryan RA, Richey JJ et al (2019) Suppression of the gut microbiome ameliorates age-related arterial dysfunction and oxidative stress in mice. J Physiol 597:2361–2378. https://doi.org/10.1113/JP277336

Brunt VE, Gioscia-Ryan RA, Casso AG et al (2020) Trimethylamine-N-oxide promotes age-related vascular oxidative stress and endothelial dysfunction in mice and healthy humans. Hypertension (Dallas, TX 1979) 76:101–112. https://doi.org/10.1161/HYPERTENSIONAHA.120.14759

Bull MJ, Plummer NT (2014) Part 1: the human gut microbiome in health and disease. Integr Med (Encinitas) 13:17–22

Buret AG, Allain T, Motta J-P, Wallace JL (2022) Effects of hydrogen sulfide on the microbiome: from toxicity to therapy. Antioxid Redox Signal 36:211–219. https://doi.org/10.1089/ars.2021.0004

Chaudhuri J, Bains Y, Guha S et al (2018) The role of advanced glycation end products in aging and metabolic diseases: bridging association and causality. Cell Metab 28:337–352. https://doi.org/10.1016/j.cmet.2018.08.014

Chen S-Y, Wang T-Y, Zhao C, Wang H-J (2022) Oxidative stress bridges the gut microbiota and the occurrence of frailty syndrome. World J Gastroenterol 28:5547–5556. https://doi.org/10.3748/wjg.v28.i38.5547

Cheng D, Xie MZ (2021) A review of a potential and promising probiotic candidate-Akkermansia muciniphila. J Appl Microbiol 130:1813–1822. https://doi.org/10.1111/jam.14911

Chi M, Ma K, Wang J et al (2021) The immunomodulatory effect of the gut microbiota in kidney disease. J Immunol Res 2021:5516035. https://doi.org/10.1155/2021/5516035

Coman V, Vodnar DC (2020) Gut microbiota and old age: modulating factors and interventions for healthy longevity. Exp Gerontol 141:111095. https://doi.org/10.1016/j.exger.2020.111095

Cryan JF, O'Riordan KJ, Cowan CSM et al (2019) The microbiota-gut-brain axis. Physiol Rev 99:1877–2013. https://doi.org/10.1152/physrev.00018.2018

Dao MC, Everard A, Aron-Wisnewsky J et al (2016) Akkermansia muciniphila and improved metabolic health during a dietary intervention in obesity: relationship with gut microbiome richness and ecology. Gut 65:426–436. https://doi.org/10.1136/gutjnl-2014-308778

Das A, Ganesan H, Sriramulu S et al (2021) A review on interplay between small RNAs and oxidative stress in cancer progression. Mol Cell Biochem 476:4117–4131. https://doi.org/10.1007/s11010-021-04228-9

Di Meo S, Reed TT, Venditti P, Victor VM (2016) Role of ROS and RNS sources in physiological and pathological conditions. Oxid Med Cell Longev 2016:1245049. https://doi.org/10.1155/2016/1245049

Dinan TG, Cryan JF (2017) The microbiome-gut-brain axis in health and disease. Gastroenterol Clin N Am 46:77–89. https://doi.org/10.1016/j.gtc.2016.09.007

Dodig S, Čepelak I, Pavić I (2019) Hallmarks of senescence and aging. Biochem Med 29:30501. https://doi.org/10.11613/BM.2019.030501

Ducarmon QR, Zwittink RD, Hornung BVH et al (2019) Gut microbiota and colonization resistance against bacterial enteric infection. Microbiol Mol Biol Rev 83:e00007–e00019. https://doi.org/10.1128/MMBR.00007-19

Dumitrescu L, Popescu-Olaru I, Cozma L et al (2018) Oxidative stress and the microbiota-gut-brain axis. Oxid Med Cell Longev 2018:2406594. https://doi.org/10.1155/2018/2406594

Flint HJ, Scott KP, Louis P, Duncan SH (2012) The role of the gut microbiota in nutrition and health. Nat Rev Gastroenterol Hepatol 9:577–589. https://doi.org/10.1038/nrgastro.2012.156

Franceschi C, Capri M, Monti D et al (2007) Inflammaging and anti-inflammaging: a systemic perspective on aging and longevity emerged from studies in humans. Mech Ageing Dev 128:92–105. https://doi.org/10.1016/j.mad.2006.11.016

Gebrayel P, Nicco C, Al Khodor S et al (2022) Microbiota medicine: towards clinical revolution. J Transl Med 20:111. https://doi.org/10.1186/s12967-022-03296-9

Ghosh TS, Shanahan F, O'Toole PW (2022a) Toward an improved definition of a healthy microbiome for healthy aging. Nat Aging 2:1054–1069. https://doi.org/10.1038/s43587-022-00306-9

Ghosh TS, Shanahan F, O'Toole PW (2022b) The gut microbiome as a modulator of healthy ageing. Nat Rev Gastroenterol Hepatol 19:565–584. https://doi.org/10.1038/s41575-022-00605-x

Gonzales MM, Krishnamurthy S, Garbarino V et al (2021) A geroscience motivated approach to treat Alzheimer's disease: senolytics move to clinical trials. Mech Ageing Dev 200:111589. https://doi.org/10.1016/j.mad.2021.111589

Grajeda-Iglesias C, Durand S, Daillère R et al (2021) Oral administration of Akkermansia muciniphila elevates systemic antiaging and anticancer metabolites. Aging (Albany NY) 13:6375–6405. https://doi.org/10.18632/aging.202739

Guo Y, Wang B, Wang T et al (2021) Biological characteristics of IL-6 and related intestinal diseases. Int J Biol Sci 17:204–219. https://doi.org/10.7150/ijbs.51362

Guo J, Huang X, Dou L et al (2022) Aging and aging-related diseases: from molecular mechanisms to interventions and treatments. Signal Transduct Target Ther 7:391. https://doi.org/10.1038/s41392-022-01251-0

Hajam YA, Rani R, Ganie SY et al (2022) Oxidative stress in human pathology and aging: molecular mechanisms and perspectives. Cells 11:552. https://doi.org/10.3390/cells11030552

Heintz-Buschart A, Wilmes P (2018) Human gut microbiome: function matters. Trends Microbiol 26:563–574. https://doi.org/10.1016/j.tim.2017.11.002

Honda K, Littman DR (2016) The microbiota in adaptive immune homeostasis and disease. Nature 535:75–84. https://doi.org/10.1038/nature18848

Hu C-K, Brunet A (2018) The African turquoise killifish: a research organism to study vertebrate aging and diapause. Aging Cell 17:e12757. https://doi.org/10.1111/acel.12757

Hu T, Dong Y, He C et al (2020) The gut microbiota and oxidative stress in autism spectrum disorders (ASD). Oxid Med Cell Longev 2020:8396708. https://doi.org/10.1155/2020/8396708

Jiang C, Li G, Huang P et al (2017) The gut microbiota and Alzheimer's disease. J Alzheimers Dis 58:1–15. https://doi.org/10.3233/JAD-161141

Juárez-Fernández M, Porras D, García-Mediavilla MV et al (2020) Aging, gut microbiota and metabolic diseases: management through physical exercise and nutritional interventions. Nutrients 13:16. https://doi.org/10.3390/nu13010016

Ke Y, Li D, Zhao M et al (2018) Gut flora-dependent metabolite Trimethylamine-N-oxide accelerates endothelial cell senescence and vascular aging through oxidative stress. Free Radic Biol Med 116:88–100. https://doi.org/10.1016/j.freeradbiomed.2018.01.007

Khaltourina D, Matveyev Y, Alekseev A et al (2020) Aging fits the disease criteria of the international classification of diseases. Mech Ageing Dev 189:111230. https://doi.org/10.1016/j.mad.2020.111230

Kim S-K, Guevarra RB, Kim Y-T et al (2019) Role of probiotics in human gut microbiome-associated diseases. J Microbiol Biotechnol 29:1335–1340. https://doi.org/10.4014/jmb.1906.06064

Koropatkin NM, Cameron EA, Martens EC (2012) How glycan metabolism shapes the human gut microbiota. Nat Rev Microbiol 10:323–335. https://doi.org/10.1038/nrmicro2746

López-Otín C, Blasco MA, Partridge L et al (2023) Hallmarks of aging: an expanding universe. Cell 186:243–278. https://doi.org/10.1016/j.cell.2022.11.001

Luo J, Mills K, le Cessie S et al (2020) Ageing, age-related diseases and oxidative stress: what to do next? Ageing Res Rev 57:100982. https://doi.org/10.1016/j.arr.2019.100982

Mariat D, Firmesse O, Levenez F et al (2009) The Firmicutes/Bacteroidetes ratio of the human microbiota changes with age. BMC Microbiol 9:123. https://doi.org/10.1186/1471-2180-9-123
Marquez-Exposito L, Tejedor-Santamaria L, Valentijn FA et al (2022) Oxidative stress and cellular senescence are involved in the aging kidney. Antioxidants (Basel, Switzerland) 11:301. https://doi.org/10.3390/antiox11020301
Martin P, Kelly N, Kahana B et al (2015) Defining successful aging: a tangible or elusive concept? Gerontologist 55:14–25. https://doi.org/10.1093/geront/gnu044
Martínez de Toda I, Ceprián N, Díaz-Del Cerro E, De la Fuente M (2021) The role of immune cells in oxi-inflamm-aging. Cells 10:2974. https://doi.org/10.3390/cells10112974
Mayer EA, Nance K, Chen S (2022) The gut-brain axis. Annu Rev Med 73:439–453. https://doi.org/10.1146/annurev-med-042320-014032
Moldogazieva NT, Mokhosoev IM, Mel'nikova TI et al (2019) Oxidative stress and advanced lipoxidation and glycation end products (ALEs and AGEs) in aging and age-related diseases. Oxid Med Cell Longev 2019:3085756. https://doi.org/10.1155/2019/3085756
Morris BJ, Willcox BJ, Donlon TA (2019) Genetic and epigenetic regulation of human aging and longevity. Biochim Biophys Acta Mol Basis Dis 1865:1718–1744. https://doi.org/10.1016/j.bbadis.2018.08.039
Mossad O, Batut B, Yilmaz B et al (2022) Gut microbiota drives age-related oxidative stress and mitochondrial damage in microglia via the metabolite N(6)-carboxymethyllysine. Nat Neurosci 25:295–305. https://doi.org/10.1038/s41593-022-01027-3
Ni Q, Zhang P, Li Q, Han Z (2022) Oxidative stress and gut microbiome in inflammatory skin diseases. Front Cell Dev Biol 10:849985. https://doi.org/10.3389/fcell.2022.849985
North BJ, Sinclair DA (2012) The intersection between aging and cardiovascular disease. Circ Res 110:1097–1108. https://doi.org/10.1161/CIRCRESAHA.111.246876
Palmas V, Pisanu S, Madau V et al (2022) Gut microbiota markers and dietary habits associated with extreme longevity in healthy Sardinian centenarians. Nutrients 14:2436. https://doi.org/10.3390/nu14122436
Parker HA, Forrester L, Kaldor CD et al (2021) Antimicrobial activity of neutrophils against mycobacteria. Front Immunol 12:782495. https://doi.org/10.3389/fimmu.2021.782495
Ratajczak W, Rył A, Mizerski A et al (2019) Immunomodulatory potential of gut microbiome-derived short-chain fatty acids (SCFAs). Acta Biochim Pol 66:1–12. https://doi.org/10.18388/abp.2018_2648
Rea IM, Gibson DS, McGilligan V et al (2018) Age and age-related diseases: role of inflammation triggers and cytokines. Front Immunol 9:586. https://doi.org/10.3389/fimmu.2018.00586
Renson A, Mullan Harris K, Dowd JB et al (2020) Early signs of gut microbiome aging: biomarkers of inflammation, metabolism, and macromolecular damage in young adulthood. J Gerontol A Biol Sci Med Sci 75:1258–1266. https://doi.org/10.1093/gerona/glaa122
Rizzatti G, Lopetuso LR, Gibiino G et al (2017) Proteobacteria: a common factor in human diseases. Biomed Res Int 2017:9351507. https://doi.org/10.1155/2017/9351507
Rodrigues VF, Elias-Oliveira J, Pereira ÍS et al (2022) Akkermansia muciniphila and gut immune system: a good friendship that attenuates inflammatory bowel disease, obesity, and diabetes. Front Immunol 13:934695. https://doi.org/10.3389/fimmu.2022.934695
Roger L, Tomas F, Gire V (2021) Mechanisms and regulation of cellular senescence. Int J Mol Sci 22:13173. https://doi.org/10.3390/ijms222313173
Rungratanawanich W, Qu Y, Wang X et al (2021) Advanced glycation end products (AGEs) and other adducts in aging-related diseases and alcohol-mediated tissue injury. Exp Mol Med 53:168–188. https://doi.org/10.1038/s12276-021-00561-7
Saint-Criq V, Lugo-Villarino G, Thomas M (2021) Dysbiosis, malnutrition and enhanced gut-lung axis contribute to age-related respiratory diseases. Ageing Res Rev 66:101235. https://doi.org/10.1016/j.arr.2020.101235
Schirmer M, Smeekens SP, Vlamakis H et al (2016) Linking the human gut microbiome to inflammatory cytokine production capacity. Cell 167:1125–1136.e8. https://doi.org/10.1016/j.cell.2016.10.020

Schmidt TSB, Raes J, Bork P (2018) The human gut microbiome: from association to modulation. Cell 172:1198–1215. https://doi.org/10.1016/j.cell.2018.02.044

Schuermann D, Mevissen M (2021) Manmade electromagnetic fields and oxidative stress-biological effects and consequences for health. Int J Mol Sci 22:3772. https://doi.org/10.3390/ijms22073772

Shandilya S, Kumar S, Kumar Jha N et al (2022) Interplay of gut microbiota and oxidative stress: perspective on neurodegeneration and neuroprotection. J Adv Res 38:223–244. https://doi.org/10.1016/j.jare.2021.09.005

Shanmugham M, Bellanger S, Leo CH (2023) Gut-derived metabolite, trimethylamine-N-oxide (TMAO) in cardio-metabolic diseases: detection, mechanism, and potential therapeutics. Pharmaceuticals (Basel) 16:504. https://doi.org/10.3390/ph16040504

Sidhu M, van der Poorten D (2017) The gut microbiome. Aust Fam Physician 46:206–211

Smith P, Willemsen D, Popkes M et al (2017) Regulation of life span by the gut microbiota in the short-lived African turquoise killifish. Elife 6:e27014. https://doi.org/10.7554/eLife.27014

Socała K, Doboszewska U, Szopa A et al (2021) The role of microbiota-gut-brain axis in neuropsychiatric and neurological disorders. Pharmacol Res 172:105840. https://doi.org/10.1016/j.phrs.2021.105840

Speer H, McKune AJ (2021) Aging under pressure: the roles of reactive oxygen and nitrogen species (RONS) production and aging skeletal muscle in endothelial function and hypertension-from biological processes to potential interventions. Antioxidants (Basel, Switzerland) 10:1247. https://doi.org/10.3390/antiox10081247

Tabibian JH, O'Hara SP, Splinter PL et al (2014) Cholangiocyte senescence by way of N-ras activation is a characteristic of primary sclerosing cholangitis. Hepatology 59:2263–2275. https://doi.org/10.1002/hep.26993

Taguchi K, Fukami K, Elias BC, Brooks CR (2021) Dysbiosis-related advanced glycation endproducts and trimethylamine N-oxide in chronic kidney disease. Toxins (Basel) 13:361. https://doi.org/10.3390/toxins13050361

Tan BL, Norhaizan ME, Liew W-P-P, Sulaiman Rahman H (2018) Antioxidant and oxidative stress: a mutual interplay in age-related diseases. Front Pharmacol 9:1162. https://doi.org/10.3389/fphar.2018.01162

Tazume S, Umehara K, Matsuzawa H et al (1991) Effects of germfree status and food restriction on longevity and growth of mice. Jikken Dobutsu 40:517–522. https://doi.org/10.1538/expanim1978.40.4_517

Thursby E, Juge N (2017) Introduction to the human gut microbiota. Biochem J 474:1823–1836. https://doi.org/10.1042/BCJ20160510

Tower J (2015) Programmed cell death in aging. Ageing Res Rev 23:90–100. https://doi.org/10.1016/j.arr.2015.04.002

Twarda-Clapa A, Olczak A, Białkowska AM, Koziołkiewicz M (2022) Advanced glycation endproducts (AGEs): formation, chemistry, classification, receptors, and diseases related to AGEs. Cells 11:1312. https://doi.org/10.3390/cells11081312

Wilkie SE, Borland G, Carter RN et al (2021) Hydrogen sulfide in ageing, longevity and disease. Biochem J 478:3485–3504. https://doi.org/10.1042/BCJ20210517

Wilmanski T, Diener C, Rappaport N et al (2021) Gut microbiome pattern reflects healthy ageing and predicts survival in humans. Nat Metab 3:274–286. https://doi.org/10.1038/s42255-021-00348-0

Wu R, Li S, Hudlikar R et al (2022) Redox signaling, mitochondrial metabolism, epigenetics and redox active phytochemicals. Free Radic Biol Med 179:328–336. https://doi.org/10.1016/j.freeradbiomed.2020.12.007

Zhang Y, Zhang S, Li B et al (2022) Gut microbiota dysbiosis promotes age-related atrial fibrillation by lipopolysaccharide and glucose-induced activation of NLRP3-inflammasome. Cardiovasc Res 118:785–797. https://doi.org/10.1093/cvr/cvab114

8 The Gut Microbiome, Microbial-Produced Pro-inflammatory Neurotoxins, and Neurological Disorders

Monisha Prasad, Ponnulakshmi Rajagopal,
and Selvaraj Jayaraman

Abstract

The gut microbiome is now recognized as a crucial determinant of human health and diseases, and it has been increasingly implicated in the pathogenesis of neurological disorders. This complex ecosystem of microorganisms produces a diverse range of metabolites, some of which can act as pro-inflammatory agents and potentially disrupt neuronal activity, leading to neurological dysfunction. This chapter focuses on the role of pro-inflammatory neurotoxins produced by gut microorganisms and their potential contribution to the development of neurological disorders. The chapter discusses the mechanisms by which these neurotoxins interact with the immune system and affect neurological function, and reviews the evidence linking dysbiosis of the gut microbiome to neurological disorders such as Alzheimer's disease, Parkinson's disease, and multiple sclerosis. Additionally, the chapter explores the therapeutic implications of targeting the gut microbiome and microbial-produced neurotoxins as a novel approach to treat neurological disorders. Overall, this chapter emphasizes the crucial role of the gut microbiome and microbial-produced neurotoxins in regulating neurological function and suggests that targeting these pathways may lead to the development of effective therapies for neurological disorders.

M. Prasad · S. Jayaraman (✉)
Department of Biochemistry, Centre of Molecular Medicine and Diagnostics (COMManD), Saveetha Dental College and Hospital, Saveetha Institute of Medical and Technical Sciences, Saveetha University, Chennai, India
e-mail: selvarajj.sdc@saveetha.com

P. Rajagopal
Department of Central Research Laboratory, Meenakhsi Ammal Dental College and Hospitals, Meenakhsi Academy of Higher Education and Research, Chennai, India

S. Pathak, A. Banerjee (eds.), *Gut Microbiome and Brain Ageing*,
https://doi.org/10.1007/978-981-99-8803-7_8

Keywords

Gut microbiome · Microbial-produced neurotoxins · Neurological disorders · Multiple sclerosis · Dysbiosis · Alzheimer's disease · Parkinson's disease · Immune system

8.1 Introduction

Microscopic living organisms, also known as microorganisms or microbes, such as fungi viruses, and, bacteria, are present in the human body in enormous numbers, primarily on and in the skin and intestines. Symbiotic bacteria residing in the gut provide a secure, nutrient-rich home to these microbes and offer many benefits to the host, such as making indigestible nutrients available to the body, producing energy, vitamins, and other essential chemicals, and forming a protective biofilm to restrict pathogenic microbes from entering the gut tissue (Qin et al. 2010; McVey Neufeld et al. 2016). The gut microbiome (GM), comprising 100 billion bacteria and 1000 different species, resides in the cecum, a pocket of the human large intestine, and possesses three million genes, which is 150 times more genetic diversity than the human body (Cho and Blaser 2012). The GM residing in the gastrointestinal tract (GIT) plays a critical role in the metabolism of drugs and nutrients, modulation of the immune system, pathogen defense, and genetic variations. By carrying out various physiological functions such as maintaining gut integrity, controlling host immune, protecting against pathogens, generating energy, and supplying nutrients like vitamins, the GM benefits human health in many ways (Donaldson et al. 2016). The connection between the foetus and vaginal microbiota during birth triggers a sequence of events that culminate in this symbiotic connection. The promotion of healthy growth and development in the human body is greatly aided by gut flora (Hills Jr et al. 2019).

Research on GM has developed a highly active part of study in current years, aided by next-generation sequencing techniques used in large-scale projects such as the Human Microbiome Project (HMP) and the Metagenomics of the Human Intestinal Tract (MetaHIT) project. Some types of cancer, metabolic disorders, inflammatory bowel diseases (IBD), and even disorders of the central nervous system (CNS) have all been linked to biomarkers found in the GM (Tirosh et al. 2016; Forslund et al. 2015). The gut is connected to the CNS through a variety of factors, including microbes, lymphocytes, neurotransmitters, immunological mediators, and microbial metabolites. Growing evidence indicates that the microbiome has a significant impact on immune and brain. However, under non-homeostatic conditions, such as stress, aging, disease, or diets rich in cholesterol, these microbiotas have the potential to transform into harmful "enterotoxigenic" microorganisms that produce potent pro-inflammatory neurotoxins (Carroll et al. 2011).

Bacteria have released molecular species that include some of the most potent neurotoxins known to cause inflammation. These neurotoxins can easily cross ageing or damaged plasma membranes and break cell-cell adhesion, which allows them to enter the brain, central nervous system (CNS), and peripheral nervous system (PNS). For instance, research has demonstrated that a variety of microbial lipoprotein glycoconjugates, such as Gram-negative bacteria-derived lipopolysaccharide (LPS), bacterial amyloids, and recently identified small non-coding RNA (sncRNA) microbial-derived neurotoxins, are present in the brain cells and CNS tissues of elderly Alzheimer's disease (AD) patients (Lukiw 2020). There is bidirectional contact between the brain and gut as a result of the microbiota's involvement in humoral, neurological, and immunological pathogenic pathways. It is crucial to understand that the CNS controls gut motility, secretion, mucosal immunity, and visceral nerves, affecting the intestinal environment (Collins et al. 2012; Neunlist et al. 2013).

Microorganisms that live in the human body, especially in the gut, are crucial for maintaining health because they perform a number of physiological functions. However, under certain conditions, such as stress, ageing, illness, or poor diets, these microbes can transform into harmful enterotoxigenic microorganisms that produce potent pro-inflammatory neurotoxins. These neurotoxins can pass through plasma membranes and reach the brain, where they can cause inflammation and contribute to the onset of a variety of mental illnesses, such as pain, depression, anxiety, autism, Alzheimer's disease, Parkinson's disease, and stroke. The importance of the gut microbiome in the treatment of brain diseases is highlighted in this book chapter by Zhu et al. (2020). Changes in the permeability of the blood-brain barrier, brain vascular physiology, and brain anatomy also contribute to the development of these disorders.

8.2 Gut Microbiome

The gut microbiome is a complex environmental factor composed of members from the three domains of life on Earth, namely, Bacteria, Archaea, and Eukarya (Finegold et al. 1983). Although the highest biomass is in the colon, distinct microbial communities are present in the more proximal regions of the gastrointestinal tract (Guarner and Malagelada 2003). Bacteria make up the majority of this ecosystem, with over 90% of the identified phylotypes belonging to this group. While bacteria have received the most research attention, there is increasing evidence that nonbacterial components such as viruses, fungi, Archaea, and microeukaryotes may also play a role in health and disease (Sonnenburg and Sonnenburg 2014). Gram-positive bacteria, including Clostridium, Eubacterium, Roseburia, and Ruminococcus, and the Gram-negative genera Bacteroides and Prevotella, which belong to the Bacteroidetes division, are among the most studied. Actinobacteria, such as Bifidobacterium, Collinsella, and Atopobium, which are high GC-content Gram-positive bacteria, are also important members of the gut microbial community (Turroni et al. 2008; van der Waaij et al. 2005). Additionally, other members of the human GM, although not dominant, are found in seven phyla—Spirochaetes, Fusobacteria, Cyanobacteria, TM7, Lentisphaerae, Verrucomicrobia, and Proteobacteria.

The process of colonization of the microbiota starts at birth and is influenced by various factors such as the delivery method, feeding, and antibiotic use, which can affect the composition of the infant GM (Penders et al. 2006). During the first few years of life, the microbiota rapidly diversifies and plays a vital role in the growth and development of the immune system, protecting against harmful microorganisms and bacterial growth (Milani et al. 2017; Roswall et al. 2021). This complex process is affected by various factors such as host genetics, physiology, environmental exposures, age, diet, and the composition and function of the microbiota. In adults, the GM plays an essential role in controlling intestinal and systemic hormonal activity, altering and disposing of toxins and drugs, regulating bone density, and enhancing intestinal barrier function (Haiser et al. 2013; Perler et al. 2023). The GM and host maintain a symbiotic relationship through a complex network of interactions that involve metabolic, immunological, and neuroendocrine system cross-talk. Microbial metabolites produced by the GM can function as signaling molecules that regulate the host's neuro-immune-inflammatory axis, potentially connecting the gut with other organ systems (Kho and Lal 2018).

8.3 Brain Role in Gut Microbiome

Despite being well protected by the blood-brain barrier (BBB), the immune-friendly tissue system known as the central nervous system (CNS) can still experience inflammatory reactions. According to evidence, the endocrine, neurological, metabolic, and immunological systems are connected with the brain and gut microbiomes. Neurotransmitters, hormones, and metabolites, which can either suppress or enhance inflammatory responses, are the primary means of communication between the CNS and enteric nervous system (ENS). The latter can cause tissue damage, altered synapse formation, and decreased CNS maintenance. This has been demonstrated by various studies, such as those conducted by Silva et al. (2020) and Park and Kim (2021). A brain microbiome, like the gut microbiome, would be made up of a collection of microbes that live on in the brain, though they may not always be actively procreating. In contrast, encephalopathies or brain abscesses are conditions in which the proliferation of (typically particular) bacteria are undeniably involved. A cerebral microbiome would undoubtedly contain orders of magnitude less bacteria than the gut and would need to be maintained by either a low degree of BBB breaching by microbes balanced by their active removal or the continued existence of mostly inactive microbes. The brain is most likely a "dead end" habitat for any bacteria that may exist there, even though there is evidence that the host's gut microbiome has co-evolved with it (Link 2021) (Fig. 8.1).

Both intrinsic and extrinsic factors play a role in the development and maturation of the central nervous system (CNS) in humans. In animals treated with broad-spectrum antibiotics or in germ-free (GF) animals, studies have demonstrated that specific microbiota can impact CNS physiology and neurochemistry. In the absence of associated microbiota, GF mice show neurological impairments in learning, memory, recognition, and emotional behaviors compared to control mice, with differences observed in key neurotransmitters such as serotonin (5-HT), *N*-methyl-D-aspartate (NMDA), and brain-derived neurotrophic factor (BDNF) (Smith 2015;

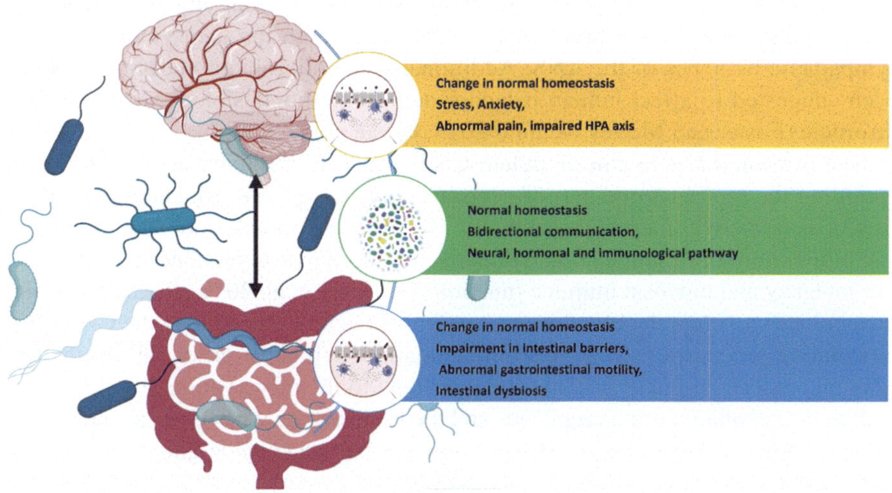

Fig. 8.1 Brain-gut microbiome axis. Neuronal, hormonal, and immunological mechanisms are involved in the two-way communication between the brain and the gut microbiome. Emotional emotions and other variables can have an impact on the composition and diversity of the gut microbiome, which in turn can have an impact on brain function and behavior

Foster et al. 2017). In humans, gastrointestinal disease has been linked to neuropsychiatric disorders such as anxiety, depression, and autism, indicating interaction between the two systems. GF mice have also been shown to exhibit structural changes in brain regions such as the hippocampus, amygdala, and prefrontal cortex, including increased volume, dendritic hypertrophy, hypermyelination, and altered gene expression (Hoban et al. 2016; Luczynski et al. 2016). The CNS is influenced by the GM, which in turn affects the immune, circulatory, and nervous systems. In addition, the GM has been linked to several other neurodevelopmental processes, such as myelination, neurogenesis, blood-brain barrier formation and stability, and microglial maturation and function (Ma et al. 2019).

8.4 Role of Microbiome in Gut Brain Axis (GBA)

The enteric nervous system (ENS), the autonomic nervous system (ANS), and the hypothalamic-pituitary-adrenal (HPA) axis are all connected via a communication pathway called the gut-brain axis. It implies that there is a physiological link between the brain and the gastrointestinal system that is essential for preserving emotional and cognitive functions and fostering gastrointestinal equilibrium (Rhee et al. 2009; Cryan and O'Mahony 2011). Recent research has focused on the GM's impact on the gut-brain axis, and the term "microbiota-gut-brain axis" has emerged, emphasizing the role of the microbiome in controlling communication between the gut and the brain. This connection has been linked to various psychiatric, neurological, and immunological disorders. Moreover, the GM's function and significance have broadened the scope of the gut-brain axis. The afferent and efferent neural impulses that link the gut-brain axis are driven by both the parasympathetic and sympathetic branches of the ANS. Additionally, the microbiota-gut-brain axis has been suggested to affect human behavior and the pathophysiology of psychiatric disorders (Foster and McVey Neufeld 2013; Cryan et al. 2020).

The gut microbiome can modulate CNS signaling through various pathways, including the biotransformation of neurotoxicants by gut bacteria, altered production of neuroactive microbial metabolites in response to environmental stressors, and bidirectional communication within the gut-brain axis to regulate intestinal barrier integrity and mucosal immune function (Dempsey et al. 2019). These microbial metabolites can enter the bloodstream and impact gene expression in the CNS, thus modulating processes such as neuroinflammation, cell survival, or cell death. The ENS and enteric immune system also contribute to the preservation of gut flora, as well as the regulation of macrophages and their interaction with mucosal and extrinsic nerve fibers (Gabanyi et al. 2016). Some gut bacteria can de-glucuronidate catecholamines, which may impact leukocytes in the gut (Asano et al. 2012; Yoo and Mazmanian 2017). The ENS also regulates gut permeability and integrity through receptors such as mechanoreceptors, intrinsic primary afferent neurons, and tension receptors, which respond to various stimuli. Moreover, substances produced by gut bacteria, including short-chain fatty acids, can activate receptors in neurons that control gastrointestinal motility (Cherbut et al. 1998).

The hypothalamic-pituitary-adrenal (HPA) axis controls the release of cortisol from the adrenal gland when the body is under stress. As it controls GI activities, communicates with the brain and other metabolic organs, and serves as a target of brain-to-gut communication, the gut microbiome is essential for bidirectional communication between the gut and other parts of the body (Carabotti et al. 2015; Fu and Cui 2017; Sharon et al. 2016; Skonieczna-Żydecka et al. 2018; Zhu et al. 2017).

8.5 Microbiota in Immune Homeostasis

The growth and operation of the mammalian immune system are significantly influenced by the microbiota in the gut. This mutually beneficial interaction between the immune system and the microbiota controls the host's homeostasis and encourages tissue survival and regeneration. T cells and B cells are important immune cells that support the integrity of the gut mucosa barrier and inhibit immune responses to innocuous antigens in order to preserve immunological homeostasis (Westerberg et al. 2008). However, dysbiosis, a GM imbalance, can cause both local and systemic T cell dysfunction, which can result in a number of immunological disorders. As a result, creating therapeutic options for treating inflammatory illnesses and enhancing cancer immunotherapy may be aided by understanding the processes that distinguish between pathogenic and homeostatic microbiota-host interactions (Honda and Littman 2016).

The innate and adaptive immune systems can be influenced by particular bacteria that live in particular niches in the gut, which can have systemic effects that are frequently far from the site of colonization. For instance, segmented filamentous bacteria (SFB) can lead to autoimmune arthritis in mice and trigger the development of T helper 17 (TH17) cells in the small intestine. The regulatory T (Treg) cells produced by some Bifidobacterium species can have systemic anti-inflammatory effects, and some Bifidobacterium species can boost the T-cell-dependent anti-tumor efficacy by blocking the PD-1 pathway (Kau et al. 2015; Sivan et al. 2015).

The passage of bacterial products through the birth canal and subsequent consumption of maternal milk expose the immune system to them for the first time. Given the lack of viable germs and the likelihood of placental sample contamination at birth, controversial research has suggested the possibility of in utero exposure to microbes. However, this is likely restricted to microbial metabolites or bacterial DNA. These early interactions are crucial for determining how the mucosal and systemic immune systems will function as adults. The development of immune tolerance in early life may be influenced by various factors such as the immature immune system and the regulatory environment during growth and development. Toll-like receptors (TLRs), which are innate immune receptors that recognize microbial patterns, play a crucial role in the differential recognition of commensal and pathogenic microbes. They are also essential for the proper development of the intestinal mucosal immune system, which is critical for recognizing important microbial pathogens that can affect infants, such as group B streptococcus, Listeria monocytogenes, and respiratory syncytial virus (RSV) (Roubalová et al. 2020) (Fig. 8.2).

TLR-mediated cytokine production by mononuclear cells in vitro reported lower levels of interferon (IFN)-α, IFN-γ, and interleukin (IL)-12 subunit p70 (IL-12-p70) expression in newborn cells compared to adult cells, but these levels rise between birth and 1–2 years of age (PrabhuDas et al. 2011; Theis et al. 2019). Antigens from the typical intestinal microbiota trigger TLR signal through preventing direct epithelial damage, protecting the epithelial barrier, and inhibiting inflammatory responses, suggesting a significant role in maintaining intestinal homeostasis. Moreover, inflammasomes are created when nucleotide oligomerization domain (Nod)-like receptors (NLRs) recognize various chemicals that are unique to particular microbes. These molecules can serve as sensors for the molecular patterns associated with microbes (MAMPs). NLR insufficiency is linked to behavior problems and changes in immune response (PrabhuDas et al. 2011; Pusceddu et al. 2019). The colonization of the GI tract by the early-life microbiota affects how T cell populations develop into different T helper cell (Th) subtypes, such as Th1, Th2, and Th17 or regulatory T cells (Tregs) (Ivanov et al. 2009). As an instance, germ-free (GF) mice have decreased Th1/Th2 ratios with a bias towards Th2 responses, absence of Th17 cells, and less Treg cell induction. By the bacterial product polysaccharide A-dependent route, Bacteroides fragilis colonization restored the establishment of the Th1-associated immune response. Furthermore, B. fragilis inhibited Th17 development by TLR-2 and increased Treg accumulation (Round et al. 2011; Fung et al. 2017). But still, under non-homeostatic circumstances, with stress, with ageing and disease, or with diets high in fat, these same bacteria have the potential of evolving into fatal "enterotoxigenic" microbes that create some of the most potent

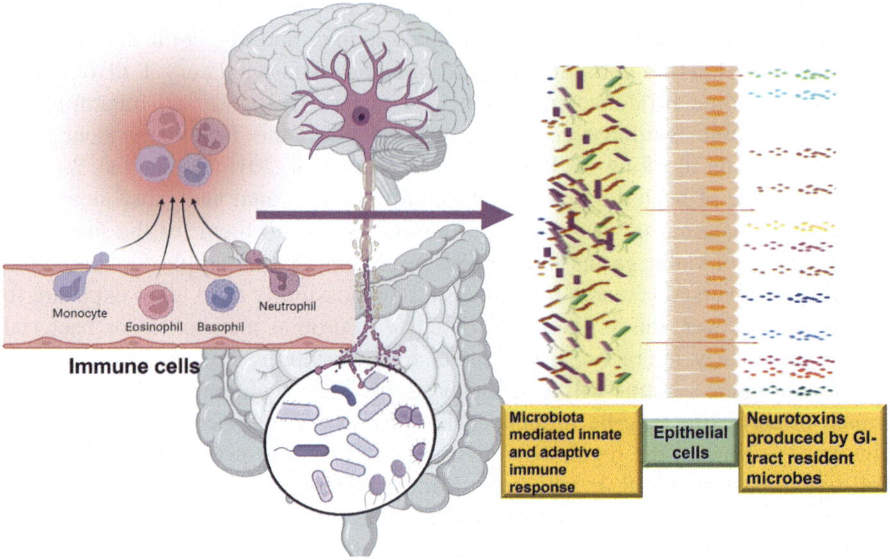

Fig. 8.2 Diagram of the intricate interplay between the immune system, GM, and nervous system in the Gut-Brain axis. The vast majority of the bacteria that make up the GM are found in the gut. Physical barriers that keep dangerous substances out of the bloodstream include the gut epithelium. Immune cells such as T cells, B cells, macrophages, and dendritic cells are all present in the gut-associated lymphoid tissue (GALT), which helps to keep the immune system in balance. The makeup and diversity of the microbial population are impacted by the GALT's interactions with the GM. Through a number of channels, including the vagus nerve, neurotransmitters, and immune cells, the gut and brain are able to communicate with one another. Dysbiosis, a disorder of the GM balance, can damage the gut-brain axis and cause inflammatory reactions, oxidative stress, and neuroinflammation

pro-inflammatory neurotoxins. It is now more well studied that the metabolic and neuro-immune systems can be affected by neurotoxins produced by GI-tract resident microbes in both health and disease (Foster et al. 2016; Allaband et al. 2019; Rutsch et al. 2020).

8.6 Gut Microbiome-Derived Neurotoxins

Many bacteria produce toxins, which are strong chemicals, to engage in interactions with their hosts. These microbial toxins damage host tissues and impair the immune system, which aids in the spread of disease and infection. For instance, gram-negative bacteria create lipopolysaccharides (LPS) or lipooligosaccharides (LOS), both of which are endotoxins. The botulinum toxin, which is mostly produced by Clostridium botulinum and less frequently by other Clostridium species, is the deadliest toxin known to man (Kessler and Benecke 1997). The targeted action of the clostridial neurotoxins on neuronal cells results in neurological diseases with distinct symptoms. It might also carry out a number of additional tasks and be able to distinguish between various cell types, including brain cells. The enteric nervous system is the only organ that these enterotoxins exclusively affect. For instance, they might excite afferent neurons or trigger the release of neurotransmitters from enterochromaffin cells, which either causes vomiting, exacerbates diarrhoea, or starts an inflammatory reaction in the colon. Moreover, some toxins have the potential to penetrate the blood-brain barrier and specifically target a particular neuron (Popoff and Poulain 2010). Many neurotoxins block the flow of nerve impulses. They inhibit or impair cholinergic receptor, acetylcholinesterase, or ion channel activity directly binding to them. They particularly bind to ion channels at the presynaptic level to perform their function. As a result, the permeability of the neuronal plasma membrane to specific ions is significantly altered. This effectively inhibits neuroexocytosis and inhibits the transmission of nerve signals (Montecucco and Schiavo 1995) (Table 8.1).

Neurotoxins are classified into two forms based on their mode of action: presynaptic and postsynaptic. Presynaptic neurotoxins stimulate the release of neuromodulators by affecting nerve ending plasmatic membranes and producing interterminal signals, resulting in neuromuscular blockade by preventing the release of acetylcholine. On the other hand, postsynaptic neurotoxins bind to the nicotinic acetylcholine receptor, leading to asphyxiation-related mortality. The gut-residing microbiota is known to produce various neurotoxins, such as BF-LPS and BFT fragilysin from Bacteroides fragilis, which can disrupt barriers and promote inflammation (Choi et al. 2016). The biggest family of gram-negative obligate bacteria in the human GI tract, the Bacteroidetes, release a complex mixture of amyloids, lipopolysaccharides (LPSs), enterotoxins, neurotoxins, and short non-coding RNAs (sncRNA) (Kim 2017). The structure and integrity of the GI tract and blood-brain barrier, as well as the homeostasis and equilibrium of the systemic, central nervous system (CNS), and peripheral nervous system (PNS), as well as the development of inflammatory degeneration within the human nervous system, are just a few examples of how these exudates from the microbiome can negatively impact microbiome-host interactions (Lukiw et al. 2021).

The toxin fragilysin produced by Bacteroides fragilis has been found to increase the permeability of the intestinal epithelium by rupturing the tight junctions between epithelial cells and inducing inflammatory signaling in human brain cells (Sears et al. 2014). When combined with exudates from other anaerobic microbes, these toxins can pass the blood-brain and GI tract barriers to trigger proinflammatory responses in the central nervous system. This might eventually give rise to NF-κB signaling, vascular permeability, immunogenicity, and reactive oxygen species generation. Further, these neuropathogenic signals can encourage the aggregation of amyloid and inflammation that are present in age-related neurologic disorders such as Alzheimer's disease (AD) and other conditions that have deficient amyloid beta 42 (Aβ42) peptide clearance pathways and increasing amyloidogenesis. The majority of localized cranial abscesses are also brought on by B. fragilis-derived toxins (Troletti et al. 2016). The GI tract's microbial sources of amyloid, LPS, and other endotoxins are a significant element of the human microbiome and can influence the load of amyloid in the CNS and systemically in each of these compartments. The GI

Table 8.1 Possible associations between neurological diseases, neurotoxins produced by GM

S. no	GM	Neurotoxin	Potential links to neurological diseases	Reference
1	Bacteroides fragilis	Polysaccharide A (PSA)	Improvements in ASD symptoms in mouse models, possible reduction in neuroinflammation	Hsiao et al. (2013); Hsiao (2018)
2	Escherichia coli	Lipopolysaccharides (LPS)	Increased inflammation and oxidative stress, potential role in AD and PD	Zhao et al. (2019); Kim et al. (2018)
3	Clostridium difficile	Toxin A and B	Possible links to cognitive impairment and dementia in elderly patients	Sadowsky (2016); De Wolfe et al. (2018)
4	Prevotella spp.	Proteolytic enzymes	Increased blood-brain barrier permeability, potential role in neurodegenerative diseases	Erny et al. (2015); Kelly et al. (2017)
5	Fusobacterium nucleatum	Lipopolysaccharides (LPS) and adhesins	Possible links to neuroinflammation and AD	Bhattacharjee and Lukiw (2013)
6	Enterobacteriaceae	Amyloid proteins	Possible role in AD pathogenesis	Jiang et al. (2017)
7	Clostridium botulinum	Botulinum toxin	Botulism	Hill et al. (2013)
8	Clostridium tetani	Tetanospasmin	Tetanus	Brin (2015)
9	Vibrio cholerae	Cholera toxin	Guillain-Barré syndrome	Kim et al. (2017)
10	Escherichia coli	Shiga toxin	Hemolytic uremic syndrome (HUS), encephalopathy	Garg et al. (2003)
11	Streptococcus pneumoniae	Pneumolysin	Meningitis, encephalitis, stroke	Yarandi et al. (2016)
12	Bacillus cereus	Cereulide	Cereulide-induced neurological syndrome	Zorzoli et al. (2017)
13	Staphylococcus aureus	Staphylococcal enterotoxin	Toxic shock syndrome, staphylococcal enteritis	Zinkevich and Beech (2018)

tract and blood-brain barriers grow more permeable with age, which may increase the contribution of harmful, proinflammatory chemicals from the GI tract microbiota (Lukiw 2016).

8.7 Gut Microbiome and Neurological Disorders

The GM is capable of detecting and regulate signaling in the central nervous system (CNS) using a number of different methods. The transformation of neurotoxins by intestinal bacteria, modifications in the synthesis of microbial metabolites brought on by environmental stressors, bidirectional communication between the gut and the brain system, and control of mucosal immune function are a few examples of these. Microbial metabolites can enter the bloodstream and alter how genes are expressed in the central nervous system (CNS), which can have an impact on processes like neuroinflammation, cell survival, and cell death (Dempsey et al. 2019). Consequently, a problem with the GM may have an adverse effect on the body as a whole and aid in the development or beginning of specific illnesses or ailments, particularly those that influence neurological function. Given that the microbiota can affect CNS function through various immunological pathways, including NF-κB, IFN-I, and inflammasome (Ma et al. 2019), there may be a link between the development of neurological disorders and the microbiota.

8.8 Alzheimer's Disease

Alzheimer's disease (AD) is a neurological condition that worsens with time and typically affects the elderly. It causes memory loss and cognitive decline. Memory loss and poor cognition are symptoms of the condition, which is characterized by the buildup of atypical amyloid beta (A) and tau protein (t-protein) in the interstitial space of the brain tissue (Hu et al. 2016). Language problems, confusion, mood swings, motivational loss, and difficulty managing self-care are some of the symptoms of AD that can get severe over time. According to research, microbial diseases such Chlamydia pneumonia, fungal infections, and spirochaetes may alter the species composition of the GM and raise levels of pro-inflammatory cytokines (Querfurth and LaFerla 2010; Fülöp et al. 2018). Bairamian et al. (2022) suggest that the presence of pro-inflammatory bacteria, such as Escherichia coli and Shigella, and anti-inflammatory bacteria, such as Escherichia coli, can dysregulate the microbiota, leading to neurodegeneration in individuals with cognitive impairment and brain amyloidosis. This dysbiosis can also result in imbalances in the GM and increased permeability of the gut and blood-brain barrier (BBB), which may impact the development of Alzheimer's disease (AD) and other neurological conditions. Additionally, the gut microbiome can produce a significant amount of LPSs and amyloids, which can disrupt signaling pathways and promote the release of pro-inflammatory cytokines that may contribute to the onset of AD (Jiang et al. 2017).

Studies have suggested that neuroinflammation and inhibition of Aβ clearance by activated microglia may contribute to the pathogenesis of AD. Downregulation of synapsin I and PSD-95 expression and upregulation of Aβ-peptide, Tau protein, COX-2, and CD11b deposition have been observed in transgenic mice models of AD (Cai et al. 2014; Sun et al. 2019). Alterations in the GM of AD patients have also been identified, with decreased abundance of butyrate-producing bacteria and increased levels of taxa associated with neurological disorders and inflammation. Furthermore, AD patients have been found to have dysregulated gut homeostasis and reduced anti-inflammatory *p*-glycoprotein in vitro, which is indicative of GM dysbiosis (Honarpisheh et al. 2020). These findings suggest that the gut microbiome may contribute to AD pathogenesis through dysbiosis-induced pro-inflammatory signaling, and that beneficial microbial metabolites or bacteria may have potential as AD treatments.

8.9 Multiple Sclerosis

Multiple sclerosis (MS) is a chronic inflammatory disease that harms the central nervous system (CNS) and is distinguished by demyelination, immune-mediated dysfunction, and damaged axons (Doshi and Chataway 2017). There are 2.3 million affected persons worldwide, with females being more likely to be impacted. The symptoms of MS might include impaired vision, motor dysfunction, fatigue, numbness, lack of coordination, vertigo, changes in sensation, discomfort, problems with the bladder and bowels, and even depression (Dendrou et al. 2015). Studies have suggested that the pathogenesis of Multiple Sclerosis (MS) involves a neuro-inflammatory response triggered by the formation of demyelinated plaques in the spinal cord and brain's grey or white matter, leading to neurodegeneration and oligodendrocyte demyelination. CD4+ T cells have been shown to play a crucial role in the development of MS, as demonstrated by experimental autoimmune encephalomyelitis (EAE) in a mouse model. The gut microbiome may also be implicated in the development of MS. For example, human gut colonization with Clostridium perfringens type B has been linked to microangiopathy that disrupts the blood-brain barrier and damages neurons and oligodendrocytes in MS patients (Lonchamp et al. 2010; Rumah et al. 2013). Several studies have found that MS patients have a higher prevalence of antibodies against epsilon toxin in their sera compared to controls. Additionally, MS patients' GM have been found to have higher levels of Methanobrevibacter archaea and lower levels of Butyricimonas, Lachnospiraceae, and Faecalibacterium. Faecalibacterium prausnitzii, which produces butyrate, has been shown to increase Treg cell numbers in the body. These findings suggest that alterations in GM may be linked to the onset of MS (Machiels et al. 2014).

Bacterial extracts from MS the patients GM have been shown in in vitro investigations to stimulate proinflammatory T-cell responses. Additionally, gut bacteria from a mouse model of autoimmune encephalomyelitis were transplanted into MS mice, worsening the condition, and demonstrating the role of the GM in the pathogenesis of MS. Studies have shown that MS patients have higher levels of antibodies generated in response to microorganism poisons and changes in the population of specific gut bacteria. The ability to reverse microbial alterations and have anti-inflammatory effects have been proven with the administration of a multi-species probiotic including Lactobacillus, Bifidobacterium, and Streptococcus species twice daily for 2 months (Hindson 2017). These findings highlight the critical part that the GM plays in the pathogenesis of MS, and efforts are being made to alter the

GM as a therapeutic strategy to reduce MS symptoms and relapse risk and perhaps even cure the condition.

8.10 Parkinson's Disease

Parkinson's disease is a neurological condition that occurs due to the loss of dopamine-producing neurons in the midbrain's substantia nigra, leading to a deficiency of dopamine. This condition affects around 7–10 million people globally and is more common in men than women. One of the defining features of Parkinson's disease is the deposition of α-synuclein, a protein that can travel from the gut to the brain via the vagus nerve. Symptoms of Parkinson's disease include difficulty with movement, tremors, muscle stiffness, and problems with walking and balance (Holmqvist et al. 2014). Additionally, patients may also experience cognitive impairment, depression, and autonomic dysfunction. In addition to dopamine loss, Parkinson's disease is characterized by the buildup of insoluble polymers of α-synuclein in the neurons, which form round, eosinophilic, cytoplasmic inclusions called Lewy bodies. The Lewy bodies, which are insoluble aggregates of α-synuclein protein, contribute to neurodegeneration and the death of neurons in Parkinson's disease (Xu and Pu 2016). There have been several studies that indicate changes in the GM of individuals with PD, which can serve as both biomarkers of PD and a possible trigger for the misfolding of α-synuclein, leading to neurodegeneration in these patients. Patients with PD often experience gastrointestinal dysmotility, including delayed gastric emptying and constipation, which are non-motor symptoms that are frequently observed in this condition (Hardoff et al. 2001; Noyce et al. 2012).

Parkinson's disease patients who also have constipation due to neurodegeneration of the enteric nervous system have been found to have significant changes in their GM composition. Studies have shown a decrease in Prevotellaceae species and an increase in Enterobacteriaceae in their stool samples compared to healthy individuals. Animal studies have shown that short-chain fatty acids (SCFAs) can contribute to neuroinflammation in the PD model, and fecal microbiota transplantation from PD patients to mice resulted in motor impairments and neuroinflammation. Antibiotic treatment has been found to improve behavioral symptoms in PD patients, and microbial tyrosine decarboxylases in the gut have been shown to decrease the levels of levodopa in the blood, a medication commonly used to manage PD symptoms (Scheperjans et al. 2015). High-throughput 16S rRNA sequencing of bacterial genomes has revealed that individuals with Parkinson's disease display lower levels of various bacterial strains, including Prevotellacopri, Bacteroides dorei, Ruminococcus callidus, and Christensenella minuta. These changes in GM composition can lead to inflammation in the gut, which can result in the accumulation of α-synuclein and the formation of Lewy bodies. Additionally, Parkinson's disease patients with intestinal inflammation exhibit increased expression of the bacterial endotoxin-specific ligand TLR4, CD3+ T cells, and cytokine expression in colonic biopsies and lower levels of SCFA-producing colonic bacteria compared to controls. These findings suggest that alterations in the GM play a crucial role in the pathogenesis and treatment of Parkinson's disease (Kang et al. 2021).

8.11 Rheumatoid Arthritis

Rheumatoid arthritis (RA) is an autoimmune disease that mainly affects the joints and is characterized by synovial inflammation and cartilage degeneration. The exact pathophysiology of this debilitating disease is still unknown. Environmental, microbial, and genetic factors, including HLADRB1, have been implicated in the development of RA. Prevotella copri has been found in higher concentrations in individuals at high risk for RA as well as in newly diagnosed, untreated RA patients. Another study found a significant association between three uncommon taxa (Eggerthella, Collinsella, and Faecalibacterium) and RA, with Collinsella being linked to the production of the proinflammatory cytokine IL-17A (Chen et al. 2016). RA patients in a Chinese cohort had an overabundance of Lactobacillus salivarius and reduced levels of Haemophilus spp. in their intestinal, dental, and saliva samples. Metabolites produced by the microbiome interact with several immunological pathways linked to RA, particularly short-chain fatty acids (SCFAs). IL1rn-/- mice must have microbial ligands activate TLR2 and TLR4 for them to develop T cell-mediated autoimmune arthritis. Intestinal lymphocytes respond with IL17 to dysbiotic microbiota from IL1rn-/- mice (Wang and Xu 2019). Additionally, genetically susceptible mice infected with RA patient dysbiotic microbiota have an enhanced Th17 response. Similarly, SFB injection in GF mice is sufficient to activate Th17 and cause inflammatory arthritis. Porphyromonas gingivalis, a periodontal pathobiont, as well as enteric bacteria, can trigger a TLR2- and IL-1-mediated Th17 response and exacerbate autoimmune arthritis (Zheng et al. 2020). Future research is needed to determine the impact of RA treatment on the microbiome and the relationships between microbiome changes and potential modulation of human RA.

8.12 Autism Spectrum Disorder (ASD)

Autism Spectrum Disorder (ASD) is a multifaceted neurodevelopmental disorder characterized by persistent impairments in social and communication skills, along with repetitive and restrictive behaviors. The development of ASD in children has been linked to various environmental and genetic factors, including the gut microbiome (Hallmayer et al. 2011). Recent studies have highlighted the potential impact of interactions between the gut microbiome and the brain on autism (Saurman et al. 2020). Additionally, it has been observed that about 40% of individuals with ASD experience gastrointestinal issues. Evidence suggests that the microbiome, which colonizes the gut soon after birth, can influence mood and behavior from childhood to adulthood. Therefore, any inflammation or impairment during this developmental process can lead to cognitive difficulties, changes in mood and memory, and atypical behavior (Clapp et al. 2017).

Autism Spectrum Disorder (ASD) is primarily a neurodevelopmental disorder, but there is increasing evidence to suggest that dysbiosis plays an active role in its development. Many individuals with ASD experience gastrointestinal problems, and research has shown that they have altered levels of certain gut bacteria, with decreased levels of Firmicutes, Fusobacteria, and Verrucomicrobia, but increased levels of Bacteroidetes. Additionally, specific gut microbial species, such as segmented filamentous bacteria (SFB), Bifidobacterium adolescentis, and certain E. coli isolates during pregnancy, have been linked to an increased risk of ASD in offspring. Intestinal CD11c+ DCs, which recognize microbial TLR3 ligands and produce IL-1β, IL-23, and IL-6, can induce pathogenic Th17 responses. Fecal microbiota transplant (FMT) studies have further supported the role of the GM in the onset of ASD (De Angelis et al. 2013; Kim et al. 2017). Therefore, it appears that an imbalance in the GM may contribute to the development of neurological disorders such as ASD.

8.13 Anxiety

A substantial fraction of the world's population is affected by anxiety disorder, which has a prevalence rate of between 3 and 25% and manifests clinically as anxious symptoms. Anxiety can occur as frequently as 1.4–70% among people with chronic diseases such cancer, cardiocerebrovascular disease, and irritable bowel syndrome. Apprehension and fear are common emotions in people with anxiety, and the outcomes are frequently accompanied by unpleasant physical symptoms such as sweating and palpitations. Up to 33.7% of people may at some point in their lives experience symptoms of anxiety. Depending on the person, the frequency, intensity, and number of symptoms of anxiety might vary. These symptoms include muscle tension, restlessness, weariness, and difficulties concentrating (Rynn and Brawman-Mintzer 2004). An imbalance in the intestinal microbiota has been related to anxiety, and studies have revealed that the GM can affect brain function through the gut-brain axis. Recent studies have shown that the composition and operation of the gut microbiome differ between people with social anxiety disorder (SAD) and healthy controls. SAD is a common psychiatric disease. For instance, in a study that used whole-genome shotgun sequencing to examine 49 faecal samples (31 cases and 18 age- and sex-matched controls), changes in the composition and functionality of the gut microbiome were found between SAD patients and healthy controls (Bercik et al. 2010). While Parasutterella was more prevalent in healthy controls, Anaeromassillibacillus and Gordonibacter genera were more prevalent in SAD. Additionally, it was discovered that Parasutterella excrementihominis was more common in controls than in SAD patients at the species level, but Anaeromassilibacillus sp. An250 was more common in SAD patients (Butler et al. 2023). This work shows that the makeup and function of the GM vary between SAD patients and healthy controls, indicating that the gut-brain axis may be a viable treatment target and biomarker for this chronic, early-onset disease.

8.14 Others Diseases

Emerging evidence suggests that the GM can significantly influence the development of various neurological disorders. For instance, Huntington's disease (HD), a progressive neurological condition, has been linked to gut dysbiosis in animal models, with significant differences in microbial composition, particularly an increase in Bacteroidetes and a decrease in Firmicutes. Furthermore, gut dysbiosis has been associated with motor deficits and altered white matter plasticity in HD mice, indicating the potential involvement of the gut-brain axis in the disease's pathogenesis. Similarly, a case study of a patient with Aicardi-Goutières syndrome (AGS), a rare autoimmune disease, revealed GM dysbiosis, with lower amounts of beneficial bacteria and higher levels of pathogenic bacteria, suggesting the involvement of the GM in AGS development (Huo et al. 2017). Another rare neurological disorder, Rett syndrome, has also been associated with GM dysbiosis, with reduced microbial diversity and altered microbial composition in patients. Dysbiosis in Rett syndrome has been linked to increased levels of potentially pathogenic bacteria and reduced levels of beneficial bacteria that produce short-chain fatty acids, which regulate brain function. In addition, GM dysbiosis has been linked to other neurological disorders, such as Alzheimer's disease, Parkinson's disease, and multiple sclerosis. These findings highlight the need for further research to investigate the role of the gut-brain axis in neurological disorders and potential therapeutic interventions (Wasser et al. 2020; Lim et al. 2016; de Theije et al. 2011; Kong et al. 2020) (Fig. 8.3).

8.15 Conclusion

In conclusion, there is growing evidence that neurological disorders are significantly influenced by the gut microbiome. Increased pro-inflammatory neurotoxic production and GM dysbiosis may be factors in the onset and progression of neurological diseases such as Alzheimer's disease, Parkinson's disease, and multiple sclerosis. The gut-brain axis facilitates communication between the gut and the central nervous system, and imbalances in the gut microbiome may contribute to alterations in this communication, potentially influencing the risk of neurological diseases. To completely comprehend the intricate mechanisms behind the gut-brain axis and the way they contribute to the aetiology of neurological diseases, more investigation is necessary. Future research may shed light on cutting-edge treatment approaches that prevent or treat neurological diseases by focusing on the gut microbiome.

8.16 Future Perspective

Research in the fields of the gut microbiome, pro-inflammatory neurotoxins produced by microbes, and neurological conditions holds great promise for the future. Unraveling the specific contributions of microbial strains and their metabolites to

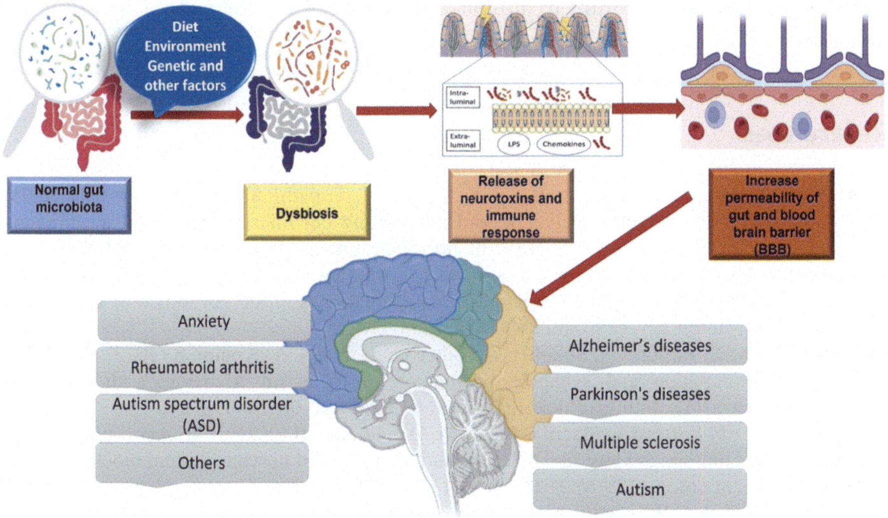

Fig. 8.3 The role of the gut flora in the development of neurotoxins and neurological diseases. The gut microbiome plays an important role for maintaining human health, and dysregulation of the gut microbiome can cause the onset of a number of neurological disorders

the development and progression of neurological conditions (e.g., Alzheimer's disease, Parkinson's disease) has the potential to revolutionize treatment strategies. These include identifying specific microbial strains and metabolic byproducts that contribute to the onset and progression of neurological diseases, which could help develop targeted treatments to restore gut microbiome balance and reduce the risk of neurological diseases. Additionally, dietary interventions and probiotics have shown promise in altering the GM and potentially preventing or treating neurological disorders. To determine the most effective dietary and probiotic therapies for these conditions, more research is necessary. Understanding the function of the gut-brain axis in neurological disorders is also crucial since a disruption of this communication link between the gut microbiome and the central nervous system can lead to the development of neurological diseases. The future prospects of research on the gut microbiome, pro-inflammatory neurotoxins produced by microbes, and their relationship with neurological disorders appear promising. To better comprehend and manage neurological diseases, further investigation is needed to identify specific microbial strains and metabolic byproducts that play a role in their onset and progression. Developing targeted therapies that aim to restore the balance of the gut microbiome and eliminate neurotoxins could be facilitated through this understanding. Research also indicates that altering the GM through dietary interventions and probiotics may be a promising preventive and treatment approach for neurological disorders, and more studies are needed to identify the most effective therapies. Additionally, research is being conducted on the potential use of faecal microbiota transplantation (FMT) as a therapy for neurological diseases, and innovative medicines targeting the gut microbiome are being developed. Overall, these findings offer hope for future treatments that may reduce the risk of neurological illnesses.

References

Allaband C, McDonald D, Vázquez-Baeza Y, Minich JJ, Tripathi A, Brenner DA et al (2019) Microbiome 101: studying, analyzing, and interpreting gut microbiome data for clinicians. Clin Gastroenterol Hepatol 17(2):218–230. https://doi.org/10.1016/j.cgh.2018.09.017

Asano Y, Hiramoto T, Nishino R, Aiba Y, Kimura T, Yoshihara K et al (2012) Critical role of GM in the production of biologically active, free catecholamines in the gut lumen of mice. Am J Physiol Gastrointest Liver Physiol 303(11):G1288–G1295

Bairamian D, Sha S, Rolhion N, Sokol H, Dorothée G, Lemere CA, Krantic S (2022) Microbiota in neuroinflammation and synaptic dysfunction: a focus on Alzheimer's disease. Mol Neurodegener 17(1):19. https://doi.org/10.1186/s13024-022-00524-3

Bercik P, Verdu EF, Foster JA, Macri J, Potter M, Huang X et al (2010) Chronic gastrointestinal inflammation induces anxiety-like behavior and alters central nervous system biochemistry in mice. Gastroenterology 139(6):2102–2112.e1

Bhattacharjee S, Lukiw WJ (2013) Alzheimer's disease and the microbiome. Front Cell Neurosci 7:153. https://doi.org/10.3389/fncel.2013.00153

Brin MF (2015) Botulinum toxin: chemistry, pharmacology, toxicity, and immunology. Muscle Nerve 52(1):6–19. https://doi.org/10.1002/mus.24643

Butler MI, Bastiaanssen TFS, Long-Smith C, Lucey S, Cryan JF, Dinan TG (2023) The gut microbiome in social anxiety disorder: evidence of altered composition and function. Transl Psychiatry 13(1):95

Cai Z, Hussain MD, Yan LJ (2014) Microglia, neuroinflammation, and beta-amyloid protein in Alzheimer's disease. Int J Neurosci 124:307–321. https://doi.org/10.3109/00207454.2013.833510

Carabotti M, Scirocco A, Maselli MA, Severi C (2015) The gut-brain axis: interactions between enteric microbiota, central and enteric nervous systems. Ann Gastroenterol 28(2):203–209

Carroll IM, Ringel-Kulka T, Keku TO, Chang YH, Packey CD, Sartor RB et al (2011) Molecular analysis of the luminal- and mucosal-associated intestinal microbiota in diarrhea-predominant irritable bowel syndrome. Am J Physiol Gastrointest Liver Physiol 301(5):G799–G807. https://doi.org/10.1152/ajpgi.00154.2011

Chen J, Wright K, Davis JM, Jeraldo P, Marietta EV, Murray J et al (2016) An expansion of rare lineage intestinal microbes characterizes rheumatoid arthritis. Genome Med 8(1):43. https://doi.org/10.1186/s13073-016-0299-7

Cherbut C, Ferrier L, Roze C, Anini Y, Blottiere H, Lecannu G et al (1998) Short-chain fatty acids modify colonic motility through nerves and polypeptide YY release in the rat. Am J Physiol 275(6):G1415–G1422

Cho I, Blaser MJ (2012) The human microbiome: at the interface of health and disease. Nat Rev Genet 13(4):260–270. https://doi.org/10.1038/nrg3182

Choi VM, Herrou J, Hecht AL, Teoh WP, Turner JR, Crosson S et al (2016) Activation of Bacteroides fragilis toxin by a novel bacterial protease contributes to anaerobic sepsis in mice. Nat Med 22:563–567

Clapp M, Aurora N, Herrera L, Bhatia M, Wilen E, Wakefield S (2017) GM's effect on mental health: the gut-brain axis. Clin Pract 7(4):987

Collins SM, Surette M, Bercik P (2012) The interplay between the intestinal microbiota and the brain. Nat Rev Microbiol 10(11):735–742. https://doi.org/10.1038/nrmicro2876

Cryan JF, O'Mahony SM (2011) The microbiome-gut-brain axis: from bowel to behavior. Neurogastroenterol Motil 23(3):187–192. https://doi.org/10.1111/j.1365-2982.2010.01664.x

Cryan JF, O'Riordan KJ, Sandhu K, Peterson V, Dinan TG (2020) The gut microbiome in neurological disorders. Lancet Neurol 19(2):179–194. https://doi.org/10.1016/S1474-4422(19)30356-4
De Angelis M, Piccolo M, Vannini L, Siragusa S, De Giacomo A, Serrazzanetti DI et al (2013) Fecal microbiota and metabolome of children with autism and pervasive developmental disorder not otherwise specified. PLoS One 8(10):e76993
de Theije CG, Wu J, da Silva SL, Kamphuis PJ, Garssen J, Kraneveld AD (2011) Pathways underlying the gut-to-brain connection in autism spectrum disorders as future targets for disease management. Eur J Pharmacol 668:S70–S80
De Wolfe TJ, Eggers JE, Barker AK, Kates AE (2018) Clostridium difficile toxins A and B are cation-dependent UDP-glucose hydrolases with differing catalytic activities. Biochemistry 57(17):2519–2531. https://doi.org/10.1021/acs.biochem.8b00137
Dempsey JL, Little M, Cui JY (2019) Gut microbiome: an intermediary to neurotoxicity. Neurotoxicology 75:41–69. https://doi.org/10.1016/j.neuro.2019.08.005
Dendrou CA, Fugger L, Friese MA (2015) Immunopathology of multiple sclerosis. Nat Rev Immunol 15:545–558
Donaldson GP, Lee SM, Mazmanian SK (2016) Gut biogeography of the bacterial microbiota. Nat Rev Microbiol 14(1):20–32. https://doi.org/10.1038/nrmicro3552
Doshi A, Chataway J (2017) Multiple sclerosis, a treatable disease. Clin Med 17:530–536
Erny D, Hrabe de Angelis AL, Jaitin D, Wieghofer P, Staszewski O, David E et al (2015) Host microbiota constantly control maturation and function of microglia in the CNS. Nat Neurosci 18(7):965–977. https://doi.org/10.1038/nn.4030
Finegold SM, Sutter VL, Mathiesen GE (1983) Microflora composition and development. In: Hentges DJ (ed) Human intestinal microflora in health and disease. Academic, New York, pp 3–119
Forslund K, Hildebrand F, Nielsen T, Falony G, Le Chatelier E, Sunagawa S et al (2015) Disentangling type 2 diabetes and metformin treatment signatures in the human GM. Nature 528(7581):262–266. https://doi.org/10.1038/nature15766
Foster J, McVey Neufeld K (2013) Gut–brain axis: how the microbiome influences anxiety and depression. Trends Neurosci 36(5):305–312
Foster JA, Lyte M, Meyer E, Cryan JF (2016) GM and brain function: an evolving field in neuroscience. Int J Neuropsychopharmacol 19(5):pyv114. https://doi.org/10.1093/ijnp/pyv114
Foster JA, Rinaman L, Cryan JF (2017) Stress & the gut-brain axis: regulation by the microbiome. Neurobiol Stress 7:124–136. https://doi.org/10.1016/j.ynstr.2017.03.001
Fu ZD, Cui JY (2017) Remote sensing between liver and intestine: importance of microbial metabolites. Curr Pharmacol Rep 3(3):101–113
Fülöp T, Itzhaki RF, Balin BJ, Miklossy J, Barron AE (2018) Role of microbes in the development of Alzheimer's disease: state of the art—an international symposium presented at the 2017 IAGG congress in San Francisco. Front Genet 9:362. https://doi.org/10.3389/fgene.2018.00362
Fung TC, Olson CA, Hsiao EY (2017) Interactions between the microbiota, immune and nervous systems in health and disease. Nat Neurosci 20(2):145–155. https://doi.org/10.1038/nn.4476
Gabanyi I, Muller PA, Feighery L, Oliveira TY, Costa-Pinto FA, Mucida D (2016) Neuro-immune interactions drive tissue programming in intestinal macrophages. Cell 164(3):378–391
Garg AX, Suri RS, Barrowman N, Rehman F, Matsell D, Rosas-Arellano MP (2003) Long-term renal prognosis of diarrhea-associated hemolytic uremic syndrome: a systematic review, meta-analysis, and meta-regression. JAMA 290(10):1360–1370. https://doi.org/10.1001/jama.290.10.1360
Guarner F, Malagelada JR (2003) Gut flora in health and disease. Lancet (London, England) 361(9356):512–519. https://doi.org/10.1016/S0140-6736(03)12489-0
Haiser HJ, Gootenberg DB, Chatman K, Sirasani G, Balskus EP, Turnbaugh PJ (2013) Predicting and manipulating cardiac drug inactivation by the human gut bacterium Eggerthella lenta. Science (New York, N.Y.) 341(6143):295–298. https://doi.org/10.1126/science.1235872
Hallmayer J, Cleveland S, Torres A, Phillips J, Cohen B, Torigoe T et al (2011) Genetic heritability and shared environmental factors among twin pairs with autism. Arch Gen Psychiatry 68(11):1095–1102

Hardoff R, Sula M, Tamir A et al (2001) Gastric emptying time and gastric motility in patients with Parkinson's disease. Mov Disord 16(6):1041–1047
Hill KK, Smith TJ, Arnon SS (2013) Botulinum neurotoxin: origin, structure, pharmacology, and current status. Pharmacol Rev 65(3):334–371. https://doi.org/10.1124/pr.112.005108
Hills RD Jr, Pontefract BA, Mishcon HR, Black CA, Sutton SC, Theberge CR (2019) Gut microbiome: profound implications for diet and disease. Nutrients 11(7):1613. https://doi.org/10.3390/nu11071613
Hindson J (2017) A possible link between multiple sclerosis and GM. Nat Rev Neurol 13(12):705
Hoban AE, Stilling RM, Ryan FJ, Shanahan F, Dinan TG, Claesson MJ et al (2016) Regulation of prefrontal cortex myelination by the microbiota. Transl Psychiatry 6(4):e774. https://doi.org/10.1038/tp.2016.42
Holmqvist S, Chutna O, Bousset L, Aldrin-Kirk P, Li W, Björklund T et al (2014) Direct evidence of Parkinson pathology spread from the gastrointestinal tract to the brain in rats. Acta Neuropathol 128:805–820
Honarpisheh P, Reynolds CR, Conesa MPB, Manchon JFM, Putluri N, Bhattacharjee MB et al (2020) Dysregulated gut homeostasis observed prior to the accumulation of the brain amyloid-β in Tg2576 mice. Int J Mol Sci 21:1711. https://doi.org/10.3390/ijms21051711
Honda K, Littman DR (2016) The microbiota in adaptive immune homeostasis and disease. Nature 535(7610):75–84
Hsiao EY (2018) Gastrointestinal issues in autism spectrum disorder. Harv Rev Psychiatry 26(5):259–273
Hsiao EY, McBride SW, Hsien S, Sharon G, Hyde ER, McCue T et al (2013) Microbiota modulate behavioral and physiological abnormalities associated with neurodevelopmental disorders. Cell 155(7):1451–1463. https://doi.org/10.1016/j.cell.2013.11.024
Hu X, Wang T, Jin F (2016) Alzheimer's disease and GM. Sci China Life Sci 59(10):1006–1023. https://doi.org/10.1007/s11427-016-5083-9
Huo H, Li T, Wang S, Lv Y, Zuo Y, Yang L (2017) Prediction of presynaptic and postsynaptic neurotoxins by combining various Chou's pseudo components. Sci Rep 7(1):5827
Ivanov II, Atarashi K, Manel N, Brodie EL, Shima T, Karaoz U et al (2009) Induction of intestinal Th17 cells by segmented filamentous bacteria. Cell 139(3):485–498. https://doi.org/10.1016/j.cell.2009.09.033
Jiang C, Li G, Huang P, Liu Z, Zhao B, The Human Microbiome Project (HMP) Consortium (2017) The GM and Alzheimer's disease. J Alzheimers Dis 58(1):1–15. https://doi.org/10.3233/JAD-161141
Kang Y, Kang X, Zhang H, Liu Q, Yang H, Fan W (2021) GM and Parkinson's disease: implications for faecal microbiota transplantation therapy. ASN Neuro 13:17590914211016217
Kau AL, Planer JD, Liu J, Rao S, Yatsunenko T, Trehan I et al (2015) Functional characterization of IgA-targeted bacterial taxa from undernourished Malawian children that produce diet-dependent enteropathy. Sci Transl Med 7(276):276ra24
Kelly JR, Minuto C, Cryan JF, Clarke G, Dinan TG (2017) Transferring the blues: depression-associated GM induces neurobehavioural changes in the rat. J Psychiatr Res 68:369–377. https://doi.org/10.1016/j.jpsychires.2015.07.007
Kessler KR, Benecke R (1997) Botulinum toxin: from poison to remedy. Neurotoxicology 18(3):761–770
Kho ZY, Lal SK (2018) The human gut microbiome—a potential controller of wellness and disease. Front Microbiol 9:1835. https://doi.org/10.3389/fmicb.2018.01835
Kim KS (2017) Microbial antigen mimics and microbial pathogenesis. Microb Pathog 106:5–13. https://doi.org/10.1016/j.micpath.2017.01.019
Kim S, Kim H, Yim YS, Ha S, Atarashi K, Tan TG et al (2017) Maternal gut bacteria promote neurodevelopmental abnormalities in mouse offspring. Nature 549(7673):528–532
Kim S, Kwon SH, Kam TI, Panicker N, Karuppagounder SS, Lee S et al (2018) Lipopolysaccharide from GM modulates α-synuclein aggregation and alters its biological function. Proc Natl Acad Sci 115(36):E8615–E8624. https://doi.org/10.1073/pnas.1800655115

Kong G, Cao KL, Judd LM, Li S, Renoir T, Hannan AJ (2020) Microbiome profiling reveals gut dysbiosis in a transgenic mouse model of Huntington's disease. Neurobiol Dis 135:104268
Lim ES, Wang D, Holtz LR (2016) The bacterial microbiome and virome milestones of infant development. Trends Microbiol 24(10):801–810
Link CD (2021) Is there a brain microbiome? Neurosci Insights 16:26331055211018709. https://doi.org/10.1177/26331055211018709
Lonchamp E, Dupont JL, Wioland L, Courjaret R, Mbebi-Liegeois C, Jover E et al (2010) Clostridium perfringens epsilon toxin targets granule cells in the mouse cerebellum and stimulates glutamate release. PLoS One 5(9):e13046. https://doi.org/10.1371/journal.pone.0013046
Luczynski P, Whelan SO, O'Sullivan C, Clarke G, Shanahan F, Dinan TG et al (2016) Adult microbiota-deficient mice have distinct dendritic morphological changes: differential effects in the amygdala and hippocampus. Eur J Neurosci 44(9):2654–2666. https://doi.org/10.1111/ejn.13291
Lukiw WJ (2016) The microbiome, microbial-generated proinflammatory neurotoxins, and Alzheimer's disease. J Sport Health Sci 5(4):393–396
Lukiw WJ (2020) Gastrointestinal (GI) tract microbiome-derived neurotoxins-potent neuro-inflammatory signals from the GI tract via the systemic circulation into the brain. Front Cell Infect Microbiol 10:22. https://doi.org/10.3389/fcimb.2020.00022
Lukiw WJ, Arceneaux L, Li W, Bond T, Zhao Y (2021) Gastrointestinal (GI)-tract microbiome derived neurotoxins and their potential contribution to inflammatory neurodegeneration in Alzheimer's disease (AD). J Alzheimers Dis Parkinsonism 11(6):525
Ma Q, Xing C, Long W, Wang HY, Liu Q, Wang RF (2019) Impact of microbiota on central nervous system and neurological diseases: the gut-brain axis. J Neuroinflammation 16(1):53. https://doi.org/10.1186/s12974-019-1434-3
Machiels K, Joossens M, Sabino J, De Preter V, Arijs I, Eeckhaut V et al (2014) A decrease of the butyrate-producing species Roseburia hominis and Faecalibacterium prausnitzii defines dysbiosis in patients with ulcerative colitis. Gut 63(8):1275–1283. https://doi.org/10.1136/gutjnl-2013-304833
McVey Neufeld KA, Luczynski P, Seira Oriach C, Dinan TG, Cryan JF (2016) What's bugging your teen?—the microbiota and adolescent mental health. Neurosci Biobehav Rev 70:300–312. https://doi.org/10.1016/j.neubiorev.2016.06.005
Milani C, Duranti S, Bottacini F, Casey E, Turroni F, Mahony J et al (2017) The first microbial colonizers of the human gut: composition, activities, and health implications of the infant GM. Microbiol Mol Biol Rev 81(4):e00036–e00017. https://doi.org/10.1128/MMBR.00036-17
Montecucco C, Schiavo G (1995) Structure and function of tetanus and botulinum neurotoxins. Q Rev Biophys 28(4):423–472. https://doi.org/10.1017/s0033583500003292
Neunlist M, Van Landeghem L, Mahé MM, Derkinderen P, des Varannes SB, Rolli-Derkinderen M (2013) The digestive neuronal-glial-epithelial unit: a new actor in gut health and disease. Nat Rev Gastroenterol Hepatol 10(2):90–100. https://doi.org/10.1038/nrgastro.2012.221
Noyce AJ, Bestwick JP, Silveira-Moriyama L, Hawkes CH, Giovannoni G, Lees AJ et al (2012) Meta-analysis of early nonmotor features and risk factors for Parkinson disease. Ann Neurol 72(6):893–901. https://doi.org/10.1002/ana.23687
Park J, Kim CH (2021) Regulation of common neurological disorders by gut microbial metabolites. Exp Mol Med 53(12):1821–1833. https://doi.org/10.1038/s12276-021-00703-x
Penders J, Thijs C, Vink C, Stelma FF, Snijders B, Kummeling I et al (2006) Factors influencing the composition of the intestinal microbiota in early infancy. Pediatrics 118(2):511–521. https://doi.org/10.1542/peds.2005-2824
Perler BK, Friedman ES, Wu GD (2023) The role of the GM in the relationship between diet and human health. Annu Rev Physiol 85:449–468. https://doi.org/10.1146/annurev-physiol-031522-092054
Popoff MR, Poulain B (2010) Bacterial toxins and the nervous system: neurotoxins and multipotential toxins interacting with neuronal cells. Toxins (Basel) 2(4):683–737. https://doi.org/10.3390/toxins2040683

PrabhuDas M, Adkins B, Gans H, King C, Levy O, Ramilo O et al (2011) Challenges in infant immunity: implications for responses to infection and vaccines. Nat Immunol 12(3):189–194. https://doi.org/10.1038/ni0311-189

Pusceddu MM, Barboza M, Keogh CE, Schneider M, Stokes P, Sladek JA et al (2019) Nod-like receptors are critical for gut-brain axis signalling in mice. J Physiol 597(24):5777–5797. https://doi.org/10.1113/JP278640

Qin J, Li R, Raes J, Arumugam M, Burgdorf KS, Manichanh C et al (2010) A human gut microbial gene catalogue established by metagenomic sequencing. Nature 464(7285):59–65. https://doi.org/10.1038/nature08821

Querfurth HW, LaFerla FM (2010) Alzheimer's disease. N Engl J Med 362:329–344. https://doi.org/10.1056/NEJMra0909142

Rhee SH, Pothoulakis C, Mayer EA (2009) Principles and clinical implications of the brain–gut–enteric microbiota axis. Nat Rev Gastroenterol Hepatol 6:306–314

Roswall J, Olsson LM, Kovatcheva-Datchary P, Nilsson S, Tremaroli V, Simon MC et al (2021) Developmental trajectory of the healthy human GM during the first 5 years of life. Cell Host Microbe 29(5):765–776.e3. https://doi.org/10.1016/j.chom.2021.02.021

Roubalová R, Procházková P, Papežová H, Smitka K, Bilej M, Tlaskalová-Hogenová H (2020) Anorexia nervosa: GM-immune-brain interactions. Clin Nutr 39(3):676–684

Round JL, Lee SM, Li J, Tran G, Jabri B, Chatila TA et al (2011) The Toll-like receptor 2 pathway establishes colonization by a commensal of the human microbiota. Science 332(6032):974–977. https://doi.org/10.1126/science.1206095

Rumah KR, Linden J, Fischetti VA, Vartanian T (2013) Isolation of Clostridium perfringens type B in an individual at first clinical presentation of multiple sclerosis provides clues for environmental triggers of the disease. PLoS One 8:e76359

Rutsch A, Kantsjö JB, Ronchi F (2020) The gut-brain axis: how microbiota and host inflammasome influence brain physiology and pathology. Front Immunol 11:604179. https://doi.org/10.3389/fimmu.2020.604179

Rynn M, Brawman-Mintzer O (2004) Generalized anxiety disorder: acute and chronic treatment. CNS Spectr 9(10):716–723

Sadowsky JM (2016) Alzheimer's disease: a role for gut bacteria? J Alzheimers Dis 52(4):1277–1278. https://doi.org/10.3233/JAD-160999

Saurman V, Margolis KG, Luna RA (2020) Autism spectrum disorder as a brain-gut-microbiome axis disorder. Dig Dis Sci 65:818–828

Scheperjans F, Aho V, Pereira PAB, Koskinen K, Paulin L, Pekkonen E et al (2015) GM are related to Parkinson's disease and clinical phenotype. Mov Disord 30:350–358

Sears CL, Geis AL, Housseau F (2014) Bacteroides fragilis subverts mucosal biology: from symbiont to colon carcinogenesis. J Clin Investig 124(10):4166–4172

Sharon G, Sampson TR, Geschwind DH, Mazmanian SK (2016) The central nervous system and the gut microbiome. Cell 167(4):915–932. https://doi.org/10.1016/j.cell.2016.10.027

Silva YP, Bernardi A, Frozza RL (2020) The role of short-chain fatty acids from GM in gut-brain communication. Front Endocrinol 11:25. https://doi.org/10.3389/fendo.2020.00025

Sivan A, Corrales L, Hubert N, Williams JB, Aquino-Michaels K, Earley ZM et al (2015) Commensal Bifidobacterium promotes antitumor immunity and facilitates anti-PD-L1 efficacy. Science 350(6264):1084–1089

Skonieczna-Żydecka K, Grochans E, Maciejewska D, Szkudłapski D, Głąbska D (2018) The comparison of GM composition in obese and non-obese individuals in a population of the Eastern European country. Arch Med Sci 14(1):157–164. https://doi.org/10.5114/aoms.2016.63298

Smith PA (2015) The tantalizing links between gut microbes and the brain. Nature 526(7573):312–314. https://doi.org/10.1038/526312a

Sonnenburg ED, Sonnenburg JL (2014) Starving our microbial self: the deleterious consequences of a diet deficient in microbiota-accessible carbohydrates. Cell Metab 20(5):779–786. https://doi.org/10.1016/j.cmet.2014.07.003

Sun J, Xu J, Ling Y, Wang F, Gong T, Yang C et al (2019) Fecal microbiota transplantation alleviated Alzheimer's disease-like pathogenesis in APP/PS1 transgenic mice. Transl Psychiatry 9:189. https://doi.org/10.1038/s41398-019-0522-7

Theis KR, Romero R, Winters AD, Greenberg JM, Gomez-Lopez N, Alhousseini A et al (2019) Does the human placenta delivered at term have a microbiota? Results of cultivation, quantitative real-time PCR, 16S rRNA gene sequencing, and metagenomics. Am J Obstet Gynecol 220(3):267.e1–267.e39

Tirosh I, Izar B, Prakadan SM, Wadsworth MH, Treacy D, Trombetta JJ et al (2016) Dissecting the multicellular ecosystem of metastatic melanoma by single-cell RNA-seq. Science (New York, N.Y.) 352(6282):189–196. https://doi.org/10.1126/science.aad0501

Troletti CD, de Goede P, Kamermans A, de Vries HE (2016) Molecular alterations of the blood-brain barrier under inflammatory conditions: the role of endothelial to mesenchymal transition. Biochim Biophys Acta 1862(3):452–460

Turroni F, Ribbera A, Foroni E, van Sinderen D, Ventura M (2008) Human GM and bifidobacteria: from composition to functionality. Antonie Van Leeuwenhoek 94(1):35–50. https://doi.org/10.1007/s10482-008-9232-4

van der Waaij LA, Harmsen HJ, Madjipour M, Kroese FG, Zwiers M, van Dullemen HM et al (2005) Bacterial population analysis of human colon and terminal ileum biopsies with 16S rRNA-based fluorescent probes: commensal bacteria live in suspension and have no direct contact with epithelial cells. Inflamm Bowel Dis 11(10):865–871. https://doi.org/10.1097/01.mib.0000179212.80778.d3

Wang Q, Xu R (2019) Data-driven multiple-level analysis of gut-microbiome-immune-joint interactions in rheumatoid arthritis. BMC Genomics 20:124

Wasser CI, Mercieca EC, Kong G, Hannan AJ, McKeown SJ, Glikmann-Johnston Y et al (2020) Gut dysbiosis in Huntington's disease: associations among GM, cognitive performance and clinical outcomes. Brain. Communications 2(1):fcaa110

Westerberg LS, Klein C, Snapper SB (2008) Breakdown of T cell tolerance and autoimmunity in primary immunodeficiency-lessons learned from monogenic disorders in mice and men. Curr Opin Immunol 20(6):646–654

Xu L, Pu J (2016) Alpha-synuclein in Parkinson's disease: from pathogenetic dysfunction to potential clinical application. Parkinsons Dis 2016:1–10

Yarandi SS, Peterson DA, Treisman GJ, Moran TH, Pasricha PJ (2016) Modulatory effects of GM on the central nervous system: how gut could play a role in neuropsychiatric health and diseases. J Neurogastroenterol Motil 22(2):201–212. https://doi.org/10.5056/jnm15146

Yoo BB, Mazmanian SK (2017) The enteric network: interactions between the immune and nervous systems of the gut. Immunity 46(6):910–926

Zhao Y, Dua P, Lukiw WJ (2019) LPS-induced neuroinflammation and neurodegeneration may be associated with Alzheimer's disease. Biochem Biophys Res Commun 513(1):195–199. https://doi.org/10.1016/j.bbrc.2019.03.013

Zheng D, Liwinski T, Elinav E (2020) Interaction between microbiota and immunity in health and disease. Cell Res 30(6):492–506. https://doi.org/10.1038/s41422-020-0332-7

Zhu X, Han Y, Du J, Liu R, Jin K, Yi W (2017) Microbiota-gut-brain axis and the central nervous system. Oncotarget 8(32):53829–53838

Zhu S, Jiang Y, Xu K, Cui M, Ye W, Zhao G et al (2020) The progress of gut microbiome research related to brain disorders. J Neuroinflammation 17(1):25. https://doi.org/10.1186/s12974-020-1705-z

Zinkevich V, Beech IB (2018) Screening of dairy propionibacteria for adhesion and antagonistic properties against Staphylococcus aureus. Foodborne Pathog Dis 15(2)

Zorzoli A, Urdiain M, Ugarte-Ruiz M, Cui Y, Liu J, Su X et al (2017) Isolation of Bacillus cereus group from the microbiota of nursing home residents. PLoS One 12(5):e0176280. https://doi.org/10.1371/journal.pone.0176280

Gut Microbiota and Altered Behaviour: Target on Neuroimmune Interplays

9

Selvaraj Jayaraman and Shiny Paul

Abstract

There are over 100 trillion microflorae in the human gut which engage in multiple interactions with human health during the whole life span. They influence human physiology, nutrition, metabolism and immunity. There is a two-way connection between the gut and brain known as gut-brain axis through which it influences the physiological, behavioural and cognitive functions of brain. This research focuses on the gut-brain axis, a bidirectional communication pathway between the gut and the brain that impacts brain function through vagus nerve and HPA axis and the metabolites such as neurotransmitters and SCFAs. The disruption in composition and relative abundance of gut microbiota known as dysbiosis leads to pathogenesis of several neurological disorders such as Alzheimer's disease, Parkinson's disease and autism. The individuals with neurological disorders presented with altered composition of gut microbes and its metabolites which varies from the normal individuals. This altered behaviour of gut microbes can be influenced or changed by supplementing with a bioactive live microorganisms known as probiotics. This study explores the potential of manipulating gut microbiota through probiotics to improve neurological function and potentially treat these disorders.

Keywords

Gut microbiota · Altered behaviour · Neuroimmune disorder · Microbiota metabolites

S. Jayaraman (✉) · S. Paul
Department of Biochemistry, Centre of Molecular Medicine and Diagnostics (COMManD), Saveetha Dental College and Hospital, Saveetha Institute of Medical and Technical Sciences, Saveetha University, Chennai, India
e-mail: selvarajj.sdc@saveetha.com

S. Pathak, A. Banerjee (eds.), *Gut Microbiome and Brain Ageing*,
https://doi.org/10.1007/978-981-99-8803-7_9

9.1 Introduction

The peripheral immune system serves as a vital defence mechanism safeguarding the body against diverse threats, such as pathogens and tissue damage. Consisting of various cell types, including leukocytes and lymphocytes, originating from haematopoietic stem cells, these immune cells fulfil specific roles essential for system functionality. Travelling through blood and lymphatic vessels, these mobile cells actively scan the body, recognising and responding to adverse conditions by producing immune factors. The orchestrated collaboration of immune cells in executing multicellular behaviours plays a key role in implementing recovery and repair programs, thereby alleviating the impact of adverse conditions on the body (Gruol, 2023). Brain is recognised to possess its own innate immune system, the neuroimmune system, working alongside the peripheral immune system to safeguard against adverse conditions and protect brain function (Cerrato 2020). This is because there is a no possible way for the peripheral immune system to gain access into the CNS to elicit immune reaction. Hence, the neuroimmune system serves vital functions in maintaining a balanced, homeostatic state crucial for normal physiology and development of the brain (Hickman et al. 2018). However, when dysregulated due to adverse conditions, it can contribute to negative effects on the brain. For instance, numerous studies have been established related to the viral infection (HIV) affecting the brain and counterattack by the neuroimmune system. When the body encounters the HIV infection, the peripheral immune system takes action in the removal of the toxic effects of that virus. However, when the infection reaches into the CNS, there is a marked change, namely cognitive dysfunction, memory loss, disturbed behaviour, and retardation of motor function. This is when the neural immune system gets into action in order to remove the infection and protect the brain from getting damaged. In response to the viral infection, microglia produce immune factors that alter the biology of various brain cells, such as microglia, astrocytes, and neurons, resulting in changes in overall brain function (Garden 2002).

Traditionally, it is noted that the glial cells residing in the brain act as a primary defence system of the brain. Within adults, the neuroimmune system primarily consists of microglia and astrocytes as the main cell types along with two other glial cell types present in the brain, namely oligodendrocytes and NG2 glia (also known as oligodendrocyte precursor cells), contributing to the customary functions of the neuroimmune system (Zeis et al. 2016). Microglia, functioning under physiological conditions, regulate brain development and ageing, maintain neuronal networks, and oversee synaptic function (Grabert et al. 2016). They play a crucial role in homeostasis by clearing pathogens, cell debris, and abnormal proteins, while also participating in the repair process (Grabert et al. 2016). Under adverse conditions, dysregulated functions of microglia can contribute to pathology, despite their initial role in combating the effects and repairing damage. Astrocytes constitute one of the predominant cell types within the brain (Garland et al. 2022). Traditionally considered support cells, astrocytes play pivotal roles in promoting brain health by providing metabolic support, maintaining ion balance, clearing the synaptic environment, and regulating the blood-brain barrier (Pannasch and Rouach 2013). Contemporary insights highlight the diverse contributions of astrocytes in establishing and sustaining functional cognitive and control systems throughout both developmental and mature stages of the brain (Pannasch and Rouach 2013).

The human gut nurses distinct variety of complex microorganisms which plays a vital role on human health. There are more than 100 species of bacteria, 1000 kinds of micro-organisms and 5000 strains of microbes which encode around 150 times as human genome (Wang and Wang 2016). Based on HTS 16S rRNA study shows the presence of 2 dominant bacterial phyla of adult gut, *Firmicutes* and *Bacteriodetes*, and other phyla such as *Actinobacteria, Proteobacteria, Verrucomicrobia* and *Fusobacteria* appear in lower proportions. Also, few butyrate-producing bacteria such as *Bacteroides uniformis, Roseburia intestinalis* and *Faecalibacterium prausnitzii* represent to be key members of gut microbiota (Eckburg et al. 2005; Tremaroli and Bäckhed 2012; Qin et al. 2010).

The microbiota of gut performs multiple interactions with human health throughout the life span. It has an effective influence on human well-being which functions on physiological, immunological, nutritional and metabolic factors which can be either beneficial or harmful (Guinane and Cotter 2013). It is evident through recent studies that microbiota plays vital role on physiological, behaviour and cognitive functions. One composition of gut microbiota depends on delivery mode, age, nutrition, stress, infection, use of antibiotics and other factors. The normal gut microbiota plays crucial roles in host functions, including nutrient and drug metabolism, maintaining the gut mucosal barrier, immunomodulation, and defending against pathogens (Jandhyala et al. 2015). From an immunological standpoint, microorganisms are typically recognised as pathogens by the host immune system, triggering their elimination. Despite this, a significant portion of gut bacteria is non-pathogenic, forming a symbiotic relationship with enterocytes (Sekirov et al. 2010). These commensals play key roles in nutrient and drug metabolism, preventing colonisation by pathogenic microorganisms, and supporting the intestinal barrier function. Simultaneously, the immune system has evolved to collaborate with the healthy microbiota while effectively combating invasive pathogenic microorganisms (Sekirov et al. 2010). The composition and structure of the gut microbiota are influenced by various factors. The reciprocal relationship exists between gut and brain functions and psychological behaviour (Petra et al. 2015). The GI tract is regulated by two factors, namely intrinsic and extrinsic. Intrinsic factor comprises of intestinal neural system with 200–600 million neurons and is called as second brain (Anglin et al. 2015). Hence, gut microbiota and its association with neuroimmunity and mental well-being is an important concept which is the main focus of this review.

9.2 Gut Microbiota

Gut microbiota includes bacteria, virus, protozoa, etc. which are inhabited in the human intestine. The bacteria present in human gut has both beneficial and deleterious effects on human health. Under symbiosis condition, there exists a balance between good and pathogenic bacteria which leads to homeostasis. Under dysbiosis processes, the homeostasis is disturbed between the bacteria leading gut inflammation and invasiveness (Matamoros et al. 2013).

There are around 1000 different species of symbiotic bacteria in the gut where *Bacteroidetes* and *Firmicutes* are two main phyla which account for major portion in the gut. Smaller number of bacteria includes Verucomicrobia, Actinomyces, Fusobacterium and Proteobacteria (Eckburg et al. 2005; Lay et al. 2005; Diamant et al. 2011). However, the composition of gut microbes depends on mode of delivery, breast fed and early weaning, use of antibiotics, dietary and lifestyle factors and environmental factors (Mayer 2011). The gut microbiota forms a symbiotic bond with the gut mucosa, delivering substantial metabolic, immunological, and gut-protective functions in a healthy individual. Operating as an autonomous organ, the gut microbiota draws nutrients from the host's diet and shed epithelial cells, illustrating robust metabolic capabilities and remarkable functional adaptability (Sonnenburg et al. 2005). The roles of gut microbiota in normal physiology of the body are nutrient metabolism, immune modulation, drug metabolism, and maintenance of gut integrity. Dietary carbohydrates constitute the main nutrient source for the gut microbiota. Colonic organisms, such as Bacteroides, Roseburia, Bifidobacterium, Faecalibacterium, and Enterobacteria, ferment undigested carbohydrates and indigestible oligosaccharides. This fermentation produces short-chain fatty acids (SCFA), notably butyrate, propionate, and acetate, which act as plentiful energy sources for the host (Macfarlane and Macfarlane 2003). The gut microbiota positively influences lipid metabolism by preventing the suppression of lipoprotein lipase activity in adipocytes and enhancing lipid hydrolysis efficiency through increased colipase expression by Bacteroides thetaiotaomicron (Hooper et al. 2001). Synthesis of various vitamins (namely vitamin K, Vitamin B, etc.) has been seen to be associated with gut microbiota. They are also involved in the metabolism of polyphenols (such as flavanols, flavanones, flavan-3-ols, anthocyanidins, isoflavones, flavones, tannins, lignans, and chlorogenic acids), derivatives of plants and their products (Marín et al. 2015). A mounting body of evidence underscores the substantial role of the gut microbiota in xenobiotic metabolism, with potential implications for shaping future therapies across diverse diseases. Research has demonstrated that the gut microbial metabolite p-cresol can diminish the liver's ability to metabolise acetaminophen by competitively inhibiting hepatic sulfotransferases (Clayton et al. 2009). Different microbes and its abundance in various neuroimmune disorders are discussed later in this chapter.

9.3 Gut-Brain Axis

There is a bidirectional route of communication that exists between the gut and brain termed as microbiota gut-brain axis (Baj and Moro 2019). The connection between microbes and the brain is not only through gut-brain axis but also through endocrine, metabolic and immune systems. Few evidential studies show that gut microbiota plays a chief role in the behaviour of brain and on cognitive developments through its products such as hormones, immune factors and metabolites, and by modifying these products, the microbiota may improve or also cure the diseases of brain. The gut microbiota interacts with central nervous system through different pathways and mechanisms such as sympathetic and parasympathetic branches of autonomic nervous system, neuroimmune and neuroendocrine system (Rieder et al. 2017). The gut-brain axis involves critical parts of central nervous system (CNS) along with autonomic nervous system (ANS), enteric nervous system (ENS), hypothalamic-pituitary-adrenal (HPA) axis and immune system. The bidirectional communication between gut and brain is through ENS in gut and ANS and vagal nerve in the spinal cord. The direct connection between the gut, brain and spinal cord is the vagal nerve which mediates signals for stress, satiety and mood changes (Abdel-Haq et al. 2019; Wang and Wang 2016).

A system of inter-kingdom communication exists between prokaryotes and eukaryotes which is evident by synthesis of several neuro/immune active compounds such as gamma aminobutyric acid (GABA), glutamate, serotonin, dopamine and noradrenaline by both the species (Baj and Moro 2019). These neuroactive compounds affect CNS by penetrating into blood brain barrier once it crosses gut mucosa (Petra et al. 2016). The alterations in the secretion of these compounds have been diagnosed in mentally ill patients (Skonieczna-Zydecka et al. 2020).

Few experimental evidence show the relationship between gut microbiota and altered behaviour: Mice lacking intestinal and other microbiota exhibited exaggerated and defective response to stress which can be normalised by inducing intestinal recolonisation. It was also demonstrated that intestinal transplant of microbiota normalised the impaired behaviour of BALB/c mice (Bruce-Keller et al. 2018). Sudo et al. (2004) found that recolonisation of *Bifidobacterium infantis* through stool transplant, normalises the behavioural alterations of mice that exhibited the changes in stress induced environment (Sudo et al. 2004).

Several studies have identified different mechanisms of gut microorganisms on nervous system including stimulation of vagal nerve (Bonaz et al. 2018; Bravo et al. 2011), functions of stress associated hypothalamic-pituitary-adrenal (HPA) axis (Bonaz et al. 2018), along with permeability of blood brain barrier (Braniste et al. 2014). Through studies it is evident that the glutameric pathway modulates the inter kingdom communication both in physiological and disease conditions (Filpa et al. 2016). Glutamate is derived from dietary proteins and from Glu containing food products and synthesised by gut bacteria (Briguglio et al. 2018). Glutamate mainly acts as excitatory neurotransmitter in CNS and enteric nervous system (ENS) where neurons and glial cells are involved in synthesis of amino acids (Fig. 9.1). In CNS, spinal and vagal afferent synapse co-ordinates emotional response along gut-brain axis which interacts with higher brain region which includes emotional motor system along with limbic and paralimbic system (Tremaroli and Bäckhed 2012; Mulak and Bonaz 2004). The main stress axis that participates in bidirectional communication is hypothalamic-pituitary adrenal axis which releases corticotrophin releasing factor (CRF) which in turn promotes adrenal and pituitary secretions such as adrenocorticotrophic hormone (ACTH). Glutamate regulates several gut and brain functions via glutamate receptor activation which functions through activation either

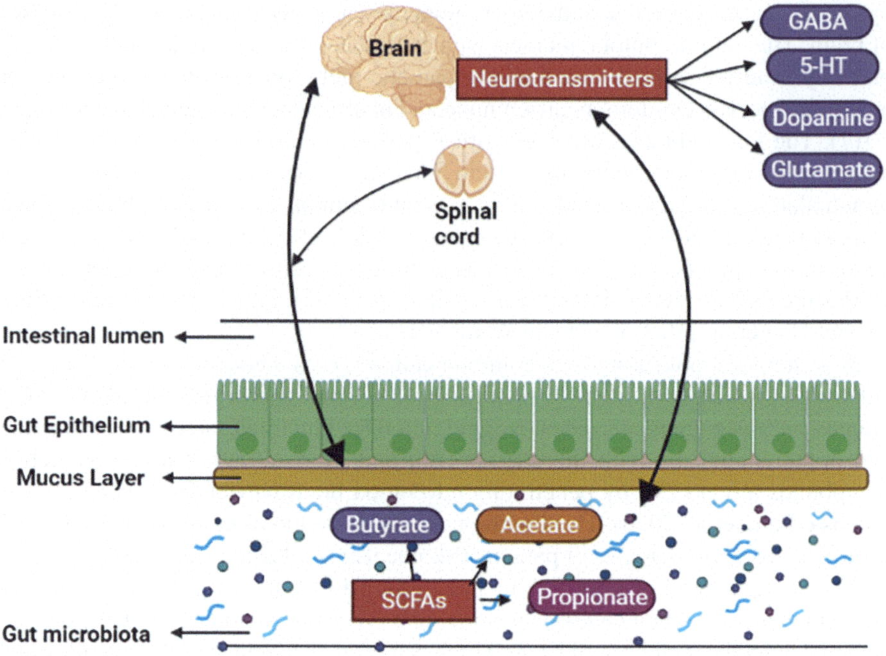

Fig. 9.1 The schematic representation of gut microbiota metabolites and its interactions with gut-brain axis

vagal, pelvic or splanchin afferent (Furness et al. 2014; Tremaroli and Bäckhed 2012; Mulak and Bonaz 2004).

Endocrine system is also involved in gut brain signalling axis which regulates through secretion of several hormones such as somatostatin, serotonin and ghrelin. These endocrine components exert action on physiological and homeostasis of gastrointestinal tract (Skonieczna-Zydecka et al. 2020).

9.4 Neuroimmune Metabolites and Mental Health

Astrocytes and microglia primarily generate and release small signalling proteins, typically within the range of 6–70 kDa, categorised as 'cytokines.' These signalling factors, belonging to a sizable superfamily of soluble proteins, hold significant importance in the functions of the neuroimmune system, contributing to both healthy and pathological states (Gruol 2023). Within the cytokine superfamilies, one can find representatives from various groups, including chemokines, interleukins (IL), interferons (IFN), colony-stimulating factors (CSF), transforming growth factors (TGF), and tumour necrosis factors (TNF) (Dinarello 2007). The IL-1 family includes 11 cytokines, IL-6 family consists of 8 cytokines, TNF superfamily has 19 members, IFN superfamily comprises over 20 members, the CSF family has 4 members, and the TGF-β family includes 33 members (Gruol 2023). Despite the identification of a considerable number of cytokines, only a relatively modest percentage has been observed in the brain, likely due to the neuroimmune system being an emerging field with limited available information. This is because, only a few immune markers from the peripheral innate system enters the CNS. In both normal and pathological conditions, the cerebellum has been demonstrated to express major cytokines at the mRNA or protein levels. These include IL-1β, IL-6, IL-9, IL-10, IL-15, IL-8, TNFα, CCL2, CXCL12, CXCL14, IP-10, MIP1α, GM-CSF, and TGFβ (Gruol 2023). Ordinarily, the quantification of neuroimmune factors in the brain is carried out using immunoassays like enzyme-linked immunoassay (ELISAs). ELISAs can assess the level of an individual protein, while multiplex analysis systems enable the simultaneous determination of multiple cytokine levels (Liu et al. 2021).

The gut microbes communicate with central nervous system through vagus nerve and different metabolites such as neurotransmitter and neurotropic factors (Rieder et al. 2017). The gut microbiota synthesises numerous bioactive metabolites which enter into the circulation and regulate host immune system along with activity of some organs like brain. Persistent activation of glial cells and excessive production of neuroimmune factors are prevalent characteristics in neurodegenerative diseases, potentially resulting in adverse effects on both brain structure and function (Stephenson et al. 2018). In experiments, inducing inflammation and promoting the production of neuroimmune factors is commonly achieved by administering lipopolysaccharide (LPS), a bacterial toxin found in the outer membrane of gram-negative bacteria (Llorens et al. 2014). These metabolites include small chain fatty acids, serotonin, Gamma aminobutyric acid and many more (Lucerne et al. 2021).

Since the gut microbes are able to produce the metabolites such as neurotransmitter, their homeostasis can have an impact on neurodegenerative disorders.

9.5 Small Chain Fatty Acids (SCFAs)

The studies have found that there is a direct connection between microbiota derived metabolites and human immunity. The metabolites such as small chain fatty acids (SCFAs), butyrate, propionates and acetates which are produced during intestinal fermentation, exhibits neuroactive properties by stimulating sympathetic and autonomous nervous system through G-protein receptors (Bruce-Keller et al. 2018). SCFAs stimulates ENS to produce neuropeptides and are also permeable to blood brain barrier which directly regulates microglial morphology and maturity (Erny et al. 2015). SCFAs are produced through high fibre diet and they exhibit anti-inflammatory effects on intestinal mucosa (Furusawa et al. 2013).

Butyrate is an important SCFAs as it supplies energy to enterocytes and acts as anti-depressant which can be used in treatment of Parkinson's and Huntington's disease (Laurent et al. 2013; Ferrante et al. 2003). G-protein receptors activates butyrate which regulates homeostasis between pathogen and bacterial immune balance (Thangaraju et al. 2009). Nuclear factor kappa B and Histone deacetylase inhibitor (HDAC) which are responsible for anticancer action of dietary fibre is downregulated by butyrate (Louis and Flint 2009; Fung et al. 2012). The synthesis of neurotransmitters such as dopamine, epinephrine and norepinephrine is regulated by butyrate which exerts its action by altering the expression tyrosine hydroxylase gene (DeCastro et al. 2005).

Propionic acid (PPA) induces the production of pro-inflammatory cytokines and inhibits certain cellular antioxidants such as GSH and superoxide dismutase (Wajner et al. 2004). High dosage of PPA showed hyperactivity and abnormal motor movements in mice which resembled the ASD in humans (MacFabe et al. 2007).

Inducing SCFAs may alter microbiota and therefore gut-brain axis and hence SCFAs can be used in the treatment of neuroimmune disorders (Hahnen et al. 2008). Few animal studies have shown that SCFAs such as omega-3 polyunsaturated fatty acids (*n*-3-PUFA) and its derivatives such as docosahexaenoic acid (DHA) and eicosapentaenoic acid (EPA) which are required for proper brain and visual functioning and developments and behavioural regulations (Kidd 2007) improve neurological and neuroimmune disorders by enhancing GI homeostasis (Lombardi and De Meirleir 2018).

Studies on animal models suggest that administration of SCFAs such as sodium butyrate and propionic acid reduces the progression of Parkinson's and Huntington's diseases (Laurent et al. 2013; Ferrante et al. 2003).

9.6 Gamma Aminobutyric Acid (GABA)

GABA is an inhibitory neurotransmitter of CNS which is synthesised by the bacteria such as *Bifidobacteria* and *Lactobacilli* (Barrett et al. 2012). It regulates neuronal excitability and its dysfunction leads to several chronic neurological disorder (Sun et al. 2012). GABA produced by gut microbes are permeable to blood brain barrier and its synthesis depends on factors such as high fat food diet and exposure to antibiotics (Takanaga et al. 2001; Shyamaladevi et al. 2002; Fujisaka et al. 2018; Bienenstock et al. 2010). GABA shunt is the major pathway where GABA converts to succinate to enter TCA cycle (Feehily and Karatzas 2013). GABA signalling which is associated with depression and anxiety, mediates with gut microbiota through vagus nerve. Later, an improvement was shown in anxiety and depression related mood and reduced visceral pain when administered with the strains of probiotics which produce GABA (Bercik et al. 2011; Bravo et al. 2011; Janik et al. 2016).

9.7 Serotonin

Serotonin is known as 5-hydroxytrytamine (5-HT) and plasma tryptophan is a precursor of 5-HT. Serotonin is a neurotransmitter which functions on both gut and CNS and plays vital role in controlling mood and cognition (Filosa et al. 2018). Bacteria such as *Streptococcus*, *Candida*, *Enterococcus* and *Escherichia coli* secrete serotonin (Vitetta et al. 2014) and are also synthesised by certain specialised cells such as enterochromaffin cells, mesenteric neurons and mucosal mast cells (Yano et al. 2015). Microbes such as *Escherichia coli* regulate serotonergic system which is associated with the production and availability of serotonin that acts as neurotransmitter of both CNS and enteric nervous system (Pistollato et al. 2016).

Of the total serotonin in the body, 90% is synthesised in the intestine and it regulates various functions such as cardiac function, regulation of immune response, platelet aggregation and regulates motor and sensory reflexes (Yano et al. 2015). It is also involved in various physiological functions such as respiration, vasoconstriction, peristalsis of GI tract, behavioural and neurological functions (Cloez-Tayarani and Changeux 2007). Change in serotonin levels either due to tryptophan dysregulation or bacterial dysbiosis leads to disorders of brain and GI tract (Clarke et al. 2013). Serum serotonin is a main biomarker of autism syndrome disorder (DeCastro et al. 2005). Increased serotonin levels in serum showed abnormal social and repetitive behaviour and non-standard communication in mice (Veenstra-VanderWeele et al. 2012).

9.8 Dopamine

Dopamine is a neurotransmitter and a precursor of catecholamine such as adrenaline and noradrenaline. Tyrosine is a precursor of dopamine and gut microbiota regulates the enzymes involved in conversion of tyrosine to dopamine (Matsumoto et al. 2017). It is produced by certain bacteria such as *Bacillus* and *Serattia* (Vitetta et al. 2014). Dopamine plays vital role in regulating several neurological processes such as reward and learning, motor control and cognition (Chaudhuri et al. 2006). Hence it is called as reward-motivated behavioural neurotransmitter (Feehily and Karatzas 2013). The decreased levels of dopamine was reported in Parkinson's disease and it serves as biomarker of Parkinson's disease (Goldstein et al. 2018). It was reported that dopamine and its metabolites were associated with corresponding profusion of bacterial genus like *Coprococcus* which was correlated with physical and mental behaviour (Valles-Colomer et al. 2019).

9.9 Neuroimmune Disorders and Gut Microbiota Interplay

The gut-brain axis plays a profound role in neurological diseases and disorders such as Alzheimer's, Parkinson's, multiple sclerosis and Autism (Borsom et al. 2020). The studies have confirmed that mental disorders show the existence of altered composition of intestinal microbiota which significantly functions through bacterial produced neurotransmitters and neuroimmune metabolites.

9.9.1 Alzheimer's Disease

Alzheimer's disease (AD) is characterised by memory loss, confusion, paranoia and aggression in patients which was first observed by a psychiatrist and neuropathologist named Alois Alzheimer in 1906 (Hippius and Neundörfer 2003). Alzheimer's is a most common dementia in United States affecting 5.8 million people and is predicted to rise to 13.5 million by 2050 (Alzheimer's Association 2020). Through recent studies it is found out that intestinal microbiota and gut-brain axis contributes majorly to the aetiology and pathogenesis of AD. The intestinal microbes act on neuroinflammation in AD through synthesis of bacterial metabolites and proinflammatory cytokines which exerts its action on brain immune cells. These proinflammatory cytokines includes IL-6, IL-1β, IFN-γ and TNF-α, and metabolites such as trimethylamine *N*-oxide (TMAO) and SCFA such as butyric acid and propionic acid (Giau et al. 2018; Domingues et al. 2017; Koeth et al. 2013).

A study performed on AD patients revealed a decline in phyla Fermicutes and raise in genus *Blautia* and Bacteriodetes. It was noticed that there was a positive correlation between Bacteriodes and CSF marker, suggesting that Bacteriodes may increase the level of neuroinflammation (Vogt et al. 2017). High load of bacteria such as Actinobacteria, specifically Propionibacterium acnes, was seen in post mortem brain of AD patient (Emery et al. 2017). Mice orally induced with *Porphyromonas gingivalis* exposed increased amyloidosis which correlated with brain colonisation. This data showing translocation of bacteria from the gut indicate that microbes are capable of crossing both gut epithelial barrier and blood brain barrier (Dominy et al. 2019). Escherichia/Singella isolated from patients stool sample showed direct activation of pro-inflammatory condition and accumulation of amyloid-β (Borsom et al. 2020).

A study conducted on mouse model of AD showed that the oral administration of *Bifidobacterium breve* strain reversed the cognitive impairment, and other studies on AD mice showed that gut microbiota automatically changes over the time and promotes infiltration of brain immune cells which causes cognitive impairment, microglial activation and Aβ amyloidosis. Kobayashi et al. compared AD and non-AD animal models and transcribed the hippocampus and identified 305 variably expressed genes in AD mice which were involved in immune response. They found that only two genes differed in both the models which were the potent genes which determines AD associated immune response suggesting that microbes play an effective role in pathology of AD (Kobayashi et al. 2017; Wang et al. 2019).

Through these studies, it is evident that composition of gut microbes significantly modulates pathogenesis of AD. The gut-brain axis regulates processes such as maturation of brain immune cells, synthesis of metabolites such as SCFAs, production of proinflammatory cytokines which are permeable to BBB and migrate to brain causing neuroinflammation and eventually increasing the severity of AD related pathologies (Borsom et al. 2020).

9.9.2 Parkinson's Disease

Parkinson's disease (PD) is a second most common neuro degenerative disease affecting around 2% of aged people of 65 years and above (Roversi et al. 2021). It is a chronic neuro degenerative disorder caused due to loss of dopamine neurons and under major loss of neurons, patients experience shuffling gait, corporal instability leading to loss of balance and tremors in upper or lower limbs (Vila and Przedborski 2004). Before motor symptoms, the onset of several non-motor symptoms such as dementia, depression, rapid eye movement, apathy and constipation are seen in PD patient (Chaudhuri et al. 2006; Fasano et al. 2015).

The non-motor symptoms such as constipation associated with inflammatory bowel disease (IBD) and Crohn's disease could be present even before 10 years in PD patients (Zhu et al. 2022). This prolonged GI dysfunction leads to severe inflammations which is evident through significant increase in inflammatory markers such as Tumor Necrosis Factor (TNF) and several proinflammatory cytokines in PD patients. This inflammatory cytokine profile in PD patients is very similar to the cytokine profile in IBD patients indicating that IBD may be an early symptom for PD (Dumitrescu et al. 2021).

PD patients showed increased number of genera such as *Bifidobacterium, Lactobacillus* and *Akkermansia* and decreased dominant taxa such as *Faecalibacterium, Butyriciccocaceae* and *Ruminococcaceae*, which are part of core community of gut microbes involved in synthesis of several SCFAs including butyrate and energy metabolisms (Geirnaert et al. 2014; Vacca et al. 2020).

It was reported that PD patients showed decreased level of beneficial organisms such as *Roseburia, Blaudia* and *coprococcus* which are involved in neuronal inflammatory signalling and BBB permeability and integrity (Keshavarzian et al. 2015), and increase in members of *Enterobacteriaceae* which are involved in gait, rigidity and stability in PD patients (Scheperjans et al. 2015).

The deficiency of dietary fibres also leads to increased colonisation and infiltration of opportunistic pathogens (Desai et al. 2016). Increase in Akkermansia in PD patients also co-localises opportunistic pathogens leading to increased risk of PD pathogenesis (Wallen et al. 2020). This shows that the dysbiosis of gut microbiota may lead to increased risk and progression of disease. The GI dysfunction and inflammation leading to gut dysbiosis is a major contribution for progression of a disease.

9.9.3 Autism Spectrum Disorder

Autism Spectrum Disorder (ASD) is a neurodevelopmental disorder characterised by impaired social and communicational interactions, stereotyped and repetitive behaviour with cognitive disturbances. It is also associated with GI dysfunction such as chronic diarrhoea and abdominal pain and discomfort (Horvath et al. 1999).

Recent studies have found out that gut microbiota plays a significant role in neuroimmune health and diseases which is correlated with its distinct composition in

ASD patients than in normal individuals (Li et al. 2017). The gut microbiota interacts with brain through gut-brain axis and through production of many neurotransmitters such as serotonin and small chain fatty acids such as butyrate. Under gut dysbiosis condition, certain bacterial products such as inflammatory cytokines pass BBB and act on brain neurons contributing to the pathologies of the disease.

The report showed the abundant presence of *Lactobacillus, Bacteriodes, Cloustridium* and *Proteobacteria* and lower levels of *Blautia, Dialister, Turicibacter* and *Bafidobacterium* in autistic patients than compared to normal individuals (Liu et al. 2019). Autism patients also showed the enriched presence of *Bacteriodes* and *Cloustridium* which mainly synthesises of propionates and that may be associated with ASD symptoms (Navarro et al. 2016; Finegold et al. 2002). A study on mice showed that the administration of propionic acid resulted in hyperactivity, abnormal motor movements and repetitive behaviour of mice which featured the similar activity in individuals with ASD (MacFabe et al. 2007).

A study of autistic children showed a relatively high affluence of *Akkermansia, Sarcina, Alistipes, Enterobacteriaceae and Clostridium* genera and low levels of *Coprococcus, Prevotella* and unclassified *Veillonallaceae* which contribute to altered composition of organic compounds such as free amino acids (De Angelis et al. 2015).

Children with ASD showed high level of *Clostridium histolyticum* which produces neurotoxins and contributes to aetiology of autism and a significant improvement in ASD children were observed by lowering the *Clostridium* yields (Parracho et al. 2005; Sandler et al. 2000).

This shows that gut microbiota plays a vital role in ASD and potential microbial modulation therapies could be effective in treating ASD.

9.9.4 Multiple Sclerosis

Multiple Sclerosis (MS) is an autoimmune disorder characterised by chronic inflammation of CNS leading progressive demyelination and deterioration of neurological functions (Hemmer et al. 2015; Mahad et al. 2015). The survey-based study reported that most of the MS patients experienced GI issues such as diarrhoea, faecal incontinence and constipation (Levinthal et al. 2013).

It was found that the phylum Fermicutes, butyrate producing bacteria were diminished in MS patients implicating its pathogenesis. Butyrate inhibits the pro-inflammatory pathway and protects from exposure to gut antigens (Canani et al. 2011). *Anaerostipes, Lacnospiraceae* and *Feacalibacterium* found to be reduced in MS which produces butyrate, supresses CNS demyelination through activation of G protein receptor and histone deacetylase (Chen et al. 2019). MS patients showed increased abundance of *Akkermansia, Bafidobacterium and Streptococcus* and relatively low portion of *Butyricimonas, Haemophilus, Slackia, Anaerostipes, Feacalibacterium and Bacteriodes* (Miyake et al. 2015; Berer et al. 2017).

9.10 Dysbiosis of Gut Microbiota

Dysbiosis means an alteration in relative composition and abundance of gut microbiota which causes dysfunctioning of neuroimmunological pathways leading to implications of pathogenesis of several neurological and neurodegenerative disorders in humans (Padhi et al. 2022). Normobiosis means the homeostasis between microbes which enables basal physiological immune response by immune effector site alteration in CNS. Bacterial infection which leads to microbial dysbiosis can restrict the immunoregulatory activity of neuroglial cells which aggravates neuroinflammation and neurodegeneration (Braniste et al. 2014; Erny et al. 2015; Rothhammer et al. 2016, 2018; Powell et al. 2017; Wekerle 2017).

Inflammation within the gastrointestinal (GI) tract imposes stress on the microbiome as cytokines and neurotransmitters are released. Concurrently, with the rise in intestinal permeability, these molecules circulate systemically. Elevated blood levels of cytokines such as TNF-α and MCP (monocyte chemoattractant protein) heighten the permeability of the blood-brain barrier, amplifying the impact of molecules escaping the permeable gut. This release affects brain function, contributing to symptoms like anxiety, depression, and memory loss (Biesmans et al. 2015). During dysbiosis, the pathogenic microbes producing pro-inflammatory cytokines increases and beneficial microbes which produces anti-inflammatory cytokines decreases (Maes et al. 2019; Capuco et al. 2020). In neurological disorders, the altered gut microbiota and its metabolites causes imbalance in nerve signalling through gut-brain axis which provokes terrible inflammations. This shows that the altered composition of gut microbes and its functional dysregulation leads to pathogenesis of neuroimmune disorders (Foster and McVey Neufeld 2013). Many studies have shown the evidence of gut microbes dysbiosis and its implications in several neurological disorders such as Alzheimer's and Parkinson's disease (Padhi et al. 2022). In autistic patients, the gut dysbiosis led to intestinal permeability known as leaky gut which resulted in secretion and spread of proinflammatory endotoxin like Lipopolyssacharide (LPS) which modulates the CNS activity on emotions and behaviour (Haba et al. 2012). In multiple sclerosis patients, the dysbiosis caused reduced abundance of Butyricimonas, the butyrate synthesising bacteria. This dysbiosis may increase the pro-inflammation of autoreactive T cells in the blood (Matusevicius et al. 1999). The multiple sclerosis patients also showed decreased abundance of Clostridia XIVa and IV clusters (Miyake et al. 2018). The interconnection of endocrine, neural, and immune pathways is demonstrated through the relationship between brain-derived neurotrophic factor (BDNF) mRNA in the hippocampal dentate gyrus and the stress response in germ-free mice. BDNF is vital for the development of neurons and synapses, influencing emotional and cognitive regulation. In male germ-free mice, an elevated stress response is linked to decreased hippocampal BDNF levels, which can be reversed by recolonisation with Bifidobacteria species. Furthermore, Bifidobacteria influence GABA receptor mRNA expression and decrease serum cortisol. Notably, these effects are absent after vagotomies, emphasising the significance of the parasympathetic nervous system in mediating the bacteria's impact on the stress response (Cryan and Dinan 2012).

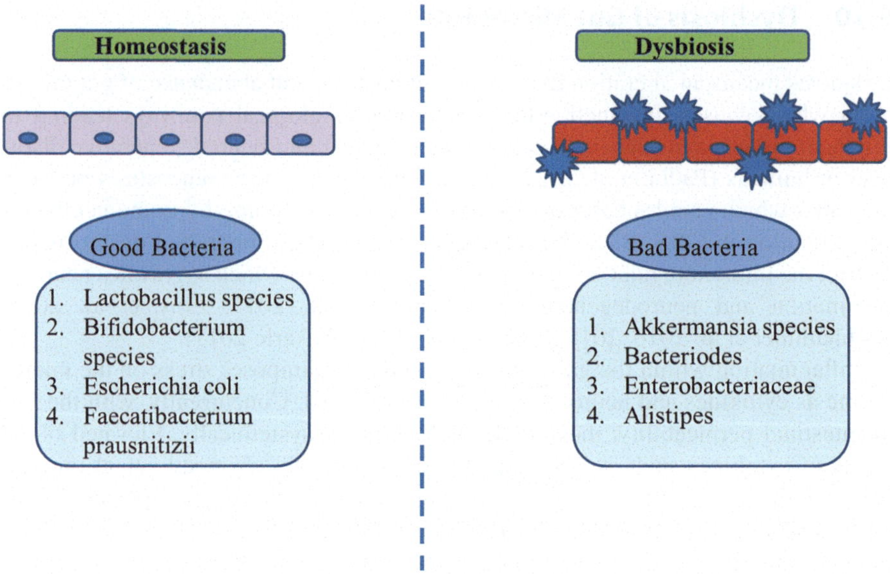

Fig. 9.2 The schematic representation of gut microbiota involved in homeostasis and dysbiosis

Therefore, the composition of the gut microbiota regulates the homeostasis and equilibrium between the microbes and the dysbiosis of microbes is associated with brain dysfunctioning. It is possible to reverse the disease condition completely or partially by creating the rebalance of disturbed composition of gut microbiota (Fig. 9.2). This can be achieved by treating with probiotics or by developing the microbiota-based therapies (Filosa et al. 2018).

9.11 Probiotics and Neuroimmune Interplay

Modulating the composition of the microflora provides an opportunity to enhance immune function, maintain homeostasis, and alleviate gut inflammation (Marlow et al., 2015). Probiotics are living microorganisms which are known to offer health benefits when taken in required amount (Gareau et al. 2010). The increased accessibility and resurgence in food trends favouring ancient food preparation techniques may lead patients to prefer natural sources of probiotics. Recent research indicates that incorporating fermented foods into diets does provide gastrointestinal and cognitive benefits (Selhub et al., 2014). Probiotics are lactic acid producing bacteria such as *Lactobacilli*, *Lactococcin*, *Saccharomycetes* and *Bafidobacterium* which are capable of treating various disease in humans (Verna and Lucak 2010; Verma and Shukla 2014; Sharma and Shukla 2016; Valsecchi et al. 2016). Probiotics regulates composition of intestinal microbes, maintains the integrity of intestinal barrier, regulates neuroinflammatory reactions and avoids bacterial migration (Rios et al. 2017). Many studies have proved that consumption of probiotics aids in several

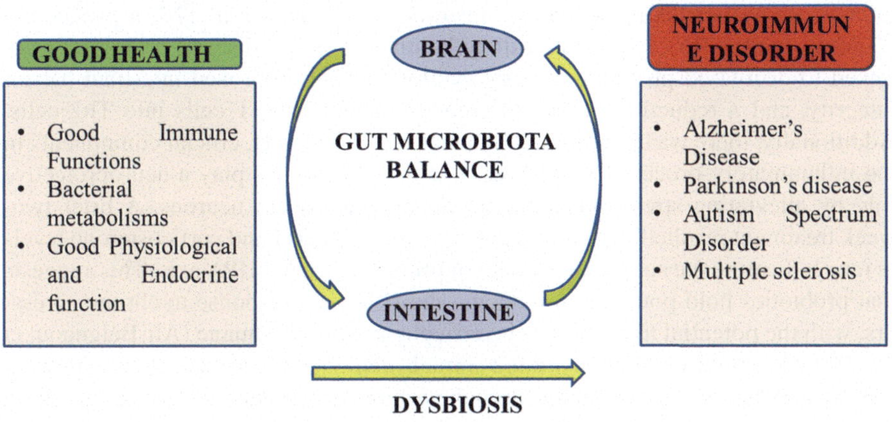

Fig. 9.3 The schematic diagram depicting the relationship of gut microbiota in health and neuro-immune disorder

health benefits by providing favourable conditions and modifications for gut microbiota (Fig. 9.3). Probiotics modifies the composition and functioning of the existing gut microbes and interacts directly with the host immune system (Navarro et al. 2016). Probiotics acts through gut-brain axis and controls CNS and the behaviour (Sampson and Mazmanian 2015). Even with the administration of multiple antidepressants employing various mechanisms of action, approximately 20% of patients do not exhibit improvements in the reduction of anxiety or depressive symptoms (Holtzheimer and Mayberg 2011). Soluble fibres like fructo-oligosaccharides and galacto-oligosaccharides, known as prebiotics, are utilised to promote the existing gut microbiota. Recent studies indicate that prebiotics, similar to probiotics, exhibit anxiolytic and antidepressant effects by mitigating stress-induced alterations in the colonic microbiota. Additionally, they contribute to the stabilisation of Bifidobacteria and Lactobacilli populations (Burokas et al. 2017). In a recent study, it was shown that the probiotics such as *Lactobacillus farciminis* inhibit permeability by reconstructing the colonic tight junction and improve the activity of HPA axis and neuroinflammatory changes caused by stress (Kelly et al. 2015). The study reported that the rat under stress supplemented with the probiotic *Lactobacillus helveticus* reduced cognitive dysfunction and improved stress-induced anxiety and depression (Liang et al. 2017). A *Lactobacillus rhamnosus*, a probiotic effectively enhanced GABA activity and improved their response to stress through vagus nerve stimulation (Bravo et al. 2011). Mice under stress showed improved anxiety and decreased level of plasma inflammatory cytokines when treated with *Lactobacillus plantarum* (Liu et al. 2016). The study showed that when children with ASD supplemented with probiotics containing *Lactobacillus*, *Streptococci* and *Bifidobacteria* normalised the ratio of *Firmicutes*/*Bacteroidetes* and the abundance of *Desulfovibrio* spp. Was found to be similar in both ASD and non-autistic children indicating the improved action of probiotics on ASD individual (Tomova et al. 2015). Numerous studies have documented positive outcomes in individuals with chronic inflammation after taking probiotics, resulting in a decrease in TNF-α production (D'Mello et al. 2015). In patients with inflammatory bowel disease, probiotics were linked to decreased pro-inflammatory cytokine levels, improved intestinal barrier integrity, and a reduction in the differentiation of CD4+ T cells into Th2 cells. Additionally, there was inhibition of nuclear factor kappa B, crucial components in the inflammatory process (Buckley et al. 2014). Probiotics play a neuroprotective role by alleviating stress-induced synaptic dysfunction in neurons. A brief two-week treatment resulted in a significant decrease in ACTH and corticosterone levels in rats, indicating the suppressive effect of probiotics on the HPA axis. This suggests that probiotics hold potential in reducing the HPA axis response to chronic stressors, with the potential to prevent or reverse physiological damage (Ait-Belgnaoui et al. 2012). Studies in both humans and animals demonstrate consistent reductions in anxiety and depressive symptoms through probiotic interventions. In a human study focusing on chronic stress, a three-week treatment with probiotics containing Bifidobacteria species proved notably effective, particularly benefiting individuals initially ranked in the bottom third of the elated/depressed scale (Benton et al. 2007).

In consideration, there are several evidence that support the beneficial effects of probiotics on mental health and altered behaviour. The interplay between gut microbiota and altered behaviour can be modulated with the supplementation of certain amount and composition of probiotics. This needs more studies to compel the efficacy and limitations regarding strains, doses, treatment duration and its benefits.

9.12 Conclusion

This summarises about the gut microbiota and its effect on neuroimmune disorders and altered behaviour. Specific findings about the gut-brain axis and neuroimmunity, Investigating the potential of personalized medicine approaches based on individual gut microbiome profiles on pathophysiology of neuroimmune disorders and its treatment using probiotics. Future research is necessary to explore the potential of modulatory approaches targeting the gut microbiota for broader health promotion and disease prevention. Unraveling the complex interplay between the gut microbiome and the brain holds immense promise for developing novel therapeutic strategies and improving the lives of individuals with neuroimmune disorders.

References

Abdel-Haq R, Schlachetzki JCM, Glass CK, Mazmanian SK (2019) Microbiome-microglia connections via the gut-brain axis. J Exp Med 216:41–59

Ait-Belgnaoui A, Durand H, Cartier C et al (2012) Prevention of gut leakiness by a probiotic treatment leads to attenuated HPA response to an acute psychological stress in rats. Psychoneuroendocrinology 37:1885–1895

Alzheimer's Association (2020) 2020 Alzheimer's disease facts and figures. Alzheimers Dement 5:234–270

Anglin R, Surette M, Moayyedi P, Bercik P (2015) Lost in translation: the gut microbiota in psychiatric illness. Can J Psychiatry 60(10):460–463. https://doi.org/10.1177/070674371506001007
Baj A, Moro E (2019) Glutamatergic signaling along the microbiota-gut-brain axis. Int J Mol Sci 20:1482. https://doi.org/10.3390/ijms20061482
Benton D, Williams C, Brown A (2007) Impact of consuming a milk drink containing a probiotic on mood and cognition. Eur J Clin Nutr 61:355–361
Barrett E, Ross RP, O'Toole PW, Fitzgerald GF, Stanton C (2012) γ-Aminobutyric acid production by culturable bacteria from the human intestine. J Appl Microbiol 113:411–417
Bercik P, Park AJ, Sinclair D, Khoshdel A, Lu J, Huang X, Deng Y, Blennerhassett PA, Fahnestock M, Moine D, Berger B, Huizinga JD, Kunze W, McLean PG, Bergonzelli GE, Collins SM, Verdu EF (2011) The anxiolytic effect of Bifidobacterium longum NCC3001 involves vagal pathways for gut-brain communication. Neurogastroenterol Motil 23(12):1132–1139. https://doi.org/10.1111/j.1365-2982.2011.01796.x
Berer K, Gerdes LA, Cekanaviciute E, Jia X, Xiao L, Xia Z, Liu C, Klotz L, Stauffer U, Baranzini SE, Kümpfel T, Hohlfeld R, Krishnamoorthy G, Wekerle H (2017) Gut microbiota from multiple sclerosis patients enables spontaneous autoimmune encephalomyelitis in mice. Proc Natl Acad Sci U S A 114(40):10719–10724. https://doi.org/10.1073/pnas.1711233114
Bienenstock J, Forsythe P, Karimi K, Kunze W (2010) Neuroimmune aspects of food intake. Int Dairy J 20:253–258
Bonaz B, Bazin T, Pellissier S (2018) The vagus nerve at the interface of the microbiota-gut-brain axis. Front Neurosci 12:49. https://doi.org/10.3389/fnins.2018.00049
Borsom EM, Lee K, Cope EK (2020) Do the bugs in your gut eat your memories? Relationship between gut microbiota and Alzheimer's disease. Brain Sci 10(10):814. https://doi.org/10.3390/brainsci10110814
Braniste V, Al-Asmakh M, Kowal C, Anuar F, Abbaspour A, Tóth M, Korecka A, Bakocevic N, Ng LG, Kundu P, Gulyás B, Halldin C, Hultenby K, Nilsson H, Hebert H, Volpe BT, Diamond B, Pettersson S (2014) The gut microbiota influences blood-brain barrier permeability in mice. Sci Transl Med 6:263ra158
Bravo JA, Forsythe P, Chew MV, Escaravage E, Savignac HM, Dinan TG, Bienenstock J, Cryan JF (2011) Ingestion of Lactobacillus strain regulates emotional behavior and central GABA receptor expression in a mouse via the vagus nerve. Proc Natl Acad Sci U S A 108:16050–16055
Briguglio M, Dell'Osso B, Panzica G, Malgaroli A, Banfi G, Dina CZ, Galentino R, Porta M (2018) Dietary neurotransmitters: a narrative review on current knowledge. Nutrients 10:591
Bruce-Keller AJ, Salbaum JM, Berthoud HR (2018) Harnessing gut microbes for mental health: getting from here to there. Biol Psychiatry 83(3):214–223. https://doi.org/10.1016/j.biopsych.2017.08.014
Buckley MM, O'Mahony SM, O'Malley D (2014) Convergence of neuro-endocrine-immune pathways in the pathophysiology of irritable bowel syndrome. World J Gastroenterol 20:8846–8858
Canani RB, Costanzo MD, Leone L, Pedata M, Meli R, Calignano A (2011) Potential beneficial effects of butyrate in intestinal and extraintestinal diseases. World J Gastroenterol 17(12):1519–1528. https://doi.org/10.3748/wjg.v17.i12.1519
Capuco A, Urits I, Hasoon J, Chun R, Gerald B, Wang JK, Kassem H, Ngo AL, Abd-Elsayed A, Simopoulos T, Kaye AD, Viswanath O (2020) Current perspectives on gut microbiome dysbiosis and depression. Adv Ther 37(4):1328–1346. https://doi.org/10.1007/s12325-020-01272-7
Cerrato V (2020) Cerebellar astrocytes: much more than passive bystanders in ataxia pathophysiology. J Clin Med 9:757
Chaudhuri KR, Healy DG, Schapira AH, National Institute for Clinical Excellence (2006) Non-motor symptoms of Parkinson's disease: diagnosis and management. Lancet Neurol 5(3):235–245
Chen T, Noto D, Hoshino Y, Mizuno M, Miyake S (2019) Butyrate suppresses demyelination and enhances remyelination. J Neuroinflammation 16:165
Clarke G, Grenham S, Scully P, Fitzgerald P, Moloney RD, Shanahan F, Dinan TG, Cryan JF (2013) The microbiome-gut-brain axis during early life regulates the hippocampal serotonergic system in a sex-dependent manner. Mol Psychiatry 18:666–673

Clayton TA, Baker D, Lindon JC, Everett JR, Nicholson JK (2009) Pharmacometabonomic identification of a significant host-microbiome metabolic interaction affecting human drug metabolism. Proc Natl Acad Sci U S A 106:14728–14733

Cloez-Tayarani I, Changeux JP (2007) Nicotine and serotonin in immune regulation and inflammatory processes: a perspective. J Leukoc Biol 81:599–606. https://doi.org/10.1189/jlb.0906544

De Angelis M, Francavilla R, Piccolo M, De Giacomo A, Gobbetti M (2015) Autism spectrum disorders and intestinal microbiota. Gut Microbes 6(3):207–213

DeCastro M, Nankova BB, Shah P, Patel P, Mally PV, Mishra R, La Gamma EF (2005) Short chain fatty acids regulate tyrosine hydroxylase gene expression through a cAMP-dependent signaling pathway. Brain Res Mol Brain Res 142:28–38. https://doi.org/10.1016/j.molbrainres.2005.09.002

Desai MS, Seekatz AM, Koropatkin NM, Kamada N, Hickey CA, Wolter M, Pudlo NA, Kitamoto S, Terrapon N, Muller A, Young VB, Henrissat B, Wilmes P, Stappenbeck TS, Núñez G, Martens EC (2016) A dietary fiber-deprived gut microbiota degrades the colonic mucus barrier and enhances pathogen susceptibility. Cell 167(5):1339–1353.e21. https://doi.org/10.1016/j.cell.2016.10.043

Diamant M, Blaak EE, de Vos WM (2011) Do nutrient-gut-microbiota interactions play a role in human obesity, insulin resistance and type 2 diabetes? Obes Rev 12(4):272–281. https://doi.org/10.1111/j.1467-789X.2010.00797.x

D'Mello C, Ronaghan N, Zaheer R et al (2015) Probiotics improve inflammation-associated sickness behavior by altering communication between the peripheral immune system and the brain. J Neurosci 35:10821–10830

Domingues C, da Cruz E Silva OAB, Henriques AG (2017) Impact of cytokines and chemokines on Alzheimer's disease neuropathological hallmarks. Curr Alzheimer Res 14(8):870–882

Dominy SS, Lynch C, Ermini F, Benedyk M, Marczyk A, Konradi A, Nguyen M, Haditsch U, Raha D, Griffin C, Holsinger LJ, Arastu-Kapur S, Kaba S, Lee A, Ryder MI, Potempa B, Mydel P, Hellvard A, Adamowicz K, Hasturk H et al (2019) *Porphyromonas gingivalis* in Alzheimer's disease brains: evidence for disease causation and treatment with small-molecule inhibitors. Sci Adv 5(1):eaau3333. https://doi.org/10.1126/sciadv.aau3333

Dumitrescu L, Marta D, Danau A, Lefter A, Tulba D, Cozma L, Manole E, Gherghiceanu M, Ceafalan LC, Popescu BO (2021) Serum and fecal markers of intestinal inflammation and intestinal barrier permeability are elevated in Parkinson's disease. Front Neurosci 15:689723

Eckburg PB, Bik EM, Bernstein CN, Purdom E, Dethlefsen L, Sargent M, Gill SR, Nelson KE, Relman DA (2005) Diversity of the human intestinal microbial flora. Science 308(5728):1635–1638. https://doi.org/10.1126/science.1110591

Emery DC, Shoemark DK, Batstone TE, Waterfall CM, Coghill JA, Cerajewska TL, Davies M, West NX, Allen SJ (2017) 16S rRNA next generation sequencing analysis shows bacteria in Alzheimer's post-mortem brain. Front Aging Neurosci 9:195

Erny D, Hrabe de Angelis AL, Jaitin D, Wieghofer P, Staszewski O, David E, Keren-Shaul H, Mahlakoiv T, Jakobshagen K, Buch T, Schwierzeck V, Utermöhlen O, Chun E, Garrett WS, McCoy KD, Diefenbach A, Staeheli P, Stecher B, Amit I, Prinz M (2015) Host microbiota constantly control maturation and function of microglia in the CNS. Nat Neurosci 18(7):965–977. https://doi.org/10.1038/nn.4030

Fasano A, Visanji NP, Liu LW, Lang AE, Pfeiffer RF (2015) Gastrointestinal dysfunction in Parkinson's disease. Lancet Neurol 14(6):625–639

Feehily C, Karatzas KA (2013) Role of glutamate metabolism in bacterial responses towards acid and other stresses. J Appl Microbiol 114:11–24

Ferrante RJ, Kubilus JK, Lee J, Ryu H, Beesen A, Zucker B, Smith K, Kowall NW, Ratan RR, Luthi-Carter R, Hersch SM (2003) Histone deacetylase inhibition by sodium butyrate chemotherapy ameliorates the neurodegenerative phenotype in Huntington's disease mice. J Neurosci 23:9418–9427

Filosa S, Di Meo F, Crispi S (2018) Polyphenols-gut microbiota interplay and brain neuromodulation. Neural Regen Res 13(12):2055–2059. https://doi.org/10.4103/1673-5374.241429

Filpa V, Moro E, Protasoni M, Crema F, Frigo G, Giaroni C (2016) Role of glutamatergic neurotransmission in the enteric nervous system and brain-gut axis in health and disease. Neuropharmacology 111:14–33

Finegold SM, Molitoris D, Song Y, Liu C, Vaisanen ML, Bolte E, McTeague M, Sandler R, Wexler H, Marlowe EM, Collins MD, Lawson PA, Summanen P, Baysallar M, Tomzynski TJ, Read E, Johnson E, Rolfe R, Nasir P, Shah H et al (2002) Gastrointestinal microflora studies in late-onset autism. Clin Infect Dis 35(Suppl 1):S6–S16. https://doi.org/10.1086/341914

Foster JA, McVey Neufeld KA (2013) Gut-brain axis: how the microbiome influences anxiety and depression. Trends Neurosci 36(5):305–312. https://doi.org/10.1016/j.tins.2013.01.005

Fujisaka S, Avila-Pacheco J, Soto M, Kostic A, Dreyfuss JM, Pan H, Ussar S, Altindis E, Li N, Bry L, Clish CB, Kahn CR (2018) Diet, genetics, and the gut microbiome drive dynamic changes in plasma metabolites. Cell Rep 22(11):3072–3086. https://doi.org/10.1016/j.celrep.2018.02.060

Fung KY, Cosgrove L, Lockett T, Head R, Topping DL (2012) A review of the potential mechanisms for the lowering of colorectal oncogenesis by butyrate. Br J Nutr 108:820–831

Furness JB, Callaghan BP, Rivera LR, Cho HJ (2014) The enteric nervous system and gastrointestinal innervation: integrated local and central control. Adv Exp Med Biol 817:39–71. https://doi.org/10.1007/978-1-4939-0897-4_3

Furusawa Y, Obata Y, Fukuda S, Endo TA, Nakato G, Takahashi D, Nakanishi Y, Uetake C, Kato K, Kato T, Takahashi M, Fukuda NN, Murakami S, Miyauchi E, Hino S, Atarashi K, Onawa S, Fujimura Y, Lockett T, Clarke JM, Topping DL, Tomita M, Hori S, Ohara O, Morita T, Koseki H, Kikuchi J, Honda K, Hase K, Ohno H (2013) Commensal microbe-derived butyrate induces the differentiation of colonic regulatory T cells. Nature 504:446–450. https://doi.org/10.1038/nature12721

Garden GA (2002) Microglia in human immunodeficiency virus-associated neurodegeneration. Glia 40:240–251

Grabert K, Michoel T, Karavolos MH, Clohisey S, Baillie JK, Stevens MP, Freeman TC, Summers KM, McColl BW (2016) Microglial brain region-dependent diversity and selective regional sensitivities to aging. Nat Neurosci 19:504–516

Gareau MG, Sherman PM, Walker WA (2010) Probiotics and the gut microbiota in intestinal health and disease. Nat Rev Gastroenterol Hepatol 7(9):503–514. https://doi.org/10.1038/nrgastro.2010.117

Garland EF, Hartnell IJ, Boche D (2022) Microglia and astrocyte function and communication: what do we know in humans? Front Neurosci 16:824888

Geirnaert A, Steyaert A, Eeckhaut V, Debruyne B, Arends JB, Van Immerseel F, Boon N, Van de Wiele T (2014) Butyricicoccus pullicaecorum, a butyrate producer with probiotic potential, is intrinsically tolerant to stomach and small intestine conditions. Anaerobe 30:70–74. https://doi.org/10.1016/j.anaerobe.2014.08.010

Giau V, Wu S, Jamerlan A, An S, Kim S, Hulme J (2018) Gut microbiota and their neuroinflammatory implications in Alzheimer's disease. Nutrients 10(11):1765

Goldstein DS, Holmes C, Lopez GJ, Wu T, Sharabi Y (2018) Cerebrospinal fluid biomarkers of central dopamine deficiency predict Parkinson's disease. Parkinsonism Relat Disord 50:108–112

Gruol DL (2023) The neuroimmune system and the cerebellum. Cerebellum. https://doi.org/10.1007/s12311-023-01624-3

Guinane CM, Cotter PD (2013) Role of the gut microbiota in health and chronic gastrointestinal disease: understanding a hidden metabolic organ. Ther Adv Gastroenterol 6(4):295–308. https://doi.org/10.1177/1756283X13482996

Haba R, Shintani N, Onaka Y, Wang H, Takenaga R, Hayata A, Baba A, Hashimoto H (2012) Lipopolysaccharide affects exploratory behaviors toward novel objects by impairing cognition and/or motivation in mice: possible role of activation of the central amygdala. Behav Brain Res 228(2):423–431. https://doi.org/10.1016/j.bbr.2011.12.027

Hahnen E, Hauke J, Trankle C, Eyupoglu IY, Wirth B, Blumcke I (2008) Histone deacetylase inhibitors: possible implications for neurodegenerative disorders. Expert Opin Investig Drugs 17:169–184

Hemmer B, Kerschensteiner M, Korn T (2015) Role of the innate and adaptive immune responses in the course of multiple sclerosis. Lancet Neurol 14(4):406–419. https://doi.org/10.1016/S1474-4422(14)70305-9

Hickman S, Izzy S, Sen P, Morsett L, El Khoury J (2018) Microglia in neurodegeneration. Nat Neurosci 21:1359–1369

Hippius H, Neundörfer G (2003) The discovery of Alzheimer's disease. Dialogues Clin Neurosci 5(1):101–108

Hooper LV, Wong MH, Thelin A, Hansson L, Falk PG, Gordon JI (2001) Molecular analysis of commensal host-microbial relationships in the intestine. Science. 291:881–884

Horvath K, Papadimitriou JC, Rabsztyn A, Drachenberg C, Tildon JT (1999) Gastrointestinal abnormalities in children with autistic disorder. J Pediatr 135(5):559–563

Janik R, Thomason LAM, Stanisz AM, Forsythe P, Bienenstock J, Stanisz GJ (2016) Magnetic resonance spectroscopy reveals oral lactobacillus promotion of increases in brain GABA, N-acetyl aspartate and glutamate. NeuroImage 125:988–995. https://doi.org/10.1016/j.neuroimage.2015.11.018

Jandhyala SM, Talukdar R, Subramanyam C, Vuyyuru H, Sasikala M, Nageshwar Reddy D (2015) Role of the normal gut microbiota. World J Gastroenterol 21(29):8787–8803. https://doi.org/10.3748/wjg.v21.i29.8787

Kelly JR, Kennedy PJ, Cryan JF, Dinan TG, Clarke G, Hyland NP (2015) Breaking down the barriers: the gut microbiome, intestinal permeability and stress-related psychiatric disorders. Front Cell Neurosci 9:392. https://doi.org/10.3389/fncel.2015.00392

Keshavarzian A, Green SJ, Engen PA, Voigt RM, Naqib A, Forsyth CB, Mutlu E, Shannon KM (2015) Colonic bacterial composition in Parkinson's disease. Mov Disord 30(10):1351–1360. https://doi.org/10.1002/mds.26307

Kidd PM (2007) Omega-3 DHA and EPA for cognition, behavior, and mood: clinical findings and structural-functional synergies with cell membrane phospholipids. Altern Med Rev 12:207–227

Kobayashi Y, Sugahara H, Shimada K, Mitsuyama E, Kuhara T, Yasuoka A, Kondo T, Abe K, Xiao JZ (2017) Therapeutic potential of Bifidobacterium breve strain A1 for preventing cognitive impairment in Alzheimer's disease. Sci Rep 7:13510

Koeth RA, Wang Z, Levison BS, Buffa JA, Org E, Sheehy BT, Britt EB, Fu X, Wu Y, Li L, Smith JD, DiDonato JA, Chen J, Li H, Wu GD, Lewis JD, Warrier M, Brown JM, Krauss RM, Tang WH et al (2013) Intestinal microbiota metabolism of L-carnitine, a nutrient in red meat, promotes atherosclerosis. Nat Med 19(5):576–585. https://doi.org/10.1038/nm.3145

Laurent RS, O'Brien LM, Ahmad ST (2013) Sodium butyrate improves locomotor impairment and early mortality in a rotenone-induced Drosophila model of Parkinson's disease. Neuroscience 246:382–390

Lay C, Sutren M, Rochet V, Saunier K, Doré J, Rigottier-Gois L (2005) Design and validation of 16S rRNA probes to enumerate members of the Clostridium leptum subgroup in human faecal microbiota. Environ Microbiol 7(7):933–946. https://doi.org/10.1111/j.1462-2920.2005.00763.x

Levinthal DJ, Rahman A, Nusrat S, O'Leary M, Heyman R, Bielefeldt K (2013) Adding to the burden: gastrointestinal symptoms and syndromes in multiple sclerosis. Mult Scler Int 2013:319201. https://doi.org/10.1155/2013/319201

Li Q, Han Y, Dy ABC, Hagerman RJ (2017) The gut microbiota and autism spectrum disorders. Front Cell Neurosci 11:120

Liang L, Zhou H, Zhang S, Yuan J, Wu H (2017) Effects of gut microbiota disturbance induced in early life on the expression of extrasynaptic GABA-A receptor $\alpha5$ and δ subunits in the hippocampus of adult rats. Brain Res Bull 135:113–119. https://doi.org/10.1016/j.brainresbull.2017.09.014

Liu YW, Liu WH, Wu CC, Juan YC, Wu YC, Tsai HP, Wang S, Tsai YC (2016) Psychotropic effects of Lactobacillus plantarum PS128 in early life-stressed and naïve adult mice. Brain Res 1631:1–12. https://doi.org/10.1016/j.brainres.2015.11.018

Liu F, Li J, Wu F, Zheng H, Peng Q, Zhou H (2019) Altered composition and function of intestinal microbiota in autism spectrum disorders: a systematic review. Transl Psychiatry 9(1):43

Lombardi VC, De Meirleir KL (2018) Nutritional modulation of the intestinal microbiota; future opportunities for the prevention and treatment of neuroimmune and neuroinflammatory disease. J Nutr Biochem 61:1–16

Louis P, Flint HJ (2009) Diversity, metabolism and microbial ecology of butyrate-producing bacteria from the human large intestine. FEMS Microbiol Lett 294:1–8

Lucerne KE, Osman A, Meckel KR, Kiraly DD (2021) Contributions of neuroimmune and gut-brain signaling to vulnerability of developing substance use disorders. Neuropharmacology 192:108598. https://doi.org/10.1016/j.neuropharm.2021.108598

MacFabe DF, Cain DP, Rodriguez-Capote K, Franklin AE, Hoffman JE, Boon F, Taylor AR, Kavaliers M, Ossenkopp KP (2007) Neurobiological effects of intraventricular propionic acid in rats: possible role of short chain fatty acids on the pathogenesis and characteristics of autism spectrum disorders. Behav Brain Res 176:149–169. https://doi.org/10.1016/j.bbr.2006.07.025

Macfarlane S, Macfarlane GT (2003) Regulation of short-chain fatty acid production. Proc Nutr Soc 62:67–72

Maes M, Simeonova D, Stoyanov D, Leunis JC (2019) Upregulation of the nitrosylome in bipolar disorder type 1 (BP1) and major depression, but not BP2: increased IgM antibodies to nitrosylated conjugates are associated with indicants of leaky gut. Nitric Oxide 91:67–76. https://doi.org/10.1016/j.niox.2019.07.003

Mahad DH, Trapp BD, Lassmann H (2015) Pathological mechanisms in progressive multiple sclerosis. Lancet Neurol 14(2):183–193. https://doi.org/10.1016/S1474-4422(14)70256-X

Marín L, Miguélez EM, Villar CJ, Lombó F (2015) Bioavailability of dietary polyphenols and gut microbiota metabolism: antimicrobial properties. Biomed Res Int 2015:905215

Matamoros S, Gras-Leguen C, Le Vacon F, Potel G, de La Cochetiere MF (2013) Development of intestinal microbiota in infants and its impact on health. Trends Microbiol 21(4):167–173. https://doi.org/10.1016/j.tim.2012.12.001

Matsumoto M, Ooga T, Kibe R, Aiba Y, Koga Y, Benno Y (2017) Colonic absorption of low-molecular-weight metabolites influenced by the intestinal microbiome: a pilot study. PLoS One 12(2):e0169207

Matusevicius D, Kivisäkk P, He B, Kostulas N, Ozenci V, Fredrikson S, Link H (1999) Interleukin-17 mRNA expression in blood and CSF mononuclear cells is augmented in multiple sclerosis. Mult Scler 5(2):101–104. https://doi.org/10.1177/135245859900500206

Mayer EA (2011) Gut feelings: the emerging biology of gut-brain communication. Nat Rev Neurosci 12(8):453–466. PMID: 21750565

Miyake S, Kim S, Suda W, Oshima K, Nakamura M, Matsuoka T, Chihara N, Tomita A, Sato W, Kim SW, Morita H, Hattori M, Yamamura T (2015) Dysbiosis in the gut microbiota of patients with multiple sclerosis, with a striking depletion of species belonging to clostridia XIVa and IV clusters. PLoS One 10(9):e0137429. https://doi.org/10.1371/journal.pone.0137429

Miyake Y, Tanaka K, Okubo H, Sasaki S, Furukawa S, Arakawa M (2018) Soy isoflavone intake and prevalence of depressive symptoms during pregnancy in Japan: baseline data from the Kyushu Okinawa Maternal and Child Health Study. Eur J Nutr 57(2):441–450. https://doi.org/10.1007/s00394-016-1327-5

Mulak A, Bonaz B (2004) Irritable bowel syndrome: a model of the brain-gut interactions. Med Sci Monitor 10(4):RA55–RA62

Navarro F, Liu Y, Rhoads JM (2016) Can probiotics benefit children with autism spectrum disorders? World J Gastroenterol 22(46):10093–10102

Padhi P, Worth C, Zenitsky G, Jin H, Sambamurti K, Anantharam V, Kanthasamy A, Kanthasamy AG (2022) Mechanistic insights into gut microbiome dysbiosis-mediated neuroimmune dysregulation and protein misfolding and clearance in the pathogenesis of chronic neurodegenerative disorders. Front Neurosci 16:836605. https://doi.org/10.3389/fnins.2022.836605

Pannasch U, Rouach N (2013) Emerging role for astroglial networks in information processing: from synapse to behavior. Trends Neurosci 36:405–417

Parracho HM, Bingham MO, Gibson GR, McCartney AL (2005) Differences between the gut microflora of children with autistic spectrum disorders and that of healthy children. J Med Microbiol 54:987–991

Petra AI, Panagiotidou S, Hatziagelaki E, Stewart JM, Conti P, Theoharides TC (2015) Gut-microbiota-brain axis and its effect on neuropsychiatric disorders with suspected immune dysregulation. Clin Ther 37(5):984–995. https://doi.org/10.1016/j.clinthera.2015.04.002

Petra N, Petra CG, Zhang Z, Constantinescu EM, Anitescu M (2016. Nov 4) A Bayesian approach for parameter estimation with uncertainty for dynamic power systems. IEEE Trans Power Syst 32(4):2735–2743

Pistollato F, Cano SS, Elio I, Vergara MM, Giampieri F, Battino M (2016) Role of gut microbiota and nutrients in amyloid formation and pathogenesis of Alzheimer's disease. Nutr Rev 74(10):624–634. https://doi.org/10.1093/nutrit/nuw023

Powell N, Walker MM, Talley NJ (2017) The mucosal immune system: master regulator of bidirectional gut-brain communications. Nat Rev Gastroenterol Hepatol 14(3):143–159. https://doi.org/10.1038/nrgastro.2016.191

Qin J, Li R, Raes J, Arumugam M, Burgdorf KS, Manichanh C, Nielsen T, Pons N, Levenez F, Yamada T, Mende DR, Li J, Xu J, Li S, Li D, Cao J, Wang B, Liang H, Zheng H, Xie Y, Tap J, Lepage P, Bertalan M, Batto JM, Hansen T, Le Paslier D, Linneberg A, Nielsen HB, Pelletier E, Renault P, Sicheritz-Ponten T, Turner K, Zhu H, Yu C, Li S, Jian M, Zhou Y, Li Y, Zhang X, Li S, Qin N, Yang H, Wang J, Brunak S, Doré J, Guarner F, Kristiansen K, Pedersen O, Parkhill J, Weissenbach J, MetaHIT Consortium, Bork P, Ehrlich SD, Wang JA (2010) Human gut microbial gene catalogue established by metagenomic sequencing. Nature 464(7285):59–65. https://doi.org/10.1038/nature08821

Rieder R, Wisniewski PJ, Alderman BL, Campbell SC (2017) Microbes and mental health: a review. Brain Behav Immun 66:9–17

Rios AC, Maurya PK, Pedrini M, Zeni-Graiff M, Asevedo E, Mansur RB, Wieck A, Grassi-Oliveira R, McIntyre RS, Hayashi MAF, Brietzke E (2017) Microbiota abnormalities and the therapeutic potential of probiotics in the treatment of mood disorders. Rev Neurosci 28(7):739–749. https://doi.org/10.1515/revneuro-2017-0001

Rothhammer V, Mascanfroni ID, Bunse L, Takenaka MC, Kenison JE, Mayo L, Chao CC, Patel B, Yan R, Blain M, Alvarez JI, Kébir H, Anandasabapathy N, Izquierdo G, Jung S, Obholzer N, Pochet N, Clish CB, Prinz M, Prat A et al (2016) Type I interferons and microbial metabolites of tryptophan modulate astrocyte activity and central nervous system inflammation via the aryl hydrocarbon receptor. Nat Med 22(6):586–597. https://doi.org/10.1038/nm.4106

Rothhammer V, Borucki DM, Tjon EC, Takenaka MC, Chao CC, Ardura-Fabregat A, de Lima KA, Gutiérrez-Vázquez C, Hewson P, Staszewski O, Blain M, Healy L, Neziraj T, Borio M, Wheeler M, Dragin LL, Laplaud DA, Antel J, Alvarez JI, Prinz M et al (2018) Microglial control of astrocytes in response to microbial metabolites. Nature 557(7707):724–728. https://doi.org/10.1038/s41586-018-0119-x

Roversi K, Callai-Silva N, Roversi K, Griffith M, Boutopoulos C, Prediger RD, Talbot S (2021) Neuro-immunity and gut dysbiosis drive Parkinson's disease-induced pain. Front Immunol 12:759679

Sampson TR, Mazmanian SK (2015) Control of brain development, function, and behavior by the microbiome. Cell Host Microbe 17(5):565–576. https://doi.org/10.1016/j.chom.2015.04.011

Sandler RH, Finegold SM, Bolte ER, Buchanan CP, Maxwell AP, Väisänen ML, Nelson MN, Wexler HM (2000) Short-term benefit from oral vancomycin treatment of regressive-onset autism. J Child Neurol 15(7):429–435. https://doi.org/10.1177/088307380001500701

Scheperjans F, Aho V, Pereira PA, Koskinen K, Paulin L, Pekkonen E, Haapaniemi E, Kaakkola S, Eerola-Rautio J, Pohja M, Kinnunen E, Murros K, Auvinen P (2015) Gut microbiota are related to Parkinson's disease and clinical phenotype. Mov Disord 30(3):350–358. https://doi.org/10.1002/mds.26069

Sekirov I, Russell SL, Antunes LC, Finlay BB (2010) Gut microbiota in health and disease. Physiol Rev 90:859–904

Sharma M, Shukla G (2016) Metabiotics: one step ahead of probiotics; an insight into mechanisms involved in anticancerous effect in colorectal cancer. Front Microbiol 7:1940. https://doi.org/10.3389/fmicb.2016.01940

Shyamaladevi N, Jayakumar AR, Sujatha R, Paul V, Subramanian EH (2002) Evidence that nitric oxide production increases γ-amino butyric acid permeability of blood-brain barrier. Brain Res Bull 57:231–236

Skonieczna-Zydecka K, Jakubczyk K, Maciejewska-Markiewicz D, Janda K, Kaźmierczak-Siedlecka K, Kaczmarczyk M, Łoniewski I, Marlicz W (2020) Gut biofactory—neurocompetent metabolites within the gastrointestinal tract. A scoping review. Nutrients 12:3369. https://doi.org/10.3390/nu12113369

Sonnenburg JL, Xu J, Leip DD, Chen CH, Westover BP, Weatherford J, Buhler JD, Gordon JI (2005) Glycan foraging in vivo by an intestine-adapted bacterial symbiont. Science. 307:1955–1959

Sudo N, Chida Y, Aiba Y, Sonoda J, Oyama N, Yu XN, Kubo C, Koga Y (2004) Postnatal microbial colonization programs the hypothalamic-pituitary-adrenal system for stress response in mice. J Physiol 558(Pt 1):263–275. https://doi.org/10.1113/jphysiol.2004.063388

Sun X, Meng X, Zhang J, Li Y, Wang L, Qin X, Sui N, Zhang Y (2012) GABA attenuates amyloid toxicity by downregulating its endocytosis and improves cognitive impairment. J Alzheimers Dis 31:635–649

Takanaga H, Ohtsuki S, Hosoya K-I, Terasaki T (2001) GAT2/BGT-1 as a system responsible for the transport of gamma-aminobutyric acid at the mouse blood-brain barrier. J Cereb Blood Flow Metab 21:1232–1239

Thangaraju M, Cresci GA, Liu K, Ananth S, Gnanaprakasam JP, Browning DD, Mellinger JD, Smith SB, Digby GJ, Lambert NA, Prasad PD, Ganapathy V (2009) GPR109A is a G-protein-coupled receptor for the bacterial fermentation product butyrate and functions as a tumor suppressor in colon. Cancer Res 69:2826–2832. https://doi.org/10.1158/0008-5472.CAN-08-4466

Tomova A, Husarova V, Lakatosova S, Bakos J, Vlkova B, Babinska K, Ostatnikova D (2015) Gastrointestinal microbiota in children with autism in Slovakia. Physiol Behav 138:179–187. https://doi.org/10.1016/j.physbeh.2014.10.033

Tremaroli V, Bäckhed F (2012) Functional interactions between the gut microbiota and host metabolism. Nature 489(7415):242–249. https://doi.org/10.1038/nature11552

Vacca M, Celano G, Calabrese FM, Portincasa P, Gobbetti M, De Angelis M (2020) The controversial role of human gut Lachnospiraceae. Microorganisms 8(4):573. https://doi.org/10.3390/microorganisms8040573

Valles-Colomer M, Falony G, Darzi Y, Tigchelaar EF, Wang J, Tito RY, Schiweck C, Kurilshikov A, Joossens M, Wijmenga C, Claes S, Van Oudenhove L, Zhernakova A, Vieira-Silva S, Raes J (2019) The neuroactive potential of the human gut microbiota in quality of life and depression. Nat Microbiol 4(4):623–632. https://doi.org/10.1038/s41564-018-0337-x

Valsecchi C, Carlotta Tagliacarne S, Castellazzi A (2016) Gut microbiota and obesity. J Clin Gastroenterol 50:S157–S158. https://doi.org/10.1097/MCG.0000000000000715

Veenstra-VanderWeele J, Muller CL, Iwamoto H, Sauer JE, Owens WA, Shah CR, Cohen J, Mannangatti P, Jessen T, Thompson BJ, Ye R, Kerr TM, Carneiro AM, Crawley JN, Sanders-Bush E, McMahon DG, Ramamoorthy S, Daws LC, Sutcliffe JS, Blakely RD (2012) Autism gene variant causes hyperserotonemia, serotonin receptor hypersensitivity, social impairment and repetitive behavior. Proc Natl Acad Sci U S A 109(14):5469–5474. https://doi.org/10.1073/pnas.1112345109

Verma A, Shukla G (2014) Synbiotic (Lactobacillus rhamnosus+Lactobacillus acidophilus+inulin) attenuates oxidative stress and colonic damage in 1,2 dimethylhydrazine dihydrochloride-induced colon carcinogenesis in Sprague-Dawley rats: a long-term study. Eur J Cancer Prev 23(6):550–559. https://doi.org/10.1097/CEJ.0000000000000054

Verna EC, Lucak S (2010) Use of probiotics in gastrointestinal disorders: what to recommend? Ther Adv Gastroenterol 3(5):307–319. https://doi.org/10.1177/1756283X10373814

Vila M, Przedborski S (2004) Genetic clues to the pathogenesis of Parkinson's disease. Nat Med 10(Suppl):S58–S62

Vitetta L, Bambling M, Alford H (2014) The gastrointestinal tract microbiome, probiotics, and mood. Inflammopharmacology 22:333–339. https://doi.org/10.1007/s10787-014-0216-x

Vogt NM, Kerby RL, Dill-McFarland KA, Harding SJ, Merluzzi AP, Johnson SC, Carlsson CM, Asthana S, Zetterberg H, Blennow K, Bendlin BB, Rey FE (2017) Gut microbiome alterations in Alzheimer's disease. Sci Rep 7(1):13537. https://doi.org/10.1038/s41598-017-13601-y
Wajner M, Latini A, Wyse AT, Dutra-Filho CS (2004) The role of oxidative damage in the neuropathology of organic acidurias: insights from animal studies. J Inherit Metab Dis 27(4):427–448
Wallen ZD, Appah M, Dean MN, Sesler CL, Factor SA, Molho E, Zabetian CP, Standaert DG, Payami H (2020) Characterizing dysbiosis of gut microbiome in PD: evidence for overabundance of opportunistic pathogens. NPJ Parkinsons Dis 6:11. https://doi.org/10.1038/s41531-020-0112-6
Wang HX, Wang YP (2016) Gut microbiota-brain axis. Chin Med J 129(19):2373–2380. https://doi.org/10.4103/0366-6999.190667
Wang X, Sun G, Feng T, Zhang J, Huang X, Wang T, Xie Z, Chu X, Yang J, Wang H, Chang S, Gong Y, Ruan L, Zhang G, Yan S, Lian W, Du C, Yang D, Zhang Q, Lin F et al (2019) Sodium oligomannate therapeutically remodels gut microbiota and suppresses gut bacterial amino acids-shaped neuroinflammation to inhibit Alzheimer's disease progression. Cell Res 29(10):787–803. https://doi.org/10.1038/s41422-019-0216-x
Wekerle H (2017) Brain autoimmunity and intestinal microbiota: 100 trillion game changers. Trends Immunol 38(7):483–497. https://doi.org/10.1016/j.it.2017.03.008
Yano JM, Yu K, Donaldson GP, Shastri GG, Ann P, Ma L, Nagler CR, Ismagilov RF, Mazmanian SK, Hsiao EY (2015) Indigenous bacteria from the gut microbiota regulate host serotonin biosynthesis. Cell 161(2):264–276. https://doi.org/10.1016/j.cell.2015.02.047
Zhu Y, Yuan M, Liu Y, Yang F, Chen WZ, Xu ZZ, Xiang ZB, Xu RS (2022) Association between inflammatory bowel diseases and Parkinson's disease: systematic review and meta-analysis. Neural Regen Res 17(2):344–353

Customised Microbiome Restoration Approaches in Older People: Perspectives and Therapeutic Prospects

10

Rajalakshmi Manikkam, Vijayalakshmi Periyasamy, and Indu Sabapathy

Abstract

The microbiome performs a crucial role in the development, function, and regulation of the host immune system from birth to old age. In exchange, millions of helpful microorganisms living inside of us have helped the immune system co-evolve while developing effective defences against pathogen invasion. Age-related major compositional and functional changes in the immune system and gut microbiome are associated with heightened vulnerability to infectious diseases and decreased immunisation responses. Recent research suggests that addressing age-related dysbiosis may increase longevity and enhance health, in part by lowering systemic chronic inflammation and immunosenescence. Our knowledge of microbial colonisation, maturation, dysbiosis in health- and disease-related subgroups has improved as a result of investigations into the human gut microbiome. There are new opportunities to use gut microorganisms as therapeutic agents to treat human diseases as well as their enormous metabolic capacity and role in maintaining human health. Microbiome treatments aim to engineer the gut microbiome by applying native or synthetic bacteria, antibiotics, bacteriophages, and bacteriocins in additive, subtractive, or modulatory therapy.

R. Manikkam (✉)
PG & Research Department of Zoology, Holy Cross College (Autonomous), Tiruchirappalli, Tamil Nadu, India

PG & Research Department of Biotechnology & Bioinformatics, DBT-BIF Centre, Holy Cross College (Autonomous), Affiliated to Bharathidasan University, Tiruchirappalli, Tamil Nadu, India
e-mail: rajalakshmi@hcctrichy.ac.in

V. Periyasamy · I. Sabapathy
PG & Research Department of Biotechnology & Bioinformatics, DBT-BIF Centre, Holy Cross College (Autonomous), Affiliated to Bharathidasan University, Tiruchirappalli, Tamil Nadu, India

S. Pathak, A. Banerjee (eds.), *Gut Microbiome and Brain Ageing*,
https://doi.org/10.1007/978-981-99-8803-7_10

The global medicines market has demonstrated its enormous economic potential. In this review we discussed about the role of gut microbiome in ageing and their therapeutic application.

Keywords

Gut-microbiota · Ageing · Additive therapy · Subtractive therapy · Probiotics

10.1 Introduction

The term 'gut microbiota', also referred to as the 'human gastrointestinal microbiota', refers to all of the microorganisms present in human digestive tracts. There are also reports of several microorganisms being harboured inside the gastrointestinal tract by other species like insects. Archaea and bacteria are the main players in the gut microbiota. According to Saxena and Sharma (2016), the term 'human gastrointestinal metagenome' refers to the compilation of all the genomes found in the gut microbiota. Anaerobes make up nearly all of the total amount of bacteria in the gut. However, the cecum contains a significant amount of aerobic bacteria (Sherwood et al. 2013).

When compared to other species, the gut microbiota has a large number of different bacteria, just like in humans (Quigley 2013). In the body, the gut microbial community develops 2–3 years after birth. During this stage, the gastrointestinal musical barrier that it emits and the gut epithelium both expand. The entire pathogenic organism is kept at bay by this development, which is tolerable to the gut flora (Sommer and Bäckhed 2013). O'Hara and Shanahan (2006) referred to the gut microbiota as a 'forgotten organ'. The Human Microbiome Project has been launched with the goal of defining the function of gut flora in human physiology and disorders (Turnbaugh et al. 2007).

Other reviews contain additional research on the interaction between the microbiome and the immune system and its function in health and disease. According to a theory, the immune system and gut microbiota cooperate to generate signals that promote immune cell maturity and the emergence of immune system reactions. Numerous investigations into the human gut microbiota are being conducted to see how it may affect human health (Shen and Wong 2016).

Dysbiosis, which is defined by disturbances in the gut's microbial ecosystem, has recently been recognised as a leading global cause of mortality and disease. It also plays a role in a variety of age-related disorders. The majority of age-related dysbiosis, also known as microbial ageing in humans, was characterised by a decline in the population of Clostridiales and Bifidobacterium and an increase in Proteobacteria. Additionally, this resulted in pathobionts such Enterobacteriaceae becoming overpopulated (Xu et al. 2019).

According to López-Otín et al. (2013), ageing is linked to a reduction in a number of physiological processes, metabolic dysregulation, and an increase in the prevalence of non-communicable diseases and disorders. These factors, as well as the ensuing modifications in lifestyle habits, such as a decline in the standard of

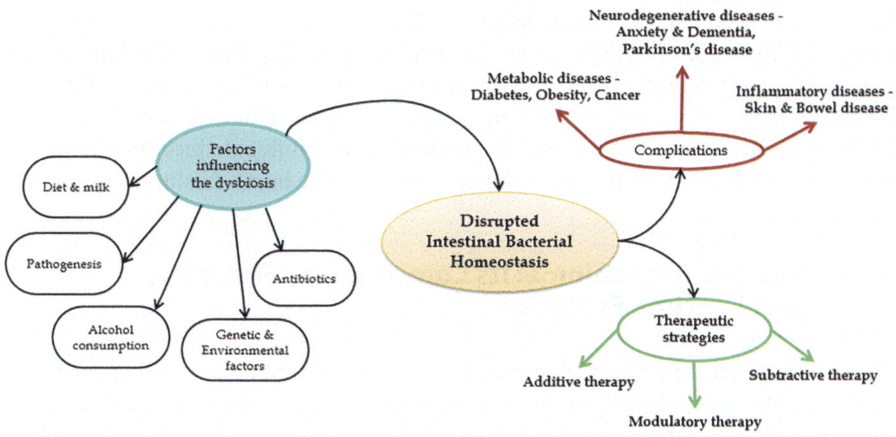

Fig. 10.1 Impact of microbiome enhancement in maintaining of good health

eating and physical activity and an increase in medication use, have a significant impact on the composition of the gut microbiome (DeJong et al. 2020). Recent investigations have showed these microbiome changes could also be linked to the risk or onset of a number of age-related diseases (Raza et al. 2019). Sarcopenia, inflamm-aging, the steady deterioration of physiological processes, and the emergence of several diseases are all characteristics of ageing. The shift in the gut microbiome that comes with ageing may also increase vulnerability to a number of diseases. These changes, which are hastened by lifestyle variables including medicine, limited mobility, and restricted nutrition, include a loss of commensals and an increase in disease-associated pathobionts. Figure 10.1 represents the impact of microbiome enhancement in maintaining of good health.

10.2 The Gut Microbiome's Classification

According to Qin et al. (2010), the gut microbiota is thought to contain a gene content that is more than 100 times more than that of the human genome. Normal microorganisms, opportunistic microbes, and pathogenic bacteria are the three broad categories for the gut microbiome. The species occurring in the stomach were not extensively investigated because of their survival restrictions. The bacteria are not capable of being grown outside of their hosts (Sears 2005). The populations of microorganisms differ significantly from person to person, despite the fact that only a small number of fundamental species are shared by all individuals (Tap et al. 2009). If we focus on a single person, the microorganisms in their bodies proliferate and remain stable throughout time, despite potential changes brought on by factors such as ageing, nutrition, and lifestyle changes (O'Hara and Shanahan 2006). Firmicutes, Bacteroidetes, Proteobacteria, and Actinobacteria are the four bacterial phyla that are most commonly observed in the human gut (Khanna and Tosh 2014).

According to Guarner and Malagelada (2003), the majority of bacteria belong to the genera Clostridium, Eubacterium, Peptococcus, Faecalibacterium, and Ruminococcus, among many others. Other genera including Lactobacillus and Escherichia are also present, albeit in smaller numbers (Guarner and Malagelada 2003). The species from the genus Bacteroides are significant since they make up over 30% of the gut bacteria alone and are crucial to the host's health (Sears 2005).

10.3 The Gut Microbiome: Its Composition, Function, and Lifetime Evolution

The vast, intricate, and varied community of microbes that inhabit our intestines is known as the gut microbiota. It is composed of microorganisms that evolved together with humans, including bacteria, archaea, viruses, and some single-celled eukaryotes. The gut microbiota is the biggest and most intense collection of microorganisms in the human body, with levels of 1014 cells in the colon. The so-called 'intestinal microbiome' is a massive genetic database of this bacteria. The human microbiome is currently thought to contain over ten million distinct genes, making it a significant genetic resource. It has been estimated that there are between 1000 and 1500 different bacterial species present in the human gut, with each individual host home to about 150 different species that collectively contain more genes than the human genome. Our second genome, the microbiome, is frequently referred to as our second genome5 and has a significant impact on our physiology and metabolism.

Deciphering the role of the microbiota in preserving human health is of growing interest to the scientific community, and several initiatives have been started with this goal all around the world. More thorough research on the gut microbiome's makeup and functionality has been made possible in recent years thanks to the advent of high-throughput analytical methods and 'meta-omics' technology. Despite these developments, it is still unknown exactly what a 'healthy' gut microbiome looks like.

10.4 Health and Disease Outcomes of the Gut Microbiota

Animals are home to complex populations of gut bacteria that evolved together with their multicellular hosts and display host uniqueness in both composition and activity (Ley et al. 2008). According to estimates, there are 2000 different bacterial species living in the human gut, which may be more than the entire amount of cells in the human body as well as the total number of different coding genes (Sender et al. 2016). Culture-independent techniques revealed that patients with a variety of illnesses had microbiomes that are different from those of healthy people, which increased interest in the human gut microbiome. However, few mechanisms have been demonstrated to connect pathophysiology to specific microbial metabolites that are either added to the microbiome associated with the disease or subtracted from the healthy microbiome (Brown and Hazen 2015). This problem is made more complex by the difficulties in identifying a healthy microbiome. Despite these obstacles, there are a number of broad categories that may be used to think about how microbiome-health interactions might become dysfunctional as people age and what the consequences might be for the host.

10.5 Age-Related Influences on the Microbiome

Homeostasis gradually disappears as we become older, and we become less functional and more susceptible to death. Infectious, neoplastic, metabolic, and degenerative diseases associated with frailty and cognitive loss are some examples of age-related diseases. However, they are brought on by changes in the microbiome, that in turns impact how quickly things get worse with age (López-Otín et al. 2013). The basic markers of ageing in animals have been found at the molecular and cellular levels.

The microbiome alterations associated with ageing are extremely varied and regulated by internal and external factors related to the environment. In this respect, the gut microbiota would unavoidably be impacted by the gradual deterioration in the physiological functioning of the alimentary tract. In addition to these changes, the genome is unstable, cells (and mitochondria) fail, proteostasis is reduced, and there is epigenetic dysregulation, which leads to the emergence of chronic diseases, metabolic conditions, and altered gut-brain communication (Pellanda et al. 2020). The consequences for host behaviour and lifestyle (more frailty, prescription use, surgery, decreased physical activity, and diet quality) may worsen the effects on the gut flora. Although there is a chance for positive behavioural change, age-associated modifications within the microbiome are also influenced by choices made during one's lifetime, notably regarding food. Less research has been done on themes such as how society treats its elderly and how social interactions affect the microbiota's composition.

According to growing body of research, it is beneficial to one's health for members of a social group to share microbes (Stein et al. 2016). Even though research on strain monitoring have revealed that commensals and mutualistic bacteria are also distributed through social networks, the spread of illnesses there has garnered a lot of interest (Sarkar et al. 2020). Comparatively to the microbiomes of residents of other families, those who share a residence likely to have similar compositions (Lax et al. 2014). Pets kept in the home can spread bacteria by serving as transmission vectors. Additionally, those with more social circles also appears to possess a wider range of gut microbiomes. The term 'social microbiome' refers to the collective microbial metacommunity of a family and describes how it changes over time, with elderly people experiencing a sometimes rapid loss in their capacity to acquire microbes. Whether the people live alone or under institutional care will also affect the situation.

10.6 Changes in Gut Microbiome with Ageing

Longitudinal research can identify the microbiome components linked with both normal and unhealthy ageing by observing people throughout an extended period of time and tying the composition of the gut microbiome determined at intermediary points in time to their finalised physiological or medical condition. A popular alternative is the use of a cross-sectional research design, which categorises older individuals according to measures of both healthy and unhealthy ageing and identifies the associated taxonomic markers. The microbiome has been associated with conditions such as less exercise, cardiometabolic illnesses, cognitive impairment, migraine, decreased bone mass density, weight gain, metabolic syndrome along with multiple medical conditions, chronic renal failure, and pre-mortality, according to studies (Langsetmo et al. 2019; Luan et al. 2020). An increase in pathobionts, such as Ruminococcus torques, Eggerthella, Desulfovibrio, and members of the Enterobacteriaceae family, as well as disease-associated Clostridium species, is the main distinction between these two groups. While the third group of naturally occurring microbial markers (including Akkermansia, Odoribacter, Butyricimonas, Butyrivibrio, Oscillospira, Christensenellaceae, and Barnesiellaceae) decrease in certain age-related disorders or conditions associated with unhealthy ageing, they increase in overall ageing research. These species are examples of taxonomic 'milestones' that become more prevalent during healthy ageing but disappear after physiological deterioration.

10.7 Ageing and the Gut Microbiome

As it incorporates and reacts to signals from the environment throughout time, the gut microbiome experiences dynamic alterations (O'Toole and Jeffery 2015). The makeup and operation of the gut microbiome are continuously shaped by factors such as diet, drug use, physical exercise, and social environment (Yatsunenko et al. 2012). Dysbiosis, an imbalance in the composition of the microbial population, is frequently associated with age-related diseases, which together account for the majority of disability and mortality cases globally. The combination of bacterial species and metabolic activity are altered as part of the gut microbiome's dynamic adaptation to an environment that is always changing. The host-immune system meticulously regulates this process by allowing beneficial bacteria to prosper and occupy mucosal areas while removing or neutralising pathogenic microbes, which is essential for the formation of the gut microbiome. Immunological wellness progressively decreases with age, which has significant functional implications for host well-being and immunity, making it challenging to monitor this dynamic host-microbial relationship.

A reduction in Clostridiales and Bifidobacterium, an enrichment in Proteobacteria, and an overrepresentation of pathobionts such Enterobacteriaceae are the hallmarks of age-related dysbiosis in humans, which is here referred to as microb-aging (Xu et al. 2019). The current understanding of age-related changes to the composition, function, and diversity of the gut microbiota has been extensively evaluated globally (Odamaki et al. 2016). Environmental factors, dietary choices, and medications all have a significant impact on the composition and operation of the intestinal microbiome (Zmora et al. 2019). However, host organ activity may also influence how the gut's microbial population is put together. It is likely that changes in the mucosal niche cause dysbiotic states as we age because the intestine is susceptible to age-related changes to the function of tissues and integrity, including changes in regenerating ability, epithelial barrier formation, mucus layer composition, and peristalsis (Mitchell et al. 2017; Pentinmikko and Katajisto 2020).

10.8 Microbiome as a Therapeutically Important Factor

Traditional therapies are believed to be a factor in the development of microbial resistance to antibiotics and chemotherapy, as well as non-responsiveness to treatment and insufficient specificity. These symptoms pose a major threat to the general public's health. The shortcomings of contemporary medicine are addressed through microbial therapy (Greene and Reid 2012). Microbes, which are frequent visitors to the human body, boost the body's ability to recover without causing any harm. Microorganisms may also be genetically altered to improve their potency and safety (Ogunrinola et al. 2020). Microbes are beneficial for the growth of microbiota therapies as they can stop the transmission of illnesses by interacting with the host (Thaiss and Elinav 2017). It is well recognised that *Christensenella* sp. lessens depressive and anxiety-like symptoms. According to de la Cuesta-Zuluaga et al. (2017), Akkermansia muciniphila cures metabolic diseases, works in conjunction with metformin to treat cancer, and protects against atherosclerosis by lowering intestinal permeability and reducing inflammation (Li et al. 2016; Cheema et al. 2016). *Lactobacillus johnsonii* guards against cancer. According to Joossens et al. (2012), Bifidobacterium longum improves the integrity of the mucus layer that has been damaged by a high-fat diet and lessens the severity of Crohn's disease. By maintaining the homeostasis of oxalic acid, Oxalibacterium formigenes prevents kidney stones (Jalanka-Tuovinen et al. 2011). Adiposity is protected from by *Bacteroides* spp. The field of therapeutics expanded with new opportunities for illness diagnostics, test processes, and techniques for acquiring and modifying data as more was learned about the potential of gut microbes. To introduce bacteria into the host, live biotherapeutics are being developed (McCarville et al. 2020). For the past 10 years, the addition of exogenous bacteria has been an interesting way to modify the microbiome. Small-molecule therapies that can alter the host microbiota are urgently needed. To stop the origin of diseases, the tiny chemicals ought to be able to change how bacteria behave (Cully 2019).

10.9 Implementation of Microbiome Therapeutics

There has been an effort to utilise the advantages of the host-microbiome relationship for the creation of microbiome therapeutics (Mimee et al. 2016). For microbiome therapies, additive therapy, subtractive therapy, and modulatory therapy are often employed techniques. Subtractive therapy seeks to destroy the fatal pathogens that are known to cause the start of a certain disease, while additive therapy involves the addition of microbial strains or microbial consortia (Marchesi et al. 2016). Modulatory treatment uses certain non-living substances to alter or manipulate the host-microbiome relationship for specific purposes (Claesen and Fischbach 2015).

10.10 Additive Therapy

To achieve the health-promoting effects of a specific strain or microbial consortia, either as probiotics or by FMT, is known as additive therapy. Both naturally occurring and genetically modified microbes have the potential to create therapeutic compounds in additive therapy (Khoruts and Sadowsky 2016).

FMT frequently entails the delivery of medicinal microorganisms. FMT is an effective way to switch out harmful bacteria for good ones. Through a variety of delivery methods, it involves the transmission of beneficial microorganisms from donors to receivers. Strict procedures must be followed when screening the donor (32). Within 4 weeks prior to transplantation, the necessary stool and blood tests must be performed to decrease the risk of infection spreading from the donor (Cammarota et al. 2017). To ensure that the receiver is immune to the donor's germs, it is preferable for the donor to be a close relative of the recipient (Bakken et al. 2011). While a genetic illness such as inflammatory bowel disease should be treated with an unrelated donor (Kelly et al. 2015). It was discovered that treating *Clostridioides difficile* infection (CDI) with just faecal filtrate, which contains bacterial debris, metabolites, DNA, etc., was sufficient (Ott et al. 2017). Recurrent CDI was treated with a pure culture of gut bacteria from a single healthy donor (Petrof et al. 2013). Antibiotic-induced dysbiosis caused by *C. difficile* infection has been successfully treated with FMT (Quraishi et al. 2017). With a remarkable recovery rate, FMT has been successfully utilised to treat *C. difficile* recurrent infections (Baktash et al. 2018). Additionally, studies are still being conducted to see if FMT may be used to treat other disorders (with success rates substantially lower than those shown with CDI) (Allegretti et al. 2019). The transfer of faecal microorganisms from lean to obese mice resulted in the restoration of butyrate-producing bacteria and increased insulin production (Vrieze et al. 2012). Following FMT, the prevalence of antibiotic-resistant bacteria linked to recurrent urinary tract infections significantly decreased (Tariq et al. 2017). A lack of mucosa-associated invariant T-cells is linked to certain diseases such as cirrhosis and alcoholic hepatitis. After FMT, patients' T-cell counts were restored (Gao et al. 2018). Similar to this, after FMT, the loss of Bacteroidetes caused by alcohol was recovered. Several FMT clinical trials have been conducted with positive results in the treatment of alcoholic hepatitis, fibrosis progression, hepatic encephalopathy, and liver diseases. Numerous neurological illnesses, including autism, sclerosis, and Parkinson's disease, have been effectively examined with FMT (Finegold et al. 2002). The response to immune checkpoint inhibitors was improved when healthy individuals' feces were transferred to cancer patients (Sweis et al. 2016). The heredity of microorganisms after transfer depends on the host immune system-related genes, which in turn depends on the efficacy of the FMT technique (Hall et al. 2017). Although there is a potential of remission, FMT has demonstrated efficacy in the treatment of ulcerative colitis (Paramsothy et al. 2019). In recurrent Crohn's disease, the FMT treatment was also found to be ineffective. As was documented in the case of the transmission of -lactam-resistant *Escherichia coli*, immune-compromised individuals are more vulnerable to the negative effects of faecal transfer. Researchers have proposed a replacement for FMT involving the construction of artificial stools, in which commensals from stools could be fermented and developed in vitro under environments equivalent to the mammalian gut before being encapsulated in the living form (Kump et al. 2018). This is due to the negative effects of FMT and the risk associated with constipation (Petrof and Khoruts 2014). To recover the patient's faeces and re-establish the natural gut microbes in cases of severe disease.

Probiotics Probiotic therapy is among the most useful kind of addition therapy. The monotherapy use of either naturally occurring or genetically created therapeutic microbes forms the foundation for the use of probiotics. The definition of probiotics is 'Live microorganisms that when administered in adequate amounts confer a health benefit to the host' in accordance with guidelines provided by the FAO/WHO working group in 2002. According to Guslandi et al. (2003), the optimal probiotic must be specific to a species pathogen-free, helpful to the host, and capable of thriving within the human body. A variety of disorders have been successfully treated with Lactobacilli, Bifidobacteria, and *E. coli* (Cuello-Garcia et al. 2015). By creating bacteriocins, competing for attachment sites and resources, modifying pathogen activities, and enhancing the host's immune-stimulatory and nutritional status, probiotics work through competition to remove pathogens (Kesarcodi-Watson et al. 2008). Recognised health advantages include pathogen defence, host immune system stimulation, and the production of potentially effective antimicrobials inside the human body (Varankovich et al. 2015). The improvement in gut-microbiome composition is one of probiotics' most important benefits. In order to enhance the makeup of the intestinal microbiome, intestinal microorganisms are currently used as next-generation probiotics (Depommier et al. 2019). The probiotic microorganisms have a favourable effect on the human gut microbial composition, consequently enhancing nutritional health and status because the gut microbes respond to the live microbes consumed orally. Probiotics have been successfully utilised to treat a number of illnesses, including diarrhoea, Crohn's disease, ulcerative colitis, and cancer (Coqueiro et al. 2019). The gut barrier is damaged during inflammatory bowel illness, ulcers, or fistula. Pathogens and pH changes are more likely to affect the lining of a leaky gut (Podolsky 1999). These probiotics are currently being modified to express specific bio therapeutics. For the detection and therapy of diseases, engineered probiotics are being created. This technique re-modulates the diseased environment to healthy environmental conditions by adding an external functional element.

10.11 Subtractive Therapy

In the field of microbiome engineering, subtractive therapy has become an intriguing technique (Lu and Collins 2009). With the assistance of bacteriocins and bacteriophages' antibacterial properties, this treatment tries to decrease the number of harmful microorganisms in the microbiome. In addition to the emergence of antibiotic resistance within gut bacteria, bacteriocins and bacteriophages have been used to target pathogenic organisms with minimal effect on the microbiome as a whole. Antibiotics were typically used to eradicate undesired pathogens. Bacteriocins are peptides produced by ribosomes that have antibacterial properties. Bacteriocins combat pathogens in a variety of ways, including by rupturing membranes, producing toxins, impeding the respiratory system, and generally lysing cells (Mahlapuu et al. 2016). According to Ołdak and Zielińska (2017), bacteriocins can be either lanthionine- or non-lanthionine-containing. While lanthionine-containing enterocin and nisin have antibacterial activity against *Bacillus cereus*, *Geobacillus stearothermophilus*, and *Clostridium botulism*, non-lanthionine-containing bacteriocins act against Clostridium, Enterococcus, Pediococcus, Lactobacillus, and Leuconostoc (Egan et al. 2016). Bacteriocins are known to be produced by Firmicutes, Bacteroidetes, Proteobacteria, and Actinobacteria. Commensals rely on bacteriocin to thrive in the gut's competitive niche (Kommineni et al. 2015). They suppress the defensins, stop pathogen colonisation, and boost host immunity all around. Dairy products have been preserved using bacteriocins (Perez et al. 2014; Gautam and Sharma 2009). Bacteriocins are used to preserve meat, vegetables, drinks, and other foods. Commercial food preservatives pediocin and nisin are available. Peptic ulcers are treated with bacteriocins produced by Pediococcus acidilactici BA28 (Kaur et al. 2012). Due to its spermicidal and antibacterial properties, ferrmenticin HV6b, which is generated by Limosilactobacillus fermentum HV6b, is utilised in vaginal creams (Kaur et al. 2013). To control microbial diseases, veterinary professionals also utilise bacteriocins such as nisin. Similar to this, ESL5, which is produced by Enterococcus faecalis SL-5, is applied topically to prevent acne lesions brought on by Propionibacterium acnes (Kang et al. 2009). Bacteriocins are also employed in dental hygiene. For mouthwash and sustaining dental health, *S. macedonicus*' macedocin is utilised (Zoumpopoulou et al. 2013). Similar to how bacteria can become resistant to antibiotics, bacteriocin resistance can also emerge as a result of environmental adaptation or bacteriocin breakdown (Dicks et al. 2018). An attempt should be made to boost bacteriocin potency in order to progress towards treatments.

10.12 Modulatory Therapy

For the benefit of human health, modulatory therapy may involve modifying gut microorganisms or the interactions they have with the human host. It takes into account rebuilding the depleted microbiome and altering current bacteria to create a healthier microbiome. Through a variety of modifications, including those to food, exercise, and antibiotic use, the gut microbiota can be restored or modulated (Bhalodi et al. 2019). Because the microbiome supports our nutrition, changing our diets is a key strategy for altering the gut microbiome. The gut microbiome is greatly impacted by dietary changes. Short-chain fatty acid (SCFA) production and a healthy microbiota are both correlated with exercise (Allen et al. 2018). Sportspeople eat more proteins that affect the gut microbiome. The rise in Veillonella improving exercise endurance was tested in marathon runners (Birchenough et al. 2019). A crucial element in determining the microbiota's composition is vitamin D. Dietary changes may aid in the continuous development of good bacteria and enhance the synthesis of microbial metabolites such SCFAs (Bindels et al. 2015). Dietary changes that are combined with other forms of diabetes treatment enhance the glycemic index. Children with severe epilepsy who were given a ketogenic diet had fewer gut Eubacterium rectale, Bifidobacteria, and Dialister bacteria (Lindefeldt et al. 2019). Thus, a drug-resistant form of epilepsy was treated with a ketogenic diet. Long-chain fatty acid supplementation relieved the pathological circumstances and restored the Lactobacillus in ethanol-induced liver damage. Similar to how butyrate concentration was reduced by giving glycerol tributyrate to patients, this had a good impact on their health (Cresci et al. 2014). The antioxidant tempol was used to treat obesity by changing the bacterial makeup to prevent obese situations. Prebiotics, such as fibre and galactooligosaccharides that boost the abundance of Bifidobacterium, boost the good microorganisms and eliminate the infections. Drug use, smoking, and alcohol intake also have an impact on the composition of the gut microbiota. Some medications/drugs may alter the composition of the gut microbiota and may possibly heighten antibiotic-induced resistance. According to Meroni et al. (2019), drinking alcohol increases gram-negative bacterial concentration, reduces SCFA production, and makes the intestinal wall more permeable. Alcohol decreases the quantity of Proteobacteria and increases the amount of Lactobacilli and Bacteroidetes (Barr et al. 2018). The abundance of Proteobacteria increases and Faecalibacterium decreases in the human stool as a result of higher alcohol intake. Smoking also affects the composition of the gut, airway, and oral microbiomes. Quitting smoking changes the composition of the gut microbiome by favouring Firmicutes and Actinobacteria while simultaneously depleting Bacteroidetes and Proteobacteria. Dysbiosis, smoking, and the beginning of a disease are all related. The development and severity of the disease are influenced by the increased abundance of Bacteroidetes in CD patients. In addition to having a deleterious impact on the gut flora, antibiotics also cause changes in metabolic activity, the development of antibiotic resistance, and antibiotic-associated diarrhoea and CDI infections (Ramirez et al. 2020).

10.13 Limitations in the Field of Microbiome-Based Therapies

Microbiome therapy creates a native gut microbial ecology that helps improve gut health and prevent dysregulation. The goal of using microorganisms as therapy is to promote host survival by influencing physiological, metabolic, and nutritional processes. Although microbiome treatments show promise, they frequently face a few difficulties. Identification of the bacteria to address disease complexity is the main obstacle in the field of microbiome therapies. Depending on their capacity to survive inside the body, various microbial strains are suited for specific treatment methods. The *Lactobacillus* sp. and *E. coli* Nissle successfully enrich within the small intestine, *Bacteroides* sp. successfully colonises the colon and caecum. According to Donaldson et al. (2016), *Lactobacillus lactis* cannot colonise the gut. Therefore, the appropriateness of the probiotic utilised for treatment is contingent upon the biogeography of the disease. Prior to choosing microorganisms for treatment, it is necessary to properly characterise the bacteria according to their practical advantages. It has been difficult for a long time and in a variety of situations to determine whether microbiome treatments are effective. Furthermore, since the majority of the research on the medical use of the microbiome has been conducted using rodent models, efforts continue to be required to carry out human trials. The durability and stability of the therapeutically important microbial strains enable effective microbiome therapy. To understand the environmental difficulties that bacteria encounter and how their interactions with one another affect their functionality, chemostats must be constructed (Auchtung et al. 2015). Similarly, investigations into the interactions between hosts and probiotics have been carried out utilising organoids, gut-on-a-chip models, and 3D intestinal scaffolds. Similar to that, studies exploring the interactions between hosts and probiotics have been conducted using organoids, gut-on-a-chip models, and 3D intestinal scaffolds. It is necessary to assess the safety of customised probiotics for sustained therapeutic efficacy. There are issues with the horizontal transfer of recombinant DNA from the intended microbiome to the natural microbiome (Smillie et al. 2011). Recombinant probiotics that are released into the environment might also be dangerous. Auxotrophic microorganisms that become inactive in the absence of a certain substrate must be used as therapies because they are unable to colonise their environment. Synchronised research and regulatory mechanisms must be employed for a safe treatment strategy as well as therapeutic maintenance because changed phages may lead them to lose function (Gladstone et al. 2012). In order to ensure the long-term viability of the drugs, further measures should be done to lessen the burden on cellular therapy.

10.14 Conclusion

The human body has a diverse range of various gut bacteria, which can be separated and examined to learn more about their methods of action and determine whether they might substantially influence advanced medical therapies. There is a still need to investigate the human microbiota in a novel way to improve our understanding of the functionality and mechanisms associated with these microbes and move beyond these simple associations to more complex analysis. In addition, it is essential to characterized the role of the microbiota, which includes viruses, phage, fungus, and archaea, in human health and disease. Innovations in this field could have a real impact on medicine and human health. We can explore the bacteria in the human gut as well as those in various body locations by using various metagenomic techniques study the uncultivated bacterial population. Probiotics may potentially provide therapeutic approaches to maintain the microbiome's equilibrium in the gut, preventing dysbiosis, It is crucial to emphasise the importance of studying the optimal number of probiotic strains required to achieve specific objectives with proven effectiveness. Future probiotic research must concentrate on stratifying clinical trials according to sex, nutrition, ethnicity, location, and the functional aspects of each participant's microbiome. Probiotic strains and their products play a significant role in the improvement of global health problems as natural and safe therapeutic therapies. Clinical research have shown that there are a number of health advantages, including improvements in bowel function, diarrhoea, vaginal dysbiosis, respiratory infections, bone mineralisation, and body weight. Exploring these findings has piqued the curiosity of researchers all across the world, raising the demand for accurate and understandable scientific communications. By taking advantage of how diet and food choices affect the microbiome's composition, these gut microbial populations may be suggested as possible therapeutic targets. Probiotics and suggestions for a healthier lifestyle and diet may be crucial in reducing illness susceptibility and enhancing human health. This requires thorough extensive communication and a greater comprehension of the populations of human gut microbes and their link with the host's health through complex analyses that can significantly impact human wellness and disease.

References

Allegretti JR, Mullish BH, Kelly C et al (2019) The evolution of the use of faecal microbiota transplantation and emerging therapeutic indications. Lancet 394:420–431

Allen JM, Mailing LJ, Niemiro GM et al (2018) Exercise alters gut microbiota composition and function in lean and obese humans. Med Sci Sports Exerc 50:747–757

Auchtung JM, Robinson CD, Britton RA (2015) Cultivation of stable, reproducible microbial communities from different fecal donors using minibioreactor arrays (MBRAs). Microbiome 3:42

Bakken JS, Borody T, Brandt LJ et al (2011) Treating Clostridium difficile infection with fecal microbiota transplantation. Clin Gastroenterol Hepatol 9:1044–1049

Baktash A, Terveer EM, Zwittink RD et al (2018) Mechanistic insights in the success of fecal microbiota transplants for the treatment of Clostridium difficile infections. Front Microbiol 9:1242

Barr T, Sureshchandra S, Ruegger P et al (2018) Concurrent gut transcriptome and microbiota profiling following chronic ethanol consumption in nonhuman primates. Gut Microbes 9:338–356

Bhalodi AA, van Engelen TSR, Virk HS et al (2019) Impact of antimicrobial therapy on the gut microbiome. J Antimicrob Chemother 74:i6–i15

Bindels LB, Delzenne NM, Cani PD et al (2015) Towards a more comprehensive concept for prebiotics. Nat Rev Gastroenterol Hepatol 12:303–310

Birchenough G, Schroeder BO, Bäckhed F et al (2019) Dietary destabilisation of the balance between the microbiota and the colonic mucus barrier. Gut Microbes 10:246–250

Brown JM, Hazen SL (2015) The gut microbial endocrine organ: bacterially derived signals driving cardiometabolic diseases. Annu Rev Med 66:343–359

Cammarota G, Ianiro G, Tilg H et al (2017) European consensus conference on faecal microbiota transplantation in clinical practice. Gut 66:569–580

Cheema AK, Maier I, Dowdy T et al (2016) Chemopreventive metabolites are correlated with a change in intestinal microbiota measured in A-T mice and decreased carcinogenesis. PLoS One 11:e0151190

Claesen J, Fischbach MA (2015) Synthetic microbes as drug delivery systems. ACS Synth Biol 4:358–364

Coqueiro AY, Raizel R, Bonvini A et al (2019) Probiotics for inflammatory bowel diseases: a promising adjuvant treatment. Int J Food Sci Nutr 70:20–29

Cresci GA, Bush K, Nagy LE (2014) Tributyrin supplementation protects mice from acute ethanol-induced gut injury. Alcohol Clin Exp Res 38:1489–1501

Cuello-Garcia CA, Brożek JL, Fiocchi A et al (2015) Probiotics for the prevention of allergy: a systematic review and meta-analysis of randomized controlled trials. J Allergy Clin Immunol 136:952–961

Cully M (2019) Microbiome therapeutics go small molecule. Nat Rev Drug Discov 18:569–567

de la Cuesta-Zuluaga J, Mueller NT, Corrales-Agudelo V et al (2017) Metformin is associated with higher relative abundance of mucin-degrading Akkermansia muciniphila and several short-chain fatty acid-producing microbiota in the gut. Diabetes Care 40:54–62

DeJong EN, Surette MG, Bowdish DME (2020) The gut microbiota and unhealthy aging: disentangling cause from consequence. Cell Host Microbe 28:180–189

Depommier C, Everard A, Druart C et al (2019) Supplementation with Akkermansia muciniphila in overweight and obese human volunteers: a proof-of-concept exploratory study. Nat Med 25:1096–1103

Dicks LMT, Dreyer L, Smith C et al (2018) A review: the fate of bacteriocins in the human gastro-intestinal tract: do they cross the gut–blood barrier? Front Microbiol 9:2297

Donaldson GP, Lee SM, Mazmanian SK (2016) Gut biogeography of the bacterial microbiota. Nat Rev Microbiol 14:20–32

Egan K, Field D, Rea MC et al (2016) Bacteriocins: novel solutions to age old spore-related problems? Front Microbiol 7:461

Finegold SM, Molitoris D, Song Y et al (2002) Gastrointestinal microflora studies in late-onset autism. Clin Infect Dis 35:S6–S16

Gao B, Ma J, Xiang X (2018) MAIT cells: a novel therapeutic target for alcoholic liver disease? Gut 67:784–786

Gautam N, Sharma N (2009) Bacteriocin: safest approach to preserve food products. Indian J Microbiol 49:204–211

Gladstone EG, Molineux IJ, Bull JJ (2012) Evolutionary principles and synthetic biology: avoiding a molecular tragedy of the commons with an engineered phage. J Biol Eng 6:13

Greene S, Reid A (2012) Moving targets: fight the evolution of resistance in infections, pests and cancer. American Academy of Microbiology Colloquium Report, Washington, DC

Guarner F, Malagelada J (2003) Gut flora in health and disease. Lancet 361:512–519

Guslandi M, Giollo P, Testoni PA (2003) A pilot trial of Saccharomyces boulardii in ulcerative colitis. Eur J Gastroenterol Hepatol 15:697–698

Hall AB, Tolonen AC, Xavier RJ (2017) Human genetic variation and the gut microbiome in disease. Nat Rev Genet 18:690–699

Jalanka-Tuovinen J, Salonen A, Nikkilä J et al (2011) Intestinal microbiota in healthy adults: temporal analysis reveals individual and common core and relation to intestinal symptoms. PLoS One 6:e23035

Joossens M, De Preter V, Ballet V et al (2012) Effect of oligofructose-enriched inulin (OF-IN) on bacterial composition and disease activity of patients with Crohn's disease: results from a double-blinded randomised controlled trial. Gut 61:958

Kang BS, Seo JG, Lee GS et al (2009) Antimicrobial activity of enterocins from Enterococcus faecalis SL-5 against Propionibacterium acnes, the causative agent in acne vulgaris, and its therapeutic effect. J Microbiol 47:101–109
Kaur B, Garg N, Sachdev A et al (2012) Isolation and molecular characterization of anti-Helicobacter pylori bacteriocin producing Pediococcus acidilactici BA28. Sci Rep 1:323
Kaur B, Balgir PP, Mittu B et al (2013) Biomedical applications of fermenticin HV6b isolated from Lactobacillus fermentum HV6b MTCC10770. Biomed Res Int 2013:168438
Kelly CR, Kahn S, Kashyap P et al (2015) Update on fecal microbiota transplantation 2015: indications, methodologies, mechanisms, and outlook. Gastroenterology 149:223–237
Kesarcodi-Watson A, Kaspar H, Lategan MJ et al (2008) Probiotics in aquaculture: the need, principles and mechanisms of action and screening processes. Aquaculture 274:1–14
Khanna S, Tosh PK (2014) A clinician's primer on the role of the microbiome in human health and disease. Mayo Clin Proc 89:107–114
Khoruts A, Sadowsky M (2016) Understanding the mechanisms of faecal microbiota transplantation. Nat Rev Gastroenterol Hepatol 13:508–516
Kommineni S, Bretl DJ, Lam V et al (2015) Bacteriocin production augments niche competition by *Enterococci* in the mammalian gastrointestinal tract. Nature 526:719–722
Kump P, Wurm P, Gröchenig HP et al (2018) The taxonomic composition of the donor intestinal microbiota is a major factor influencing the efficacy of faecal microbiota transplantation in therapy refractory ulcerative colitis. Aliment Pharmacol Ther 47:67–77
Langsetmo L et al (2019) The association between objectively measured physical activity and the gut microbiome among older community dwelling men. J Nutr Health Aging 23:538–546
Lax S et al (2014) Longitudinal analysis of microbial interaction between humans and the indoor environment. Science 345:1048–1052
Ley RE et al (2008) Evolution of mammals and their gut microbes. Science 320:1647–1651
Li J, Lin S, Vanhoutte PM et al (2016) Akkermansia muciniphila protects against atherosclerosis by preventing metabolic endotoxemia-induced inflammation in Apoe−/− mice. Circulation 133:2434–2446
Lindefeldt M, Eng A, Darban H et al (2019) The ketogenic diet influences taxonomic and functional composition of the gut microbiota in children with severe epilepsy. NPJ Biofilms Microbiomes 5:5
López-Otín C, Blasco MA, Partridge L, Serrano M, Kroemer G (2013) The hallmarks of aging. Cell 153(6):1194–1217
Lu TK, Collins JJ (2009) Engineered bacteriophage targeting gene networks as adjuvants for antibiotic therapy. Proc Natl Acad Sci U S A 106:4629–4634
Luan Z et al (2020) Metagenomics study reveals changes in gut microbiota in centenarians: a cohort study of Hainan centenarians. Front Microbiol 11:1474
Mahlapuu M, Håkansson J, Ringstad L et al (2016) Antimicrobial peptides: an emerging category of therapeutic agents. Front Cell Infect Microbiol 6:194
Marchesi JR, Adams DH, Fava F et al (2016) The gut microbiota and host health: a new clinical frontier. Gut 65:330–339
McCarville JL, Chen GY, Cuevas VD et al (2020) Microbiota metabolites in health and disease. Annu Rev Immunol 38:147–170
Meroni M, Longo M, Dongiovanni P (2019) Alcohol or gut microbiota: who is the guilty? IJMS 20:4568
Mimee M, Citorik RJ, Lu TK (2016) Microbiome therapeutics—advances and challenges. Adv Drug Deliv Rev 105:44–54
Mitchell EL, Davis AT, Brass K, Dendinger M, Barner R, Gharaibeh R et al (2017) Reduced intestinal motility, mucosal barrier function, and inflammation in aged monkeys. J Nutr Health Aging 21:354–361
O'Hara AM, Shanahan F (2006) The gut flora as a forgotten organ. EMBO Rep 7:688–693
O'Toole PW, Jeffery IB (2015) Gut microbiota and aging. Sci Am Assoc Adv Sci 350:1214–1215

Odamaki T, Kato K, Sugahara H, Hashikura N, Takahashi S, Xiao J-Z et al (2016) Age-related changes in gut microbiota composition from newborn to centenarian: a cross-sectional study. BMC Microbiol 16:90–12

Ogunrinola GA, Oyewale JO, Oshamika OO et al (2020) The human microbiome and its impacts on health. Int J Microbiol 2020:8045646–8045647

Ołdak A, Zielińska D (2017) Bacteriocins from lactic acid bacteria as an alternative to antibiotics. Postepy Hig Med Dosw (Online) 71:328–338

Ott SJ, Waetzig GH, Rehman A et al (2017) Efficacy of sterile fecal filtrate transfer for treating patients with Clostridium difficile infection. Gastroenterology 152:799–811.e7

Paramsothy S, Nielsen S, Kamm MA et al (2019) Specific bacteria and metabolites associated with response to fecal microbiota transplantation in patients with ulcerative colitis. Gastroenterology 156:1440–54.e2

Pellanda P, Ghosh TS, O'Toole PW (2020) Understanding the impact of age-related changes in the gut microbiome on chronic diseases and the prospect of elderly-specific dietary interventions. Curr Opin Biotechnol 70:48–55

Pentinmikko N, Katajisto P (2020) The role of stem cell niche in intestinal aging. Mech Ageing Dev 191:111330

Perez RH, Zendo T, Sonomoto K (2014) Novel bacteriocins from lactic acid bacteria (LAB): various structures and applications. Microb Cell Factories 13:S3

Petrof EO, Khoruts A (2014) From stool transplants to next-generation microbiota therapeutics. Gastroenterology 146:1573–1582

Petrof EO, Gloor GB, Vanner SJ et al (2013) Stool substitute transplant therapy for the eradication of Clostridium difficile infection: "RePOOPulating" the gut. Microbiome 1:3

Podolsky DK (1999) Mucosal immunity and inflammation. V. Innate mechanisms of mucosal defense and repair: the best offense is a good defense. Am J Phys 277:G495–G499

Qin J, Li R, Raes J, Arumugam M, Burgdorf KS, Manichanh C, Nielsen T, Pons N et al (2010) A human gut microbial gene catalogue established by metagenomic sequencing. Nature 464:59–65

Quigley EM (2013) Gut bacteria in health and disease. Gastroenterol Hepatol 9:560–569

Quraishi MN, Widlak M, Bhala N et al (2017) Systematic review with meta-analysis: the efficacy of faecal microbiota transplantation for the treatment of recurrent and refractory *Clostridium difficile* infection. Aliment Pharmacol Ther 46:479–493

Ramirez J, Guarner F, Bustos Fernandez L et al (2020) Antibiotics as major disruptors of gut microbiota. Front Cell Infect Microbiol 10:572912

Raza M, Gul K, Arshad A, Riaz N, Waheed U, Rauf A, Aldakheel F, Alduraywish S, Rehman M, Abdullah M et al (2019) Microbiota in cancer development and treatment. J Cancer Res Clin Oncol 145:49–63

Sarkar A et al (2020) Microbial transmission in animal social networks and the social microbiome. Nat Ecol Evol 4:1020–1035

Saxena R, Sharma VK (2016) A metagenomic insight into the human microbiome: its implications in health and disease. In: Kumar D, Antonarakis S (eds) Medical and health genomics medical and health genomics. Elsevier Science, Amsterdam, pp 107–119. https://doi.org/10.1016/B978-0-12-420196-5.00009-5

Sears CL (2005) A dynamic partnership: celebrating our gut flora. Anaerobe 11:247–251

Sender R, Fuchs S, Milo R (2016) Are we really vastly outnumbered? Revisiting the ratio of bacterial to host cell humans. Cell 164:337–340

Shen S, Wong CH (2016) Bugging inflammation: role of the gut microbiota. Clin Transl Immunol 5:e72

Sherwood L, Willey J, Woolverton C (2013) Prescott's microbiology, 9th edn. McGraw Hill, New York, pp 713–721

Smillie CS, Smith MB, Friedman J et al (2011) Ecology drives a global network of gene exchange connecting the human microbiome. Nature 480:241–244

Sommer F, Bäckhed F (2013) The gut microbiota–masters of host development and physiology. Nat Rev Microbiol 11:227–238

Stein MM et al (2016) Innate immunity and asthma risk in Amish and Hutterite farm children. N Engl J Med 375:411–421

Sweis RF, Spranger S, Bao R et al (2016) Molecular drivers of the non-T-cell-inflamed tumor microenvironment in urothelial bladder cancer. Cancer Immunol Res 4:563–568

Tap J, Mondot S, Levenez F, Pelletier E, Caron C, Furet JP, Ugarte E, Muñoz-Tamayo R et al (2009) Towards the human intestinal microbiota phylogenetic core. Environ Microbiol 11:2574–2584

Tariq R, Pardi DS, Tosh PK et al (2017) Fecal microbiota transplantation for recurrent *Clostridium difficile* infection reduces recurrent urinary tract infection frequency. Clin Infect Dis 65:1745–1747

Thaiss CA, Elinav E (2017) The remedy within: will the microbiome fulfill its therapeutic promise? J Mol Med (Berl) 95:1021–1027

Turnbaugh PJ, Ley RE, Hamady M, Fraser-Liggett CM, Knight R, Gordon JI (2007) The human microbiome project. Nature 449:804–810

Varankovich NV, Nickerson MT, Korber DR (2015) Probiotic-based strategies for therapeutic and prophylactic use against multiple gastrointestinal diseases. Front Microbiol 6:685

Vrieze A, Van Nood E, Holleman F et al (2012) Transfer of intestinal microbiota from lean donors increases insulin sensitivity in individuals with metabolic syndrome. Gastroenterology 143:913–6.e7

Xu C, Zhu H, Qiu P (2019) Aging progression of human gut microbiota. BMC Microbiol 19:236–210

Yatsunenko T, Rey FE, Manary MJ, Trehan I, Dominguez-Bello MG, Contreras M et al (2012) Human gut microbiome viewed across age and geography. Nature 486:222–227

Zmora N, Suez J, Elinav E (2019) You are what you eat: diet, health and the gut microbiota. Nat Rev Gastroenterol Hepatol 16:35–56

Zoumpopoulou G, Pepelassi E, Papaioannou W et al (2013) Incidence of bacteriocins produced by food-related lactic acid bacteria active towards oral pathogens. Int J Mol Sci 14:4640–4654

Gut-Microbiota as a Therapeutic Intervention for Cognitive Damage

11

Hadeer M. Hamdalla, Manal Abdul-Hamid, and Mohamed Y. Zaky

Abstract

Elderly mortality is significantly influenced by cognitive impairment. The dysregulation of the gut microbiota is believed to be a contributing factor in the pathogenesis of several gastrointestinal and extraintestinal disorders, and it is showing signs of evolving into a significant disease-controlling factor. Through the increasingly well-defined gut microbiota–brain axis (GMBA), an altered gut microbiome (GM) has more recently been linked to mild cognitive impairment (MCI) and Alzheimer's disease (AD). It is known that the gut microbiota yields vitamins, intestinal toxins, and neurotransmitters like acetylcholine, dopamine, and serotonin as well as modulates nerve signaling, with a focus on the vagus nerve. Furthermore, gut dysbiosis impairs the synthesis of signaling proteins, which impacts metabolic processes important for the onset of AD. Corrections made to the gut microflora composition may have a pleiotropic effect on dementia pathology because the gut microbiota has numerous connections to important metabolic and inflammatory pathways. In this book chapter, we provide a brief overview of the function of GM homeostasis in both diseased and healthy brains. On the foundation of these findings, we proceed to discuss how dysbiosis might be used as a new diagnostic tool in both early and advanced disease stages. This book chapter explores the potential of gut dysbiosis as a diagnostic tool for cognitive decline, and examines dietary modifications, probiotics, and fecal

H. M. Hamdalla · M. Abdul-Hamid
Cell Biology, Histology and Genetics Division, Department of Zoology, Faculty of Science, Beni-Suef University, Beni-Suef, Egypt

M. Y. Zaky (✉)
Molecular Physiology Division, Department of Zoology, Faculty of Science, Beni-Suef University, Beni-Suef, Egypt
e-mail: mohamedzaki448@science.bsu.edu.eg

S. Pathak, A. Banerjee (eds.), *Gut Microbiome and Brain Ageing*,
https://doi.org/10.1007/978-981-99-8803-7_11

microbiota transplantation (FMT) as promising therapeutic interventions for these disorders.

Keywords

Gut microbiota · Gut-brain axis · Biomarker · Probiotics · Diet · Fecal microbiota transplantation · Alzheimer's disease

Abbreviations

AD	Alzheimer's disease
Aβ	Beta amyloid
FMT	Fecal microbiota transplantation
GBA	Gut-brain axis
GM	Gut microbiome
GMBA	Gut microbiota brain axis
MCI	Mild cognitive impairment
NDs	Neurodegenerative diseases
PGNs	Peptidoglycans
SAMP8	Senescence-accelerated mouse prone 8
SAMR1	Senescence-accelerated mouse resistant 1
SCFAs	Synthesizing short-chain fatty acids
TLRs	Toll-like receptors

11.1 Introduction

Healthcare expenditures are rising and caretakers are being devastated by an epidemic of age-associated cognitive loss, which is most frequently attributed to neurodegenerative diseases (NDs). Cognitive disorders include dementia (e.g., Alzheimer's disease), amnesia, and delirium (Berryhill et al. 2012). In these disorders, patients are no longer fully oriented to time and space. Each cognitive illness has a varied course of treatment depending on the precise diagnosis, but they are never curable. Therapies instead work to minimize symptoms experienced by patients. The goal of current research is to increase understanding of the underlying causes of cognitive problems and to create effective therapies for them.

Intestinal flora refers to the microbes that reside in the gut and includes eukaryotes and archaea in addition to bacteria. According to Li et al. (2022), the gut microbiota (GM) is thought of as a separate organ system that contributes to a variety of essential roles such as immunity, digestion, and nutrient absorption. It is essential for the brain's behavioral and cognitive processes as well. It has been clearly established for the past 20 years that the brain and gut communicate in both directions

(Varesi et al. 2022). One of the main factors for the emergence of a number of mental diseases is the close relationship between the GM and the host brain (Varesi et al. 2022).

The gut-brain axis' (GBA) existence and how intestinal bacteria affect brain function have drawn a lot of research over the past decade. The infectious hypothesis that underlies NDs like Alzheimer's disease has been postulated in light of growing data. According to this theory, gut flora may be beneficial both as a protective measure and as a treatment (Kim et al. 2020; Long-Smith et al. 2020). This book chapter seeks to identify whether altering the gut microbiota can offer any benefits in terms of enhancing cognitive performance in individuals experiencing cognitive impairment.

11.2 Gut-Brain Axis

GBA is a signaling pathway that connects the central nervous system (CNS) and gastrointestinal tract, enabling two-way communication between the two systems (Varesi et al. 2022). Its main tasks include monitoring and assimilating intestinal processes as well as establishing connections between emotional and cognitive brain regions and peripheral intestinal processes like immune activation, intestinal permeability, enteric reflex, and enteroendocrine signaling via immune and neuroendocrine mediators. In this integrated world, signals from the stomach affect brain activity while the brain regulates the sensory, secretory, and movement activities of the gut. Because of this, maintaining gut homeostasis depends heavily on this relationship, which has been associated with the etiology of a number of metabolic and mental (psychiatric and neurological) dysfunctions and disorders (Kaur et al. 2021). Several potential routes of contact between the GM and the brain have been proposed. It might be through the main modulatory pathway, which is made up of the branches of the vagus nerve (Han et al. 2022). As a result of the microbiota modulating transmitters like serotonin and acetylcholine as well as the creation of bioactive peptides and metabolites like short-chain fatty acids by the GM (Chen et al. 2021). Cortisol may affect intestinal motility, integrity, and mucus production as well as alter the composition of the GM when it is released by the hypothalamic-pituitary-adrenal axis (HPA) in response to stress. This alteration may then have an effect on CNS by modulating stress hormones (Heijtz et al. 2011). It also might be through the production of pro-inflammatory cytokines and chemokines (Cryan 2021). Also, immunity is a key factor. Toll-like receptors (TLRs) and peptidoglycans (PGNs) specifically serve as sensors of microbial components to mediate the immune response to microbes (Banks and Erickson 2010). Through various pathways, a local immune activation can result in an immune activation in various organs, including the brain (Silver et al. 2014). The pathophysiology of some types of depression and neurodegenerative diseases like AD and Parkinson's disease (PD) have been linked to this low-grade immune activation (Margolis et al. 2021). Given this intricate interplay, it is not surprising that the GBA, and by extension the GM as its primary element, either directly or indirectly influences neuropsychiatric illnesses (Generoso et al. 2021).

11.3 Correlation Between Gut Microbiota and Cognitive Dysfunction

Patients with dementia generally experience a decline in memory and other cognitive skills that is so severe that it impairs their ability to perform daily duties. A consistent and steady loss in cognitive function is one of its hallmarks. People with impairments commonly forget things and have little to no awareness of their inadequacies (Cipriani et al. 2020). The control of human behavior and the impact on mental processes such as mood and cognition, and the brain's physiological processes like myelination, neurogenesis, and microglial activation are all attributed to the gut microbiota (Guzzetta et al. 2022). Although many routes are considered when relating GM to cognitive function, little is understood about the precise mechanisms behind dementia in the elderly. Additionally, GM is extremely sensitive to outside lifestyle factors, including diet, lack of sleep, circadian rhythm disturbances, chronic noise exposure, and sedentary behavior, which are also thought to be risk factors for some NDs (Banks and Erickson 2010; Cryan 2021; Heijtz et al. 2011). In order to prevent neurodevelopmental abnormalities and NDs, the GM is essential for preserving the microglia's healthy functional state (Silver et al. 2014).

11.4 Gut Microbiota and Alzheimer's Disease

AD, the most typical dementia-causing factor, affects roughly 50,000,000 individuals globally and is a serious global health issue (Mullane and Williams 2020). Beta amyloid (Aβ) plaques and tangles of hyperphosphorylated tau neurofibrils, which cause neuroinflammation and gradual cognitive impairment, are the disease's hallmarks (Den et al. 2020). Any infections or traumatic events that affect the brain (traumatic brain injury) can interfere with central immune homeostasis and hasten the disease's progression. Recent research papers assert that changes in the GM have a direct influence on cognitive impairment and actively contribute to the development and progression of AD. Unbalanced GM could contribute to immunosenescence, oxidative stress, cytokine production, and neuroinflammation in the early stages of AD pathogenesis, according to a growing body of research (Ausó and Gómez-Vicente 2020; Leblhuber et al. 2020; Varesi et al. 2022).

Kim et al. (2020) shed new light on the relationship between the gut, immune cell population blood, and brain (pathology) axis and AD by demonstrating that microbiota-mediated intestinal and systemic immune aberrations contribute to the pathogenesis of AD in Aβ and neurofibrillary tangles ($ADLP^{APT}$) transgenic mouse model. $ADLP^{APT}$ animals displayed persistent intestinal and systemic inflammation as well as a loss of epithelial barrier integrity. The development of amyloid plaques and neurofibrillary tangles, glial reactivity, and cognitive impairment were all reduced by repeated transfer and transplant of the fecal microbiota from wild-type mice into $ADLP^{APT}$ animals.

11.5 Gut Microbiota as Biomarkers for Cognitive Disorders

One of the main objectives of several investigations is the search for prognostic, precise, non-invasive, and approachable biomarkers for early illness diagnosis. Although it is believed that an early diagnosis of AD may be vital for taking action during the earliest stages of a disease, effective and repeatable biomarkers have not yet found clinical use (Ausó and Gómez-Vicente 2020). Although fluid biomarkers for early disease detection are the subject of numerous studies, we have yet to develop an assay that is reliable and efficient enough to be used in clinical settings (Mullane and Williams 2020). In recent years, biomedical research has focused on the GMBA as a potential therapeutic target for disorders of the CNS, including AD (Kim et al. 2020; Long-Smith et al. 2020). Hence, GM-related biomarkers may be a promising substitute or additional tool to assess illness states.

In both human and animal research, dysbiosis of GM has been linked to neuropsychiatric illnesses. Recent research revealed that there were substantial differences in fecal bacteria diversity and composition between AD patients and healthy controls (Zhuang et al. 2018). In contrast to the lower concentration of bacteria capable of synthesizing short-chain fatty acids (SCFAs), a larger concentration of bacteria that cause pro-inflammatory conditions was present in the GM of patients with AD. In a mouse model of AD, elevated verrucomicrobia and proteobacteria and lower levels of ruminococcus and butyricicoccus were detected (Zhang et al. 2017). Also, our earlier research showed that fecal microbiota transplantation (FMT) could lessen the symptoms of AD in the amyloid precursor protein/presenilin 1 (APP/PS1) animal model (Sun et al. 2019). A recent research showed that the distinctive GM, particularly Enterobacteriaceae, may be able to predict PSCI in post-stroke patients and may be employed as clinical biomarkers of patients with ischemic stroke (Ling et al. 2020). In addition, recently, GM-derived biomarkers have been considered for psychological and NDs, such as bipolar disorder (Lucidi et al. 2021), multiple sclerosis (Navarro-López et al. 2022), and PD (Qian et al. 2020).

11.6 Gut Microbiome-Based Therapeutic Approaches

High-fiber diets, probiotics, and FMT are potential therapeutic approaches that target the gut to treat or prevent the symptoms of dementia and other forms of cognitive impairment.

11.7 Diet

One of the most crucial lifestyle choices for long term, including a person's physical and mental well-being is diet. An expanding body of research indicates that dietary elements or nutrients have an impact on different biological processes, including brain activity (Kim et al. 2020). The emerging idea of maintaining brain health through dietary interventions is thus actively being researched. Food may be a controllable risk factor for cognitive deterioration. A complicated signaling system comprising several cells, mediators, and receptors is how food communicates with the brain (Ettinger 2022). Dietary ingredients affect the microbial populations' diversity and abundance, protect the gut barrier's integrity, and control how easily microorganisms and their metabolites enter the bloodstream to influence the immune system and communicate with other organs, including the brain. The microbiota regulates the communication between the gut and the brain as previously mentioned. Microbes not only contribute significantly to food digestion and fermentation, supplying nutrients and bioactive metabolites, but they also reflect the kind and quantity of food consumed as well as exposure to food-borne toxins (Ettinger 2022).

As previously mentioned, one of the primary elements that affects the GM composition during the course of a person's life is nutrition (Flint 2012). In turn, bacteria mediate interactions between a regular diet and a host organism's numerous activities, including cognitive function (Shabbir et al. 2021). In this context, the GM may interact with dietary components by influencing energy balance, metabolic risk factors, and systemic inflammatory response through food metabolites, as well as through influencing the availability of nutrients that are crucial for brain function (Varesi et al. 2022). Together, studies demonstrate that foods heavy in saturated fats and simple carbohydrates, poor in fiber, and associated with the Western diet include cognitive impairment and an increased risk of dementia.

Recent research has shown a close connection between the composition of GM (then influenced by dietary habits) and the onset of AD. This connection often results from neuroinflammation brought on by bacterial brain migration or bacterial products, which frequently happens to influence the regulation of brain synaptogenesis and development as well as the evolution of mood and cognition (Lin et al. 2018). Accordingly, gut microbiota dysregulations, which are frequently reported in AD patients, can be used to research novel treatment and diagnostic options for this debilitating condition.

Intake of fiber is linked to a decreased risk of AD in older persons. SCFA generation by gut bacteria is encouraged by the consumption of plant-based diets high in soluble fiber (butyrate, acetate, and propionate). Administration of butyrate provides anti-inflammatory effects, whereas propionate encourages neuroinflammation. Even in the prodromal phases of the disease, gut microbiota dysbiosis is a typical symptom in AD patients. It is unclear if the neuroprotective effects of fiber consumption depend on changes to the gut microbiota and particular actions of SCFAs in brain cells (Cuervo-Zanatta et al. 2023). It is challenging to demonstrate a cause-and-effect relationship between these variables, however, because there is comparatively little clinical research exploring the relationships between GM, diet, and neurodegeneration.

11.8 Diet Approach-Related Challenges

Our understanding of the timeframe required for a dietary change to have a significant impact on the gut flora and, consequently, health, is constrained by a number of factors. First, it is difficult to determine if a diet-induced modulation of the gut microbiota persists without long-term human research or even follow-ups of short-term dietary interventions, but wash-out periods of cross-sectional studies can offer some indirect information (Leeming et al. 2019). Second, some studies have demonstrated that a host's microbiota may respond or not to a dietary intervention depending on the presence or lack of specific bacteria (such as bacteria that degrade fiber) within their core microbial community, leading to varied results. Sonnenburg et al. (2016) demonstrated that a decline or extinction of these fiber-degrading bacteria results from a decrease in fiber over multiple generations in mouse.

Third, research has typically focused on the analysis of general dietary indices or single and mixed nutrients rather than foods in the context of specific dietary patterns. These approaches ignore the synergistic interaction seen in food matrices. The challenge of accurately expressing complexity of diet itself is still one that is far from being solved. This is mostly due to the substantial individual variation in dietary intake, which makes it challenging to collect and combine data that can be used in statistical modeling. At last, existing gut microbiome techniques occasionally undervalue how transit time affects the diversity and makeup of GM. Although feces richness is seen as a sign of gut health, the fecal microbiota's composition really captures the level of ecological development more accurately than it does social cohesion (Falony et al. 2018).

11.9 Probiotics

The use of probiotics is one potential gut-targeted therapy to reduce cognitive impairment symptoms (Sun et al. 2020). When given in sufficient proportions, probiotics are characterized as live bacteria that help the host's health (Reid 2016). Probiotics can influence gut dysbiosis in a variety of ways (Ng et al. 2009); nonetheless, in the case of cognitive dysfunction, the anti-inflammatory effects could be crucial. Probiotics may improve cognitive impairment by lowering inflammation linked to tryptophan serves as a substrate for the synthesis of a number of bioactive chemicals and is an essential amino acid. The metabolites in the tryptophan-kynurenine pathway are implicated in inflammation, immunological response, and neurotransmission and may have either neuroprotective or neurodegenerative effects. The majority of tryptophan is metabolized through the kynurenine pathway (Cervenka et al. 2017). For instance, the composition of the intestinal microbiota can control the formation of indole compounds, which can be either advantageous or poisonous (Zelante et al. 2013).

According to several clinical and preclinical intervention studies, probiotics are effective in treating a number of age-related diseases, including AD (Athari Nik Azm et al. 2018; Bonfili et al. 2022; Nimgampalle and Kuna 2017), neuroinflammation (Musa et al. 2017), and vascular dementia (Liu et al. 2015), despite the fact that the effects of probiotics on cognitive impairment are less well understood. Kobayashi and colleagues found that giving amyloid-β injection-induced AD mice oral Bifidobacterium breve strain A1 for 6 days reversed the impairment of alternation behavior in the Y maze test and decreased latency time in the passive avoidance test (Kobayashi et al. 2017).

Additionally, the consumption of probiotics reduced the expression of immune-reactive and pro-inflammatory genes in the hippocampal region (Kobayashi et al. 2017). Probiotics have been shown to improve cognitive deficits in gross behavior and spatial learning tasks in other studies in rodents with an AD model (Athari Nik Azm et al. 2018; Nimgampalle and Kuna 2017), to increase the expression of neuronal proteolytic pathways (Bonfili et al. 2022), and to decrease neurodegeneration (Athari Nik Azm et al. 2018). There isn't much evidence from humans to support this idea, despite encouraging results in preclinical studies. Probiotic supplementation's impact on AD patients was investigated in two studies. According to research by Akbari and colleagues, probiotic therapy given to AD's patients for 12 weeks with a supplement containing *Lactobacillus acidophilus*, *Lactobacillus casei*, *Bifidobacterium bifidum*, and *Lactobacillus fermentum* improved their cognitive function and metabolic homeostasis (Akbari et al. 2016).

Probiotics have recently been suggested as a secure and effective method of transport for other therapeutic substances. GM probiotic are being recognized as a viable therapeutic technique because they offer accuracy and a greater degree of site specificity than standard medication regimens (Kumar et al. 2016).

11.10 Probiotic Therapeutics' Challenges

Parameters that must be controlled on both the probiotic and host sides when planning intervention trials are as follows:

Considering the probiotic:

1. **Strain**: Depending on the strain, a probiotic can have a positive impact on health. Some strains have a good impact on cognitive function, whereas others do not. Probiotics from a single strain or several strains are both used in the research. In studies involving humans, Lv et al. (2021) found that a single strain of probiotics performed better. However, diversity could be essential for the elderly's cognitive health; the majority of studies haven't stared at how probiotic supplementation affects the microbiota.
2. **Dose**: Information on the dose-response properties of probiotics is not supported by sufficient data. The majority of studies have not contrasted the various doses.

The most frequently used doses ranged from 108 to 1011, but additional trustworthy evidence from different dosages is required, particularly outside of these typical doses (Den et al. 2020).

3. **Period**: Regarding how long probiotics should be consumed, there are still many questions. The majority of the studies' interventions lasted 12 weeks. Despite the fact that these studies used middle-aged adults as participants, Lv et al. (2021) contend that a time length of less than 12 weeks was more effective. Long-term research studies are lacking, particularly in regard to the elderly population.

Considering the host:

1. **Host's gut microbiome**: The hosts' diet and lifestyle, age, sex, geographic location, concurrent disease, exposure to antibiotics, and baseline microbiota composition could all affect the intervention's success. Accordingly, all of these elements ought to be under control. Despite the fact that diet has an impact on the gut microbiome, most studies analyze diet only superficially and did not take into account the individuals' baseline gut microbiota. More closely monitored randomized controlled trials must be conducted in order to determine the precise effects of dietary changes on the gut microbiota (Johnson et al. 2022).
2. **Cognitive dysfunction level**: studies comparing people who have different levels of cognitive functioning do not exist. When analyzing the findings between mild cognitive impairment (MCI) or early dementia and cognitively normal individuals, Kobayashi et al. (2019) found a beneficial effect of probiotics in the low-score subgroup but not in the high-score subgroup (indicating favorable cognitive performance). Although in one study with cognitively healthy participants, Lv et al. (2021) demonstrated that the effects of probiotic supplementation were greater in the case of cognitively impaired individuals than in healthy ones. Additionally, the current Randomized Controlled Trial (RCT), which includes patients with cognitive impairment, focuses more on inflammatory and oxidative biomarkers than cognitive function, so they used the mini-mental state examination (MMSE) or test your memory (TYM), which are both measures of inflammation and oxidative stress.
3. **Probiotic side effects**: Along with the benefits that have been noticed, probiotics' side effects should also be taken into account. Hwang et al.'s (2019) findings indicate that it is safe to give MCI patients *Lactiplantibacillus plantarum* C29-fermented soybean to improve their cognitive function. As mild adverse events, the following adverse events were noted: headaches, abdominal pain, gastritis, erectile dysfunction, and seborrheic dermatitis (Dore et al. 2019).

11.11 Fecal Microbiota Transplantation

FMT is an additional promising treatment that was first developed to treat gastrointestinal illnesses and may be a potential method to treat cognitive decline (Kim and Gluck 2019). FMT involves inserting "healthy" donor feces into the intestinal tract of a recipient who is presumed to have an altered or dysbiotic colonic microbiome in an attempt to directly change the recipient's microbial composition (Vindigni and Surawicz 2017; Zhan et al. 2018). The goal of this treatment is to restore the recipient's gastrointestinal tract's normal gut microbiota and colonization resistance (Vindigni and Surawicz 2017).

Retention enemas were the most popular method of FMT administration in the 1980s. Throughout time, the routes of delivery have altered. Other delivery methods were then employed, such as fecal material instillation through a nasogastric tube, administration during colonoscopies and upper endoscopies in the 2000s, and self-administered enemas in 2010. Recently, concentrated cryopreserved fecal-derived bacteria has emerged as a therapy option for recurrent *Clostridium difficile* disease and has shown promising results (Aroniadis and Brandt 2013; Hirsch et al. 2015).

FMT increases the diversity of fecal bacterial communities and fosters long-term engraftment in recipients, and it has been successfully used for almost 60 years to treat *Clostridium difficile* infection (Ventham et al. 2013). For instance, FMT has been demonstrated to have therapeutic potential in the treatment of inflammatory bowel disease (Lopez and Grinspan 2016), and type 2 diabetes (Wang et al. 2020). Senescence-accelerated mouse prone 8 (SAMP8) mice had considerably worse cognitive performance than senescence-accelerated mouse resistant 1 (SAMR1) mice in a preclinical paradigm of accelerated aging, which was associated with altered gut microbiota composition. Germ-free mice who received fecal transplants from SAMR1 mice but not from SAMP8 mice showed enhanced behavior and microbiota diversity metrics (Zhan et al. 2018).

Sun et al. (2019) proposed that FMT could be a potential treatment for AD. They reported that FMT therapy could help APPswe/PS1dE9 transgenic (Tg) mice with their cognitive deficits and amyloid-β (Aβ) brain deposition. These improvements were accompanied by a decline in tau protein phosphorylation as well as Aβ-40 and Aβ-42 concentrations. There was also an increased synaptic plasticity in the Tg mice, which demonstrated that the expression of PSD-95 and synapsin I increased following FMT. Moreover, after FMT, a drop in the levels of COX-2 and CD11b in Tg mice was observed. Additionally, FMT treatment reversed the changes in SCFAs and gut microbiota.

Only two case studies with encouraging outcomes in humans have been undertaken thus far. After receiving FMT from an 85-year-old woman (the recipient's wife), Hazan (2020) showed that an 82-year-old man's AD symptoms (cognitive function, memory, and mood) had improved An increase in cognitive function, microbial diversity, and SCFA production was also seen in a second case-study involving a 90-year-old woman with AD and severe *C. difficile* infection who received FMT from a 27-year-old healthy male (Park and Seo 2021). Many restrictions still exist for FMT, despite the possibility of using it to treat AD. Standardization of the therapeutic regimens, time and duration of administration, short- and long-term dangers, and inclusion criteria are all issues that need to be taken into account and addressed (Varesi et al. 2022).

In a study by Park et al. (2022), cognitive improvement is observed following FMT to treat *Clostridioides difficile* infection (CDI), indicating a critical link between the gut microbiome and neural function. Although there has been research linking the gut microbiome to cognitive function, it is unclear whether fecal microbiota transplantation can help patients with cognitive decline. After fecal transplant, FMT seems to be safe and well tolerated except for minor adverse events, including transient diarrhea, abdominal cramps or pain, low-grade fever, bloating, flatulence, and constipation during short-term follow-up (Colman and Rubin 2014; Ettinger 2022; Park and Seo 2021). However, in a prospective registry, of the 156 participants who had a 6-month follow-up following FMT, 2 patients received new diagnoses of irritable bowel syndrome and 2 patients received new diagnoses of ulcerative colitis (Kelly et al. 2021).

FMT could therefore be used as a modulation strategy to address cognitive impairment. However, in order to establish a reliable FMT causality relationship, well-designed experiments must be quickly developed before making claims about the positive effects of FMT. For the purpose of selecting donors for FMT, larger long-term follow-up studies are required to find any potential long-term adverse outcomes (Sbahi and Di Palma 2016). Furthermore, there are still many unanswered questions regarding FMT therapy in human subjects, including the route of administration, frequency of application, donor microbiota screening, antibiotic administration, and other issues (Sun et al. 2020).

11.12 Conclusion and Future Directions

The exponential increase in research into the function of the GM in human health has occurred over the past 20 years. The emergence of GMBA as a crucial integrative physiological regulator via the bidirectional microbiota-gut-brain communication has sparked a great deal of excitement both within and outside the scientific community. Interventions that target the microbiota are intriguing and prospective therapeutic methods for the prevention and treatment of cognitive decline, especially in consideration of the plasticity of the microbiome. However, gut microbiota as a therapy still faces a number of methodological and experimental challenges, despite some reported evidence about the potential applicability of some novel therapeutic approaches. Preclinical studies account for the majority of the data currently available linking the microbiome to chronic diseases. Before they can be translated into useful advice, the few human studies still need to demonstrate their applicability. For instance, a growing body of research indicates that probiotics, particularly the GMP strain, may help to alter the GBA, making them a promising treatment option for cognitive impairment because they deliver precise medications with high levels of site specificity. Even though it has been hypothesized that cognitive impairment may share some of the pathways connected to gut dysbiosis, suggesting that it may be a strategy to restore intestinal homeostasis in this condition, the potential effectiveness of FMT is still based on theoretical presumptions. In fact, the efficacy of this therapeutic approach to treat this age-related condition has not yet been demonstrated in preclinical or clinical studies. Further research using both animal models and human intervention trials will be essential in establishing the efficacy of these approaches and paving the way for their clinical application.

References

Akbari E, Asemi Z, Daneshvar Kakhaki R, Bahmani F, Kouchaki E, Tamtaji OR et al (2016) Effect of probiotic supplementation on cognitive function and metabolic status in Alzheimer's disease: a randomized, double-blind and controlled trial. Front Aging Neurosci 8:256. https://doi.org/10.3389/fnagi.2016.00256

Aroniadis OC, Brandt LJ (2013) Fecal microbiota transplantation: past, present and future. Curr Opin Gastroenterol 29(1):79–84. https://doi.org/10.1097/MOG.0b013e32835a4b3e

Athari Nik Azm S, Djazayeri A, Safa M, Azami K, Ahmadvand B, Sabbaghziarani F et al (2018) Lactobacilli and bifidobacteria ameliorate memory and learning deficits and oxidative stress in β-amyloid (1-42) injected rats. Appl Physiol Nutr Metab 43(7):718–726. https://doi.org/10.1139/apnm-2017-0648

Ausó E, Gómez-Vicente V (2020) Biomarkers for Alzheimer's disease early diagnosis. Pers Med 10(3):114. https://doi.org/10.3390/jpm10030114

Banks WA, Erickson MA (2010) The blood–brain barrier and immune function and dysfunction. Neurobiol Dis 37(1):26–32

Berryhill ME, Peterson D, Jones K, Tanoue R (2012) Cognitive disorders. In: Ramachandran VS (ed) Encyclopedia of human behavior, 2nd edn. Academic Press, San Diego, pp 536–542

Bonfili L, Cecarini V, Berardi S, Scarpona S, Suchodolski JS, Nasuti C et al (2022) Microbiota modulation counteracts Alzheimer's disease progression influencing neuronal proteolysis and gut hormones plasma levels. Sci Rep 7(1):2426. https://doi.org/10.1038/s41598-017-02587-2

Cervenka I, Agudelo L, Ruas J (2017) Kynurenines: tryptophan's metabolites in exercise, inflammation, and mental health. Science 357:eaaf9794

Chen Y, Xu J, Chen Y (2021) Regulation of neurotransmitters by the gut microbiota and effects on cognition in neurological disorders. Nutrients 13(6):2099. https://doi.org/10.3390/nu13062099

Cipriani G, Danti S, Picchi L, Nuti A, Fiorino MD (2020) Daily functioning and dementia. Dement Neuropsychol 14(2):93–102. https://doi.org/10.1590/1980-57642020dn14-020001

Colman RJ, Rubin DT (2014) Fecal microbiota transplantation as therapy for inflammatory bowel disease: a systematic review and meta-analysis. J Crohns Colitis 8(12):1569–1581. https://doi.org/10.1016/j.crohns.2014.08.006

Cryan J (2021) The microbiome as a key regulator of stress & neuroinflammation across the lifespan. Psychoneuroendocrinology 131:105571. https://doi.org/10.1016/j.psyneuen.2021.105571

Cuervo-Zanatta D, Syeda T, Sánchez-Valle V, Irene-Fierro M, Torres-Aguilar P, Torres-Ramos MA, Shibayama-Salas M, Silva-Olivares A, Noriega LG, Torres N, Tovar AR, Ruminot I, Barros LF, García-Mena J, Perez-Cruz C (2023) Dietary fiber modulates the release of gut bacterial products preventing cognitive decline in an Alzheimer's mouse model. Cell Mol Neurobiol 43(4):1595–1618. https://doi.org/10.1007/s10571-022-01268-7

Den H, Dong X, Chen M, Zou Z (2020) Efficacy of probiotics on cognition, and biomarkers of inflammation and oxidative stress in adults with Alzheimer's disease or mild cognitive impairment—a meta-analysis of randomized controlled trials. Aging 12(4):4010–4039. https://doi.org/10.18632/aging.102810

Dore MP, Bibbò S, Fresi G, Bassotti G, Pes GM (2019) Side effects associated with probiotic use in adult patients with inflammatory bowel disease: a systematic review and meta-analysis of randomized controlled trials. Nutrients 11(12):2913. https://doi.org/10.3390/nu11122913. PMID: 31810233; PMCID: PMC6950558

Ettinger S (2022) Gut microbiome, and cognitive decline. Curr Nutr Rep 11(4):643–652. https://doi.org/10.1007/s13668-022-00435-y

Falony G, Vieira-Silva S, Raes J (2018) Richness and ecosystem development across faecal snapshots of the gut microbiota. Nat Microbiol 3(5):526–528. https://doi.org/10.1038/s41564-018-0143-5

Flint HJ (2012) The impact of nutrition on the human microbiome. Nutr Rev 70(Suppl 1):S10–S13. https://doi.org/10.1111/j.1753-4887.2012.00499.x

Generoso JS, Giridharan VV, Lee J, Macedo D, Barichello T (2021) The role of the microbiota-gut-brain axis in neuropsychiatric disorders. Braz J Psychiatry 43(3):293–305. https://doi.org/10.1590/1516-4446-2020-0987

Guzzetta KE, Cryan JF, O'Leary OF (2022) Microbiota-gut-brain axis regulation of adult hippocampal neurogenesis. Brain Plast 8(1):97–119. https://doi.org/10.3233/BPL-220141

Han Y, Wang B et al (2022) Vagus nerve and underlying impact on the gut microbiota-brain axis in behavior and neurodegenerative diseases. J Inflamm Res 15:6213–6230. https://doi.org/10.2147/jir.s384949

Hazan S (2020) Rapid improvement in Alzheimer's disease symptoms following fecal microbiota transplantation: a case report. J Int Med Res 48(6):300060520925930. https://doi.org/10.1177/0300060520925930

Heijtz RD, Wang S, Anuar F, Qian Y, Björkholm B, Samuelsson A et al (2011) Normal gut microbiota modulates brain development and behavior. Proc Natl Acad Sci 108(7):3047–3052

Hirsch BE, Saraiya N, Poeth K, Schwartz RM, Epstein ME, Honig G (2015) Effectiveness of fecal-derived microbiota transfer using orally administered capsules for recurrent Clostridium difficile infection. BMC Infect Dis 15:191. https://doi.org/10.1186/s12879-015-0930-z

Hwang YH, Park S, Paik JW, Chae SW, Kim DH, Jeong DG, Ha E, Kim M, Hong G, Park SH, Jung SJ, Lee SM, Na KH, Kim J, Chung YC (2019) Efficacy and safety of *Lactobacillus plantarum* C29-fermented soybean (DW2009) in individuals with mild cognitive impairment: a 12-week, multi-center, randomized, double-blind, placebo-controlled clinical trial. Nutrients 11(2):305. https://doi.org/10.3390/nu11020305. PMID: 30717153; PMCID: PMC6412773

Johnson AJ, Zheng JJ, Kang JW, Saboe A, Knights D, Zivkovic AM (2022) A guide to diet-microbiome study design. Front Nutr 7:79. https://doi.org/10.3389/fnut.2020.00079

Kaur G, Behl T, Bungau S, Kumar A, Uddin MS, Mehta V, Zengin G, Mathew B, Shah MA, Arora S (2021) Dysregulation of the gut-brain axis, dysbiosis and influence of numerous factors on gut microbiota associated Parkinson's disease. Curr Neuropharmacol 19(2):233–247. https://doi.org/10.2174/1570159X18666200606233050. PMID: 32504503; PMCID: PMC8033978

Kelly CR, Yen EF, Grinspan AM, Kahn SA, Atreja A, Lewis JD, Moore TA, Rubin DT, Kim AM, Serra S, Nersesova Y, Fredell L, Hunsicker D, McDonald D, Knight R, Allegretti JR, Pekow J, Absah I, Hsu R, Vincent J, Khanna S, Tangen L, Crawford CV, Mattar MC, Chen LA, Fischer M, Arsenescu RI, Feuerstadt P, Goldstein J, Kerman D, Ehrlich AC, Wu GD, Laine L (2021) Fecal microbiota transplantation is highly effective in real-world practice: initial results from the FMT National Registry. Gastroenterology 160(1):183–192.e3. https://doi.org/10.1053/j.gastro.2020.09.038

Kim KO, Gluck M (2019) Fecal microbiota transplantation: an update on clinical practice. Clin Endosc 52(2):137–143. https://doi.org/10.5946/ce.2019.009

Kim MS, Kim Y, Choi H, Kim W, Park S, Lee D et al (2020) Transfer of a healthy microbiota reduces amyloid and tau pathology in an Alzheimer's disease animal model. Gut 69(2):283–294. https://doi.org/10.1136/gutjnl-2018-317,431

Kobayashi Y, Sugahara H, Shimada K, Mitsuyama E, Kuhara T, Yasuoka A, Kondo T, Abe K, Xiao JZ (2017) Therapeutic potential of Bifidobacterium breve strain A1 for preventing cognitive impairment in Alzheimer's disease. Sci Rep 7(1):13510. https://doi.org/10.1038/s41598-017-13368-2

Kobayashi Y, Kuhara T, Oki M, Xiao JZ (2019) Effects of *Bifidobacterium breve* A1 on the cognitive function of older adults with memory complaints: a randomised, double-blind, placebo-controlled trial. Benef Microbes 10(5):511–520. https://doi.org/10.3920/BM2018.0170. Epub 2019 May 15. PMID: 31090457

Kumar M, Yadav AK, Verma V, Singh B, Mal G, Nagpal R, Hemalatha R (2016) Bioengineered probiotics as a new hope for health and diseases: an overview of potential and prospects. Future Microbiol 4:585–600. https://doi.org/10.2217/fmb.16.4

Leblhuber F, Steiner K, Geisler S, Fuchs D, Gostner JM (2020) On the possible relevance of bottom-up pathways in the pathogenesis of Alzheimer's disease. Curr Top Med Chem 20(15):1415–1421. https://doi.org/10.2174/1568026620666200514090359

Leeming ER, Johnson AJ, Spector TD, Le Roy CI (2019) Effect of diet on the gut microbiota: rethinking intervention duration. Nutrients 11(12):2862. https://doi.org/10.3390/nu11122862. PMID: 31766592; PMCID: PMC6950569

Li X, Chen LM, Kumar G, Zhang SJ, Zhong QH, Zhang HY et al (2022) Therapeutic interventions of gut-brain axis as novel strategies for treatment of alcohol use disorder associated cognitive and mood dysfunction. Front Neurosci 16:820106. https://doi.org/10.3389/fnins.2022.820106

Lin L, Zheng LJ, Zhang LJ (2018) Neuroinflammation, gut microbiome, and Alzheimer's disease. Mol Neurobiol 55(11):8243–8250. https://doi.org/10.1007/s12035-018-0983-2

Ling Y, Gong T, Zhang J, Gu Q, Gao X, Weng X, Liu J, Sun J (2020) Gut microbiome signatures are biomarkers for cognitive impairment in patients with ischemic stroke. Front Aging Neurosci 12:511562. https://doi.org/10.3389/fnagi.2020.511562

Liu J, Sun J, Wang F, Yu X, Ling Z, Li H, Zhang H, Jin J, Chen W, Pang M, Yu J, He Y, Xu J (2015) Neuroprotective effects of Clostridium butyricum against vascular dementia in mice via metabolic butyrate. Biomed Res Int 2015:412946. https://doi.org/10.1155/2015/412946

Long-Smith C, O'Riordan KJ, Clarke G, Stanton C, Dinan TG, Cryan JF (2020) Microbiota-gut-brain axis: new therapeutic opportunities. Annu Rev Pharmacol Toxicol 60:477–502. https://doi.org/10.1146/annurev-pharmtox-010919-023628

Lopez J, Grinspan A (2016) Fecal microbiota transplantation for inflammatory bowel disease. Gastroenterol Hepatol (N Y) 12(6):374–379

Lucidi L, Pettorruso M, Vellante F, Di Carlo F, Ceci F, Santovito MC, Di Muzio I, Fornaro M, Ventriglio A, Tomasetti C, Valchera A, Gentile A, Kim YK, Martinotti G, Fraticelli S, Di Giannantonio M, De Berardis D (2021) Gut microbiota and bipolar disorder: an overview on a novel biomarker for diagnosis and treatment. Int J Mol Sci 22(7):3723. https://doi.org/10.3390/ijms22073723

Lv T, Ye M, Luo F, Hu B, Wang A, Chen J, Yan J, He Z, Chen F, Qian C, Zhang J, Liu Z, Ding Z (2021) Probiotics treatment improves cognitive impairment in patients and animals: a systematic review and meta-analysis. Neurosci Biobehav Rev 120:159–172. https://doi.org/10.1016/j.neubiorev.2020.10.027. Epub 2020 Nov 4. PMID: 33157148

Margolis KG, Cryan JF, Mayer EA (2021) The microbiota-gut-brain axis: from motility to mood. Gastroenterology 160(5):1486–1501. https://doi.org/10.1053/j.gastro.2020.10.066

Mullane K, Williams M (2020) Alzheimer's disease beyond amyloid: can the repetitive failures of amyloid-targeted therapeutics inform future approaches to dementia drug discovery? Biochem Pharmacol 177:113945. https://doi.org/10.1016/j.bcp.2020.113945

Musa NH, Mani V, Lim SM, Vidyadaran S, Abdul Majeed AB, Ramasamy K (2017) Lactobacilli-fermented cow's milk attenuated lipopolysaccharide-induced neuroinflammation and memory impairment in vitro and in vivo. J Dairy Res 84(4):488–495. https://doi.org/10.1017/S0022029917000620

Navarro-López V, Méndez-Miralles MÁ, Vela-Yebra R, Fríes-Ramos A, Sánchez-Pellicer P, Ruzafa-Costas B, Núñez-Delegido E, Gómez-Gómez H, Chumillas-Lidón S, Picó-Monllor JA, Navarro-Moratalla L (2022) Gut microbiota as a potential predictive biomarker in relapsing-remitting multiple sclerosis. Genes (Basel) 13(5):930. https://doi.org/10.3390/genes13050930

Ng SC, Hart AL, Kamm MA, Stagg AJ, Knight SC (2009) Mechanisms of action of probiotics: recent advances. Inflamm Bowel Dis 15(2):300–310. https://doi.org/10.1002/ibd.20602

Nimgampalle M, Kuna Y (2017) Anti-Alzheimer properties of probiotic, Lactobacillus plantarum MTCC 1325 in Alzheimer's disease induced albino rats. J Clin Diagn Res 11(8):KC01–KC05. https://doi.org/10.7860/JCDR/2017/26106.10428

Park SY, Seo GS (2021) Fecal microbiota transplantation: is it safe? Clin Endosc 54(2):157–160. https://doi.org/10.5946/ce.2021.072

Park SH, Lee JH, Kim JS, Kim TJ, Shin J, Im JH et al (2022) Fecal microbiota transplantation can improve cognition in patients with cognitive decline and Clostridioides difficile infection. Aging 14(16):6449–6466. https://doi.org/10.18632/aging.204230

Qian Y, Yang X, Xu S, Huang P, Li B, Du J, He Y, Su B, Xu LM, Wang L, Huang R, Chen S, Xiao Q (2020) Gut metagenomics-derived genes as potential biomarkers of Parkinson's disease. Brain 143(8):2474–2489. https://doi.org/10.1093/brain/awaa201

Reid G (2016) Probiotics: definition, scope and mechanisms of action. Best Pract Res Clin Gastroenterol 30(1):17–25. https://doi.org/10.1016/j.bpg.2015.12.001

Sbahi H, Di Palma JA (2016) Faecal microbiota transplantation: applications and limitations in treating gastrointestinal disorders. BMJ Open Gastroenterol 3:e000087. https://doi.org/10.1136/bmjgast-2016-000087

Shabbir U, Arshad MS, Sameen A, Oh DH (2021) Crosstalk between gut and brain in Alzheimer's disease: the role of gut microbiota modulation strategies. Nutrients 13(2):690. https://doi.org/10.3390/nu13020690

Silver J, Schwab ME, Popovich PG (2014) Central nervous system regenerative failure: role of oligodendrocytes, astrocytes, and microglia. Cold Spring Harb Perspect Biol 7(3):a020602. https://doi.org/10.1101/cshperspect

Sonnenburg ED, Smits SA, Tikhonov M, Higginbottom SK, Wingreen NS, Sonnenburg JL (2016) Diet-induced extinctions in the gut microbiota compound over generations. Nature 529(7585):212–215. https://doi.org/10.1038/nature16504

Sun J, Xu J, Ling Y, Wang F, Gong T, Yang C, Ye S, Ye K, Wei D, Song Z, Chen D, Liu J (2019) Fecal microbiota transplantation alleviated Alzheimer's disease-like pathogenesis in APP/PS1 transgenic mice. Transl Psychiatry 9(1):189. https://doi.org/10.1038/s41398-019-0525-3

Sun Y, Baptista LC, Roberts LM, Jumbo-Lucioni P, McMahon LL, Buford TW, Carter CS (2020) The gut microbiome as a therapeutic target for cognitive impairment. J Gerontol A Biol Sci Med Sci. 75(7):1242–1250. https://doi.org/10.1093/gerona/glz281

Varesi A, Pierella E, Romeo M, Piccini GB, Alfano C, Bjørklund G, Oppong A, Ricevuti G, Esposito C, Chirumbolo S, Pascale A (2022) The potential role of gut microbiota in Alzheimer's disease: from diagnosis to treatment. Nutrients 14(3):668. https://doi.org/10.3390/nu14030668

Ventham NT, Kennedy NA, Nimmo ER, Satsangi J (2013) Beyond gene discovery in inflammatory bowel disease: the emerging role of epigenetics. Gastroenterology. 145(2):293–308. https://doi.org/10.1053/j.gastro.2013.05.050

Vindigni SM, Surawicz CM (2017) Fecal microbiota transplantation. Gastroenterol Clin North Am 46(1):171–185. https://doi.org/10.1016/j.gtc.2016.09.012

Wang H, Lu Y, Yan Y, Tian S, Zheng D, Leng D, Wang C, Jiao J, Wang Z, Bai Y (2020) Promising treatment for type 2 diabetes: fecal microbiota transplantation reverses insulin resistance and impaired islets. Front Cell Infect Microbiol 9:455. https://doi.org/10.3389/fcimb.2019.00455

Zelante T, Iannitti RG, Cunha C, De Luca A, Giovannini G, Pieraccini G, Zecchi R, D'Angelo C, Massi-Benedetti C, Fallarino F, Carvalho A, Puccetti P, Romani L (2013) Tryptophan catabolites from microbiota engage aryl hydrocarbon receptor and balance mucosal reactivity via interleukin-22. Immunity 39(2):372–385. https://doi.org/10.1016/j.immuni.2013.08.003

Zhan G, Yang N, Li S, Huang N, Fang X, Zhang J, Zhu B, Yang L, Yang C, Luo A (2018) Abnormal gut microbiota composition contributes to cognitive dysfunction in SAMP8 mice. Aging (Albany NY) 10(6):1257–1267. https://doi.org/10.18632/aging.101464

Zhang L, Wang Y, Xiayu X, Shi C, Chen W, Song N, Fu X, Zhou R, Xu YF, Huang L, Zhu H, Han Y, Qin C (2017) Altered gut microbiota in a mouse model of Alzheimer's disease. J Alzheimers Dis 60(4):1241–1257. https://doi.org/10.3233/JAD-170020

Zhuang ZQ, Shen LL, Li WW, Fu X, Zeng F, Gui L, Lü Y, Cai M, Zhu C, Tan YL, Zheng P, Li HY, Zhu J, Zhou HD, Bu XL, Wang YJ (2018) Gut microbiota is altered in patients with Alzheimer's disease. J Alzheimers Dis 63(4):1337–1346. https://doi.org/10.3233/JAD-180176

Pharmacological and Nutritional Approaches to Modulate Microglial Polarization in Cognitive Senescence

12

Bhuvaneswari Ponnusamy, Ponnulakshmi Rajagopal, and Selvaraj Jayaraman

Abstract

Brain aging, characterized by age-related changes, significantly impacts cognitive function. Microglia cells are resident macrophages in the CNS that are responsible for maintaining brain homeostasis. Normally older people exhibit a high inflammatory condition in brain. Thus any small insult will trigger the microglia activation and polarize into pro-inflammatory (M1) and anti-inflammatory (M2) states. Microglia acts as a double-end sword by both promoting and ameliorating cognitive senescence. Cognitive impairment is a common condition seen in elderly people. Microglia holds an upper hand in pathophysiology of diseases, like Alzheimer's, that impair cognitive health. Thus understanding the action of microglia is essential for avoiding further depletion of cognitive activities. In this book chapter, we have discussed a detailed view on the senescence of brain cells and the role of microglial cells senescence in the impairment of cognitive functioning. Moreover we also discuss about various pharmacological and nutritional approaches involved in the modulating microglial function thereby restoring the cognitive functions.

B. Ponnusamy · S. Jayaraman (✉)
Department of Biochemistry, Centre of Molecular Medicine and Diagnostics (COMManD), Saveetha Dental College and Hospital, Saveetha Institute of Medical and Technical Sciences, Saveetha University, Chennai, India
e-mail: selvarajj.sdc@saveetha.com

P. Rajagopal
Department of Central Research Laboratory, Meenakshi Ammal Dental College and Hospitals, Meenakshi Academy of Higher Education and Research, Chennai, India

S. Pathak, A. Banerjee (eds.), *Gut Microbiome and Brain Ageing*,
https://doi.org/10.1007/978-981-99-8803-7_12

Keywords

Microglia · Ageing · Senescence · Inflammation · Cytokine · M1 and M2 polarization

12.1 Introduction

Nervous system is an intertwined web of network constantly functioning to ease the living system to communicate with its environment (Ludwig et al. 2022). Broadly, nervous system is divided into central nervous system (CNS) and peripheral nervous system (PNS). Brain is a versatile organ that controls the entire functioning of the organism through responses, senses, emotions, movement, language, communication, thinking and memory (Maldonado and Alsayouri 2022). Each division of brain is involved in a number of functions in order to maintain a balanced cognitive stimuli and responses. Furthermore, brain is distinguished as two parts namely the white and grey matter. The grey matter holds up majority of brain composition and forms the cerebral cortex. The grey matter mainly consists of neuronal cell bodies, dendrites, both myelinated and unmyelinated axons, glial cells, neuronal junctions (synapsis) and blood vessels (Maldonado and Alsayouri 2022). The area underlying, referred to as subcortical region, constitutes the white matter, majorly made up of myelinated axons, with a fewer neuronal cell bodies and makes up the spinal cord.

Neurons and glial cells are the functional units of central nervous system (CNS). Neuronal cells are mainly involved in the propagation of chemical and electrical impulses. It is said that human brain is made up of approximately 100 billion neurons with more or less 100 trillion synapses (Peters and Connor 2014). Rudolf Virchow was the first to discover a new type of cell apart from neurons in the brain. He found that these cells are involved in holding the neurons in their place and named these cells as glial cells. However, advances in the glial cell research proved that these cells possess various essential functioning namely, increasing synaptic fidelity, providing trophic support and contributing to the immune response within the brain. Thus glial cells are important type of cell contributing to CNS homeostasis. Glial cells are non-conductive type of cells that are further divided into three types namely, astrocytes, oligodendrocytes and microglial cells. Among the glial cell population, astrocytes accounts for about 20–50% (Sofroniew and Vinters 2010). They play a vital role in metabolism, development of blood brain barrier, maintaining a balanced osmolarity in brain and act as storage site of glycogen and scavenging of unwanted debris (Peters and Connor 2014). Oligodendrocytes are involved in providing insulation to axons present inside the CNS. An extension of oligodendrocyte plasma membrane is labelled as myelin sheath. Microglia cells are unique type of immune cells localized in brain. They act as a macrophage in the brain playing a major role in exerting inflammatory properties and scavenging activity thereby providing innate immunity.

After a course of development every individual organism is prone to a natural time-based deregulation of physiological roles that are essential for existence of that individual. This so-called process is referred to as ageing. The aging happening in the brain is a complex process and has a special interest for researchers. On course of ageing, brain undergoes structural, chemical and functional changes that have an impact on normal functioning of an individual (Mattson and Arumugam 2018). Cognitive pattern refers to the ability of conceptual reasoning, memory, processing speed and vocabulary. Deterioration of cognitive functions may be due to natural process as well as pathological condition. Growing researches in neuroscience have pointed that cognitive impairments may be due to decrease in the grey and white matter content, and decrease in levels of neurotransmitters contributing to age-related cognitive changes.

12.2 Senescence

Hayflick and Moorhead were the first to define the term "cellular senescence" in the year 1961 (Hayflick and Moorhead 1961). The growth and division of cells can be arrested upon various triggers. This type of programmed cell growth arrest is called as senescence; occurs mainly due to presence of unique phenotypic alterations and inflammatory cytokines release (Kuilman et al. 2010). Senescence is a natural physiological process occurring in the cells in order to regulate the cell proliferation and division, thus acting as a blockade to cancer induction (Hanahan and Weinberg 2011). In simple terms, senescence cells are those cells that do not divide and proliferate, but exhibit stable metabolic and biochemical functions. This action comes upon different signalling cascade, stimuli and stress factors that trigger senescence gene. Possible factors that are involved in triggering the senescence process are alterations in the telomere morphology, telomeric arm shortening, oncogenic signals, activation of the mitogen related cascade, oxidative stress, radiation exposure, epigenetic modifications, deformities of the chromatin structure, inflammation, malfunctioning of mitochondria, perturbed proteostasis, xenobiotics and deficiency of essential nutrients (Kuilman et al. 2010; Kumari and Jat 2021). Thus the extent of stress imposed on a cell determines the fate of that cell by entering repair mechanisms or entering the apoptotic pathway or cellular senescence (Surova and Zhivotovsky 2013). Senescence cells exhibit unique characteristic features namely excess of lysosome secretion, damage to the systemic macromolecules and initiation of SASP (senescence-associated secretory phenotype) (Gorgoulis et al. 2018; Saleh et al. 2018; Lee et al. 2006). These cells also show significant morphological and structural changes with enlargement of cells, poly-nucleated, large vacuoles, modified plasma membrane makeup and marked nucleus enlargement (Kumari and Jat 2021). These changes in the senescent cells are the major reason for eliciting growth arrest in cells and engaged in the formation of SASP secretome. This process has both beneficiary roles in the maintenance of normal physiology and has an equal disadvantage during pathological conditions. Every cell has a limit to divide and when its limit is attained, the cell will stop to divide and initiate ageing.

12.3 Ageing

Ageing is a natural process that can be described as a systematic progression in the decrease of functionality (McHugh and Gil 2018). Thus ageing is characterized by the gradual deterioration of every organ system in the body thereby leading to loss of tissue function. This opens up the reason for increased risk for cardiovascular disease, diabetes, kidney dysfunction, neuron degeneration and many more in the aged people. On a broad classification, ageing is classified into three categories namely primary, antagonistic and integrative (McHugh and Gil 2018). Primary ageing refers to the naturally occurring age-related deterioration, antagonistic refers to the deterioration due to damage or injury or stress and integrative refers to the combined effect of primary and secondary ageing that results in the loss of age-related cell homeostasis (Table 12.1). There are number theories available for the better understanding of mechanism of ageing. They are genetic programme theory, wear and tear theory, telomere theory, endocrine theory, DNA damage hypothesis, error catastrophe theory, mitochondrial theory, free radical theory and rate of living theory (Park and Yeo 2013).

The process of ageing begins right from the birth of offspring. Various genetic factors and environmental process are involved in determining the rate of ageing. As ageing progresses there are a number of physiological and anatomical variations seen in the living system. Moreover risk factors such as smoking, tobacco, alcohol, hypertension, diabetes, stress etc. may also result in pathological ageing condition. In terms of senescence theory, during the ageing process two types of actions occur namely the senescent cells' accumulation in tissues result in loss of tissue physiology and senescent cells are involved in hindering the regenerative potential of stem cells thereby losing the normal physiology of the cells (van Deursen 2014). During ageing the senescent cells are found to emit a unique type of signals called paracrine signals. These signals play an important role in affecting the nearby normal cells and converting them to a senescent cell. The paracrine signals are involved in recruiting inflammatory markers to the site (Di Micco et al. 2021). Though the senescence plays a positive role in the control of cancer cell growth, it also has a negative impact on the stem cells by hindering their proliferation and division (Paul et al. 2022). Apart from ageing, various pathological conditions are associated with cellular senescence namely osteoarthritis, pulmonary fibrosis, atherosclerosis, Alzheimer's disease, cancer and frailty (McHugh and Gil 2018).

Table 12.1 List of various theories related to ageing

Theories	Description	References
Genetic program theory	According to this theory, ageing in each and every individual organism in a species is preprogrammed as they have their own average life span The theory suggests that there is an improvement in the average life span of humans with not much change in the maximum life span of human This shows a clear statement that genetic mutations play an important role in the extension of life span This theory proposes that senescence gene controls age associated decline in the biological functions	Troen (2003)
Wear and tear theory	This theory believes that living system is encountered with a number of injuries due to various reasons like accidents, diseases, radiation, toxic substances and food However this theory is not accepted, because there is no change in the maximum life span changes even if an organism is protected from damages	Harman (1981)
Telomere theory	Telomere shortening is linked with ageing process. According to this theory the life span of an individual is pre-programmed in accordance with the length of the telomere	Wong and Collins (2003)
Endocrine theory	This theory states that hormones have a major impact on the life span expectancy as it proves that after menopause, andropause and somatopause organisms showed an extended life time when provided with hormonal supplements	Tatar et al. (2003)
DNA damage hypothesis	This hypothesis states that the loss of DNA repair mechanism and increased susceptibility to DNA damage are directly linked with the progression of ageing	Bohr and Anson (1995)
Error catastrophe theory	This theory states that mutations in the central dogma can lead to generation of defected genes, mRNA and proteins which ultimately lead to the defective and damaged cells formation However, further researches are needed in this field	Weinert and Timiras (2003)
Mitochondrial theory	This theory proposes that mitochondria is the major site that releases a number of free radicles because it is majorly prone to oxidative stress and damage. Thus free radical-induced damaged mitochondria can result in decreased functions which result in increased rate of ageing	Trifunovic et al. (2004)
Free radical theory	This theory believes that generation of free radicles like superoxide (O_2^-) and hydroxyl (OH^-) anions, nitric oxide (NO), peroxynitrite ($ONOO^-$) are directly linked with ageing of the individual	Beckman and Ames (1998)
Rate of living theory	This theory suggests that rate of energy expenditure is inversely proportional to lifetime. Thus increased energy release results in short life span	Park and Yeo (2013)

12.4 Microglial Cells and Inflammation

Microglial cells are immune cells present in the central nervous system characterized as mono nuclear phagocytic cells. They account for only 10% of the cells in the brain. They evolve from early myeloid progenitor cells which are involved in the building of neural tube that retains throughout the lifetime of the organism (Ginhoux et al. 2010). In the process of development, these cells are involved in the regulation of the neural circuit and synapse formation (Colonna and Butovsky 2017). However when there is a damage or injury to the neuronal cells, microglial cells are actively engaged in the clearing of the damaged cells via phagocytosis or by releasing chemo-attractants, cytokines and neurotropic factors thereby providing an immense response against the pathogens and stimulating tissue repair in the cells (Colonna and Butovsky 2017). Thus microglial cell brings about cell-mediated immunity in maintaining the normal physiology of the brain. It is now very well known that microglial cells actively participate in precipitating a number of neurological diseases like Alzheimer's disease, brain injury and psychiatric diseases. Homeostasis of microglial cells mainly depend on the signalling molecules namely colony stimulating factor 1 receptor (CSF1R), protein kinase B and extracellular signal regulated kinases (ERK) (Dai et al. 2002). The CSF1 ligand shares a similar structure with Interleukin (IL)-34, though the basic amino acid makeup differs, binds with CSF1R to elicit its action (Chitu et al. 2016). Upon mutation or damage or knockout of this receptor, there is a marked reduction in the release of microglial cells. However there are two more important proteins that are involved in maintaining the number of microglia, called the adapter protein DAP12 and interferon regulatory factor (IRF)-8. Upon DAP12 gene and IRF-8 gene mutation, there is a marked reduction in the microglial cell population. Table 12.2 enlists various pathological condition in which microglia play an important role.

Microglial cells express wide range receptors for the binding of various molecules as well as neurotransmitters to elicit its functions. They express a number of pattern-recognition receptors (PRRs), scavenger receptors, immune receptors and receptors for neurotransmitters (Table 12.3).

As the microglia cells mature and age, they are more prone to proinflammatory triggers which are referred to as microglia priming (Norden and Godbout 2013). They have the ability to elicit a prolonged and persistent action upon stimulus. These types of primed microglia are characterized with a larger soma and small dendritic arbours. They respond to cytokines namely TNF-α, IL-1β, IL-6, IL-10 and TGF-β; and express certain specific markers such as MHC-II, AXL, Lgals3 and CD11C (Holtman et al. 2015). Condition in which downregulation of inhibitory neuronal ligands expression namely, fractalkine, CD47, CD200 and siglecs occurs is also involved in microglia priming.

Table 12.2 Role of microglia under various clinical conditions

S. no.	Clinical condition	Role of microglia	Factors involved	References
1.	Multiple sclerosis	Microglia and macrophages are the two most terrorizing elements that are involved in the progression of MS. Studies revealed that persistent inflammatory condition due to microglia action acts as one of the main causes for the adverse clinical symptom. Thus a microglia knockout animal model showed a decrease in the inflammation and an improved clinical condition in the experimental rats However microglia is also noted for its beneficial role in MS, by promoting the scavenging of myelin debris and apoptotic cells which in hand promotes the tissue regeneration. This act of microglia cells is noted for its ameliorating potential and providing a better clinical outcome	Fibrinogen activation may play an important role in recruiting microglia cells in the brain MS displays a marked degeneration of axons which could possibly trigger the activation of microglia cells Another clinical feature in identification of microglial clusters which could act as an antigen presenting cell for T lymphocytes	Bogie et al. (2014)
2.	Stroke	Stroke can lead to the degeneration of neurons, recruiting microglia cells and other immune cells to the site. Upon ischemic condition, the brain filters a number of immune cells and microglia in the periphery of the ischemic tissue, thereby promoting phagocytosis and creating an inflammatory microenvironment	Production of reactive oxidative species at the site of ischemia and cytokine production lead to the breakdown of the blood brain barrier that could promote the diffusion of immune cells into the CNS	Fumagalli et al. (2015)
3.	Alzheimer's disease	In this condition the amyloid plaques surrounded with a number of inflammatory cytokines are considered as the basic pathology in AD. Single nucleotide polymorphism is considered as a risk factor of precipitation of AD. A number of genes related to microglia are found to be associated with SNPs, therefore resulting in expression of proteins that are expressed on microglia. A recent study by Zhang et al. showed that microglia-induced phagocytosis and immune reactions are greatly linked to AD pathology	SNPs encode proteins namely TREM2, CD33, CR1, ABCA7, SHIP1 and APOE that are related to microglia Microglia-mediated secretion of cytokines (IL-1β, IL-6, TNF-α, TGF-β) reactive species, NLRP3 inflammasome and other chemokines are noted in AD	Zhang et al. (2013), Heneka et al. (2013)

(continued)

Table 12.2 (continued)

S. no.	Clinical condition	Role of microglia	Factors involved	References
4.	Amyotrophic lateral sclerosis	At an early stage of ALS there is marked concentration of microglia at the site and it plays a beneficial role by increasing the expression of BNDF (brain-derived neurotrophic factor) and phagocytosis of debris. However as the disease progresses to later stages the microglia cells are found to secrete high-mobility group box 1 (HMGB1), a neurotoxin, that promotes more number of proinflammatory cytokines	TREM2, granulin and profilin 1 are some factors that are responsible for reduced microglial phagocytosis and altered immune response	Brites and Vaz (2014)
5.	Parkinson disease	PD is characterized by the presence of Lewy bodies and α-synuclein in parts of brain. Primarily microglia cells are involved in the clearance of these characteristic elements from the CNS by phagocytosis	IL-1 and IL-18 are the most noted cytokines in PD. Apart from them the NLRP3 inflammasome is also widely studied	Machado et al. (2016)

Table 12.3 Various receptors expressed by microglia cells and their function

Name	Receptor	Function	References
Pattern-recognition receptors	• Pathogen associated molecular patterns (PAMPs) • Tissue damage-associated molecular patterns (DAMPs) • Toll-like receptors (TLRs) • NOD-like receptors (NLRs) • Nucleic acid receptors • C-type lectin receptors	These receptors are actively involved in the identification of specific molecular patterns or structures in the host that occur due to pathogen or tissue damage	Saijo et al. (2013), Sancho and Reis e Sousa (2012), Heneka et al. (2015)
Scavenger receptors	• CD36 • SR1 • MARCO • LDL receptor family members such as LDLR, ApoER2 and VLDL • Three receptor tyrosine kinases—Tyro3, Axl and Mertk (TAM) • Chemokine receptor—CX3CR1 and CXCR4	These receptors are involved in the clearance of unwanted particles from the site by either phagocytosis or endocytosis	Areschoug and Gordon (2009), Mizutani et al. (2012)
Immune receptors	• Immunoglobulin superfamily (Ig-SF) • TREM2 (triggering receptor expressed on myeloid cells 2) • CD33 • CD200R1 • SIRPA • Signalling lymphocytic activation molecule (SLAM)	This group of receptors are mainly involved in the regulatory action of microglial cells by taking control over the duration of any type of reaction activation or inhibition	Colonna and Wang (2016)
Receptors for neurotransmitters and neuropeptides	• α-Amino-3-hydroxy-5methyl-4-isoxazolepropionic acid (AMPA) • *N*-methyl-D-aspartate (NMDA) receptors • Glutamate metabotropic receptors, such as mGluR2	These receptors are mainly involved in the neuron and glial cells cross-talks. Moreover they are also involved in linking the microglial cells for supervising the neuronal functionality, removing the damage cells, promoting regeneration of new cells replacing the damaged ones and also for regulatory purpose	Pocock and Kettenmann (2007)

12.5 Microglia Polarization and Cognitive Senescence

Microglia cells are either in active state or resting state. Microglia is believed to be found in two phenotypes during active state namely M1 (classic path) and M2 (alternate path) phase, in which the M1 phase is the active phase involved in eliciting the inflammatory action and the M2 phase is involved in exhibiting the anti-inflammatory action of microglia cells, thus playing a regulatory role. Therefore the M1 phenotype is involved in the precipitation of inflammatory neurodegeneration and the M2 phenotype is involved in the neuroprotection action and tissue regeneration processes. Liposaccharides and interferon-γ are found to stimulate M1 microglia; and interleukin-4 (IL-4)/IL-13 are found to stimulate M2 phenotype (Kwon and Koh 2020). Specific markers namely, IL-1β, IL-6 and inducible nitric oxide synthase; and arginase 1 (ARG1), chitinase-3-like-3 (CHI3L3/YM1), and found in inflammatory zone 1 (FIZZ1) are used to identify the presence of M1 and M2 phenotypes, respectively (Jin et al. 2018). However the inflammatory action of microglia cells is regulated by M2 phenotype and various other factors. PGC-1-related coactivator (PRC) is largely expressed on M2 microglia cells. Moreover the excessive expression of PRC gene is involved in increasing the expression of other markers expressed on M2 cells namely, ARG1, FIZZ1 and YM1 (Mou et al. 2015). Trehalose6,60-dibehenate (TDB) is another regulatory molecule that is involved in decreasing protein markers expression on M1 cells (Mohanraj et al. 2019). Ageing is an important factor that determines the expression of either of microglia phenotypes. Studies reveal that upon ageing there is a marked increase in the expression of M1 phenotype protein markers, thereby increasing the proinflammatory condition and decreased expression of M2 protein markers (Zhang et al. 2020).

The most common type of senescence affecting cognition is seen in Alzheimer's disease. In AD there is a marked increase in the expression of proinflammatory protein marker on the M1 in the brain. Most importantly gene encoding TNF-α is found predominantly on the frontal cortex in the AD patients. Contrastingly they also expressed protein markers, namely YM1, in the cortex region of the brain. This shows that both classic and alternate pathways of microglia are activated (Colton et al. 2006). There are three main pathophysiological processes that are happening in AD, firstly, interaction of Amyloid-β (Aβ) and microglia cells, secondly, interaction with tau proteins and thirdly, microglia-mediated synaptic dysfunction. Upon aggregation of Aβ in the CNS, the activation and recruitment of microglia occurs (Glass et al. 2010), which further promotes the generation of a proinflammatory cytokine (iNOS, TNF-α, IL-6, IL-1β) wave. On the other hand, they are involved in the clearance of plaque aggregates via phagocytosis and proteases-mediated degradation (Kettenmann et al. 2011). Similarly tau proteins are also involved in the accumulation of microglia cells in the vicinity. The activation of microglia in this condition is mediated by NLRP3 inflammasome which is responsible for the release of inflammatory cytokine IL-1β (Ising et al. 2019). Another important feature is that the synaptic dysfunction plays a major role in the AD precipitation. This is mainly due to neuronal degeneration and death due to the production of inflammatory cytokines at specific sites. TNF-α and IL-1β are the two main cytokines found increased

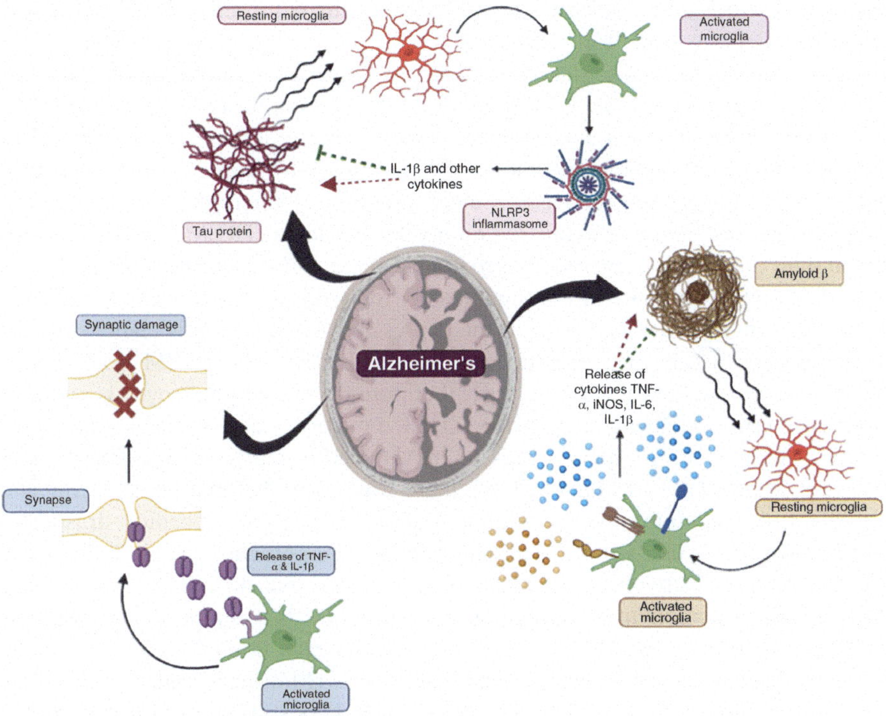

Fig. 12.1 Illustration on the role of microglia in Alzheimer's disease progression and amelioration

in the AD and responsible for the synaptic damage in AD patients (Xiao et al. 2020). Thus in AD, the microglia cells play a dual role in both promotion of the disease condition and also amelioration process (Fig. 12.1).

12.6 Pharmacological Approach in Cognitive Senescence

It is important to understand the pathology underlying in AD and microglia to draw a definite pharmacological target design in order to bring a better amelioration process. Thus an efficient pharmacological approach in treating AD patients should be the potential target that would compromise the ill microglia and not the beneficial one. Thus the drug should be designed with more specificity to alter the microglial phagocytosis. From the above information it is now evident that M1 activation and its phagocytic action are greatly stimulated by the presence of Aβ and tau protein aggregates. But on the other hand the inflammatory conditions generated by the microglia cells itself have an impact on the formation or progression of Aβ and tau aggregates. Moreover there is extensive damage to the neurons and synaptic clefts on account of inflammation produced by the microglia cells in AD condition. Thus targeting the molecular pathways in which microglia cells activate will aid in formulating a definite target for AD treatment.

Many other markers like the CD40 gene, responsible for recruiting of TNF-α, is found to be upregulated and a M2 marker CD11 is found to be upregulated in the AD patients (Kamphuis et al. 2016). Evidences prove that there is a major part of microglia involved in pathophysiology of AD. A study by Nordengen et al. showed that there was a marked increase in the expression of microglia marker namely, soluble triggering receptor expressed on myeloid cells 2 (sTREM2) in the cerebrospinal fluid of patients diagnosed at preclinical stage, mild impairment in the cognition in AD when compared to normal healthy individuals (Nordengen et al. 2019). Expression of TREM is linked with the expression Arg1, which is responsible for the phagocytic action of microglia cells. This adds point that TREM deficiency is linked to decreased expression of Arg1, therefore decreased phagocytosis and thus progression of AD (Zhang et al. 2018). CD33, a majorly expressed gene in microglia and macrophage, is largely expressed in AD condition promoting inflammation. Whereas when CD33 was knocked out, there was anti-inflammation occurring and formation of inflammasome (Griciuc et al. 2013). P2X7 receptor drastically increases in AD patients in the later stages with no significant change in the early stages. They are actively involved in neurodegenerative process and promoting inflammatory microenvironment. Upon activation of this receptor, the microglia is instructed to move towards the senile plaques whereas upon deactivation of the receptor, there was marked increase in phagocytosis (Martínez-Frailes et al. 2019). Another potential target is TLRs knockout. TLR knockout showed a transitional change from M1 to M2 phenotype in AD patients (Liu et al. 2012). Thus targeting these protein biomarkers in AD patient would be efficient in the management and amelioration of the disease.

12.7 Nutritional Approach in Cognitive Senescence

There are various agents involved in producing damage to the CNS. Though some injury and damage are inevitable some modifications can be made to protect the brain from injuries. In this context, diet and lifestyle modifications are two main aspects that are related to brain health. Deterioration of cognition is only evident when there is more burden on CNS (Clare et al. 2017). Normally ageing is a resultant of a number of factors like oxidative stress, inflammation and vascular diseases. These factors are also seen in neurological diseases development, thus upon unfavourable conditions these factors are involved in the aggravation of pathological states. Studies reveal that amount of protein intake is related to cognition health with visuospatial function, verbal fluency, processing speed and sustained attention (Li et al. 2020; Okereke et al. 2012). Studies relate that high-fat diet consumption is strongly associated with impaired cognitive functions. One author states that high-fat diet triggers the hippocampus to generate an array of neuro-inflammatory markers upon immune challenge that could result in memory impairments (Spencer et al. 2017). Moreover high-fat diet is linked to other pathological conditions, namely insulin resistance, obesity, impaired glucose metabolism and many more. These conditions play a major role as risk factors in AD pathology (Barbagallo and Dominguez 2014). Thus diets with low fat could be ideal for the prevention of cognitive decline. Moreover intake of polyunsaturated fatty acids (PUFAs), eicosapentaenoic acid (EPA), docosahexaenoic acid (DHA) and omega-3 fatty acids is found to improve the functioning of neurons, endothelial cells and glial cells in the brain by altering neurotransmission, down-regulate neuroinflammation and induce neuronal growth (Bazinet and Layé 2014). Vitamins are recognized for their role in improving cognition. Especially B group vitamins are mainly investigated for their beneficial role in cognition health (Dangour et al. 2015). In this study the author suggests that B-group vitamins are involved in improving the cognitive functions of individuals in their mid-age. Moreover women in their postmenopausal stage exhibited mild cognitive impairment due to low folate intake which could serve as a risk factor for dementia and AD. Vitamin D is also found to play an important role in cognition homeostasis. A study suggests that individuals with low 25-hydroxyvitamin D (25(OH)D) levels in the serum is associated with impaired cognition and AD when compared with healthy individuals (Goodwill and Szoeke 2017). A systematic analysis by Etgen et al. showed that population with Vitamin D deficiency exhibited impairment in cognition (Etgen et al. 2012). Similarly Vitamin C and E supplementation also showed a marked improvement in cognitive health in older people (Basambombo et al. 2017). Apart from these vitamin supplementation, the level of iron in the blood also plays a vital role in maintaining the homeostasis of brain. Iron deficiency is also associated with cognitive impairment due to dysregulations in the neurophysiological mechanisms of CNS (Smyth et al. 2015).

12.8 Conclusion

In conclusion, cognitive senescence is an inevitable process upon ageing. However this can be manageable with various treatment approaches. It is essential to acquire in-depth knowledge on the process of cognitive senescence and the role of microglia in it. As microglia displays a dual role in both promoting and detrimentally affecting the cognitive senescence, it is very much necessary for the researcher to identify a correct target during drug development process against microglia function. Furthermore, ongoing research is exploring various pharmacological and nutritional interventions like (olyunsaturated fatty acids (PUFAs), eicosapentaenoic acid (EPA), docosahexaenoic acid (DHA) and omega-3 fatty acids, vitamin D,C and E) to improve cognitive function in individuals experiencing age-related decline.

References

Areschoug T, Gordon S (2009) Scavenger receptors: role in innate immunity and microbial pathogenesis. Cell Microbiol 11(8):1160–1169. https://doi.org/10.1111/j.1462-5822.2009.01326.x

Barbagallo M, Dominguez LJ (2014) Type 2 diabetes mellitus and Alzheimer's disease. World J Diabetes 5(6):889–893. https://doi.org/10.4239/wjd.v5.i6.889

Basambombo LL, Carmichael PH, Côté S, Laurin D (2017) Use of vitamin E and C supplements for the prevention of cognitive decline. Ann Pharmacother 51(2):118–124. https://doi.org/10.1177/1060028016673072

Bazinet RP, Layé S (2014) Polyunsaturated fatty acids and their metabolites in brain function and disease. Nat Rev Neurosci 15(12):771–785. https://doi.org/10.1038/nrn3820

Beckman KB, Ames BN (1998) The free radical theory of aging matures. Physiol Rev 78(2):547–581. https://doi.org/10.1152/physrev.1998.78.2.547

Bogie JF, Stinissen P, Hendriks JJ (2014) Macrophage subsets and microglia in multiple sclerosis. Acta Neuropathol 128(2):191–213. https://doi.org/10.1007/s00401-014-1310-2

Bohr VA, Anson RM (1995) DNA damage, mutation and fine structure DNA repair in aging. Mutat Res 338(1–6):25–34. https://doi.org/10.1016/0921-8734(95)00008-t

Brites D, Vaz AR (2014) Microglia centered pathogenesis in ALS: insights in cell interconnectivity. Front Cell Neurosci 8:117. https://doi.org/10.3389/fncel.2014.00117

Chitu V, Gokhan Ş, Nandi S, Mehler MF, Stanley ER (2016) Emerging roles for CSF-1 receptor and its ligands in the nervous system. Trends Neurosci 39(6):378–393. https://doi.org/10.1016/j.tins.2016.03.005

Clare L, Wu YT, Teale JC, MacLeod C, Matthews F, Brayne C, Woods B, CFAS-Wales Study Team (2017) Potentially modifiable lifestyle factors, cognitive reserve, and cognitive function in later life: a cross-sectional study. PLoS Med 14(3):e1002259. https://doi.org/10.1371/journal.pmed.1002259

Colonna M, Butovsky O (2017) Microglia function in the central nervous system during health and neurodegeneration. Annu Rev Immunol 35:441–468. https://doi.org/10.1146/annurev-immunol-051116-052358

Colonna M, Wang Y (2016) TREM2 variants: new keys to decipher Alzheimer disease pathogenesis. Nat Rev Neurosci 17(4):201–207. https://doi.org/10.1038/nrn.2016.7

Colton CA, Mott RT, Sharpe H, Xu Q, Van Nostrand WE, Vitek MP (2006) Expression profiles for macrophage alternative activation genes in AD and in mouse models of AD. J Neuroinflammation 3:27. https://doi.org/10.1186/1742-2094-3-27

Dai XM, Ryan GR, Hapel AJ, Dominguez MG, Russell RG, Kapp S, Sylvestre V, Stanley ER (2002) Targeted disruption of the mouse colony-stimulating factor 1 receptor gene results in osteopetrosis, mononuclear phagocyte deficiency, increased primitive progenitor cell frequencies, and reproductive defects. Blood 99(1):111–120. https://doi.org/10.1182/blood.v99.1.111

Dangour AD, Allen E, Clarke R, Elbourne D, Fletcher AE, Letley L, Richards M, Whyte K, Uauy R, Mills K (2015) Effects of vitamin B-12 supplementation on neurologic and cognitive function in older people: a randomized controlled trial. Am J Clin Nutr 102(3):639–647. https://doi.org/10.3945/ajcn.115.110775

Di Micco R, Krizhanovsky V, Baker D, d'Adda di Fagagna F (2021) Cellular senescence in ageing: from mechanisms to therapeutic opportunities. Nat Rev Mol Cell Biol 22(2):75–95. https://doi.org/10.1038/s41580-020-00314-w

Etgen T, Sander D, Bickel H, Sander K, Förstl H (2012) Vitamin D deficiency, cognitive impairment and dementia: a systematic review and meta-analysis. Dement Geriatr Cogn Disord 33(5):297–305. https://doi.org/10.1159/000339702

Fumagalli S, Perego C, Pischiutta F, Zanier ER, De Simoni MG (2015) The ischemic environment drives microglia and macrophage function. Front Neurol 6:81

Ginhoux F, Greter M, Leboeuf M, Nandi S, See P, Gokhan S, Mehler MF, Conway SJ, Ng LG, Stanley ER, Samokhvalov IM, Merad M (2010) Fate mapping analysis reveals that adult microglia derive from primitive macrophages. Science 330(6005):841–845. https://doi.org/10.1126/science.1194637

Glass CK, Saijo K, Winner B, Marchetto MC, Gage FH (2010) Mechanisms underlying inflammation in neurodegeneration. Cell 140(6):918–934. https://doi.org/10.1016/j.cell.2010.02.016

Goodwill AM, Szoeke C (2017) A systematic review and meta-analysis of the effect of low vitamin D on cognition. J Am Geriatr Soc 65(10):2161–2168. https://doi.org/10.1111/jgs.15012

Gorgoulis VG, Pefani DE, Pateras IS, Trougakos IP (2018) Integrating the DNA damage and protein stress responses during cancer development and treatment. J Pathol 246(1):12–40. https://doi.org/10.1002/path.5097

Griciuc A, Serrano-Pozo A, Parrado AR, Lesinski AN, Asselin CN, Mullin K, Hooli B, Choi SH, Hyman BT, Tanzi RE (2013) Alzheimer's disease risk gene CD33 inhibits microglial uptake of amyloid beta. Neuron 78(4):631–643. https://doi.org/10.1016/j.neuron.2013.04.014

Hanahan D, Weinberg RA (2011) Hallmarks of cancer: the next generation. Cell 144(5):646–674. https://doi.org/10.1016/j.cell.2011.02.013

Harman D (1981) The aging process. Proc Natl Acad Sci U S A 78(11):7124–7128. https://doi.org/10.1073/pnas.78.11.7124

Hayflick L, Moorhead PS (1961) The serial cultivation of human diploid cell strains. Exp Cell Res 25:585–621. https://doi.org/10.1016/0014-4827(61)90192-6

Heneka MT, Kummer MP, Stutz A, Delekate A, Schwartz S, Vieira-Saecker A, Griep A, Axt D, Remus A, Tzeng TC, Gelpi E, Halle A, Korte M, Latz E, Golenbock DT (2013) NLRP3 is activated in Alzheimer's disease and contributes to pathology in APP/PS1 mice. Nature 493(7434):674–678. https://doi.org/10.1038/nature11729

Heneka MT, Golenbock DT, Latz E (2015) Innate immunity in Alzheimer's disease. Nat Immunol 16(3):229–236. https://doi.org/10.1038/ni.3102

Holtman IR, Raj DD, Miller JA, Schaafsma W, Yin Z, Brouwer N, Wes PD, Möller T, Orre M, Kamphuis W, Hol EM, Boddeke EW, Eggen BJ (2015) Induction of a common microglia gene expression signature by aging and neurodegenerative conditions: a co-expression meta-analysis. Acta Neuropathol Commun 3:31. https://doi.org/10.1186/s40478-015-0203-5

Ising C, Venegas C, Zhang S, Scheiblich H, Schmidt SV, Vieira-Saecker A, Schwartz S, Albasset S, McManus RM, Tejera D, Griep A, Santarelli F, Brosseron F, Opitz S, Stunden J, Merten M, Kayed R, Golenbock DT, Blum D, Latz E et al (2019) NLRP3 inflammasome activation drives tau pathology. Nature 575(7784):669–673. https://doi.org/10.1038/s41586-019-1769-z

Jin MM, Wang F, Qi D, Liu WW, Gu C, Mao CJ, Yang YP, Zhao Z, Hu LF, Liu CF (2018) A critical role of autophagy in regulating microglia polarization in neurodegeneration. Front Aging Neurosci 10:378. https://doi.org/10.3389/fnagi.2018.00378

Kamphuis W, Kooijman L, Schetters S, Orre M, Hol EM (2016) Transcriptional profiling of CD11c-positive microglia accumulating around amyloid plaques in a mouse model for Alzheimer's disease. Biochim Biophys Acta 1862(10):1847–1860. https://doi.org/10.1016/j.bbadis.2016.07.007

Kettenmann H, Hanisch UK, Noda M, Verkhratsky A (2011) Physiology of microglia. Physiol Rev 91(2):461–553. https://doi.org/10.1152/physrev.00011.2010

Kuilman T, Michaloglou C, Mooi WJ, Peeper DS (2010) The essence of senescence. Genes Dev 24(22):2463–2479. https://doi.org/10.1101/gad.1971610

Kumari R, Jat P (2021) Mechanisms of cellular senescence: cell cycle arrest and senescence associated secretory phenotype. Front Cell Dev Biol 9:645593. https://doi.org/10.3389/fcell.2021.645593

Kwon HS, Koh SH (2020) Neuroinflammation in neurodegenerative disorders: the roles of microglia and astrocytes. Transl Neurodegener 9(1):42. https://doi.org/10.1186/s40035-020-00221-2

Lee BY, Han JA, Im JS, Morrone A, Johung K, Goodwin EC, Kleijer WJ, DiMaio D, Hwang ES (2006) Senescence-associated beta-galactosidase is lysosomal beta-galactosidase. Aging Cell 5(2):187–195. https://doi.org/10.1111/j.1474-9726.2006.00199.x

Li Y, Li S, Wang W, Zhang D (2020) Association between dietary protein intake and cognitive function in adults aged 60 years and older. J Nutr Health Aging 24(2):223–229. https://doi.org/10.1007/s12603-020-1317-4

Liu S, Liu Y, Hao W, Wolf L, Kiliaan AJ, Penke B, Rübe CE, Walter J, Heneka MT, Hartmann T, Menger MD, Fassbender K (2012) TLR2 is a primary receptor for Alzheimer's amyloid β peptide to trigger neuroinflammatory activation. J Immunol 188(3):1098–1107. https://doi.org/10.4049/jimmunol.1101121

Ludwig PE, Reddy V, Varacallo M (2022) Neuroanatomy, central nervous system (CNS). In: StatPearls. StatPearls Publishing, Treasure Island. https://www.ncbi.nlm.nih.gov/books/NBK442010/

Machado V, Zöller T, Attaai A, Spittau B (2016) Microglia-mediated neuroinflammation and neurotrophic factor-induced protection in the MPTP mouse model of Parkinson's disease-lessons from transgenic mice. Int J Mol Sci 17(2):151. https://doi.org/10.3390/ijms17020151

Maldonado KA, Alsayouri K (2022) Physiology, brain. In: StatPearls. StatPearls Publishing, Treasure Island. https://www.ncbi.nlm.nih.gov/books/NBK551718/

Martínez-Frailes C, Di Lauro C, Bianchi C, de Diego-García L, Sebastián-Serrano Á, Boscá L, Díaz-Hernández M (2019) Amyloid peptide induced neuroinflammation increases the P2X7 receptor expression in microglial cells, impacting on its functionality. Front Cell Neurosci 13:143. https://doi.org/10.3389/fncel.2019.00143

Mattson MP, Arumugam TV (2018) Hallmarks of brain aging: adaptive and pathological modification by metabolic states. Cell Metab 27(6):1176–1199. https://doi.org/10.1016/j.cmet.2018.05.011

McHugh D, Gil J (2018) Senescence and aging: causes, consequences, and therapeutic avenues. J Cell Biol 217(1):65–77. https://doi.org/10.1083/jcb.201708092

Mizutani M, Pino PA, Saederup N, Charo IF, Ransohoff RM, Cardona AE (2012) The fractalkine receptor but not CCR2 is present on microglia from embryonic development throughout adulthood. J Immunol 188(1):29–36. https://doi.org/10.4049/jimmunol.1100421

Mohanraj M, Sekar P, Liou HH, Chang SF, Lin WW (2019) The mycobacterial adjuvant analogue TDB attenuates neuroinflammation via mincle-independent PLC-γ1/PKC/ERK signaling and microglial polarization. Mol Neurobiol 56(2):1167–1187. https://doi.org/10.1007/s12035-018-1135-4

Mou C, Liu B, Wang M, Jiang M, Han T (2015) PGC-1-related coactivator (PRC) is an important regulator of microglia M2 polarization. J Mol Neurosci 55(1):69–75. https://doi.org/10.1007/s12031-014-0315-6

Norden DM, Godbout JP (2013) Review: microglia of the aged brain: primed to be activated and resistant to regulation. Neuropathol Appl Neurobiol 39(1):19–34. https://doi.org/10.1111/j.1365-2990.2012.01306.x

Nordengen K, Kirsebom BE, Henjum K, Selnes P, Gísladóttir B, Wettergreen M, Torsetnes SB, Grøntvedt GR, Waterloo KK, Aarsland D, Nilsson LNG, Fladby T (2019) Glial activation and inflammation along the Alzheimer's disease continuum. J Neuroinflammation 16(1):46. https://doi.org/10.1186/s12974-019-1399-2

Okereke OI, Rosner BA, Kim DH, Kang JH, Cook NR, Manson JE, Buring JE, Willett WC, Grodstein F (2012) Dietary fat types and 4-year cognitive change in community-dwelling older women. Ann Neurol 72(1):124–134. https://doi.org/10.1002/ana.23593

Park DC, Yeo SG (2013) Aging. Korean J Audiol 17(2):39–44. https://doi.org/10.7874/kja.2013.17.2.39

Paul R, Dorsey JF, Fan Y (2022) Cell plasticity, senescence, and quiescence in cancer stem cells: biological and therapeutic implications. Pharmacol Ther 231:107985. https://doi.org/10.1016/j.pharmthera.2021.107985

Peters DG, Connor JR (2014) Introduction to cells comprising the nervous system. Adv Neurobiol 9:33–45. https://doi.org/10.1007/978-1-4939-1154-7_2

Pocock JM, Kettenmann H (2007) Neurotransmitter receptors on microglia. Trends Neurosci 30(10):527–535. https://doi.org/10.1016/j.tins.2007.07.007

Saijo K, Crotti A, Glass CK (2013) Regulation of microglia activation and deactivation by nuclear receptors. Glia 61(1):104–111. https://doi.org/10.1002/glia.22423

Saleh T, Tyutynuk-Massey L, Cudjoe EK Jr, Idowu MO, Landry JW, Gewirtz DA (2018) Non-cell autonomous effects of the senescence-associated secretory phenotype in cancer therapy. Front Oncol 8:164. https://doi.org/10.3389/fonc.2018.00164

Sancho D, Reis e Sousa, C. (2012) Signaling by myeloid C-type lectin receptors in immunity and homeostasis. Annu Rev Immunol 30:491–529. https://doi.org/10.1146/annurev-immunol-031210-101352

Smyth A, Dehghan M, Anderson C, Teo K, Gao P, Sleight P et al (2015) Healthy eating and reduced risk of cognitive decline. Am Acad Neurol 84:2258–2265. https://doi.org/10.1212/WNL.0000000000001638
Sofroniew MV, Vinters HV (2010) Astrocytes: biology and pathology. Acta Neuropathol 119(1):7–35. https://doi.org/10.1007/s00401-009-0619-8
Spencer SJ, Korosi A, Layé S, Shukitt-Hale B, Barrientos RM (2017) Food for thought: how nutrition impacts cognition and emotion. NPJ Sci Food 1:7. https://doi.org/10.1038/s41538-017-0008-y
Surova O, Zhivotovsky B (2013) Various modes of cell death induced by DNA damage. Oncogene 32(33):3789–3797. https://doi.org/10.1038/onc.2012.556
Tatar M, Bartke A, Antebi A (2003) The endocrine regulation of aging by insulin-like signals. Science 299(5611):1346–1351. https://doi.org/10.1126/science.1081447
Trifunovic A, Wredenberg A, Falkenberg M, Spelbrink JN, Rovio AT, Bruder CE, Bohlooly-Y M, Gidlöf S, Oldfors A, Wibom R, Törnell J, Jacobs HT, Larsson NG (2004) Premature ageing in mice expressing defective mitochondrial DNA polymerase. Nature 429(6990):417–423. https://doi.org/10.1038/nature02517
Troen BR (2003) The biology of aging. Mt Sinai J Med 70(1):3–22
van Deursen JM (2014) The role of senescent cells in ageing. Nature 509(7501):439–446. https://doi.org/10.1038/nature13193
Weinert BT, Timiras PS (2003) Invited review: theories of aging. J Appl Physiol (1985) 95(4):1706–1716. https://doi.org/10.1152/japplphysiol.00288.2003
Wong JM, Collins K (2003) Telomere maintenance and disease. Lancet 362(9388):983–988. https://doi.org/10.1016/S0140-6736(03)14369-3
Xiao J, Yao R, Xu B, Wen H, Zhong J, Li D, Zhou Z, Xu J, Wang H (2020) Inhibition of PDE4 attenuates TNF-α-triggered cell death through suppressing NF-κB and JNK activation in HT-22 neuronal cells. Cell Mol Neurobiol 40(3):421–435. https://doi.org/10.1007/s10571-019-00745-w
Zhang B, Gaiteri C, Bodea LG, Wang Z, McElwee J, Podtelezhnikov AA, Zhang C, Xie T, Tran L, Dobrin R, Fluder E, Clurman B, Melquist S, Narayanan M, Suver C, Shah H, Mahajan M, Gillis T, Mysore J, MacDonald ME et al (2013) Integrated systems approach identifies genetic nodes and networks in late-onset Alzheimer's disease. Cell 153(3):707–720. https://doi.org/10.1016/j.cell.2013.03.030
Zhang Y, Feng S, Nie K, Li Y, Gao Y, Gan R, Wang L, Li B, Sun X, Wang L, Zhang Y (2018) TREM2 modulates microglia phenotypes in the neuroinflammation of Parkinson's disease. Biochem Biophys Res Commun 499(4):797–802. https://doi.org/10.1016/j.bbrc.2018.03.226
Zhang ZJ, Zheng XX, Zhang XY, Zhang Y, Huang BY, Luo T (2020) Aging alters Hv1-mediated microglial polarization and enhances neuroinflammation after peripheral surgery. CNS Neurosci Ther 26(3):374–384. https://doi.org/10.1111/cns.13271

Role of Gut Brain and Gut Oral Axis in Progression of Parkinson's Disease with Special Focus on Gut Microbes

13

Ashwin Rajeev and Indranil Chattopadhyay

Abstract

The function of the microbiome in human health and disease has become one of the most important issues in modern medicine. Currently, a growing number of studies are being done to explain the connection between the prevalence of Parkinson's disease and gut microbial dysbiosis. Both the central nervous system (CNS) and the enteric nervous system (ENS) are affected in a specific way by gut microbial dysbiosis, suggesting that there is a gut-microbiota-brain axis that underlies CNS abnormalities. Neuroinflammation and gut-inflammation-immune responses are suppressed by certain gut microbes, Certain gut microbes exhibit a double-edged sword effect, potentially influencing the development of Parkinson's disease in both detrimental and protective ways. The immune system, the metabolism of tryptophan, the vagus nerve and the ENS are some of the pathways by which the brain and gut microbiota communicate. These pathways also involve microbial metabolites such as short-chain fatty acids (SCFAs), branched chain amino acids and peptidoglycans. Moreover, oral *Porphyromonas gingivalis* (*P. gingivalis*) has been shown in animal models to influence the gut-brain axis, which may indicate the existence of an oral-gut-brain axis.

Keywords

Parkinson's disease · Gut microbiota · Oral microbiota · Gut-brain axis · Gut dysbiosis

A. Rajeev · I. Chattopadhyay (✉)
Department of Biotechnology, Central University of Tamil Nadu, Thiruvarur, India
e-mail: indranil@cutn.ac.in

S. Pathak, A. Banerjee (eds.), *Gut Microbiome and Brain Ageing*,
https://doi.org/10.1007/978-981-99-8803-7_13

13.1 Introduction

Loss of dopaminergic neurons in the substantia nigra pars compacta is the primary cause of Parkinson's disease (PD), a multifocal, progressive neurodegenerative condition (Kang et al. 2021). It is a common neurological degenerative condition in the elderly also known as tremor paralysis and it is characterised by substantia nigra and striatum lesions (Liepelt-Scarfone et al. 2013). After Alzheimer's disease, PD is the second most prevalent neurodegenerative disorder that affects people globally. The incidence and prevalence of PD have increased year over year due to the acceleration of the population ageing process and the decline in physical abilities of senile patients (Hirsch et al. 2016). Global PD prevalence was estimated to be between 4.1 million and 4.6 million in 2005, and it is predicted that in the most populous countries, this number will probably multiply to the range of 8.7–9.3 million by 2030 (Wang et al. 2020). PD is estimated to affect about 1% of populations over the age of 60 (Bekris et al. 2010). An intriguing phenomenon is that PD individuals who experience gastrointestinal (GI) symptoms may do so several years before they experience conventional motor symptoms (Chen et al. 2015). The disruption of the GM can lead to α-synuclein (α-syn) buildup in the enteric nerve cell, which can simultaneously initiate mucosal inflammation and oxidative stress (Chen et al. 2019). Thus, researchers gave the notion that PD may begin in the GI system and transmit to the brain through the gut-brain axis (Liu et al. 2020).

The gut-brain axis comprises the central, autonomic and enteric nervous system (ENS) modulated by the gut microbiota through immunological, neuroendocrine and direct neural mechanisms (Mulak and Bonaz 2015). It is a vastly studied arena which constitutes a communication centre between the brain and the gut microbiota of the GI tract, aiding in the maintenance of homeostasis (Bercik et al. 2012). The intricate pathways of communication of the gut-brain axis are maintained through the vagus nerve and pelvic nerve in mammals (Powell et al. 2017). Gut-derived inflammation in central nervous system (CNS) has found substantial importance in the research aiming to pinpoint the role of gut microbiota in neurodegenerative diseases (Perez-Pardo et al. 2019). The diversity and composition of the colon reaches the adult microbial profile at around 3–5 years of age, consisting of ~3.8×10^{13} microorganisms (Sender et al. 2016a, b). Neural, immune and hormonal pathways regulate brain functioning by gut microbiota (Galland 2014). Disruptions in the mucosal integrity of the intestine alters the gut homeostasis, which further leads to inflammation and generation of reactive oxygen species (ROS) in the lumen (Ha et al. 2014). In addition to the gut microbiota, the nasal and oral cavities represent two significant entry points for a potential pathogenic agent that could spread to the CNS and be linked to the development of PD (Müller et al. 2011; Pereira et al. 2017).

13.2 Gut Microbiota

The gut microbiota (GM) contains a diverse array of microorganisms (~3×10^{13}) comprising bacteria, fungi, viruses and other eukaryotic organisms of lower complexities. Nevertheless, bacteria dominate other microorganisms in the gut through its activity of anaerobic carbohydrate breakdown of substances which are incompatible for the human digestive system (Lubomski et al. 2020). Initiation of the development of GM commences from birth itself, being vital for further development of immune and ENS (Gomez de Agüero et al. 2016). It forms a protective barrier against debilitating external stimuli and is adaptive to the changes in diet, lifestyle and the surrounding environment (Zuker 2015). An alteration in the composition of GM is brought about by the changes in diet, exercise, probiotic supplements, antibiotics and colonisation of invasive pathogenic microorganisms (Sommer and Bäckhed 2013; Costello et al. 2012). GM aids in the strengthening of intestinal barrier through the production of bactericides, which inhibits the colonisation of potentially pathogenic foreign microbes (Maslowski and Mackay 2011). GM composition during adulthood remains stable in spite of constant variations in the gut environment (Lozupone et al. 2012).

The gut microbiota hosts a wide range of microorganisms comprising bacteria, archaea, fungi and viruses, comparable to the number of cells residing in the body, thereby considered recently as a new organ or even a system (Sender et al. 2016a, b). The enteric microbiota distributed along the human GI tract was found to be quite stable throughout healthy individuals in terms of relative abundance and distribution, although the microbe profile is unique for each individual and can be considered as a personal microbial fingerprint or enterotype (Costea et al. 2018). Nevertheless, the most dominant phyla in the gut with about 51% and 48% are Firmicutes and Bacteroidetes, respectively. Other phyla with relatively lower abundance are Actinobacteria (including the *Bifidobacteria* genera), Proteobacteria, Spirochaetes, Cyanobacteria, Lentisphaerae, Fusobacteria and Verrucomicrobia (Eckburg et al. 2005).

GM interacts with the gut-brain axis in a number of ways, predominantly through the modulation of intestinal barrier function (Carabotti et al. 2015). The intestinal inflammation associated with PD, which causes alteration in the gut microbiota composition, may be the reason for α-syn misfolding (Mulak and Bonaz 2015). The role of *Helicobacter pylori* was enumerated in numerous studies related to its involvement in PD pathophysiology. Although some studies were unable to prove that *H. pylori* eradication improves motor symptoms through better absorption of levodopa (Rees et al. 2011), a few others proved that *H. pylori* infection is associated with deteriorated motor functions in PD (Tan et al. 2015). Another study which compared GM with clinical PD, in which faecal microbiomes of individuals with PD were compared with healthy controls, revealed that in individuals with PD there was a reduction in Prevotellaceae, which plays a dual role of producing health-promoting short-chain fatty acids (SCFAs) and also synthesis of the vitamins thiamine and folate (Arumugam et al. 2011). In addition to this, there was found to be a positive correlation between the abundance of Enterobacteriaceae with postural instability and gait difficulty (PIGD) phenotype than tremor dominant (TD) phenotype individuals (Scheperjans et al. 2015). In another study, the mucosal microbial communities and faecal microbial communities of PD affected individuals were compared with that of healthy individuals. It was concluded that the variations in the faecal communities were more pronouncing than that of mucosal communities. A reduction in butyrate-producing bacteria belonging to the genera *Blautia*, *Coprococcus* and *Roseburia* were observed in faecal communities of PD samples, which indicates a possible explanation for gut leakiness. This might also be a valid explanation for the increase in inflammation during PD, as these genera are associated with anti-inflammatory effects. Consistent with this result, the abundance of pro-inflammatory Proteobacteria of the genera *Ralstonia* increased substantially during this period (Keshavarzian et al. 2015). Substantial reduction in SCFAs such as propionate, acetate and butyrate were observed in elderly individuals affected with PD, which was greater than age-matched controls (Unger et al. 2016).

13.3 Involvement of Gut in Parkinson's Disease

Regarding the involvement of gut in the pathogenesis of PD, particularly as an initiation point in PD pathology, a dual-hit hypothesis was put forward. This hypothesis suggests that a neurotropic pathogen, such as a virus, may enter the brain through two routes—nasal and gastric routes. The pathogen may enter the gastric compartment when nasal secretions (in which the pathogen is present) are introduced into the saliva, and ingested along with it. On reaching the GI tract, these pathogens may move into the preganglionic parasympathetic motor neurons of the vagus nerve by trans-synaptic transmission through axons of Meissner's plexus, by initially penetrating the epithelial lining. This would ultimately lead to the transport of the pathogen into the substantia nigra (Hawkes et al. 2007). The possibility of a prion-like propagation of α-syn pathology across the nervous system was also suggested regarding the PD pathogenesis from this hypothesis (Li et al. 2008). The hypothesis also supports the idea finding that α-syn aggregation may start in the gut and spread to the brain via the vagus nerve (Vuuren et al. 2020). This notion is backed by reports of α-syn inclusions being found in the ENS, including the vagal nerves, often years before the development of the first motor manifestations (Shannon et al. 2012).

The gold standard for definite diagnosis of PD is the detection of α-syn aggregation accompanied by neuronal loss in substantia nigra in brain by immunohistochemistry (Ruffmann and Parkkinen 2016). One of the hallmark GI dysfunctions associated with PD is constipation. It might precede the motor symptoms associated with PD, by over 10 years making it a litmus test for early detection of the pathological processes that lead to PD (Berg et al. 2015). A malabsorption syndrome characterised by increased bacterial density and/or presence of colonic-type species in the small intestine known as small intestine bacterial overgrowth (SIBO) is highly prevalent in individuals with PD (Gasbarrini et al. 2007; Tan et al. 2014). SIBO might lead to increased intestinal permeability, which aids in bacterial translocation, thus propagating inflammatory response (Chen and Quigley 2014).

13.4 Inflammation in Parkinson's Disease

A comparative study of the autopsies of PD patients and healthy controls revealed an increased activation of microglial cells in the substantia nigra of PD patients than those of healthy controls. Neuroinflammation serves as an initiation point, as well as propagative factor of PD (McGeer and McGeer 2004). Meta-analysis studies demonstrated an increased peripheral expression of the inflammatory cytokines, viz., interleukin (IL)-6, tumour necrosis factor (TNF), IL-1β, IL-2, IL-10, C-reactive protein (CRP) and Regulated upon Activation, Normal T Cell Expressed and Presumably Secreted (RANTES) in different groups of PD patients than in healthy controls (Qin et al. 2016). Chronic inflammatory diseases such as Crohn's disease and ulcerative colitis which causes dysbiosis of the GI tract constitute risk factors for the development of PD (Zhu et al. 2019). A higher level of inflammation (Houser et al. 2018) and increased intestinal permeability have been reported in PD (Forsyth et al. 2011). Glial cell dysregulation and its potential association with the development of inflammation have been observed in CNS of PD patients. Elevated levels of pro-inflammatory cytokines such as IL-6 and TNF-α have been observed in the cerebrospinal fluid (CSF) of PD patients (Dutta et al. 2019).

13.5 Gut Dysbiosis

Alterations to the anatomical and/or functional makeup of the gut microbiota that disturb gut homeostasis are known as gut dysbiosis. Many factors, including dietary changes, antibiotic exposure, infections, disease conditions and ageing, might affect this equilibrium (Sommer and Bäckhed 2013). According to one study, gut dysbiosis causes changes in the brain's microglial signalling and the ratio of SCFA, particularly butyrate, which result in the development of the disease and the emergence of PD-related motor symptoms (Sampson et al. 2016). Scheperjans et al.'s study was the first to examine the connection between gut dysbiosis and PD. These researchers discovered a decrease in Prevotellaceae abundance and an increase in Enterobacteriaceae abundance in PD patients. These abnormalities in microbiota species were linked to postural instability and gait disorder in PD patients (Scheperjans et al. 2015). The fact that microbial species harvested from PD patients showed lower levels of *Prevotella*, *Lactobacillus*, *Peptostreptococcus* and *Butyricicoccus* spp. and an increase in *Proteus* and *Enterobacter* spp. compared to healthy controls provides additional evidence for the involvement of gut dysbiosis in the pathogenesis of PD (Sampson et al. 2016). The alterations associated with the gut microbiota and their mechanisms are described below (Table 13.1).

Table 13.1 Alterations in gut microbiota makeup associated with PD

		Microbiota			
Comparison	Sample	Increased	Decreased	Mechanism	References
PD vs. HC	Stool	*Proteus* sp., *Bilophila* sp. and *Roseburia* sp.	Lachnospiraceae, Rikenellaceae, Peptostreptococcaceae, *Butyricicoccus* sp.	Reduction in SCFA-producing families	Scheperjans et al. (2015)
PD vs. HC	Stool	*Proteus*, *Enterobacter* spp.	*Lactobacillus*, Prevotellaceae, *Peptostreptococcus*, *Butyricicoccus* spp.	Decreased Prevotellaceae lead to increased intestinal permeability, systemic exposure of bacterial endotoxins	Sampson et al. (2016)
PD vs. HC	Stool	*Prevotella*	Firmicutes, *Faecalibacterium*	Reduction in SCFA-producing taxon	Heinzel et al. (2020)
PD vs. HC	Stool	*Ralstonia*, Proteobacteria, Enterococcaceae	*Blautia*, *Coprococcus*, *Roseburia* and *Faecalibacterium*	Bacteria producing pro-inflammatory cytokines is increased, and bacteria involved in anti-inflammatory effects decreased	Keshavarzian et al. (2015)
PD vs. HC	Stool	*Helicobacter pylori*		*H. pylori* has the capability to elicit an onset of PD pathogenesis	Çamcı and Oğuz (2016)
PD vs. HC	Stool		*Bacteroides fragilis*, *Bifidobacterium*	*Bacteroides fragilis* were related with impairing of motivation/activeness and *Bifidobacterium* was associated to hallucinations/delusions	Minato et al. (2017)
PD vs. HC	Stool	*Porphyromonas*, *Corynebacterium*, *Prevotella*, *Lactobacillus*, *Bifidobacteria*	Ruminococcaceae, Lachnospiraceae	Opportunistic pathogens were increased, SCFAs producing bacteria reduced, probiotics with carbohydrate-metabolizing increased	Wallen et al. (2020)

Comparison	Sample	Microbiota		Mechanism	References
		Increased	Decreased		
PD vs. HC	Stool	*Akkermansia*, *Catabacter*, Akkermansiaceae	*Roseburia*, *Faecalibacterium*, Lachnospiraceae	*Akkermansia* increases the permeability of intestine that results in α-syn aggregation in intestine. Butyrate-producing bacteria decrease	Nishiwaki et al. (2020)
PD vs. HC	Stool, serum	Christensenellaceae, Desulfovibrionaceae, *Bifidobacterium*, *Bilophila*, *Akkermansia*	Lachnospiraceae, Roseburia, Faecalibacterium	Carbohydrate fermentation reduced. Low butyrate synthesis capacity. Proteolytic fermentation and damaging amino acid metabolites production increased	Cirstea et al. (2020)
PD vs. HC	Stool	Enterobacteriaceae	Bacteroidetes, Prevotellaceae	SCFAs may induce ENS alterations and dysmotility of GI in PD	Unger et al. (2016)
PD vs. HC	Stool	*Staphylococci*	*Lactobacillus casei* Shirota	*Lactobacillus casei* Shirota can improve the bowel movement by decreasing the number of staphylococci in PD patients	Cassani et al. (2011)
PD mice vs. HC	Stool	Proteobacteria, Turicibacterales, Enterobacteriales	Firmicutes, Clostridiales	Faecal SCFAs concentration decrease, increases DA and 5-HT levels, reduces activation of microglia and astrocytes	Sun et al. (2018)

PD Parkinson's disease patients, *HC* healthy controls

13.6 Microbiota-Gut-Brain Axis

The GM and CNS stimulate each other through a two-way pathway which comprises neural, immune and endocrine signals (Morais et al. 2021). The microbiota-gut-brain axis (MGBA) comprises several mediators comprising neurosynaptic pathways of vagal nerve, autonomic and enteric nervous systems, along with neuro-immune and neuroendocrine pathways, which connects the brain and GI tract, thus upholding its integrity (Cryan and O'Mahony 2011; Grenham et al. 2011). The MGBA outplays most other systems as a crucial entity for the regulation of digestion and maintaining metabolic homeostasis, under physiological conditions. A plethora of mechanisms can disrupt cognitive activities, behaviour, learning and neuropsychiatric conditions (Cryan and Dinan 2012; Hsiao et al. 2013). One of the predominant predisposing factors for PD pathogenesis is the dysregulation of MGBA, leading to neurodegeneration throughout the gut-brain axis, due to the accumulation of α-syn from the gut till brain and LB formation (Braak et al. 2006). The GM is capable of synthesising numerous neurotransmitters such as dopamine, serotonin, noradrenaline and neuromodulators like γ-amino butyric acid (GABA) and SCFAs, which aid in regulating blood flow, gut motility and nutrient absorption, apart from assisting the innate immune functionality of the GI tract (Mulak and Bonaz 2015; Asano et al. 2012). The development and function of all subsidiaries of the nervous system is swayed by the gut microbiota through the ENS (Diaz Heijtz et al. 2011), an association established during the evolution of microorganisms itself (Santos et al. 2019). The co-evolution of GM with its host keeps a check on the competitive nature among the microorganisms comprising the microbiome, through the immune and nervous systems (Powell et al. 2017; Galley et al. 2014; Foster et al. 2017). A key player in the transmission of signals to the CNS from the GM via the vagus nerve is a sub-type of intestinal epithelial cells called enteroendocrine cells (Kaelberer et al. 2018; Hoffman and Lumpkin 2018). These cells express a diverse array of hormones/peptides that act as signalling molecules by binding to specific targets, and they also act as chemoreceptors by responding to a variety of intestinal stimuli (Latorre et al. 2016).

One of the paramount effects of GM on CNS is appreciated through behaviour modifications of the host (Johnson and Foster 2018; Akami et al. 2019), which depends on the fitness of the concerning bacterial population (Santos et al. 2019) (Fig. 13.1). In the host, levels of neurotransmitters such as serotonin (5-HT), norepinephrine (NE), dopamine, GABA and glutamate, which are all necessary for the expression of social behaviour, are regulated by bacteria (Liu et al. 2017; Strandwitz et al. 2019).

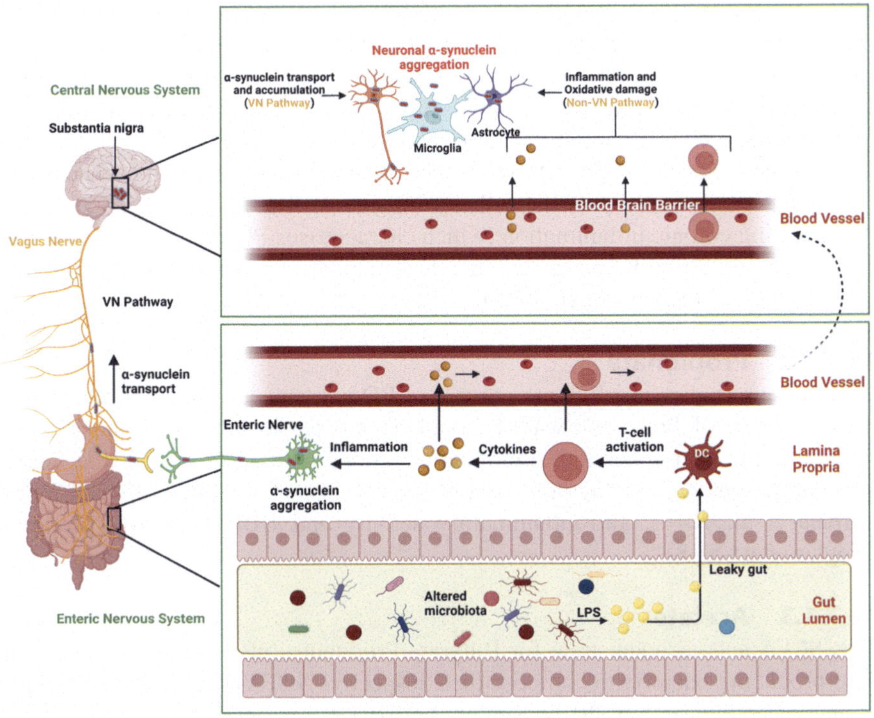

Fig. 13.1 Vagus nerve pathway and non-vagus nerve pathway involved in generating the inflammation of CNS which leads to Parkinson's disease

13.7 Bacterial-Derived Products and Potential Effects in Parkinson's Disease Pathology

13.7.1 SCFA

A decrease in the amount of SCFA-producing bacteria have been reported, particularly from the bacterial families of Prevotellaceae, Lachnospiraceae and Ruminoccocaceae, in faeces and/or intestinal mucosa of PD patients. They are bearers of potent anti-inflammatory and anti-oxidant properties (Scott et al. 2013), in addition to intestinal epithelial barrier's function (D'Souza et al. 2017). The three major forms of SCFAs related to gut integrity are:

13.7.1.1 Butyrate

Owing to the diverse properties of butyrate such as anti-inflammatory effects, regulation of the expression of genes related to cell-survival, regeneration and plasticity, it is gaining utmost importance. Apart from this, neuroprotective activity takes place through histone deacetylase's (HDAC) inhibitory activity (Bourassa et al. 2016; Arpaia et al. 2013). Experiments in mouse models of PD have shown that the DNA damage induced by α-syn can be reversed by sodium butyrate (Paiva et al. 2017). Other potential beneficial effects of butyrate in PD involves the following:

1. Strengthening the integrity of blood-brain barrier (BBB), which is disrupted in PD through HDAC's inhibiting activity (Gray and Woulfe 2015).
2. Intestinal barrier integrity is reinforced by modulating tight junction protein expression between epithelial cells (Stilling et al. 2016).
3. Inhibiting colonic inflammation (which characterises PD) by inducing apoptosis in the colonic T cells and modulating the T-regulatory cells activity (Zimmerman et al. 2012; Furusawa et al. 2013).

13.7.1.2 Propionate

In numerous studies involving the faeces of PD patients, a significant reduction in the abundance of *Prevotella* was observed. This is particularly important because, this genus is concerned with the production of propionate (Chen et al. 2017). Several other properties are synonymous to that of butyrate, such as inhibition of HDAC (Arpaia et al. 2013), anti-inflammatory activity and BBB protective properties (Hoyles et al. 2018).

13.7.1.3 Acetate

This SCFA acts as a substrate for Butyrate Producing (pBP) bacteria and propionate-producing bacteria to compose their respective SCFAs, i.e., butyrate and propionate (Fernández et al. 2016; Koh et al. 2016). An increase in the abundance of putative-acetate-producing bacteria such as *Bifidobacterium*, *Clostridium clusters*, *Lactobacillus* and *Akkermansia muciniphila* were reported in the studies (Bullich et al. 2019).

Contrary to the above-mentioned beneficial effects of SCFAs, negative impacts related to SCFAs were suggested, in which they led to inflammation, subsequently leading to α-syn aggregation and α-syn-dependent microglia activation. All these cascades into impaired motor function via amplified α-syn pathology (Sampson et al. 2016).

13.7.2 Mucin-Degrading Activity

The major mucin-degrading bacteria in the colonic compartment comprises the genera of *Ruminococcus*, *Prevotella* and *Akkermansia* (Arumugam et al. 2011). In the case of *Ruminococcus* and *Prevotella*, a reduction in their abundance indicates lower mucin synthesis in the colonic mucosal layers, thus directly leading to pronounced mucosal permeability (Brown et al. 2011). Subsequent entry of bacteria into the inner mucus layer triggers colonic inflammation through the activation of epithelial immune cells (Jakobsson et al. 2015; Schwab et al. 2014). *Akkermansia muciniphila* has been credited for the strengthening of the epithelial layer of the intestine (Reunanen et al. 2015). Based on the studies, it can be concluded that there exists an equilibrium between mucin degradation by *Akkermansia muciniphila* and mucus synthesis by the host (Derrien et al. 2011, 2017). Nevertheless, if this compensatory mucus synthesis by host is impaired, it would ultimately lead to leaky gut, and eventually inflammation (Bullich et al. 2019).

13.7.3 Hydrogen Production

In mouse and rat models of PD, the intestinal gaseous neurotransmitter hydrogen sulphide (H_2S), which is generated by *Prevotella* (Linden 2014), protects the dopaminergic neurons of the SN (Hu et al. 2010; Kida et al. 2011). Consequently, a decline in *Prevotella* is presumably going to result in intestinal H_2S underproduction, which will reduce its availability to dopaminergic neurons in the nigrostriatum (Ostojic 2018). *Prevotella* is not the only probable hydrogen-producing genus; some reviewed research also noted a decline in the population of *Roseburia* (Vogt et al. 2015). When H_2S levels are low, LPS-induced microglial inflammation that leads to neurodegeneration may become more pronounced (Peterson and Flood 2012), which implies the potential advantageous qualities of H_2S (Hu et al. 2007). During hydrogen depleted conditions, a substantial reduction in the abundance of putative hydrogen-consuming bacteria *Blautia* was observed (Bik et al. 2018).

13.7.4 Vitamin Biosynthesis

Prevotella-dominated GM enterotypes are linked to a better ability for thiamine biosynthesis (Arumugam et al. 2011). Prevotellaceae and *Prevotella* were less frequently found in PD patients' faeces in some of the studies conducted, which is consistent with the previously described thiamine deficiencies in PD patients (Jhala and Hazell 2011; Costantini et al. 2015) as well as the olfactory dysfunction linked to low levels of thiamine in the early stages of the disease (Håglin et al. 2017). Similarly, *Bacteroides* has been associated to increased riboflavin production, which has shown potential neuroprotective effects in Parkinson's disease given that riboflavin deficiency may induce an upregulation of PD pathways (Bullich et al. 2019).

13.7.5 Intestinal Ghrelin Secretion

In studies, *Prevotella* levels in PD faeces were found to be lower, whereas levels of *Bifidobacterium* and *Lactobacillus* were found to be higher. These bacterial changes have been linked to a decrease in ghrelin release in the intestine (Queipo-Ortuño et al. 2013). Likewise, regardless of the disease stage, ghrelin concentrations are low in the plasma of PD patients (Song et al. 2017), and a mouse model of PD generated by MPTP has shown that the acylated isoform of ghrelin exerts neuroprotective effects on dopaminergic neurons in the SN (Bayliss et al. 2016; Moon et al. 2009).

13.8 Correlation Between GM Alterations and Parkinson's Disease Clinical Features

To gain an insight into the relations between GM variations and PD clinical features, and to assert whether there is any impact for GM on the clinical features, several studies were required:

13.8.1 Disease Severity

According to the unified PD rating scale (UPDRS), the prevalence of *Enterococcus* and *Escherichia-Shigella* was positively correlated with disease severity, while the prevalence of *Bacteroides fragilis*, *Bifidobacterium*, *Blautia*, *Ruminococcus* and *Faecalibacterium* was negatively correlated with disease severity (Li et al. 2017; Minato et al. 2017; Qian et al. 2018). Indeed, a study revealed a notable decrease in *Faecalibacterium* in the group with severe PD as opposed to a lower decline reported in the group with mild PD (Li et al. 2017). Moreover, Prevotellaceae abundance was associated with UPDRS-III (Scheperjans et al. 2015). According to these results, GM may alter as the ailment advances, ostensibly towards a more proinflammatory bacterial profile (Keshavarzian et al. 2015; Winter and Bäumler 2014). Based on the studies using *Bifidobacterium* and *Atopobium* as models to predict worsening of total UPDRS scores, only *Bifidobacterium* was correlated with lowering of UPDRS-I (Minato et al. 2017).

13.8.2 Parkinson's Disease Duration

Bacteroidetes, *Enterococcus*, *Lactobacillus* spp., Proteobacteria and Ruminococcaceae had positive correlations with PD duration, while *Blautia*, *Clostridium* spp., *Faecalibacterium*, *Ruminococcus*, Lachnospiraceae and Firmicutes had negative correlations (Hasegawa et al. 2015; Keshavarzian et al. 2015; Li et al. 2017; Qian et al. 2018). It's interesting to note that the majority of bacteria-producing SCFAs have negative correlation with disease duration. As a result, a steady decline in these bacterial taxa could result in a depletion in SCFA production, which would eventually result in a decrease in the antioxidant, anti-inflammatory and neuroprotective properties of SCFAs as well as an increase in inflammation and intestinal permeability (Bullich et al. 2019).

13.8.3 Medication

Levodopa equivalent doses were found to be adversely associated with *Dorea* and *Phascolarctobacterium* (Qian et al. 2018), and catechol-O-methyl transferase inhibitor intake was found to be negatively correlated with Enterobacteriaceae (Scheperjans et al. 2015). Overall, these investigations indicate that GM composition changes may take effect even in patients with early-stage PD diagnosis. Longer disease duration makes these alterations worse, probably as a result of PD medication (Hill-Burns et al. 2017), constipation (Cao et al. 2017), disturbed sleep (Voigt et al. 2016) and more severe intestinal inflammation (Hooper et al. 2012). Furthermore, a recent study suggests that high levels of intestine bacterial tyrosine decarboxylase in PD patients restrict the availability of L-dopa, indicating that GM can alter drug efficacy in addition to GM composition (Van Kessel et al. 2019).

13.8.4 Motor Symptoms

Motor symptoms were related to *Aquabacterium* (Proteobacteria), *Anaerotruncus* (Clostridiaceae, Firmicutes), *Clostridium* XIVa, *Peptococcus* (Firmicutes) and Lachnospiraceae abundances (Heintz-Buschart et al. 2018; Qian et al. 2018). Moreover, another study demonstrated a positive correlation between the prevalence of Enterobacteriaceae and the severity of motor impairments in PD patients with postural instability and gait difficulty (PIGD). The PIGD phenotype may indicate a more severe α-syn disease in the colon, as demonstrated in that study (Scheperjans et al. 2015). In PD patients, *Escherichia coli* staining was found to positively correlate with α-syn staining, according to another study (Forsyth et al. 2011). Hence, a greater translocation of *E. coli* (Enterobacteriaceae) into the colonic mucosa, resulting in a more severe α-syn pathology and perhaps leading to the PIGD phenotype, may account for the correlation between the levels of Enterobacteriaceae and the aforementioned motor phenotype (Unger et al. 2016).

13.8.5 Nonmotor Symptoms

Nonmotor symptoms were typically linked to *Akkermansia* and *Anaerotruncus* (Clostridiaceae) (Heintz-Buschart et al. 2018). Constipation was attributed to the Bradyrhizobiaceae and Verrucomicrobiaceae in PD patients (Scheperjans et al. 2015). For example, an increase in *Akkermansia*, a mucin-degrading bacterium from the family Verrucomicrobiaceae, has been found in IBD patients (Png et al. 2010), suggesting that these bacteria may play a role in the emergence of GI ailments characterised by PD (Fasano et al. 2015).

13.9 Oral Microbiota

The oral microbiota (OM) harbours diverse microorganisms which comprise nearly 1000 species of bacteria, apart from viruses, microeukaryotes and archaea (Escapa et al. 2018). It holds the title of being the second most complex microbiota in the human body, right after gut microbiota. The major microenvironments of the OM consist of the mucosal epithelial surfaces (gingival sulcus, tongue, cheek, palate and lips), non-adherent hard surfaces of teeth and saliva. The composition of OM depends on lifestyle factors such as tobacco usage, dietary intake, antibiotic intake, dental hygiene etc. Synonymous to GM, the OM also inhibits colonisation of external pathogenic microorganisms, thereby preventing local infections, and promoting normal development of host tissues. Inorganic compounds such as nitrates are metabolised by OM, which are essential for vascular health. As the major point of access for ingested material, and due to its high vascularity, the oral cavity has abundant opportunities to influence the activities of other major systems of the body, including lungs and digestive tract. This is also the reason behind the fact that oral dysbiosis leads to a number of other systemic diseases (Radaic and Kapila 2021).

13.10 Oral Dysbiosis

The role of oral cavity in PD has been overlooked for so long. But recently, the importance of oral microbiota related to PD has been well studied and documented. Hyposialia and dysphagia can be taken into account as early symptoms of PD (Cersósimo et al. 2009; Noyce et al. 2012). Based on several recent studies, it was concluded that there is significant alteration in the oral microbiota of PD patients compared with healthy controls, thus aiming at a possible link between oral microbiota and PD (Pereira et al. 2017; Mihaila et al. 2019; Fleury et al. 2021). The relative abundance of Firmicutes, Negativicutes, Lactobacillaceae, Scardovia, Actinomyces, Veillonella, *Streptococcus mutans* and *Kingellaoralis* were found to be increased in PD patients, while Lachnospiraceae and *Treponema* were found to be reduced, compared to healthy controls. The elevated secretion of proinflammatory cytokine interleukin-1beta (IL-1β) in the gingival crevicular fluid of PD patients indicates onset of local inflammation (Fleury et al. 2021).

13.11 Role of *Porphyromonas gingivalis* in Forming the Oral-Gut Axis in Parkinson's Disease

Porphyromonas gingivalis (*P. gingivalis*), a gram-negative bacterium, has been linked to the development of certain inflammatory disorders and has been identified as a cause of periodontitis and gingivitis (Shanker et al. 2013; Amar and Engelke 2015; Singhrao et al. 2014; Olsen and Singhrao 2019; Dominy et al. 2019). Although it is a bacterium belonging to the buccal cavity, oral administration to animals in experiments has shown that it can also cause gut dysbiosis, decreased gut barrier function (Nakajima et al. 2015) and systemic inflammation (Arimatsu et al. 2014). Bacterial inflammagens, which are usually shed by bacteria in the form lipopolysaccharide (LPS) and lipoteichoic acid (LTA) (Kell and Pretorius 2015), are ligands for toll-like receptor-4 (TLR-4), which in turn activate inflammation (Olumuyiwa-Akeredolu et al. 2019). Novel studies, related to the proteases released from the periodontal pathogen *P. gingivalis* in PD gained much attention (Adams et al. 2019). This bacterium has the potential to cause microbial dysbiosis in the gut microbiota

(Olsen 2015). *P. gingivalis* has been detected in the brain of animals and humans diagnosed with Alzheimer's disease, where it releases its LPS and cysteine proteases known as gingipains, which is characterised as the antecedent of disease. Proteolysis is an unavoidable attribute of this bacterium, as it nourishes itself through the breakdown of proteins. This same aspect is reflected in its ability to degrade antibacterial peptides with the aid of gingipains (Guo et al. 2010), which also succour the bacterium in evading the complement system (Olsen and Singhrao 2019).

P. gingivalis produces gingipains, which are lethal bacterial proteases (Dominy et al. 2019). The virulence of this organism is greatly influenced by gingipains, which are made up of Arg-gingipain (Rgp) (RgpA and RgpB) and Lys-gingipain (Kgp) and exist in both cell-associated and secreted forms (Vuuren et al. 2020). Also detected in the brains of AD patients, gingipains have been linked to the onset of AD (Dominy et al. 2019). Recently, it was reported on gingipains, that it has a comparable hypercoagulation impact to LPS, in the blood of PD patients (Adams et al. 2019). Gingipains can also enter the body through the nasal cavity and then exit through the olfactory bulb. The postulated dual-hit mechanism, which involves anterograde disease progression from the olfactory system into the temporal lobe and retrograde pathology progression to the brainstem, may potentially be the mechanism underlying this process (Hawkes et al. 2007; Reichmann 2011).

13.12 Probiotics in Parkinson's Disease

Probiotics frequently improve the gut microbiota by replenishing the microbiota and sustaining immunological homeostasis (Reid et al. 2011). Probiotics have been demonstrated to improve intestinal epithelial integrity, safeguard against disruption of the gut barrier, modulate the immune system in the GI mucosa and limit the growth of pathogenic bacteria (Ait-Belgnaoui et al. 2012). The majority of the bacterial species in the human gut microbiome belong to the phyla Firmicutes (*Ruminococcus*, *Clostridium* and *Eubacteria*), Bacteroidetes (*Porphyromonas*, *Prevotella*) and Actinobacteria (*Bifidobacterium*). Few amounts of *Lactobacilli*, *Streptococci* and *Escherichia coli* are present in the gut (Azad et al. 2018). Through the gut-brain axis (Liu et al. 2015), probiotics have been shown to have positive modulatory effects on brain function, including reports of the normalisation of anxiety (Wallace and Milev 2017) and depressive-like behaviour (Abildgaard et al. 2017). Potential antioxidants, vitamins and bioactive compounds can indeed be produced by probiotic strains such as *Lactobacilli* and *Bifidobacteria* (LeBlanc et al. 2013). Vitamin E, vitamin D3, riboflavin and vitamin B6 have all demonstrated positive effects on PD patients (Parashar and Udayabanu 2017).

13.13 Conclusion

A deeper comprehension of the interactions between the microbiota-gut-brain axis could result in new insights into the pathophysiology of PD, enable an earlier diagnosis with a focus on peripheral ENS biomarkers, and unravel novel therapy options for PD. Studies involving the ideal sampling site, technique and suitable pathogenic targets are required in order to establish the potential relevance of intestinal α-syn as a biomarker of early PD. In PD patients or people at increased risk for the disease, dietary or pharmaceutical therapies should be designed to change the gut microbiota composition and strengthen the intestinal epithelial barrier integrity. This may affect the first stage of the subsequent cascade of neurodegeneration in PD. It will be highly clinically relevant to explain the temporal and ad hoc relationship between the gut microbiota changes and the pathogenesis of PD. Changes in bacterial taxa have been frequently demonstrated to be associated with disease across the most recent GM investigations in PD, supporting a probable biological connection between the GM and PD. Future studies must investigate such pathways that control the GM in PD as a feasible treatment approach to reduce or inhibit the spread of disease, however, as the frontier is quickly shifting. Furthermore, to get around the malfunctioning gut, researchers should look into improved dopaminergic drug delivery mechanisms. While *P. gingivalis* has been implicated in PD development, further studies using contemporary methods are needed to confirm its presence in brain tissue and elucidate its role in the disease process.

References

Abildgaard A, Elfving B, Hokland M, Wegener G, Lund S (2017) Probiotic treatment reduces depressive-like behaviour in rats independently of diet. Psychoneuroendocrinology 79:40–48. https://doi.org/10.1016/j.psyneuen.2017.02.014

Adams B, Nunes JM, Page MJ, Roberts T, Carr J, Nell TA, Kell DB, Pretorius E (2019) Parkinson's disease: a systemic inflammatory disease accompanied by bacterial inflammagens. Front Aging Neurosci 11:210. https://doi.org/10.3389/fnagi.2019.00210

Ait-Belgnaoui A, Durand H, Cartier C, Chaumaz G, Eutamene H, Ferrier L, Houdeau E, Fioramonti J, Bueno L, Theodorou V (2012) Prevention of gut leakiness by a probiotic treatment leads to attenuated HPA response to an acute psychological stress in rats. Psychoneuroendocrinology 37(11):1885–1895. https://doi.org/10.1016/j.psyneuen.2012.03.024

Akami M, Andongma AA, Zhengzhong C, Nan J, Khaeso K, Jurkevitch E, Niu CY, Yuval B (2019) Intestinal bacteria modulate the foraging behavior of the oriental fruit fly Bactrocera dorsalis (Diptera: Tephritidae). PLoS One 14(1):e0210109. https://doi.org/10.1371/journal.pone.0210109

Amar S, Engelke M (2015) Periodontal innate immune mechanisms relevant to atherosclerosis. Mol Oral Microbiol 30(3):171–185. https://doi.org/10.1111/omi.12087

Arimatsu K, Yamada H, Miyazawa H, Minagawa T, Nakajima M, Ryder MI, Gotoh K, Motooka D, Nakamura S, Iida T, Yamazaki K (2014) Oral pathobiont induces systemic inflammation and metabolic changes associated with alteration of gut microbiota. Sci Rep 4:4828. https://doi.org/10.1038/srep04828

Arpaia N, Campbell C, Fan X, Dikiy S, van der Veeken J, deRoos P, Liu H, Cross JR, Pfeffer K, Coffer PJ, Rudensky AY (2013) Metabolites produced by commensal bacteria promote peripheral regulatory T-cell generation. Nature 504(7480):451–455. https://doi.org/10.1038/nature12726

Arumugam M, Raes J, Pelletier E, Le Paslier D, Yamada T, Mende DR, Fernandes GR, Tap J, Bruls T, Batto JM, Bertalan M, Borruel N, Casellas F, Fernandez L, Gautier L, Hansen T, Hattori M, Hayashi T, Kleerebezem M, Kurokawa K et al (2011) Enterotypes of the human gut microbiome. Nature 473(7346):174–180. https://doi.org/10.1038/nature09944

Asano Y, Hiramoto T, Nishino R, Aiba Y, Kimura T, Yoshihara K, Koga Y, Sudo N (2012) Critical role of gut microbiota in the production of biologically active, free catecholamines in the gut lumen of mice. Am J Physiol Gastrointest Liver Physiol 303(11):G1288–G1295. https://doi.org/10.1152/ajpgi.00341.2012

Azad MAK, Sarker M, Li T, Yin J (2018) Probiotic species in the modulation of gut microbiota: an overview. Biomed Res Int 2018:9478630. https://doi.org/10.1155/2018/9478630

Bayliss JA, Lemus M, Santos VV, Deo M, Elsworth JD, Andrews ZB (2016) Acylated but not des-acyl ghrelin is neuroprotective in an MPTP mouse model of Parkinson's disease. J Neurochem 137(3):460–471. https://doi.org/10.1111/jnc.13576

Bekris LM, Mata IF, Zabetian CP (2010) The genetics of Parkinson disease. J Geriatr Psychiatry Neurol 23(4):228–242. https://doi.org/10.1177/0891988710383572

Bercik P, Collins SM, Verdu EF (2012) Microbes and the gut-brain axis. Neurogastroenterol Motil 24(5):405–413. https://doi.org/10.1111/j.1365-2982.2012.01906.x

Berg D, Postuma RB, Adler CH, Bloem BR, Chan P, Dubois B, Gasser T, Goetz CG, Halliday G, Joseph L, Lang AE, Liepelt-Scarfone I, Litvan I, Marek K, Obeso J, Oertel W, Olanow CW, Poewe W, Stern M, Deuschl G (2015) MDS research criteria for prodromal Parkinson's disease. Mov Disord 30(12):1600–1611. https://doi.org/10.1002/mds.26431

Bik EM, Ugalde JA, Cousins J, Goddard AD, Richman J, Apte ZS (2018) Microbial biotransformations in the human distal gut. Br J Pharmacol 175(24):4404–4414. https://doi.org/10.1111/bph.14085

Bourassa MW, Alim I, Bultman SJ, Ratan RR (2016) Butyrate, neuroepigenetics and the gut microbiome: can a high fiber diet improve brain health? Neurosci Lett 625:56–63. https://doi.org/10.1016/j.neulet.2016.02.009

Braak H, de Vos RA, Bohl J, Del Tredici K (2006) Gastric alpha-synuclein immunoreactive inclusions in Meissner's and Auerbach's plexuses in cases staged for Parkinson's disease-related brain pathology. Neurosci Lett 396(1):67–72. https://doi.org/10.1016/j.neulet.2005.11.012

Brown CT, Davis-Richardson AG, Giongo A, Gano KA, Crabb DB, Mukherjee N, Casella G, Drew JC, Ilonen J, Knip M, Hyöty H, Veijola R, Simell T, Simell O, Neu J, Wasserfall CH, Schatz D, Atkinson MA, Triplett EW (2011) Gut microbiome metagenomics analysis suggests a functional model for the development of autoimmunity for type 1 diabetes. PLoS One 6(10):e25792. https://doi.org/10.1371/journal.pone.0025792

Bullich C, Keshavarzian A, Garssen J, Kraneveld A, Perez-Pardo P (2019) Gut vibes in Parkinson's disease: the microbiota-gut-brain axis. Mov Disord Clin Pract 6(8):639–651. https://doi.org/10.1002/mdc3.12840

Çamcı G, Oğuz S (2016) Association between Parkinson's disease and Helicobacter pylori. J Clin Neurol 12(2):147–150. https://doi.org/10.3988/jcn.2016.12.2.147

Cao H, Liu X, An Y, Zhou G, Liu Y, Xu M, Dong W, Wang S, Yan F, Jiang K, Wang B (2017) Dysbiosis contributes to chronic constipation development via regulation of serotonin transporter in the intestine. Sci Rep 7(1):10322. https://doi.org/10.1038/s41598-017-10835-8

Carabotti M, Scirocco A, Maselli MA, Severi C (2015) The gut-brain axis: interactions between enteric microbiota, central and enteric nervous systems. Ann Gastroenterol 28(2):203–209

Cassani E, Privitera G, Pezzoli G, Pusani C, Madio C, Iorio L, Barichella M (2011) Use of probiotics for the treatment of constipation in Parkinson's disease patients. Minerva Gastroenterol Dietol 57(2):117–121

Cersósimo MG, Tumilasci OR, Raina GB, Benarroch EE, Cardoso EM, Micheli F, Pazo JH (2009) Hyposialorrhea as an early manifestation of Parkinson disease. Auton Neurosci 150(1–2):150–151. https://doi.org/10.1016/j.autneu.2009.04.004

Chen WC, Quigley EM (2014) Probiotics, prebiotics & synbiotics in small intestinal bacterial overgrowth: opening up a new therapeutic horizon! Indian J Med Res 140(5):582–584

Chen H, Zhao EJ, Zhang W, Lu Y, Liu R, Huang X, Ciesielski-Jones AJ, Justice MA, Cousins DS, Peddada S (2015) Meta-analyses on prevalence of selected Parkinson's nonmotor symptoms before and after diagnosis. Transl Neurodegener 4(1):1. https://doi.org/10.1186/2047-9158-4-1

Chen T, Long W, Zhang C, Liu S, Zhao L, Hamaker BR (2017) Fiber-utilizing capacity varies in Prevotella-versus bacteroides-dominated gut microbiota. Sci Rep 7(1):2594. https://doi.org/10.1038/s41598-017-02995-4

Chen QQ, Haikal C, Li W, Li JY (2019) Gut inflammation in association with pathogenesis of Parkinson's disease. Front Mol Neurosci 12:218. https://doi.org/10.3389/fnmol.2019.00218

Cirstea MS, Yu AC, Golz E, Sundvick K, Kliger D, Radisavljevic N, Foulger LH, Mackenzie M, Huan T, Finlay BB, Appel-Cresswell S (2020) Microbiota composition and metabolism are associated with gut function in Parkinson's disease. Mov Disord 35(7):1208–1217. https://doi.org/10.1002/mds.28052

Costantini A, Pala MI, Grossi E, Mondonico S, Cardelli LE, Jenner C, Proietti S, Colangeli M, Fancellu R (2015) Long-term treatment with high-dose thiamine in Parkinson disease: an open-label pilot study. J Altern Complement Med 21(12):740–747. https://doi.org/10.1089/acm.2014.0353

Costea PI, Hildebrand F, Arumugam M, Bäckhed F, Blaser MJ, Bushman FD, de Vos WM, Ehrlich SD, Fraser CM, Hattori M, Huttenhower C, Jeffery IB, Knights D, Lewis JD, Ley RE, Ochman H, O'Toole PW, Quince C, Relman DA, Shanahan F et al (2018) Enterotypes in the landscape of gut microbial community composition. Nat Microbiol 3(1):8–16. https://doi.org/10.1038/s41564-017-0072-8

Costello EK, Stagaman K, Dethlefsen L, Bohannan BJ, Relman DA (2012) The application of ecological theory toward an understanding of the human microbiome. Science 336(6086):1255–1262. https://doi.org/10.1126/science.1224203

Cryan JF, Dinan TG (2012) Mind-altering microorganisms: the impact of the gut microbiota on brain and behaviour. Nat Rev Neurosci 13(10):701–712. https://doi.org/10.1038/nrn3346

Cryan JF, O'Mahony SM (2011) The microbiome-gut-brain axis: from bowel to behavior. Neurogastroenterol Motil 23(3):187–192. https://doi.org/10.1111/j.1365-2982.2010.01664.x

D'Souza WN, Douangpanya J, Mu S, Jaeckel P, Zhang M, Maxwell JR, Rottman JB, Labitzke K, Willee A, Beckmann H, Wang Y, Li Y, Schwandner R, Johnston JA, Towne JE, Hsu H (2017) Differing roles for short chain fatty acids and GPR43 agonism in the regulation of intestinal barrier function and immune responses. PLoS One 12(7):e0180190. https://doi.org/10.1371/journal.pone.0180190

Derrien M, Van Baarlen P, Hooiveld G, Norin E, Müller M, de Vos WM (2011) Modulation of mucosal immune response, tolerance, and proliferation in mice colonized by the mucin-degrader Akkermansia muciniphila. Front Microbiol 2:166. https://doi.org/10.3389/fmicb.2011.00166

Derrien M, Belzer C, de Vos WM (2017) Akkermansia muciniphila and its role in regulating host functions. Microb Pathog 106:171–181. https://doi.org/10.1016/j.micpath.2016.02.005

Diaz Heijtz R, Wang S, Anuar F, Qian Y, Björkholm B, Samuelsson A, Hibberd ML, Forssberg H, Pettersson S (2011) Normal gut microbiota modulates brain development and behavior. Proc Natl Acad Sci U S A 108(7):3047–3052. https://doi.org/10.1073/pnas.1010529108

Dominy SS, Lynch C, Ermini F, Benedyk M, Marczyk A, Konradi A, Nguyen M, Haditsch U, Raha D, Griffin C, Holsinger LJ, Arastu-Kapur S, Kaba S, Lee A, Ryder MI, Potempa B, Mydel P, Hellvard A, Adamowicz K, Hasturk H et al (2019) Porphyromonas gingivalis in Alzheimer's disease brains: evidence for disease causation and treatment with small-molecule inhibitors. Sci Adv 5(1):eaau3333. https://doi.org/10.1126/sciadv.aau3333

Dutta SK, Verma S, Jain V, Surapaneni BK, Vinayek R, Phillips L, Nair PP (2019) Parkinson's disease: the emerging role of gut dysbiosis, antibiotics, probiotics, and fecal microbiota transplantation. J Neurogastroenterol Motil 25(3):363–376. https://doi.org/10.5056/jnm19044

Eckburg PB, Bik EM, Bernstein CN, Purdom E, Dethlefsen L, Sargent M, Gill SR, Nelson KE, Relman DA (2005) Diversity of the human intestinal microbial flora. Science 308(5728):1635–1638. https://doi.org/10.1126/science.1110591

Escapa IF, Chen T, Huang Y, Gajare P, Dewhirst FE, Lemon KP (2018) New insights into human nostril microbiome from the expanded human oral microbiome database (eHOMD): a resource for the microbiome of the human aerodigestive tract. mSystems 3(6):e00187-18. https://doi.org/10.1128/mSystems.00187-18

Fasano A, Visanji NP, Liu LW, Lang AE, Pfeiffer RF (2015) Gastrointestinal dysfunction in Parkinson's disease. Lancet Neurol 14(6):625–639. https://doi.org/10.1016/S1474-4422(15)00007-1

Fernández J, Redondo-Blanco S, Gutiérrez-del-Río I, Miguélez EM, Villar CJ, Lombó F (2016) Colon microbiota fermentation of dietary prebiotics towards short-chain fatty acids and their roles as anti-inflammatory and antitumour agents: a review. J Funct Foods 25:511–522. https://doi.org/10.1016/j.jff.2016.06.032

Fleury V, Zekeridou A, Lazarevic V, Gaïa N, Giannopoulou C, Genton L, Cancela J, Girard M, Goldstein R, Bally JF, Mombelli A, Schrenzel J, Burkhard PR (2021) Oral dysbiosis and inflammation in Parkinson's disease. J Parkinsons Dis 11(2):619–631. https://doi.org/10.3233/JPD-202459

Forsyth CB, Shannon KM, Kordower JH, Voigt RM, Shaikh M, Jaglin JA, Estes JD, Dodiya HB, Keshavarzian A (2011) Increased intestinal permeability correlates with sigmoid mucosa alpha-synuclein staining and endotoxin exposure markers in early Parkinson's disease. PLoS One 6(12):e28032. https://doi.org/10.1371/journal.pone.0028032

Foster KR, Schluter J, Coyte KZ, Rakoff-Nahoum S (2017) The evolution of the host microbiome as an ecosystem on a leash. Nature 548(7665):43–51. https://doi.org/10.1038/nature23292

Furusawa Y, Obata Y, Fukuda S, Endo TA, Nakato G, Takahashi D, Nakanishi Y, Uetake C, Kato K, Kato T, Takahashi M, Fukuda NN, Murakami S, Miyauchi E, Hino S, Atarashi K, Onawa S, Fujimura Y, Lockett T, Clarke JM et al (2013) Commensal microbe-derived butyrate induces the differentiation of colonic regulatory T cells. Nature 504(7480):446–450. https://doi.org/10.1038/nature12721

Galland L (2014) The gut microbiome and the brain. J Med Food 17(12):1261–1272. https://doi.org/10.1089/jmf.2014.7000

Galley JD, Nelson MC, Yu Z, Dowd SE, Walter J, Kumar PS, Lyte M, Bailey MT (2014) Exposure to a social stressor disrupts the community structure of the colonic mucosa-associated microbiota. BMC Microbiol 14:189. https://doi.org/10.1186/1471-2180-14-189

Gasbarrini A, Lauritano EC, Gabrielli M, Scarpellini E, Lupascu A, Ojetti V, Gasbarrini G (2007) Small intestinal bacterial overgrowth: diagnosis and treatment. Dig Dis 25(3):237–240. https://doi.org/10.1159/000103892

Gomez de Agüero M, Ganal-Vonarburg SC, Fuhrer T, Rupp S, Uchimura Y, Li H, Steinert A, Heikenwalder M, Hapfelmeier S, Sauer U, McCoy KD, Macpherson AJ (2016) The maternal microbiota drives early postnatal innate immune development. Science 351(6279):1296–1302. https://doi.org/10.1126/science.aad2571

Gray MT, Woulfe JM (2015) Striatal blood-brain barrier permeability in Parkinson's disease. J Cereb Blood Flow Metab 35(5):747–750. https://doi.org/10.1038/jcbfm.2015.32

Grenham S, Clarke G, Cryan JF, Dinan TG (2011) Brain-gut-microbe communication in health and disease. Front Physiol 2:94. https://doi.org/10.3389/fphys.2011.00094

Guo Y, Nguyen KA, Potempa J (2010) Dichotomy of gingipains action as virulence factors: from cleaving substrates with the precision of a surgeon's knife to a meat chopper-like brutal degradation of proteins. Periodontol 2000 54(1):15–44. https://doi.org/10.1111/j.1600-0757.2010.00377.x

Ha CW, Lam YY, Holmes AJ (2014) Mechanistic links between gut microbial community dynamics, microbial functions and metabolic health. World J Gastroenterol 20(44):16498–16517. https://doi.org/10.3748/wjg.v20.i44.16498

Håglin L, Johansson I, Forsgren L, Bäckman L (2017) Intake of vitamin B before onset of Parkinson's disease and atypical parkinsonism and olfactory function at the time of diagnosis. Eur J Clin Nutr 71(1):97–102. https://doi.org/10.1038/ejcn.2016.181

Hasegawa S, Goto S, Tsuji H, Okuno T, Asahara T, Nomoto K, Shibata A, Fujisawa Y, Minato T, Okamoto A, Ohno K, Hirayama M (2015) Intestinal dysbiosis and lowered serum lipopolysaccharide-binding protein in Parkinson's disease. PLoS One 10(11):e0142164. https://doi.org/10.1371/journal.pone.0142164

Hawkes CH, Del Tredici K, Braak H (2007) Parkinson's disease: a dual-hit hypothesis. Neuropathol Appl Neurobiol 33(6):599–614. https://doi.org/10.1111/j.1365-2990.2007.00874.x

Heintz-Buschart A, Pandey U, Wicke T, Sixel-Döring F, Janzen A, Sittig-Wiegand E, Trenkwalder C, Oertel WH, Mollenhauer B, Wilmes P (2018) The nasal and gut microbiome in Parkinson's disease and idiopathic rapid eye movement sleep behavior disorder. Mov Disord 33(1):88–98. https://doi.org/10.1002/mds.27105

Heinzel S, Aho VTE, Suenkel U, von Thaler AK, Schulte C, Deuschle C, Paulin L, Hantunen S, Brockmann K, Eschweiler GW, Maetzler W, Berg D, Auvinen P, Scheperjans F (2020) Gut microbiome signatures of risk and prodromal markers of Parkinson disease. Ann Neurol 88(2):320–331. https://doi.org/10.1002/ana.25788. (Retraction published Ann Neurol. 2021;90(2):329)

Hill-Burns EM, Debelius JW, Morton JT, Wissemann WT, Lewis MR, Wallen ZD, Peddada SD, Factor SA, Molho E, Zabetian CP, Knight R, Payami H (2017) Parkinson's disease and Parkinson's disease medications have distinct signatures of the gut microbiome. Mov Disord 32(5):739–749. https://doi.org/10.1002/mds.26942

Hirsch L, Jette N, Frolkis A, Steeves T, Pringsheim T (2016) The incidence of Parkinson's disease: a systematic review and meta-analysis. Neuroepidemiology 46(4):292–300. https://doi.org/10.1159/000445751

Hoffman BU, Lumpkin EA (2018) A gut feeling. Science 361(6408):1203–1204. https://doi.org/10.1126/science.aau9973

Hooper LV, Littman DR, Macpherson AJ (2012) Interactions between the microbiota and the immune system. Science 336(6086):1268–1273. https://doi.org/10.1126/science.1223490

Houser MC, Chang J, Factor SA, Molho ES, Zabetian CP, Hill-Burns EM, Payami H, Hertzberg VS, Tansey MG (2018) Stool immune profiles evince gastrointestinal inflammation in Parkinson's disease. Mov Disord 33(5):793–804. https://doi.org/10.1002/mds.27326

Hoyles L, Snelling T, Umlai UK, Nicholson JK, Carding SR, Glen RC, McArthur S (2018) Microbiome-host systems interactions: protective effects of propionate upon the blood-brain barrier. Microbiome 6(1):55. https://doi.org/10.1186/s40168-018-0439-y

Hsiao EY, McBride SW, Hsien S, Sharon G, Hyde ER, McCue T, Codelli JA, Chow J, Reisman SE, Petrosino JF, Patterson PH, Mazmanian SK (2013) Microbiota modulate behavioral and physiological abnormalities associated with neurodevelopmental disorders. Cell 155(7):1451–1463. https://doi.org/10.1016/j.cell.2013.11.024

Hu LF, Wong PT, Moore PK, Bian JS (2007) Hydrogen sulfide attenuates lipopolysaccharide-induced inflammation by inhibition of p38 mitogen-activated protein kinase in microglia. J Neurochem 100(4):1121–1128. https://doi.org/10.1111/j.1471-4159.2006.04283.x

Hu LF, Lu M, Tiong CX, Dawe GS, Hu G, Bian JS (2010) Neuroprotective effects of hydrogen sulfide on Parkinson's disease rat models. Aging Cell 9(2):135–146. https://doi.org/10.1111/j.1474-9726.2009.00543.x

Jakobsson HE, Rodríguez-Piñeiro AM, Schütte A, Ermund A, Boysen P, Bemark M, Sommer F, Bäckhed F, Hansson GC, Johansson ME (2015) The composition of the gut microbiota shapes the colon mucus barrier. EMBO Rep 16(2):164–177. https://doi.org/10.15252/embr.201439263

Jhala SS, Hazell AS (2011) Modeling neurodegenerative disease pathophysiology in thiamine deficiency: consequences of impaired oxidative metabolism. Neurochem Int 58(3):248–260. https://doi.org/10.1016/j.neuint.2010.11.019

Johnson KV, Foster KR (2018) Why does the microbiome affect behaviour? Nat Rev Microbiol 16(10):647–655. https://doi.org/10.1038/s41579-018-0014-3

Kaelberer MM, Buchanan KL, Klein ME, Barth BB, Montoya MM, Shen X, Bohórquez DV (2018) A gut-brain neural circuit for nutrient sensory transduction. Science 361(6408):eaat5236. https://doi.org/10.1126/science.aat5236

Kang Y, Kang X, Zhang H, Liu Q, Yang H, Fan W (2021) Gut microbiota and Parkinson's disease: implications for faecal microbiota transplantation therapy. ASN Neuro 13:17590914211016217. https://doi.org/10.1177/17590914211016217

Kell DB, Pretorius E (2015) On the translocation of bacteria and their lipopolysaccharides between blood and peripheral locations in chronic, inflammatory diseases: the central roles of LPS and LPS-induced cell death. Integr Biol 7(11):1339–1377. https://doi.org/10.1039/c5ib00158g

Keshavarzian A, Green SJ, Engen PA, Voigt RM, Naqib A, Forsyth CB, Mutlu E, Shannon KM (2015) Colonic bacterial composition in Parkinson's disease. Mov Disord 30(10):1351–1360. https://doi.org/10.1002/mds.26307

Kida K, Yamada M, Tokuda K, Marutani E, Kakinohana M, Kaneki M, Ichinose F (2011) Inhaled hydrogen sulfide prevents neurodegeneration and movement disorder in a mouse model of Parkinson's disease. Antioxid Redox Signal 15(2):343–352. https://doi.org/10.1089/ars.2010.3671

Koh A, De Vadder F, Kovatcheva-Datchary P, Bäckhed F (2016) From dietary fiber to host physiology: short-chain fatty acids as key bacterial metabolites. Cell 165(6):1332–1345. https://doi.org/10.1016/j.cell.2016.05.041

Latorre R, Sternini C, De Giorgio R, Greenwood-Van Meerveld B (2016) Enteroendocrine cells: a review of their role in brain-gut communication. Neurogastroenterol Motil 28(5):620–630. https://doi.org/10.1111/nmo.12754

LeBlanc JG, Milani C, de Giori GS, Sesma F, van Sinderen D, Ventura M (2013) Bacteria as vitamin suppliers to their host: a gut microbiota perspective. Curr Opin Biotechnol 24(2):160–168. https://doi.org/10.1016/j.copbio.2012.08.005

Li JY, Englund E, Holton JL, Soulet D, Hagell P, Lees AJ, Lashley T, Quinn NP, Rehncrona S, Björklund A, Widner H, Revesz T, Lindvall O, Brundin P (2008) Lewy bodies in grafted neurons in subjects with Parkinson's disease suggest host-to-graft disease propagation. Nat Med 14(5):501–503. https://doi.org/10.1038/nm1746

Li W, Wu X, Hu X, Wang T, Liang S, Duan Y, Jin F, Qin B (2017) Structural changes of gut microbiota in Parkinson's disease and its correlation with clinical features. Sci China Life Sci 60(11):1223–1233. https://doi.org/10.1007/s11427-016-9001-4

Liepelt-Scarfone I, Gauss K, Maetzler W, Müller K, Bormann C, Fruhmann Berger M, Timmers M, Streffer J, Berg D (2013) Evaluation of progression markers in the premotor phase of Parkinson's disease: the progression markers in the premotor phase study. Neuroepidemiology 41(3–4):174–182. https://doi.org/10.1159/000353560

Linden DR (2014) Hydrogen sulfide signaling in the gastrointestinal tract. Antioxid Redox Signal 20(5):818–830. https://doi.org/10.1089/ars.2013.5312

Liu X, Cao S, Zhang X (2015) Modulation of gut microbiota-brain axis by probiotics, prebiotics, and diet. J Agric Food Chem 63(36):7885–7895. https://doi.org/10.1021/acs.jafc.5b02404

Liu R, Hong J, Xu X, Feng Q, Zhang D, Gu Y, Shi J, Zhao S, Liu W, Wang X, Xia H, Liu Z, Cui B, Liang P, Xi L, Jin J, Ying X, Wang X, Zhao X, Li W et al (2017) Gut microbiome and serum metabolome alterations in obesity and after weight-loss intervention. Nat Med 23(7):859–868. https://doi.org/10.1038/nm.4358

Liu J, Xu F, Nie Z, Shao L (2020) Gut microbiota approach—a new strategy to treat Parkinson's disease. Front Cell Infect Microbiol 10:570658. https://doi.org/10.3389/fcimb.2020.570658

Lozupone CA, Stombaugh JI, Gordon JI, Jansson JK, Knight R (2012) Diversity, stability and resilience of the human gut microbiota. Nature 489(7415):220–230. https://doi.org/10.1038/nature11550

Lubomski M, Tan AH, Lim SY, Holmes AJ, Davis RL, Sue CM (2020) Parkinson's disease and the gastrointestinal microbiome. J Neurol 267(9):2507–2523. https://doi.org/10.1007/s00415-019-09320-1

Maslowski KM, Mackay CR (2011) Diet, gut microbiota and immune responses. Nat Immunol 12(1):5–9. https://doi.org/10.1038/ni0111-5

McGeer PL, McGeer EG (2004) Inflammation and neurodegeneration in Parkinson's disease. Parkinsonism Relat Disord 10(Suppl 1):S3–S7. https://doi.org/10.1016/j.parkreldis.2004.01.005

Mihaila D, Donegan J, Barns S, LaRocca D, Du Q, Zheng D, Vidal M, Neville C, Uhlig R, Middleton FA (2019) The oral microbiome of early stage Parkinson's disease and its relationship with functional measures of motor and non-motor function. PLoS One 14(6):e0218252. https://doi.org/10.1371/journal.pone.0218252

Minato T, Maeda T, Fujisawa Y, Tsuji H, Nomoto K, Ohno K, Hirayama M (2017) Progression of Parkinson's disease is associated with gut dysbiosis: two-year follow-up study. PLoS One 12(11):e0187307. https://doi.org/10.1371/journal.pone.0187307

Moon M, Kim HG, Hwang L, Seo JH, Kim S, Hwang S, Kim S, Lee D, Chung H, Oh MS, Lee KT, Park S (2009) Neuroprotective effect of ghrelin in the 1-methyl-4-phenyl-1,2,3,6-tetrahydropyridine mouse model of Parkinson's disease by blocking microglial activation. Neurotox Res 15(4):332–347. https://doi.org/10.1007/s12640-009-9037-x

Morais LH, Schreiber HL IV, Mazmanian SK (2021) The gut microbiota-brain axis in behaviour and brain disorders. Nat Rev Microbiol 19(4):241–255. https://doi.org/10.1038/s41579-020-00460-0

Mulak A, Bonaz B (2015) Brain-gut-microbiota axis in Parkinson's disease. World J Gastroenterol 21(37):10609–10620. https://doi.org/10.3748/wjg.v21.i37.10609

Müller T, Palluch R, Jackowski J (2011) Caries and periodontal disease in patients with Parkinson's disease. Spec Care Dentist 31(5):178–181. https://doi.org/10.1111/j.1754-4505.2011.00205.x

Nakajima M, Arimatsu K, Kato T, Matsuda Y, Minagawa T, Takahashi N, Ohno H, Yamazaki K (2015) Oral administration of P. gingivalis induces dysbiosis of gut microbiota and impaired barrier function leading to dissemination of enterobacteria to the liver. PLoS One 10(7):e0134234. https://doi.org/10.1371/journal.pone.0134234

Nishiwaki H, Ito M, Ishida T, Hamaguchi T, Maeda T, Kashihara K, Tsuboi Y, Ueyama J, Shimamura T, Mori H, Kurokawa K, Katsuno M, Hirayama M, Ohno K (2020) Meta-analysis of gut dysbiosis in Parkinson's disease. Mov Disord 35(9):1626–1635. https://doi.org/10.1002/mds.28119

Noyce AJ, Silveira-Moriyama L, Gilpin P, Ling H, Howard R, Lees AJ (2012) Severe dysphagia as a presentation of Parkinson's disease. Mov Disord 27(3):457–458. https://doi.org/10.1002/mds.24006

Olsen I (2015) From the Acta Prize Lecture 2014: the periodontal-systemic connection seen from a microbiological standpoint. Acta Odontol Scand 73(8):563–568. https://doi.org/10.3109/00016357.2015.1007480

Olsen I, Singhrao SK (2019) Is there a link between genetic defects in the complement cascade and Porphyromonas gingivalis in Alzheimer's disease? J Oral Microbiol 12(1):1676486. https://doi.org/10.1080/20002297.2019.1676486

Olumuyiwa-Akeredolu OO, Page MJ, Soma P, Pretorius E (2019) Platelets: emerging facilitators of cellular crosstalk in rheumatoid arthritis. Nat Rev Rheumatol 15(4):237–248. https://doi.org/10.1038/s41584-019-0187-9

Ostojic SM (2018) Inadequate production of H2 by gut microbiota and Parkinson disease. Trends Endocrinol Metab 29(5):286–288. https://doi.org/10.1016/j.tem.2018.02.006

Paiva I, Pinho R, Pavlou MA, Hennion M, Wales P, Schütz AL, Rajput A, Szego ÉM, Kerimoglu C, Gerhardt E, Rego AC, Fischer A, Bonn S, Outeiro TF (2017) Sodium butyrate rescues dopaminergic cells from alpha-synuclein-induced transcriptional deregulation and DNA damage. Hum Mol Genet 26(12):2231–2246. https://doi.org/10.1093/hmg/ddx114

Parashar A, Udayabanu M (2017) Gut microbiota: implications in Parkinson's disease. Parkinsonism Relat Disord 38:1–7. https://doi.org/10.1016/j.parkreldis.2017.02.002

Pereira PAB, Aho VTE, Paulin L, Pekkonen E, Auvinen P, Scheperjans F (2017) Oral and nasal microbiota in Parkinson's disease. Parkinsonism Relat Disord 38:61–67. https://doi.org/10.1016/j.parkreldis.2017.02.026

Perez-Pardo P, Dodiya HB, Engen PA, Forsyth CB, Huschens AM, Shaikh M, Voigt RM, Naqib A, Green SJ, Kordower JH, Shannon KM, Garssen J, Kraneveld AD, Keshavarzian A (2019) Role of TLR4 in the gut-brain axis in Parkinson's disease: a translational study from men to mice. Gut 68(5):829–843. https://doi.org/10.1136/gutjnl-2018-316844

Peterson LJ, Flood PM (2012) Oxidative stress and microglial cells in Parkinson's disease. Mediat Inflamm 2012:401264. https://doi.org/10.1155/2012/401264

Png CW, Lindén SK, Gilshenan KS, Zoetendal EG, McSweeney CS, Sly LI, McGuckin MA, Florin TH (2010) Mucolytic bacteria with increased prevalence in IBD mucosa augment

in vitro utilization of mucin by other bacteria. Am J Gastroenterol 105(11):2420–2428. https://doi.org/10.1038/ajg.2010.281
Powell N, Walker MM, Talley NJ (2017) The mucosal immune system: master regulator of bidirectional gut-brain communications. Nat Rev Gastroenterol Hepatol 14(3):143–159. https://doi.org/10.1038/nrgastro.2016.191
Qian Y, Yang X, Xu S, Wu C, Song Y, Qin N, Chen SD, Xiao Q (2018) Alteration of the fecal microbiota in Chinese patients with Parkinson's disease. Brain Behav Immun 70:194–202. https://doi.org/10.1016/j.bbi.2018.02.016
Qin XY, Zhang SP, Cao C, Loh YP, Cheng Y (2016) Aberrations in peripheral inflammatory cytokine levels in Parkinson disease: a systematic review and meta-analysis. JAMA Neurol 73(11):1316–1324. https://doi.org/10.1001/jamaneurol.2016.2742
Queipo-Ortuño MI, Seoane LM, Murri M, Pardo M, Gomez-Zumaquero JM, Cardona F, Casanueva F, Tinahones FJ (2013) Gut microbiota composition in male rat models under different nutritional status and physical activity and its association with serum leptin and ghrelin levels. PLoS One 8(5):e65465. https://doi.org/10.1371/journal.pone.0065465
Radaic A, Kapila YL (2021) The oralome and its dysbiosis: new insights into oral microbiome-host interactions. Comput Struct Biotechnol J 19:1335–1360. https://doi.org/10.1016/j.csbj.2021.02.010
Rees K, Stowe R, Patel S, Ives N, Breen K, Clarke CE, Ben-Shlomo Y (2011) Helicobacter pylori eradication for Parkinson's disease. Cochrane Database Syst Rev (11):CD008453. https://doi.org/10.1002/14651858.CD008453.pub2
Reichmann H (2011) View point: etiology in Parkinson's disease. Dual hit or spreading intoxication. J Neurol Sci 310(1–2):9–11. https://doi.org/10.1016/j.jns.2011.04.016
Reid G, Younes JA, Van der Mei HC, Gloor GB, Knight R, Busscher HJ (2011) Microbiota restoration: natural and supplemented recovery of human microbial communities. Nat Rev Microbiol 9(1):27–38. https://doi.org/10.1038/nrmicro2473
Reunanen J, Kainulainen V, Huuskonen L, Ottman N, Belzer C, Huhtinen H, de Vos WM, Satokari R (2015) Akkermansia muciniphila adheres to enterocytes and strengthens the integrity of the epithelial cell layer. Appl Environ Microbiol 81(11):3655–3662. https://doi.org/10.1128/AEM.04050-14
Ruffmann C, Parkkinen L (2016) Gut feelings about α-synuclein in gastrointestinal biopsies: biomarker in the making? Mov Disord 31(2):193–202. https://doi.org/10.1002/mds.26480
Sampson TR, Debelius JW, Thron T, Janssen S, Shastri GG, Ilhan ZE, Challis C, Schretter CE, Rocha S, Gradinaru V, Chesselet MF, Keshavarzian A, Shannon KM, Krajmalnik-Brown R, Wittung-Stafshede P, Knight R, Mazmanian SK (2016) Gut microbiota regulate motor deficits and neuroinflammation in a model of Parkinson's disease. Cell 167(6):1469–1480.e12. https://doi.org/10.1016/j.cell.2016.11.018
Santos SF, de Oliveira HL, Yamada ES, Neves BC, Pereira A Jr (2019) The gut and Parkinson's disease—a bidirectional pathway. Front Neurol 10:574. https://doi.org/10.3389/fneur.2019.00574
Scheperjans F, Aho V, Pereira PA, Koskinen K, Paulin L, Pekkonen E, Haapaniemi E, Kaakkola S, Eerola-Rautio J, Pohja M, Kinnunen E, Murros K, Auvinen P (2015) Gut microbiota are related to Parkinson's disease and clinical phenotype. Mov Disord 30(3):350–358. https://doi.org/10.1002/mds.26069
Schwab C, Berry D, Rauch I, Rennisch I, Ramesmayer J, Hainzl E, Heider S, Decker T, Kenner L, Müller M, Strobl B, Wagner M, Schleper C, Loy A, Urich T (2014) Longitudinal study of murine microbiota activity and interactions with the host during acute inflammation and recovery. ISME J 8(5):1101–1114. https://doi.org/10.1038/ismej.2013.223
Scott KP, Gratz SW, Sheridan PO, Flint HJ, Duncan SH (2013) The influence of diet on the gut microbiota. Pharmacol Res 69(1):52–60. https://doi.org/10.1016/j.phrs.2012.10.020
Sender R, Fuchs S, Milo R (2016a) Are we really vastly outnumbered? Revisiting the ratio of bacterial to host cells in humans. Cell 164(3):337–340. https://doi.org/10.1016/j.cell.2016.01.013
Sender R, Fuchs S, Milo R (2016b) Revised estimates for the number of human and bacteria cells in the body. PLoS Biol 14(8):e1002533. https://doi.org/10.1371/journal.pbio.1002533

Shanker J, Setty P, Arvind P, Nair J, Bhasker D, Balakrishna G, Kakkar VV (2013) Relationship between periodontal disease, Porphyromonas gingivalis, peripheral vascular resistance markers and coronary artery disease in Asian Indians. Thromb Res 132(1):e8–e14. https://doi.org/10.1016/j.thromres.2013.04.023

Shannon KM, Keshavarzian A, Dodiya HB, Jakate S, Kordower JH (2012) Is alpha-synuclein in the colon a biomarker for premotor Parkinson's disease? Evidence from 3 cases. Mov Disord 27(6):716–719. https://doi.org/10.1002/mds.25020

Singhrao SK, Harding A, Simmons T, Robinson S, Kesavalu L, Crean S (2014) Oral inflammation, tooth loss, risk factors, and association with progression of Alzheimer's disease. J Alzheimers Dis 42(3):723–737. https://doi.org/10.3233/JAD-140387

Sommer F, Bäckhed F (2013) The gut microbiota—masters of host development and physiology. Nat Rev Microbiol 11(4):227–238. https://doi.org/10.1038/nrmicro2974

Song N, Wang W, Jia F, Du X, Xie A, He Q, Shen X, Zhang J, Rogers JT, Xie J, Jiang H (2017) Assessments of plasma ghrelin levels in the early stages of Parkinson's disease. Mov Disord 32(10):1487–1491. https://doi.org/10.1002/mds.27095

Stilling RM, van de Wouw M, Clarke G, Stanton C, Dinan TG, Cryan JF (2016) The neuropharmacology of butyrate: the bread and butter of the microbiota-gut-brain axis? Neurochem Int 99:110–132. https://doi.org/10.1016/j.neuint.2016.06.011

Strandwitz P, Kim KH, Terekhova D, Liu JK, Sharma A, Levering J, McDonald D, Dietrich D, Ramadhar TR, Lekbua A, Mroue N, Liston C, Stewart EJ, Dubin MJ, Zengler K, Knight R, Gilbert JA, Clardy J, Lewis K (2019) GABA-modulating bacteria of the human gut microbiota. Nat Microbiol 4(3):396–403. https://doi.org/10.1038/s41564-018-0307-3

Sun MF, Zhu YL, Zhou ZL, Jia XB, Xu YD, Yang Q, Cui C, Shen YQ (2018) Neuroprotective effects of fecal microbiota transplantation on MPTP-induced Parkinson's disease mice: gut microbiota, glial reaction and TLR4/TNF-α signaling pathway. Brain Behav Immun 70:48–60. https://doi.org/10.1016/j.bbi.2018.02.005

Tan AH, Mahadeva S, Thalha AM, Gibson PR, Kiew CK, Yeat CM, Ng SW, Ang SP, Chow SK, Tan CT, Yong HS, Marras C, Fox SH, Lim SY (2014) Small intestinal bacterial overgrowth in Parkinson's disease. Parkinsonism Relat Disord 20(5):535–540. https://doi.org/10.1016/j.parkreldis.2014.02.019

Tan AH, Mahadeva S, Marras C, Thalha AM, Kiew CK, Yeat CM, Ng SW, Ang SP, Chow SK, Loke MF, Vadivelu JS, Ibrahim N, Yong HS, Tan CT, Fox SH, Lang AE, Lim SY (2015) Helicobacter pylori infection is associated with worse severity of Parkinson's disease. Parkinsonism Relat Disord 21(3):221–225. https://doi.org/10.1016/j.parkreldis.2014.12.009

Unger MM, Spiegel J, Dillmann KU, Grundmann D, Philippeit H, Bürmann J, Faßbender K, Schwiertz A, Schäfer KH (2016) Short chain fatty acids and gut microbiota differ between patients with Parkinson's disease and age-matched controls. Parkinsonism Relat Disord 32:66–72. https://doi.org/10.1016/j.parkreldis.2016.08.019

Van Kessel SP, Frye AK, El-Gendy AO, Castejon M, Keshavarzian A, van Dijk G, El Aidy S (2019) Gut bacterial tyrosine decarboxylases restrict levels of levodopa in the treatment of Parkinson's disease. Nat Commun 10(1):310. https://doi.org/10.1038/s41467-019-08294-y

Vogt SL, Peña-Díaz J, Finlay BB (2015) Chemical communication in the gut: effects of microbiota-generated metabolites on gastrointestinal bacterial pathogens. Anaerobe 34:106–115. https://doi.org/10.1016/j.anaerobe.2015.05.002

Voigt RM, Summa KC, Forsyth CB, Green SJ, Engen P, Naqib A, Vitaterna MH, Turek FW, Keshavarzian A (2016) The circadian clock mutation promotes intestinal dysbiosis. Alcohol Clin Exp Res 40(2):335–347. https://doi.org/10.1111/acer.12943

Vuuren MJV, Nell TA, Carr JA, Kell DB, Pretorius E (2020) Iron dysregulation and inflammagens related to oral and gut health are central to the development of Parkinson's disease. Biomol Ther 11(1):30. https://doi.org/10.3390/biom11010030

Wallace CJK, Milev R (2017) The effects of probiotics on depressive symptoms in humans: a systematic review. Ann Gen Psychiatry 16:14. https://doi.org/10.1186/s12991-017-0138-2

Wallen ZD, Appah M, Dean MN, Sesler CL, Factor SA, Molho E, Zabetian CP, Standaert DG, Payami H (2020) Characterizing dysbiosis of gut microbiome in PD: evidence for overabundance of opportunistic pathogens. NPJ Parkinsons Dis 6:11. https://doi.org/10.1038/s41531-020-0112-6

Wang H, Liu X, Tan C, Zhou W, Jiang J, Peng W, Zhou X, Mo L, Chen L (2020) Bacterial, viral, and fungal infection-related risk of Parkinson's disease: meta-analysis of cohort and case-control studies. Brain Behav 10(3):e01549. https://doi.org/10.1002/brb3.1549

Winter SE, Bäumler AJ (2014) Why related bacterial species bloom simultaneously in the gut: principles underlying the 'Like will to like' concept. Cell Microbiol 16(2):179–184. https://doi.org/10.1111/cmi.12245

Zhu F, Li C, Gong J, Zhu W, Gu L, Li N (2019) The risk of Parkinson's disease in inflammatory bowel disease: a systematic review and meta-analysis. Dig Liver Dis 51(1):38–42. https://doi.org/10.1016/j.dld.2018.09.017

Zimmerman MA, Singh N, Martin PM, Thangaraju M, Ganapathy V, Waller JL, Shi H, Robertson KD, Munn DH, Liu K (2012) Butyrate suppresses colonic inflammation through HDAC1-dependent Fas upregulation and Fas-mediated apoptosis of T cells. Am J Physiol Gastrointest Liver Physiol 302(12):G1405–G1415. https://doi.org/10.1152/ajpgi.00543.2011

Zuker CS (2015) Food for the brain. Cell 161(1):9–11. https://doi.org/10.1016/j.cell.2015.03.016

The Gut Microbiome-Induced Changes in Brain Immune System and Their Role in Epilepsy

14

Ramajayam Govindan, Maheshkumar Poomarimuthu, S. N. Arumugam, Senthilkumar Kalimuthu, Selvaraj Jayaraman, and C. S. Christudass

Abstract

Patients of all ages can experience unpredictable, spontaneous, and uncontrollable seizures as a result of the severe neurological condition called epilepsy. Epilepsy, a chronic neurological disorder characterized by unpredictable seizures, affects millions globally. While traditionally the brain is the focus of treatment, emerging research suggests a potential link between the gut microbiome and epilepsy. When left untreated, these seizures have a significant effect on epileptic patient's quality of life. More than 50 million people worldwide have epilepsy; nearly 80% of them live in low- and middle-income countries (WHO

R. Govindan (✉)
DHR-ICMR, Multidisciplinary Research Unit, Madurai Medical College, Government Rajaji Hospital, Madurai, Tamil Nadu, India

M. Poomarimuthu
Multidisciplinary Research Unit, Madurai Medical College, Government Rajaji Hospital, Madurai, Tamil Nadu, India

S. N. Arumugam
DHR-Multidisciplinary Research Unit, Dr. ALM PGIBMS, University of Madras, Taramani Campus, Chennai, India

S. Kalimuthu
Central Research Laboratory, Swamy Vivekanandha Medical College Hospital and Research Institute, Namakkal, Tamil Nadu, India

S. Jayaraman
Department of Biochemistry, Centre of Molecular Medicine and Diagnostics (COMManD), Saveetha Dental College and Hospital, Saveetha Institute of Medical and Technical Sciences, Saveetha University, Chennai, India

C. S. Christudass
Unit of Neurochemistry, Department of Neurological Sciences, Christian Medical College and Hospital, Ranipet Campus, Vellore, Tamil Nadu, India

S. Pathak, A. Banerjee (eds.), *Gut Microbiome and Brain Ageing*,
https://doi.org/10.1007/978-981-99-8803-7_14

report, 2017), with developing nations reporting twice as many cases each year. Usually, when we consider neurological problems, we just consider the brain and not the gut. An increasing body of studies shows that the gut-brain axis has links and is crucial.

The gut microbiome is a complex microbial community found in the human gastrointestinal (GI) tract that includes both commensal and pathogenic microorganisms. The scientific community is very interested in these gut microorganisms and their potential to enhance human health. Age-related and many intestinal symbiosis of the gut microbiome, which is frequently connected to a compromised immune system, is linked to problems. In many circumstances, especially in older people, a drop in beneficial microorganisms and a reduction in the variety of microbes in the gut can lead to disease. Probiotics, which are ingestible good bacteria, may alter the natural microbiota in the gut in a favourable way. At the neurological, hormonal, and immunological levels, the gastrointestinal system and the brain communicate in both directions. The brain-gut axis is a concept that is essential for preserving homeostasis. The growth and maturation of the immune and endocrine systems after birth are meticulously influenced by bacterial colonization of the intestine. Hence, the current chapter may review the role of gut microbium, brain immunity, and gut microbium-induced changes in epileptic seizures. This chapter delves into the potential link between the gut microbiome, brain immunity, and changes in epileptic seizures. By exploring this connection, we may uncover novel therapeutic strategies for managing epilepsy.

Keywords

Seizure · BBB · Epilepsy · CNS · GABA · GM · GI

14.1 Introduction

14.1.1 Gut Microbiome

The earth has been home to microorganisms for billions of years, and there are microbes in practically every natural ecosystem. Abundant microorganisms, such as bacteria, yeast, and viruses, have been shown to live together in different parts of the human body like oral cavity, skin, lung, and gut (Ursell et al. 2014). The term "microbiota" first appeared in the early 1900s. Furthermore, the human microbiota, also referred to as "the hidden organ," contributes more genetic data than the entire human genome—more than 150 times more (Grice and Segre 2012). Even though the terms "microbiota" and "microbiome" are frequently used interchangeably, they have some distinct meanings. Microbiota, such as the oral and intestinal microbiota, are live microorganisms that can be discovered in a certain habitat. Microbiome is the term used to describe the collection of genomes from all the microorganisms in

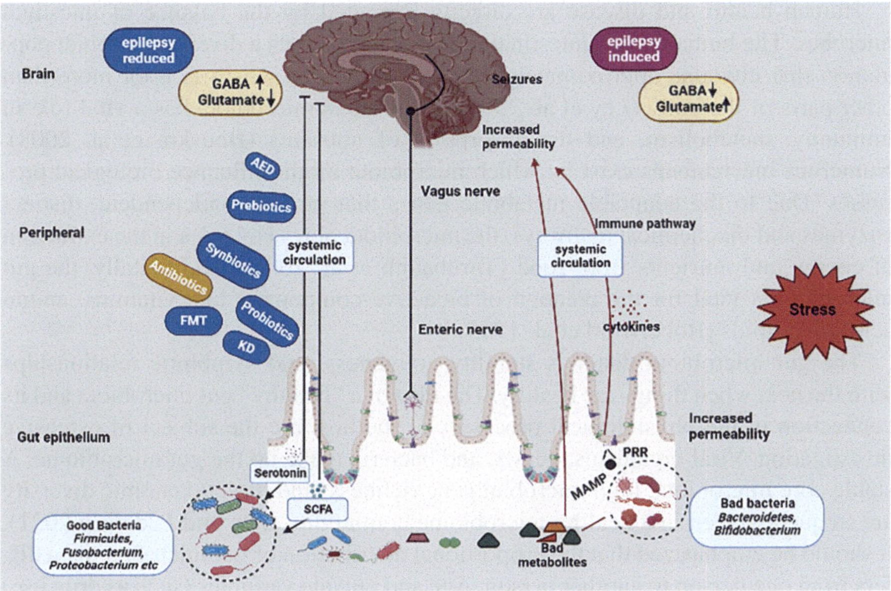

Fig. 14.1 Various factors involved in gut microbium-induced epileptic events

the environment. This includes not only the community of bacteria but also their structural components, metabolites, and environmental factors (Berg et al. 2020). The scope of the microbiome is wider than the scope of the microbiota in this regard. From one location to another, the microbiota's makeup differs (Fig. 14.1). The gut microbiota is thought to be the most important for preserving human health (Shreiner et al. 2015). The gut bacteria perform a variety of tasks, including vitamin generation, pathogen defence, immune response stimulation, and food fermentation (Hillman et al. 2017). *Firmicutes* and *Bacteroidetes* are the important types of the six phyla that make up the gut microbiota, which also includes *Actinobacteria*, *Proteobacteria*, *Fusobacteria*, and *Verrucomicrobia* (Laterza et al. 2016).

14.2 Microbiota in Healthy Conditions

Each area of the human body has a different distribution and type of glands and hair follicles. Skin regions differ physically and chemically, which influences how their bacteria are composed (Tong et al. 2018). *Actinobacteria*, *Bacteroidetes*, *Cyanobacteria*, *Firmicutes*, and *Proteobacteria* are the most common microbial species found in the skin. We primarily concentrate on how the microbiome affects human health and immune-related neurological disorders like epilepsy in this chapter. Numerous studies have focused on the connection between the microbiota and diseases like cancer, diabetes, and neurological problems in recent years. Furthermore, modifying the microbiota in the human body may be essential for the treatment of disease.

Human health and disease are directly impacted by the balance of intestinal microbes. The human gastrointestinal (GI) tract comprises a diverse microbial population that contains approximately 100 trillion germs, which is a lot more than other parts of the body (Ley et al. 2006). The human microbiota has a vital role in immunity, metabolism, and the absorption of nutrients (Bouskra et al. 2008). Numerous mechanisms exist by which microbiota might influence biological processes. Due to the adaptable metabolic genes that produce independent, distinct enzymes and biochemical pathways, the microbiota plays key roles in the extraction of energy and nutrients from food (Turnbaugh et al. 2006). Additionally, the gut microbiota is vital for the creation of bioactive compounds like vitamins, amino acids, and lipids (Roberfroid et al. 1995).

The gut microbiota displays stability, toughness, and symbiotic relationships with the host when things are healthy. The idea of a "healthy" gut microbiota and its connection to the physiological processes of the host are the subject of extensive investigation. Viral organisms, yeasts, and bacteria make up the gut microbiome. A stable core microbiota, high microbial gene richness, and high taxonomic diversity are frequently seen in a healthy microbiome community (Fan and Pedersen 2021). It should be emphasized that the proportional distribution of gut microorganism differs from one person to another person. Age and outside variables (such as drug use) can affect the gut microbiota of humans, which can vary. Additionally, the gut microbiota differs in the GI tract's various anatomical regions. For instance, *Proteobacteria* like *Enterobacteriaceae* are only found in the small intestine and not the colon. Instead, the colon is frequently home to *bacteriodetes* such as *Bacteroidaceae*, *Prevotellaceae*, and *Rikenellaceae* (Flint et al. 2012). These variances are primarily brought on by the various settings. While the colon has slower flow rates, at milder pH, larger microbial populations, especially anaerobic kinds, are frequently seen; the small intestine transit time is short, when bile concentration is high (Milani et al. 2017).

14.3 Involvement of Gut Microbiota in Pathophysiological Conditions

The gut microbiota differs by age in addition to its spatial distribution. In general, the diversity of the microbiota rises between childhood and adulthood and falls off around the age of 70 (Rinninella et al. 2019). Before the formation of a reasonably steady gut microbiota composition, in children, microbiota diversity may occur due to *Akkermansia muciniphila*, *Bacteroides*, *Veillonella*, *Clostridium coccoides* spp., and *Clostridium botulinum* spp. (Amabebe et al. 2020). The composition of the human gut microbiota may then alter as people age due to nutritional and immune system changes. Particularly, older persons frequently have higher levels of *Clostridium* and *Proteobacteria* and lower levels of *Bifidobacterium* (Guigoz et al. 2008). Due to its function in boosting the immune system, the anaerobic bacteria *Bifidobacterium*'s decline is thought to be connected to the worsening inflammatory condition.

14.4 Gut Microbiome and Immune System

The mammalian immune system comprised of innate and adaptive mechanisms in each and every tissue which plays a crucial part in host defence against various foreign antigens and endogenous disruption of homeostasis. In ecological perception, the mammals and their commensal microbes are co-evolved (Dethlefsen et al. 2007). Such mutual relationship is vital for the accurate performance of host immunity and maintaining immune tolerance to innocuous stimulus (Macpherson et al. 2005). In the first years of life, the education of host immunity is critically influenced by early-life colonization. Further, the microbial composition and their intra- and inter-individual variability influence the maturation of immune system at the age of ~3 years (Backhed et al. 2015).

The mutual relationship between the commensal microbes and host immunity is demonstrated by the use of germ-free (GF) mice. The GF mice study shows that the absence of commensal microbes leads to defects in the structure of intestinal lymphoid tissue and immune functions (Bauer et al. 1963). GF mice show substantial reduction in intra-epithelial lymphocytes (IELs) and IgA antibodies which plays crucial role in humoral mucosal immunity (Hapfelmeier et al. 2010). T helper 17 (Th17) cells are deficient in GF mice, which are stimulated by microbial colonization particularly by segmented filamentous bacteria (SFB). Further, a polysaccharide secreted from Bacteroides fragilis regulates the establishment of immune system including the development of T cells and Th1/Th2 polarization in lymphoid tissues (Ivanov et al. 2008). Similarly, extracellular signals produced from commensal microbes regulated the B-cell lineage in the intestinal mucosa and thereby influence the gut immunoglobulin repertoires (Wesemann et al. 2013).

Notably, the intestinal immune system has the ability to establish immune tolerance against numerous commensal microbes while activating immune responses against pathogenic infection. In homeostasis, the immune response towards the intestinal microbiota is decisively compartmentalized to the mucosal surface (Mowat 2018). A dense mucus layer made of hyperglycosylated mucin MUC2 segregates the intestinal epithelium from inhabitant microbes. Further, MUC2 limits the immunogenicity of intestinal antigens and maintains anti-inflammatory condition by regulating enteric dendritic cells (DCs) (Shan et al. 2013). Tight junctions play crucial role in controlling trans-epithelial permeability. Microbial signals promote the strength of the epithelial barrier by upregulating the tight junctions and associated cytoskeletal proteins. Meanwhile, the mucosal barrier function is regulated by secretory IgA antibodies and antimicrobial peptides (AMPs) (Bansal et al. 2010).

Toll-like receptors (TLRs) play crucial role in regulating host defence against pathogens, modulate the composition of commensal microbes, and sustain tissue integrity. In particular, TLR5 is involved in moulding the gut microbiota, which could be limited to a critical period at the beginning of life (Rakoff-Nahoum et al. 2004). In addition, NOD-like receptors (NLRs) are involved in shaping the gut microbiota composition. Nucleotide-binding oligomerization domain-containing protein 1 (NOD1) acts as an innate sensor and credited for the development of adaptive lymphoid tissues and preservation of intestinal homeostasis (Bouskra et al. 2008). At the same time, NOD2 protects small intestine from inflammation by restricting the growth of the commensal *Bacteroides vulgatus*. Gut epithelial stem cell survival and epithelial regeneration are promoted by the induction of NOD2 (Ramanan et al. 2014) moulding.

Recently, researchers explored the association between commensal microbes and monocytes/macrophages. Butyrate, a microbial-derived metabolite influences the differentiation of monocyte-to-macrophage through histone deacetylase 3 (HDAC3) inhibition, while the trimethylamine *N*-oxide (TMAO) can induce the polarization of murine macrophage through NLRP3 inflammasome-dependent manner (Schulthess et al. 2019; Wu et al. 2020). Moreover, the proliferation and function of group 3 innate lymphoid cells (ILCs) regulated by the microbial metabolite sensor Free fatty acid receptor 2 (Ffar2) (Chun et al. 2019). B cells are crucial regulators of gut homeostasis which secretes great number of secretory IgA antibodies. These IgA antibodies are involved in the maintenance of a heterogeneous and balanced microbiome and also regulate the differentiation and proliferation of Foxp3+ regulatory T cells (Kawamoto et al. 2014). The microbiome and their metabolites play crucial role in the differentiation and proliferation of Th17 cells. The Th17 cells stimulated by SFB are non-inflammatory, while Th17 cells elicited by Citrobacter are inflammatory which produce broad range of inflammatory cytokines (Omenetti et al. 2019).

The composition of gut microbiome is influenced by various environmental factors whose impact dictates over host genetics (Rothschild et al. 2018). The environmental factors including diet, life style, antibiotic use, etc. prominently influence the occurrence of various inflammatory and autoimmune diseases (Vojdani 2014). Antibiotics are essential for the treatment of infectious diseases, while the intake of antibiotics intensely affects the composition and function of the gut microbiota. The antibiotic-driven microbial dysbiosis in the gut affects the functions of different immune cell subsets. Use of broad spectrum antibiotics leads to reduction of microbiota-derived short-chain fatty acids (SCFAs) which consequently causes hyperactivation of intestinal macrophages and proliferation of T helper cells. Further, antibiotic treatment promotes M2 macrophage polarization and allergic airway inflammation in lungs (Becattini et al. 2016).

Recent studies explored the relationship between dietary microbiota and modulation of host immunity. Western diets extremely affect the composition of gut microbiome and have negative impact on host immunity (Christ et al. 2019). Dietary long-chain fatty acids modulate the composition of gut microbiome and metabolome which may exacerbate autoimmunity in the central nervous system (CNS) (Haghikia et al. 2015). Further, the intake of artificial sweeteners, dietary carbohydrates, emulsifiers, and certain probiotics modulate the composition of gut microbiome which leads to inflammation and dysregulation of host immunity (Christ et al. 2019). Abnormal interaction between the microbiome and the host immunity in genetically susceptible host leads to the development of various immune-mediated diseases. However, the mechanism of interaction between gut microbiome and the immune system in various human diseases is poorly understood. More detailed study addressing the modulatory effects of microbiome and their metabolites on immune system is necessary for understanding the mechanisms for prevention and treatment of immune-related disorders.

14.5 Brain Immunity

The immune coordination of the CNS control the total body. The properties of brain action in regulation of immunity are different. The impact of immune system on stress in psychological condition has been widely characterized as major in neural (sympathetic) and endocrine signals (Bowers et al. 2008; Kin and Sanders 2006). The brain immunity represents the capacity to link a specific role (immune suppression) and an exterior sign like taste (saccharine); the indication preserves and independently activates the immune response for the same. Hence, saccharine appearance alone may induce suppression of the immune system after specific conditioning (Goebel et al. 2002). In evolutionary side, the brain capacity regulates the immunity. The information-associated lifestyle including loneliness and social status may alter the immune activity (Cole et al. 2015; Lupien et al. 2009; Ben-Shaanan et al. 2017). The immunity main regulations are synchronizing effects in circadian clock or sleep (Labrecque and Cermakian 2015; Rolls et al. 2015). The brain may involved in external and internal state of physiological condition to accommodate response to adaptive and dynamic immunity. The dynamic regulation of immunity is facilitating communication of immune and CNS system through endocrine and the autonomic nervous system (ANS) (Felten 1993; Ben-Shaanan et al. 2017). Additionally, cells that have the receptors for neuropeptides and neurotransmitters (de Jonge et al. 2005; Sternberg 2006), that will give the signals to peripheral nervous system can facilitate messages from the brain and immune system. The neuronal signals control the immunity through inflammatory reflex, mediated by vagus nerve and decrease the amount of inflammatory factors including TNFα, Interleukin-1,6 and 8 (Tracey 2002; Wang et al. 2004). Therefore, vagus nerve stimulation may be suggested for the treatment to control sepsis.

In the immune system, the range of acquired T and B immune cells randomly produces V (variable), D (diversity), and J (joining) gene segments (VDJ) recombination of T-cell receptor (TCR) and genes of immunoglobulin, TCR somatic hypermutation, and immunoglobulins (Ott et al. 2018). The imbalance of immunosuppressive Treg and Breg cells (Sakaguchi et al. 2008; Rosser and Mauri 2015) affects the immune system that may result in autoimmunity and allergy. The excitatory and inhibitory neurons from CNS have the balance connecting these populations and are important for the normal activity of the neural network (Yizhar et al. 2011). The immune cells called microglia comprises majority of the cells in the brain, and others cell include myeloid cells, T, B, monocytes/macrophages, dendritic cells and natural killer (NK) cells minorly present (Korin et al. 2017). In adult naïve mice, NK and T/B cells have been identified in $CD45^{hi}$ population in CNS and are limited with approximately 10,000 per hemisphere (Pösel et al. 2016) and have potent brain function. T cells are concerned about complex brain functions including spatial learning, emotions, memory, and response to stress. CD4C T cells (helper), are recruited in the meninges and secrete interleukin (IL-4) can tilt macrophages, microglia to anti-inflammatory (M2) phenotype and then, induce secretion of brain-derived neurotrophic factor from astrocytes (Kipnis et al. 2004; Ziv et al. 2006; Derecki et al. 2010; Radjavi et al. 2014).

14.6 Gut Microbiome-Induced Brain Immunity Changes

Various microbial populations coexist in mice as well as humans. Most of these microorganisms such as archaea, bacteria, fungi, and viruses exist in human GI tract, and these are jointly called as gut "microbiota" (Ferreiro and Crook 2018). Evidence from studies suggest that microbiota is complex in the organisms' physiology and pathology, and has proposition in health and disease (Maynard et al. 2012). The physiological activity of microbiota has been identified, including immune system (Belkaid and Hand, 2014). The alterations of gut microbes in response to significant immune signalling lead to illnesses of the intestine and other distal organs (Blander et al. 2017; Roy and Trinchieri 2017). Divergent microbial flora is maternally inherited during birth and then modified due to dietary habit changes and ecological signals (Gomez de Aguero et al. 2016; Ma et al. 2019a, b). The development and maturation of CNS in humans are regulated by various factors including intrinsic and extrinsic way. Studies evidenced mostly from animals with germ-free (GF) type or animals treated from broad-spectrum antibiotics showed that specific microbiota can shock CNS neurochemistry and physiology (Smith 2015). In humans, evidence about the relationship between GI pathology and neuropsychiatric problems was reported in conditions including autism, anxiety, and depression (Foster et al. 2017; Fung et al. 2017). In addition gut microbiota was involved in the development of CNS, immune and neural signals (Tremlett et al. 2017). The immune system of human has been evolved to preserve the symbiotic connection among host and microbiota; disruption of these active interaction between immune-microbial can lead to insightful effects on health (Hooper et al. 2012). The digestive system's role during development of brain and the axis of gut-brain can be used to express the microbiota and their communication resulting in changes in the status of CNS. Dysbiosis changes in the gut microbiome can induce atypical immune signalling, imbalance of host homeostasis and disease progression of CNS (Ma et al. 2019a, b). The microbiota persuade CNS through multiple immunological signals that include NF-κB, inflammasome, and IFN-I. Microbe-associated molecular patterns (MAMPs) are generally considered by TLRs and expression in different immune cell types that promote activation of the immune cells. Activated cells produce various pro-inflammatory cytokines such as IL-1β and TNF-α, which can go into the circulation through blood-brain barrier (BBB), and result in the progress of many neurological disorders (Sampson and Mazmanian 2015). Experimental autoimmune encephalomyelitis (EAE) model in the spinal cord and intestine produces fewer amounts of pro-inflammatory cytokines, interferon (IFNγ), and IL-17A in germ-free (GF) mice. The segmented filamentous bacteria (SFB) colonization can induce Th1 and Th17 that will induce response in the spinal cord and intestine, promote EAE symptoms (Lee et al. 2011). On the other hand, colonization of *Bacteroides fragilis* and *Prevotella histicola* can repress EAE by encouraging Treg, which suggests that the gut microbiota regulates neuroinflammation through immune actions (Mangalam et al. 2017).

SFB colonization is ample to enhance symptoms of Autism spectrum disorder (ASD)-like pattern in the intestine by the regulation of Th17 cells. Conversely, blocking IL-17A with neutralizing antibodies is limiting the behavioural abnormalities (Lammert et al. 2018). Inflammasome activation can cause the maturation of apoptotic gene caspase-1, and pro-inflammatory cytokines (IL-1β and IL-18) release may promote neuroinflammation. The definite MAMPs can promote pro-inflammatory cytokine productions, and inflammasome are implicated in broad range of neurological disorders (Yang et al. 2019). In addition, caspase-1 genetic deficiency mice have less anxiety-like behaviours and depression from chronic stress (Wong et al. 2016). The components from gut microbiota cross the BBB and affect the brain immune cells' maturation and activation, including microglia and astrocytes (Fung 2020). Gut microbiota contributes to the development, homeostasis, and pathology of brain in the CNS through microglia. The microglia role by release of cytokines, phacocytosis, and complement activation in CNS (Salter and Stevens 2017). Moreover microglia and astrocytes are important cells among glial cells, involved in multiple functions, including control of BBB, regulation of CNS development and repair through the secretion of cytokines and chemokines and antigen presentation. IFN (Type 1) signalling in astrocytes, intervention by microbial tryptophan metabolites and activation of Aryl hydrocarbon receptor (AhR) can limit CNS inflammation (Rothhammer et al. 2016; Wang et al. 2014).

Research on germ-free mice, or mice with no microbiota at all, provides the strongest support for a function for the microbiome in neurodevelopment. The microbiota's composition has been demonstrated to be a key factor in these models' fundamental brain processes, including development, myelination, neurogenesis, and microglia activation. The field has advanced in determining whether the microbiota is engaged in particular brain activities, but germ-free mice represent an extreme case with a naturally faulty immune system development and minimal translation into humans (Luczynski et al. 2016).

The majority of the studies on newborn, which are the few that have been conducted, have been cross-sectional. In one study, the Mullen Scales of Early Learning were used to measure cognitive performance at 2 years old. The results showed a strong correlation between the microbiota's composition at 1 year old and cognitive ability at 2 years old (Carlson et al. 2018). In a cohort of 39 newborn, microbiota diversity (a statistic representing within-sample variability) was also linked to functional connectivity between the inferior parietal lobule and the supplementary motor region (Gao et al. 2019). It's significant to note that this functional connection was linked to cognitive outcomes at age 2 (Gao et al. 2019).

14.7 Epilepsy

Epilepsy is a brain disorder characterized by a person's propensity to experience recurring, spontaneous epileptic seizures. The aberrant paroxysmal fluctuations in neuronal electrical activity that occur during seizures are related to an imbalance between the excitatory and inhibitory brain networks. In addition to modulating brain activity, glial cells also contribute to the onset of seizures by re-establishing the equilibrium of the neurotransmitters glutamate and gamma-aminobutyric acid (GABA) (Beghi et al. 2019). Recurrent seizures are a hallmark of the neurological disorder in epilepsy, which is frequently brought on by an earlier injury to the central nervous system (CNS) (Beghi 2020). Epilepsy affects about 65 million individuals globally, making it the most common neurological condition to reduce quality of life. Mesial temporal epilepsy and lateral epilepsy are the two primary subtypes of temporal lobe epilepsy (TLE), one of the most prevalent forms (Bartolomei et al. 1999; Tatum 2012). Preclinical and clinical data have suggested in the past 10 years that the gut microbiota may be a regulator of epileptic seizures and a potential new treatment target (De Caro et al. 2019; Lum et al. 2020; Yue et al. 2021).

14.8 Gut Microbiome-Induced Factors' Role in Epilepsy

Numerous studies suggest that when the microbiota and the host interact in the intestine, immune system molecules, neurotransmitters, and microbial metabolites are released. These molecules can affect neuronal messages and possibly control brain activity and behaviour. For instance, it has been demonstrated that the intestinal microbiota can influence the enteric nervous system, a network of neurons that controls the digestive system's operations (Carabotti et al. 2015). Additionally, it has been demonstrated that the microbiota produces and reacts to a variety of neurotransmitters, including serotonin and GABA, which are involved in human behaviour and cognitive processes (Wang and Wang 2016).

We must examine the events that take place during an epileptic seizure in order to properly comprehend the function played by the gut flora. However, it is challenging to determine a specific role for the gut microbiota in epilepsy because the causes of the condition are either numerous or still unknown in primary epilepsy. The resistance of excitatory neurons to stimuli appears to decrease temporarily during an epileptic episode. This might happen as a result of altered ion channels or dysfunctional inhibitory neurons. This leads to a particular location where seizures may then start (Scharfman 2007). Following brain damage, an additional mechanism causing epilepsy may result from the "up" regulation of excitatory neuronal circuits or the "down" regulation of inhibitory circuits (Guerriero et al. 2015). "Epileptogenesis" processes cause these secondary epilepsies (Pitkänen et al. 2015). The phrase "epileptogenic threshold" is used to describe the stimulus level necessary for an attack to occur. This threshold appears to be substantially lower in epileptic patients than in the general population. A compromised blood-brain barrier

(BBB) could also be a contributing factor because it would make it possible for blood-borne chemicals to enter the brain (Rana and Musto 2018).

According to the available studies, the gut microbiome can either increase or decrease epileptic symptoms. It is possible that the gut-brain axis is the reason for this occurrence. Multiple mechanisms may be used to promote epilepsy when the so-called bad bacteria are present. The creation of a number of chemicals that can change the excitatory-inhibitory balance is carried out by gut bacteria. Cytokines and metabolites that act as neuromodulators, such as short-chain fatty acids (SCFAs), gamma-aminobutyric acid (GABA), and serotonin precursors, are examples of these (Strandwitz 2018; Bagdy et al. 2007). We are aware that an unbalanced excitatory-inhibitory balance is the cause of seizures. The threshold for the development of a seizure is lowered when the neurotransmitter GABA is decreased (Casillas-Espinosa et al. 2012).

Through a change in the BBB, a dysbiosis condition can cause an influx of toxins and cytokines. There is less SCFA and GABA production when these circumstances exist. The BBB is altered by the decrease of SCFAs, which have anti-inflammatory properties. Furthermore, the beginning of a seizure may be influenced by a decreased influx of GABA in the gut-brain axis. Furthermore, numerous studies (Lim et al. 2017) have demonstrated that epileptic manifestations can occur even with a low serotonin intake. We might therefore propose a mechanism based on changes in the gut microbiota and subsequent changes in neurotransmitters like GABA, serotonin, and glutamate that may favour epileptic episodes.

Neuroimmunity and neuroinflammation are connected with the aetiology of epilepsy (Yamanaka et al. 2021). The brain-gut axis may be implicated in immunological and inflammatory pathways that contribute to epilepsy development, according to growing data. The predominant inflammatory cells in the CNS are microglia and astrocytes, and their inflammatory condition encourages the development of epilepsy (Devinsky et al. 2013; Sanz and Garcia-Gimeno 2020). In gut immunity, 70–80% of immune cells in the body are found in the lymphoid tissue of the intestinal mucosa (Tlaskalova-Hogenova et al. 2005). Immunological cells are impacted by the gut microbiota (GM); for instance, germ-free (GF) mice exhibit immunological abnormalities, including a decline in T and B cell numbers and a drop in cytokine production (Smith et al. 2007).

Additionally, astrocyte activation, which is age- and sex-dependent (Thion et al. 2018) and appears to be one of the most crucial variables for the development of microglial cells, appears to be influenced by the GM. To control the onset of epilepsy, the GM controls inflammatory, adaptive, and innate immune responses. The blood-brain barrier (BBB) and gut mucosal barrier prohibit GM and its secretions from reaching the brain. Increased intestinal permeability, which permits bacteria, harmful metabolites, and small molecules to translocate into the circulation, is the hallmark of "leaky gut" syndrome (Obrenovich 2018). Bacteria can directly release substances into the bloodstream during gut inflammation, which activates peripheral immune cells, changes BBB integrity, and consequently transport rates, and can even cause "leaky brain" (Logsdon et al. 2018).

Lipopolysaccharides and other cytokines in the lumen enter the blood circulation, triggering toll-like receptors and releasing inflammatory cytokines that could enhance BBB permeability and harm the brain (Ait-Belgnaoui et al. 2012; McCusker and Kelley 2013). Stress can also increase intestinal mucosal permeability. The brain's astrocytes, which are the most numerous glial cells, serve a number of purposes, including as maintaining the BBB's integrity, recycling neurotransmitters, and taking part in immunological reactions. The CNS's resident macrophages, known as microglial cells, mediate the innate immune response (Ginhoux et al. 2013). The modulation of neuronal activity, inflammation, phagocytic neuron clearance, and persistent seizures can all be facilitated by microglia with a bigger, less ramified, amoeboid morphology (Bosco et al. 2020).

By releasing too many cytokines, microglia and astrocytes contribute to the pathogenesis of epilepsy (Devinsky et al. 2013). Microglia can influence astrocyte phenotype and function (Liddelow et al. 2017), and mouse microglia can control astrocyte behaviour by, for example, Vascular endothelial growth factor B (VEGF-B), which encourages the pathogenic response and inflammatory response of astrocytes, and Transforming growth factor alpha (TGF-α), which encourages the opposite. In order to regulate microglial activation, TGF-a expression, VEGF-B levels, and gut bacteria convert dietary tryptophan into aryl hydrocarbon receptor agonists and interact with their receptor (Rothhammer et al. 2018). Which in turn modifies the pathogenic activity of astrocytes. The migration, phagocytosis of apoptotic cells, and synaptic pruning of microglia are all made easier by the production of inflammatory cytokines and chemokines by astrocytes (Yang et al. 2020).

The communication between astrocytes and microglia increases the generation of pro-inflammatory cytokines, decreases BBB permeability, and infiltrates immune cells and cytokines from peripheral blood into the CNS, resulting in chronic neuroinflammation (Moradi et al. 2021). This suggests that intestinal microbial diversity is essential for microglial and CNS function (Erny et al. 2015). GF and antibiotic-treated (Abx-treated) animals also had altered microglial morphology and defects in maturation, activation, and differentiation, resulting in an inadequate immune response to a variety of pathogens, which could be repaired after GM recolonization. The morbidity of epilepsy is also influenced by peripheral immune cells that invade the brain tissue, such as T cells and monocytes, in addition to glial cells that are found in the CNS. Monocytes can mature into macrophages and occupy the brain, where they differentiate into "microglia-like cells" and contribute to epilepsy (Djukic et al. 2006). The organ (gut) with the largest population of immune cells can induce epilepsy through the innate immune pathway. In mice models, permeability of BBB increases throughout the life with decreased expression of occludin and claudin proteins in the endothelium (Braniste et al. 2014). During gut dysbiosis reductions of claudin levels may increase the permeability of the intestinal lining, which leads to the escape of microorganisms, metabolites, and toxins from the intestinal lumen. This upsurges BBB permeability and induces neuroinflammation (Welcome 2019). Both gut and brain barriers are broken, and the immune cells and molecules released by the microbiota enter the brain and induce seizures.

14.9 Neurocysticercosis and Epilepsy

The two most prevalent neurological conditions that add to the global illness burden are epilepsy and neurocysticercosis (NCC). The primary clinical sign of parenchymal NCC, acute symptomatic seizures, is brought on by an immune-inflammatory process in the host brain in reaction to the parasite's degenerative phase or death. While the local inflammatory activity persists, seizures may return over the course of several months. Once the acute process has passed, if the seizures continue, epilepsy can be identified in the patient. Epilepsy is frequently accompanied by CNS infections. Post-infectious epilepsy is frequently attributed to parasite and viral diseases (WHO 2017). The chance of a coincidental occurrence of NCC and a very prevalent brain disease in developing nations, such as epilepsy, is increased by the greater rates of T. solium infection and subsequent rates of NCC.

Innate (non-specific) and adaptive (specific) immunological responses of the host lead to the elimination of the pathogen and/or of the host when a pathogen enters an immunologically competent organism. The creation of a non-specific inflammatory event locally in the pathogen's surrounds has a significant impact on the protective or pathogenic efficiency of the innate response. The systemic, selective clonal proliferation of lymphoid cells that differentiate into effector cells of type Th1 or Th2 and produce a variety of cytokines, as well as plasma cells that produce particular antibodies, are the foundation of the adaptive immune response.

The type and rate of the immune response will affect resistance or susceptibility to the creation of metacestodes during the very first events following oncosphere ingestion. The presence of living parasites that actively avoid being destroyed by the host immune system is then followed by a strong immunological reaction. The parasite is ultimately eliminated by an inflammatory response, either throughout the course of an infection's natural development or after anti-parasitic therapy. Recent conceptual definitions for seizures related to autoimmune illnesses have been presented by the ILAE Autoimmunity and Inflammation Taskforce (Steriade et al. 2020). Despite the fact that seizures can occasionally happen weeks or even months after the resolution of immune-related encephalitis, this Taskforce recommended using the term "acute symptomatic seizures secondary to autoimmune encephalitis" to describe seizures occurring in the context of the active phase of immune-mediated encephalitis. They also proposed the term "autoimmune-associated epilepsy" to describe chronic seizures that have been determined to be secondary to autoimmune brain diseases, since autoimmune encephalitis has significant epidemiological, clinical, and therapeutic consequences (Steriade et al. 2020).

14.10 Conclusion

In this chapter we summarized how the gut microbiota function in immunological reaction and their role in brain immunity and epilepsy. It is well known that altered gut microbiota are present in epilepsy based on data from both animal models and human research. The gut flora can be altered by numerous substances in either a favourable or negative way, reducing or escalating epileptic episodes; intestinal dysbiosis is linked to chronic stress-induced epilepsy. Children with epilepsy may have a dysbiosis, according to recent investigations in human cohorts, but bigger trials with age-matched controls are required to corroborate these findings. Dysbiosis may be particularly important in some kinds of epilepsy, such as stress-induced epilepsy. Therefore, aetiology stratification may be important. If a dysbiosis is identified, microbiota-targeted approaches to treating epilepsy may be created. In developing countries 30% of epilepsy associated with taenia solium effected neurocysticercosis. Collectively the data suggest that the altered gut microbium in turn impacts immunological response. Hence the immune cell-produced cytokines and chemokines may interrupt BBB and induce epilepsy in most of the gut-altered patients. Further research is required to assess the effectiveness and safety of microbiota-targeted interventions like probiotics, prebiotics, or faecal microbiota transplantation (FMT) in the treatment or prevention of epilepsy. Targeted treatment options may need a thorough investigation of the possible involvement of gut dysbiosis in certain epilepsy subtypes such as age, location (e.g., neurocysticercosis in poor countries), and potential triggers (e.g., stress).

Acknowledgement We acknowledge DHR and ICMR, Government of India, for the partial support of this work under the scheme: Multi-Disciplinary Research Unit, Madurai Medical College, Madurai, Tamil Nadu, India.

References

Ait-Belgnaoui A, Durand H, Cartier C, Chaumaz G, Eutamene H, Ferrier L et al (2012) Prevention of gut leakiness by a probiotic treatment leads to attenuated HPA response to an acute psychological stress in rats. Psychoneuroendocrinology 37:1885–1895. https://doi.org/10.1016/j.psyneuen.2012.03.024

Amabebe E, Robert FO, Agbalalah T, Orubu ESF (2020) Microbial dysbiosis induced obesity: role of gut microbiota in homoeostasis of energy metabolism. Br J Nutr 123:1127–1137

Backhed F et al (2015) Dynamics and stabilization of the human gut microbiome during the first year of life. Cell Host Microbe 17:690–703

Bagdy G, Kecskemeti V, Riba P, Jakus R (2007) Serotonin and epilepsy. J Neurochem 100:857–873

Bansal T, Alaniz RC, Wood TK, Jayaraman A (2010) The bacterial signal indole increases epithelial-cell tight-junction resistance and attenuates indicators of inflammation. Proc Natl Acad Sci U S A 107:228–233

Bartolomei F, Wendling F, Vignal JP, Kochen S, Bellanger JJ, Badier JM, Le Bouquin-Jeannes R, Chauvel P (1999) Seizures of temporal lobe epilepsy: identification of subtypes by coherence analysis using stereo-electro encephalography. Clin Neurophysiol 110:1741–1754

Bauer H, Horowitz RE, Levenson SM, Popper H (1963) The response of the lymphatic tissue to the microbial flora. Studies on germfree mice. Am J Pathol 42:471–483

Becattini S, Taur Y, Pamer EG (2016) Antibiotic-induced changes in the intestinal microbiota and disease. Trends Mol Med 22:458–478

Beghi E (2020) The epidemiology of epilepsy. Neuroepidemiology 54:185–191

Beghi E, Giussani G, Abd-Allah FA, Abdela J, Abdelalim A, Abraha HN et al (2019) Global, Regional, and National Burden of epilepsy, 1990-2016: a systematic analysis for the global burden of disease study 2016. Lancet Neurol 18:357–375. https://doi.org/10.1016/S1474-4422(18)30454-X

Belkaid Y, Hand TW (2014) Role of the microbiota in immunity and inflammation. Cell 157:121–141. https://doi.org/10.1016/j.cell.2014.03.011
Ben-Shaanan T, Schiller M, Rolls A (2017) Studying brain-regulation of immunity with optogenetics and chemogenetics; a new experimental platform. Brain Behav Immun 65:1–8. https://doi.org/10.1016/j.bbi.2016.11.024. Epub 2016 Nov 24. PMID: 27890661
Berg G et al (2020) Microbiome definition re-visited: old concepts and new challenges. Microbiome 8:103
Blander JM, Longman RS, Iliev ID, Sonnenberg GF, Artis D (2017) Regulation of inflammation by microbiota interactions with the host. Nat Immunol 18:851–860. https://doi.org/10.1038/ni.3780
Bosco DB, Tian DS, Wu LJ (2020) Neuroimmune interaction in seizures and epilepsy: focusing on monocyte infiltration. FEBS J 287:4822–4837. https://doi.org/10.1111/febs.15428
Bouskra D et al (2008) Lymphoid tissue genesis induced by commensals through NOD1 regulates intestinal homeostasis. Nature 456:507–510
Bowers SL, Bilbo SD, Dhabhar FS, Nelson RJ (2008) Stressor-specific alterations in corticosterone and immune responses in mice. Brain Behav Immun 22:105–113
Braniste V, Al-Asmakh M, Kowal C, Anuar F, Abbaspour A, Tóth M et al (2014) The gut microbiota influences blood-brain barrier permeability in mice. Sci Transl Med 6:263ra158. https://doi.org/10.1126/scitranslmed.3009759
Carabotti M, Scirocco A, Maselli MA, Severi C (2015) The gut-brain axis: interactions between enteric microbiota, central and enteric nervous systems. Ann Gastroenterol 28:203–209
Carlson AL, Xia K, Azcarate-Peril MA et al (2018) Infant gut microbiome associated with cognitive development. Biol Psychiatry 83:148–159
Casillas-Espinosa PM, Powell KL, O'Brien TJ (2012) Regulators of synaptic transmission: roles in the pathogenesis and treatment of epilepsy. Epilepsia 53:41–58
Christ A, Lauterbach M, Latz E (2019) Western diet and the immune system: an inflammatory connection. Immunity 51:794–811
Chun E et al (2019) Metabolite-sensing receptor Ffar2 regulates colonic group 3 innate lymphoid cells and gut immunity. Immunity 51:871–884
Cole SW, Capitanio JP, Chun K, Arevalo JM, Ma J, Cacioppo JT (2015) Myeloid differentiation architecture of leukocyte transcriptome dynamics in perceived social isolation. Proc Natl Acad Sci U S A 112:15142–15147
De Caro C, Iannone LF, Citraro R, Striano P, De Sarro G, Constanti A, Cryan JF, Russo E (2019) Can we "seize" the gut microbiota to treat epilepsy? Neurosci Biobehav Rev 107:750–764
de Jonge WJ, van der Zanden EP, The FO, Bijlsma MF, van Westerloo DJ, Bennink RJ, Berthoud HR, Uematsu S, Akira S, van den Wijngaard RM et al (2005) Stimulation of the vagus nerve attenuates macrophage activation by activating the Jak2-STAT3 signaling pathway. Nat Immunol 6:844–851
Derecki NC, Cardani AN, Yang CH, Quinnies KM, Crihfield A, Lynch KR et al (2010) Regulation of learning and memory by meningeal immunity: a key role for IL-4. J Exp Med 207:1067–1080. https://doi.org/10.1084/jem.20091419
Dethlefsen L, McFall-Ngai M, Relman DA (2007) An ecological and evolutionary perspective on human-microbe mutualism and disease. Nature 449:811–818
Devinsky O, Vezzani A, Najjar S, De Lanerolle NC, Rogawski MA (2013) Glia and epilepsy: excitability and inflammation. Trends Neurosci 36:174–184. https://doi.org/10.1016/j.tins.2012.11.008
Djukic M, Mildner A, Schmidt H, Czesnik D, Brück W, Priller J et al (2006) Circulating monocytes engraft in the brain, differentiate into microglia and contribute to the pathology following meningitis in mice. Brain 129:2394–2403. https://doi.org/10.1093/brain/awl206
Erny D, Hrabě de Angelis AL, Jaitin D, Wieghofer P, Staszewski O, David E et al (2015) Host microbiota constantly control maturation and function of microglia in the CNS. Nat Neurosci 18:965–977. https://doi.org/10.1038/nn.4030
Fan Y, Pedersen O (2021) Gut microbiota in human metabolic health and disease. Nat Rev Microbiol 19:55–71

Felten DL (1993) Direct innervation of lymphoid organs: substrate for neurotransmitter signaling of cells of the immune system. Neuropsychobiology 28:110–112

Ferreiro A, Crook N, Gasparrini AJ, Dantas G (2018) Multiscale evolutionary dynamics of host-associated microbiomes. Cell 172:1216–1227. https://doi.org/10.1016/j.cell.2018.02.015

Flint HJ, Scott KP, Louis P, Duncan SH (2012) The role of the gut microbiota in nutrition and health. Nat Rev Gastroenterol Hepatol 9:577–589

Foster JA, Rinaman L, Cryan JF (2017) Stress & the gut-brain axis: regulation by the microbiome. Neurobiol Stress 7:124–136. https://doi.org/10.1016/j.ynstr.2017.03.001

Fung TC (2020) The microbiota-immune axis as a central mediator of gut-brain communication. Neurobiol Dis 136:104714

Fung TC, Olson CA, Hsiao EY (2017) Interactions between the microbiota, immune and nervous systems in health and disease. Nat Neurosci 20:145–155. https://doi.org/10.1038/nn.4476

Gao W, Salzwedel AP, Carlson AL et al (2019) Gut microbiome and brain functional connectivity in infants—a preliminary study focusing on the amygdala. Psychopharmacology 236:1641–1651

Ginhoux F, Lim S, Hoeffel G, Low D, Huber T (2013) Origin and differentiation of microglia. Front Cell Neurosci 7:45. https://doi.org/10.3389/fncel.2013.00045

Goebel MU, Trebst AE, Steiner J, Xie YF, Exton MS, Frede S, Canbay AE, Michel MC, Heemann U, Schedlowski M (2002) Behavioral conditioning of immunosuppression is possible in humans. FASEB J 16:1869–1873

Gomez de Aguero M et al (2016) The maternal microbiota drives early postnatal innate immune development. Science 351:1296–1302. https://doi.org/10.1126/science.aad2571

Grice EA, Segre JA (2012) The human microbiome: our second genome. Annu Rev Genomics Hum Genet 13:151–170

Guerriero RM, Giza CC, Rotenberg A (2015) Glutamate and GABA imbalance following traumatic brain injury. Curr Neurol Neurosci Rep 15:27

Guigoz Y, Doré J, Schiffrin EJ (2008) The inflammatory status of old age can be nurtured from the intestinal environment. Curr Opin Clin Nutr Metab Care 11:13–20

Haghikia A et al (2015) Dietary fatty acids directly impact central nervous system autoimmunity via the small intestine. Immunity 43:817–829

Hapfelmeier S et al (2010) Reversible microbial colonization of germ-free mice reveals the dynamics of IgA immune responses. Science 328:1705–1709

Hillman ET, Lu H, Yao T, Nakatsu CH (2017) Microbial ecology along the gastrointestinal tract. Microbes Environ 32:300–313

Hooper LV, Littman DR, Macpherson AJ (2012) Interactions between the microbiota and the immune system. Science 336:1268–1273. https://doi.org/10.1126/science.1223490

Ivanov II et al (2008) Specific microbiota direct the differentiation of IL-17-producing T-helper cells in the mucosa of the small intestine. Cell Host Microbe 4:337–349

Kawamoto S et al (2014) Foxp3+ T cells regulate immunoglobulin A selection and facilitate diversification of bacterial species responsible for immune homeostasis. Immunity 41:152–165

Kin NW, Sanders VM (2006) It takes nerve to tell T and B cells what to do. J Leukoc Biol 79:1093–1104

Kipnis J, Cohen H, Cardon M, Ziv Y, Schwartz M (2004) T cell deficiency leads to cognitive dysfunction: implications for therapeutic vaccination for schizophrenia and other psychiatric conditions. Proc Natl Acad Sci U S A 101:8180–8185. https://doi.org/10.1073/pnas.0402268101

Korin B, Ben-Shaanan TL, Schiller M, Dubovik T, Azulay-Debby H, Boshnak NT et al (2017) High-dimensional, single-cell characterization of the brain's immune compartment. Nat Neurosci 20:1300–1309. https://doi.org/10.1038/nn.4610

Labrecque N, Cermakian N (2015) Circadian clocks in the immune system. J Biol Rhythm 30:277–290

Lammert CR et al (2018) Cutting edge: critical roles for microbiota-mediated regulation of the immune system in a prenatal immune activation model of autism. J Immunol 201(3):845–850

Laterza L et al (2016) The gut microbiota and immune system relationship in human graft-versus-host disease. Mediterr J Hematol Infect Dis 8:2016025

Lee YK et al (2011) Proinflammatory T-cell responses to gut microbiota promote experimental autoimmune encephalomyelitis. Proc Natl Acad Sci U S A 108(Suppl 1):4615–4622

Ley RE, Turnbaugh PJ, Klein S, Gordon JI (2006) Human gut microbes associated with obesity. Nature 444:1022–1023

Liddelow SA, Guttenplan KA, Clarke LE, Bennett FC, Bohlen CJ, Schirmer L et al (2017) Neurotoxic reactive astrocytes are induced by activated microglia. Nature 541:481–487. https://doi.org/10.1038/nature21029

Lim JS, Lim MY, Choi Y, Ko G (2017) Modeling environmental risk factors of autism in mice induces IBD-related gut microbial dysbiosis and hyperserotonemia. Mol Brain 10:14

Logsdon AF, Erickson MA, Rhea EM, Salameh TS, Banks WA (2018) Gut reactions: how the blood-brain barrier connects the microbiome and the brain. Exp Biol Med (Maywood) 243:159–165. https://doi.org/10.1177/1535370217743766

Luczynski P, McVey Neufeld KA, Oriach CS, Clarke G, Dinan TG, Cryan JF (2016) Growing up in a bubble: using germ-free animals to assess the influence of the gut microbiota on brain and behavior. Int J Neuropsychopharmacol 19:pyw020

Lum GR, Olson CA, Hsiao EY (2020) Emerging roles for the intestinal microbiome in epilepsy. Neurobiol Dis 135:104576

Lupien SJ, McEwen BS, Gunnar MR, Heim C (2009) Effects of stress throughout the lifespan on the brain, behaviour and cognition. Nat Rev Neurosci 10:434–445

Ma Q, Xing C, Long W, Wang HY, Liu Q, Wang R-F (2019a) Impact of microbiota on central nervous system and neurological diseases: the gut-brain axis. J Neuroinflammation 16:1–14

Ma X, Long H, Wang L, Wang R-F (2019b) Impact of microbiota on central nervous system and neurological diseases: the gut brain axis. J Neuroinflammation 16:53

Macpherson AJ, Geuking MB, McCoy KD (2005) Immune responses that adapt the intestinal mucosa to commensal intestinal bacteria. Immunology 115:153–162

Mangalam A et al (2017) Human gut-derived commensal bacteria suppress CNS inflammatory and demyelinating disease. Cell Rep 20(6):1269–1277

Maynard CL, Elson CO, Hatton RD, Weaver CT (2012) Reciprocal interactions of the intestinal microbiota and immune system. Nature 489:231–241. https://doi.org/10.1038/nature11551

McCusker RH, Kelley KW (2013) Immune-neural connections: how the immune system's response to infectious agents influences behavior. J Exp Biol 216:84–98. https://doi.org/10.1242/jeb.073411

Milani C et al (2017) The first microbial colonizers of the human gut: composition, activities, and health implications of the infant gut microbiota. Microbiol Mol Biol Rev 81:e00036-17

Moradi K, Ashraf-Ganjouei A, Tavolinejad H, Bagheri S, Akhondzadeh S (2021) The interplay between gut microbiota and autism spectrum disorders: a focus on immunological pathways. Prog Neuro-Psychopharmacol Biol Psychiatry 6:110091. https://doi.org/10.1016/j.pnpbp.2020.110091

Mowat AM (2018) To respond or not to respond—a personal perspective of intestinal tolerance. Nat Rev Immunol 18:405–415

Obrenovich MEM (2018) Leaky gut, leaky brain? Microorganisms 6:107. https://doi.org/10.3390/microorganisms6040107

Omenetti S et al (2019) The intestine harbors functionally distinct homeostatic tissue resident and inflammatory Th17 cells. Immunity 51:77–89

Ott JA, Castro CD, Deiss TC, Ohta Y, Flajnik MF, Criscitiello MF (2018) Somatic hypermutation of T cell receptor alpha chain contributes to selection in nurse shark thymus. eLife 7:e28477. https://doi.org/10.7554/eLife.28477

Pitkänen A, Lukasiuk K, Dudek FE, Staley KJ (2015) Epileptogenesis. Cold Spring Harb Perspect Med 5:a022822

Pösel C, Möller K, Boltze J, Wagner DC, Weise G (2016) Isolation and flow cytometric analysis of immune cells from the ischemic mouse brain. J Vis Exp 108:53658. https://doi.org/10.3791/53658

Radjavi A, Smirnov I, Kipnis J (2014) Brain antigen-reactive CD4+ T cells are sufficient to support learning behavior in mice with limited T cell repertoire. Brain Behav Immun 35:58–63. https://doi.org/10.1016/j.bbi.2013.08.013

Rakoff-Nahoum S, Paglino J, Eslami-Varzaneh F, Edberg S, Medzhitov R (2004) Recognition of commensal microflora by toll-like receptors is required for intestinal homeostasis. Cell 118:229–241

Ramanan D, Tang MS, Bowcutt R, Loke P, Cadwell K (2014) Bacterial sensor Nod2 prevents inflammation of the small intestine by restricting the expansion of the commensal Bacteroides vulgatus. Immunity 41:311–324

Rana A, Musto AE (2018) The role of inflammation in the development of epilepsy. J Neuroinflammation 15:144

Rinninella E et al (2019) What is the healthy gut microbiota composition? A changing ecosystem across age, environment, diet, and diseases. Microorganisms 7:14

Roberfroid MB, Bornet F, Bouley C, Cummings JH (1995) Colonic microflora: nutrition and health. Summary and conclusions of an International Life Sciences Institute (ILSI) [Europe] workshop held in Barcelona, Spain. Nutr Rev 53:127–130

Rolls A, Pang WW, Ibarra I, Colas D, Bonnavion P, Korin B, Heller HC, Weissman IL, de Lecea L (2015) Sleep disruption impairs haematopoietic stem cell transplantation in mice. Nat Commun 6:8516

Rosser EC, Mauri C (2015) Regulatory B cells: origin, phenotype, and function. Immunity 42:607–612. https://doi.org/10.1016/j.immuni.2015.04.005

Rothhammer V et al (2016) Type I interferons and microbial metabolites of tryptophan modulate astrocyte activity and central nervous system inflammation via the aryl hydrocarbon receptor. Nat Med 22(6):586–597

Rothhammer V, Borucki DM, Tjon EC, Takenaka MC, Chao CC, ArduraFabregat A et al (2018) Microglial control of astrocytes in response to microbial metabolites. Nature 557:724–728. https://doi.org/10.1038/s41586-018-0119-x

Rothschild D et al (2018) Environment dominates over host genetics in shaping human gut microbiota. Nature 555:210–215

Roy S, Trinchieri G (2017) Microbiota: a key orchestrator of cancer therapy. Nat Rev Cancer 17:271–285. https://doi.org/10.1038/nrc.2017.13

Sakaguchi S, Yamaguchi T, Nomura T, Ono M (2008) Regulatory T cells and immune tolerance. Cell 133:775–787. https://doi.org/10.1016/j.cell.2008.05.009

Salter MW, Stevens B (2017) Microglia emerge as central players in brain disease. Nat Med 23(9):1018–1027

Sampson TR, Debelius JW, Thron T et al (2015) Gut microbiota regulate motor deficits and neuroinflammation in a model of Parkinson's disease. Cell 167:1469–1480

Sanz P, Garcia-Gimeno MA (2020) Reactive glia inflammatory signaling pathways and epilepsy. Int J Mol Sci 21(11):4096. https://doi.org/10.3390/ijms21114096

Scharfman HE (2007) The neurobiology of epilepsy. Curr Neurol Neurosci Rep 7:348–354

Schulthess J et al (2019) The short chain fatty acid butyrate imprints an antimicrobial program in macrophages. Immunity 50:432–445

Shan M et al (2013) Mucus enhances gut homeostasis and oral tolerance by delivering immunoregulatory signals. Science 342:447–453

Shreiner AB, Kao JY, Young VB (2015) The gut microbiome in health and in disease. Curr Opin Gastroenterol 31:69–75

Smith PA (2015) The tantalizing links between gut microbes and the brain. Nature 526:312–314. https://doi.org/10.1038/526312a

Smith K, McCoy KD, Macpherson AJ (2007) Use of axenic animals in studying the adaptation of mammals to their commensal intestinal microbiota. Semin Immunol 19:59–69. https://doi.org/10.1016/j.smim.2006.10.002

Steriade C, Britton J, Dale RC, Gadoth A, Irani SR, Linnoila J et al (2020) Acute symptomatic seizures secondary to autoimmune encephalitis and autoimmune associated epilepsy: conceptual definitions. Epilepsia 61:1341–1351. https://doi.org/10.1111/epi.16571

Sternberg EM (2006) Neural regulation of innate immunity: a coordinated nonspecific host response to pathogens. Nat Rev Immunol 6:318–328

Strandwitz P (2018) Neurotransmitter modulation by the gut microbiota. Brain Res 1693:128–133

Tatum WO (2012) Mesial temporal lobe epilepsy. J Clin Neurophysiol 29:356–365
Thion MS, Low D, Silvin A, Chen J, Grisel P, Schulte-Schrepping J et al (2018) Microbiome influences prenatal and adult microglia in a sex-specific manner. Cell 172:500–516.e516. https://doi.org/10.1016/j.cell.2017.11.042
Tlaskalova-Hogenova H, Tuckova L, Mestecky J, Kolinska J, Rossmann P, Stepankova R et al (2005) Interaction of mucosal microbiota with the innate immune system. Scand J Immunol 62 Suppl 1:106–113. https://doi.org/10.1111/j.1365-3083.2005.01618.x
Tong X et al (2018) Structural alteration of gut microbiota during the amelioration of human type 2 diabetes with hyperlipidemia by metformin and a traditional Chinese herbal formula: a multicenter, randomized, open label clinical trial. mBio 9:e02392-17
Tracey KJ (2002) The inflammatory reflex. Nature 420:853–859
Tremlett H, Bauer KC, Appel-Cresswell S, Finlay BB, Waubant E (2017) The gut microbiome in human neurological disease: a review. Ann Neurol 81:369–382. https://doi.org/10.1002/ana.24901
Turnbaugh PJ et al (2006) An obesity-associated gut microbiome with increased capacity for energy harvest. Nature 444:1027–1031
Ursell LK et al (2014) The intestinal metabolome: an intersection between microbiota and host. Gastroenterology 146:1470–1476
Vojdani A (2014) A potential link between environmental triggers and autoimmunity. Autoimmune Dis 2014:437231
Wang HX, Wang YP (2016) Gut microbiota-brain axis. Chin Med J Engl 129:2373–2380
Wang H, Liao H, Ochani M, Justiniani M, Lin X, Yang L, Al-Abed Y, Wang H, Metz C, Miller EJ et al (2004) Cholinergic agonists inhibit HMGB1 release and improve survival in experimental sepsis. Nat Med 10:1216–1221
Wang Y et al (2014) An intestinal commensal symbiosis factor controls neuroinflammation via TLR2-mediated CD39 signalling. Nat Commun 5:4432
Welcome MO (2019) Gut microbiota disorder, gut epithelial and blood-brain barrier dysfunctions in etiopathogenesis of dementia: molecular mechanisms and signaling pathways. NeuroMolecular Med 21:205–226. https://doi.org/10.1007/s12017-019-08547-5
Wesemann DR et al (2013) Microbial colonization influences early B-lineage development in the gut lamina propria. Nature 501:112–115
Wong ML et al (2016) Inflammasome signaling affects anxiety- and depressive-like behavior and gut microbiome composition. Mol Psychiatry 21(6):797–805
World Health Organization (2017) 10 facts about neurocysticercosis. https://www.who.int/features/factfiles/neurocysticercosis/en/. Accessed 9 Dec 2020
Wu K et al (2020) Gut microbial metabolite trimethylamine N-oxide aggravates GVHD by inducing M1 macrophage polarization in mice. Blood 136(4):501–515. https://doi.org/10.1182/blood.2019003990
Yamanaka G, Morichi S, Takamatsu T, Watanabe Y, Suzuki S, Ishida Y et al (2021) Links between immune cells from the periphery and the brain in the pathogenesis of epilepsy: a narrative review. Int J Mol Sci 22:4395. https://doi.org/10.3390/ijms22094395
Yang Y et al (2019) Recent advances in the mechanisms of NLRP3 inflammasome activation and its inhibitors. Cell Death Dis 10(2):128
Yang L, Zhou Y, Jia H, Qi Y, Tu S, Shao A et al (2020) Affective immunology: the crosstalk between microglia and astrocytes plays key role? Front Immunol 11:1818. https://doi.org/10.3389/fimmu.2020.01818
Yizhar O, Fenno LE, Prigge M, Schneider F, Davidson TJ, O'Shea DJ et al (2011) Neocortical excitation/inhibition balance in information processing and social dysfunction. Nature 477:171–178. https://doi.org/10.1038/nature10360
Yue Q, Cai M, Xiao B, Zhan Q, Zeng C (2021) The microbiota-gut-brain axis and epilepsy. Cell Mol Neurobiol 42(2):439–453. https://doi.org/10.1007/s10571-021-01130-2
Ziv Y, Ron N, Butovsky O, Landa G, Sudai E, Greenberg N et al (2006) Immune cells contribute to the maintenance of neurogenesis and spatial learning abilities in adulthood. Nat Neurosci 9:268–275. https://doi.org/10.1038/nn1629

Exploring the Interplay Between Gut Microflora and Parkinson's

15

M. Swathi, Avipsa Hazra, A. S. Dhanu, and Kanthesh M. Basalingappa

Abstract

One of the crucial elements in sustaining host health and disease is the gut microbiota. The gut microbiota is now recognized as one of the critical elements in controlling the interactions between the gut and the brain due to growing advancement of omics techniques. Age-related changes in the gut microbiome contribute to the ageing process of the brain and that further leads to different neurodegenerative diseases. The idea that the gut can affect central nervous system's (CNS) function and vice versa is becoming more widely accepted, giving rise to the microbiota-gut-brain axis as a communication pathway. A healthy body depends on maintaining a balanced gut microbiota, but dysbiosis can lead to several chronic diseases. Identifying biomarkers for chronic conditions associated with ageing will be made easier by understanding the underlying mechanisms of gut microbiota alterations and dysbiosis. Dysbiosis gut microbiota is linked to increased intestinal permeability, worsened neuroinflammation, oxidative stress and reduced neurotransmitter production, all of which are known to be crucial to the occurrence and progression of Parkinson's disease (PD). This has made the gut microbiota a promising diagnostic and therapeutic target for PD by inculcating treatments using prebiotics, probiotics and antibiotics.

This chapter summarizes gut microbiota profiles and functions related to Parkinson's disease, as well as potential roles and mechanisms of gut microbiota associated with brain ageing and its role in mitigating the disease. Understanding the underlying functions and mechanisms of the gut microbiota associated with PD will aid in reinterpreting the disease's pathogenesis and illuminating fresh treatment options.

M. Swathi · A. Hazra · A. S. Dhanu · K. M. Basalingappa (✉)
Division of Molecular Biology, School of Life Sciences, JSS Academy of Higher Education and Research, Mysore, Karnataka, India
e-mail: kantheshmb@jssuni.edu.in

S. Pathak, A. Banerjee (eds.), *Gut Microbiome and Brain Ageing*,
https://doi.org/10.1007/978-981-99-8803-7_15

Keywords

Microbiota-gut-brain axis · Neurodegenerative diseases · Dysbiosis · Short-chain fatty acids · Neuroinflammation · Parkinson's disease · Probiotics · Prebiotics

Abbreviations

AD Alzheimer's disease
ASD Autism spectrum disorder
ATP Adenosine triphosphate
BAMS Border-associated macrophages
BBB Blood-brain barrier
CD Cluster of differentiation
CML N6-carboxymethyl lysine
CNS Central nervous system
DA Dopamine
DAMPS Damage-associated molecular patterns
DC Dendritic cells
DNA Deoxyribonucleic acid
EC Epithelial cells
EDTA Ethylenediaminetetraacetic acid
FMT Faecal microbiota transplantation
GABA Gamma-aminobutyric acid
GIT Gastrointestinal tract
GPCR G-protein coupled receptor
HDAC Histone deacetylation
HE Hepatic encephalopathy
HPA Hypothalamic pituitary adrenal
ILs Interleukins
ISAPP International Scientific Association for Probiotics and Prebiotics
L-DOPA L-3,4-dihydroxyphenylalanine
LPS Lipopolysaccharide
MHC Major histocompatibility complex
MIA Maternal immune activation
MLN Mesenteric lymph nodes
MPTP 1-Methyl-4-phenyl-1,2,3,6-tetrahydropyridine
NFκ-B Nuclear factor kappa-B
NLRs Nod-like receptors
PAMPS Pathogen-associated molecular patterns
PBMCs Peripheral blood mononuclear cells
PD Parkinson's disease
QOL Quality of life
RNA Ribonucleic acid

ROS Reactive oxygen species
RSV Respiratory syncytial virus
SCFA Short-chain fatty acids
SFB Segmented filamentous bacteria
SIg Secretory immunoglobulin
TD T-cell dependent
TI T-cell independent
TLRs Toll-like receptors
TMA Trimethylamine
TNFs Tumour necrosis factor
UPDRS Unified Parkinson's disease rating scale

15.1 Introduction

Every human being at their infantile stage develops a unique environment of gut microorganisms depending upon the type of their delivery, the gestational period, the breast milk quality, the weaning period for it and the use of antibiotics by the mother when she was pregnant. This environment of microorganisms, known as the gut microbiome, helps in maintaining the nutrient metabolism of an individual by helping in the digestion of amino acids, lipids bile acids, short-chain fatty acids and vitamins; in providing structure to the mucosal barriers which prevents external bacterial attack; and in activity against pathogens by preventing their growth by either absorbing all the available nutrients or releasing bacteriocins. Though the gut's composition usually remains constant throughout the life of an individual, sometimes due to the body mass index, enterotypes, frequency of exercise and lifestyle habits there may be some composition alterations. Gut microbiome owing to differences in individual's internal and external aspects is specifically characterized for the individual (Rinninella et al. 2019). The GI tract is composed of about 100 trillion microorganisms, including yeast, viruses and bacteria of different genus, family, order and phyla, among which the colon is the most populated region with bacterial density ranging between 10^{11} and 10^{12} microorganisms per millimetre (Thursby and Juge 2017; Ley et al. 2006; Laterza et al. 2016).

Culturing the microorganisms of the gut microbiota has always been difficult; however recent technologies have given a chance to scientists to determine the constituents of the gut by analysing the DNA and RNA present in the faeces (Moore and Holdeman 1974). It is formed on the sequencing of 16S ribosomal RNA gene sequence, followed by the use of metagenomics to identify the microbial strains (Poretsky et al. 2014; Mizrahi-Man et al. 2013). Till date, about 160 species of bacteria have been detected which includes phyla like Bacteroidetes, Firmicutes, Verrucomicrobia, Actinobacteria, Proteobacteria, Fusobacteria and Actinobacteria (Laterza et al. 2016; Arumugam et al. 2011). The Firmicutes includes genera like *Clostridium*, *Lactobacillus*, *Bacillus*, *Ruminococcus* and *Enterococcus*, with *Clostridium* occupying about 95% of the Firmicutes phyla. In the Bacteroidetes phyla, *Bacteroides* and *Prevotella* are the most dominant genera. In the gut microbiome, almost 90% of the population consists of Bacteroidetes and Firmicutes. Out of all the phyla, Actinobacteria, consisting of the *Bifidobacterium* genus, are the least in number (Arumugam et al. 2011). With the advancement of technologies, to detect more of the constitution of the gut microbiome, whole-genome shotgun metagenomics in combination with MetaHit and Human Genome Project is being used to prepare a complete list of the microorganisms present in the GI tract (Hugon et al. 2015; Li et al. 2014).

These groups have a big impact on a lot of physiological processes, like how the immune system and metabolism work. The gut modulates several brain functions by delivering bacteria-derived metabolites, hormones and neuroactive substances to the central nervous system (CNS) via the vagus nerve, enteric nervous- and circulatory system and immune system (Long-Smith et al. 2020). The hypothalamic-pituitary-adrenal axis and the autonomic nervous system of the central nervous system, on the other hand, regulate gut function (Mart'yanov et al. 2021). The host's age and genetics, as well as external factors such as dietary changes and antibiotic overuse, all influence the composition and activity of the microbiota in the GI tract (David et al. 2014). As a result, even in healthy environments, each microbiome is unique and varies greatly between individuals. Dysbiosis, defined as a change in the bacterial population of the gut, has been linked to a variety of host diseases, including CNS disorders (Sudo et al. 2004). The blood-brain barrier (BBB), one that disconnects the CNS from the rest of the body, is critical for preventing pathogens and immune responses that could harm neurons (Daneman and Prat 2015). By carefully controlling the influx and efflux of molecules, ions and cells, this function is accomplished (Engelhardt and Liebner 2014). In addition to being crucial for CNS development and homeostasis, metabolites derived from bacteria are also able to cross blood-brain barrier (BBB) and contribute to the onset and progression of CNS diseases (Heijtz et al. 2011). The term "gut-brain axis" describes the two-way interaction between the central nervous system and the intestinal microbiome. Importantly, the BBB becomes increasingly less effective with age and even more so in the context of neurodegenerative diseases, allowing neurotoxic products from the gut to enter the brain where they trigger inflammatory and immune reactions; the brain endothelium gradually loses function as people age, which is connected to abnormal BBB alterations (Cai et al. 2017).

Reduced intestinal role, waning immune function and major changes in the GI tract's microbiota population are just a few of the physiological mechanisms that begin to degrade as we grow older. Many of these age-related factors are caused by dysbiosis. Individuals' microbiomes change with age, with lowered diversity, enhanced richness of bacteria linked to pro-inflammatory impacts, such as Clostridia and Enterobacteriaceae, and lowered numbers of the useful genera *Lactobacillus* and *Bifidobacterium* (O'Toole and Jeffery 2015). Diabetes, cancer, cardiovascular disease, insulin resistance and neuro-degenerative diseases such as Alzheimer's and Parkinson's are believed to be connected to age-related variations in the microbial community (Gérard and Vidal 2019).

15.2 Potential Ways Through Which the Gut Microbiome Affects the Activity of Brain

There is increasing realization and experiments that the gut microbiome plays a significant role in the interactions that occur between key nutrients and brain activity. According to a scientific theory on the ageing gut brain research, the gut microbiota and microbial metabolic products influence the gut-brain axis, hence the actions of a person, resulting in several behavioural and psychological symptoms popularly seen in the elderly struggle from a neurological condition such as Alzheimer's, Parkinson's and dementia (Lobo et al. 2000).

The physiology and operation of the gastrointestinal system change with age, influencing the microbiota that lives there (O'Toole 2012). When compared to "younger adults," frailty correlates with modifications to the community of microorganisms present in the gut. A diverse microbiota because of a diverse diet may prevent poor health and frailty (Fried et al. 2001). Correlating changes in diet composition and diversity results in a significant shift in the gut bacterial profile (Caracciolo et al. 2014). The age-related changes seen in the gut microbiome could indeed contribute to the development and progression of inflammations associated with ageing by boosting the generation of proinflammatory mediators or decreasing the production of anti-inflammatory mediators, shifting the balance towards inflammatory response (O'Toole and Jeffery 2015). A study comparing the microbial characteristics of older population to nutrient dense elderly found that bacterial species associated with inflammation are much more prevalent in the former (O'Toole 2012). A study on Alzheimer's disease (AD) found that an increase in neurotoxic proteins in brain and propagating pro-inflammatory cytokines is associated with a lower abundant supply of the butyrate-producing anti-inflammatory type of bacteria Eubacterium rectale and significantly greater rates of pro-inflammatory *Escherichia coli/Shigella*. The neuro-inflammation linked to mental decline leads to a better understanding of the connection among both neuro-inflammation and gut microbiota composition (Cattaneo et al. 2017).

The following are the mechanisms by which gut microbiota action influences the brain:

15.3 Immunological Mechanisms

15.3.1 Nonspecific Immune System (Innate)

The human system is composed of diverse innate and adaptive cells which play the major role in providing response to various events encountered by the human body. The network of these two cells acts as a regulator of host homeostasis. The entire microbiome population is commensal to the host, and it is of vital importance in the growth and development of the innate immune system (Duerkop et al. 2009). The structural microbial cell wall components, such as lipopolysaccharides, stimulate the innate immune system to generate cytokinins and interleukins, such like cytokines interleukin-1beta (IL-1β) and tumour necrosis factor alpha (TNFα) caused by bacterial cell wall lipopolysaccharides (LPS); this intestinal activation commences at the intestinal mucosal membrane and impacts every part of the body. The gut microbiome communicates with the hypothalamic-pituitary-adrenal (HPA) axis to mould an individual's normal sleep cycle (Heumann et al. 1994).

Toll-like receptors (TLRs) are innate immune pattern recognition receptors that are responsible for the normal growth and development of the intestinal mucosal immune system and play a role in the recognition of microbial pathogens associated with infants, such as group B streptococcus, respiratory syncytial virus (RSV) and Listeria monocytogenes. The stimulation of these toll-like receptors in alcoholics' peripheral blood mononuclear cells (PBMCs) by LPS and peptidoglycans was linked to elevated messenger RNA and plasma levels of IL-8, IL-1 and IL-18. The levels of IL-8 and IL-1 were found to be positively related to alcohol consumption. They divided their chronic alcoholic population into those with high and normal intestinal permeability using Cr51-EDTA (a marker when the radioisotope of Chromium is complexed with EDTA, a chelating agent) as a probe of intestinal permeability (Leclercq et al. 2014). The group with higher permeability had a greater incidence of depression, anxiety and booze hankering than the group with lower permeability, with a specific pattern of variations in the gut microbes population, including a decrease in colonial expansion with bacteria known to be have anti-inflammatory impacts (*Bifidobacterium* species and *Faecalibacterium prausnitzii*). Overall, the gut microbiome stimulates a chronic state of reduced activation of the innate immune system of human beings, which is influenced by the circadian layout of adrenal cortical activity. Because of increased intestinal permeability, such altered exposure to building structures of the microbiota composition disrupts normal neuroendocrine regulation, resulting in abnormal CNS functions and other disorders (Maes et al. 2007).

15.3.2 Specific Immune System (Adaptive)

The specific immune system is designed to react to microbiota with antibodies or antigen-specific cell-mediated immune responses that can cause CNS dysfunction via auto-immune responses caused by bacterial and self-protein molecular imitation (Berer and Krishnamoorthy 2012). The mucosal membrane of the gastrointestinal tract (GIT) is made up of closely linked cells called intestinal epithelial cells (EC) that are covered by a mucin-hydrated gel layer that contains antibacterial substances like secretory IgA (SIgA) and anti-bacterial peptides. The intestinal mucosal boundary means preventing multiple intestinal microbes from invading mucus layer. Bacteroides species (Bacteroides fragilis and Bacteroides thetaiotaomicron) can enter the mucus layer and colonize epithelial cells in the tiny bowels as well as colonic crypts, respectively (Peterson et al. 2007). Dendritic cells may present microorganisms that penetrate the mucus layer to immune cells (DC). DCs can transport a minimal quantity of commensal bacteria to the mesenteric lymph nodes (MLN) and remain there for 28–30 days. The migration of these dendritic cells from the intestine to the mesenteric lymph node is confined to lymphatic vessels, and DCs typically could indeed cross the very first deflating lymph node they approach, preventing commensal bacteria from entering the systemic immune circulation, required to ensure that these commensal bacteria can generate an efficient mucosal immune response. Furthermore, parasitized epithelial cells can be readily taken up by macrophages, allowing the antigen to be delivered and adaptive immunity to be activated (Macpherson and Uhr 2004). B cells, for example, play an important role in adaptive humoral immunity by generating a significant amount of sIgA antibodies in response to commensal bacteria. Initially, the microbes are reference immunogens that can stimulate the production of antigen-specific IgA. These sIgAs coat commensal bacteria and then prevent their growth, limiting their ability to enter the mucosal barrier, assisting the host to shield against pathogenic organisms and contributing to the immunological balance between commensal bacteria as well as host (Peterson et al. 2007).

15.4 Biochemical Mechanisms

Short-Chain Fatty Acids (SCFAs) As a result of bacterial fermentation of indigestible carbohydrates, the normal colon produces large amounts of volatile fatty acids, which are substances with carbon chain length of between two and four carbon molecules (acetate, propionate and butyrate). Increased SCFA production has indeed been associated to the health advantages of consuming a great deal of fibre (Macfarlane and Macfarlane 2011). For example, butyric acid fulfils 70% of an energy requirement of colonic epithelium (De Preter et al. 2011) and has instantaneous anti-inflammatory effects through blocking nuclear factor kappa-B (NFkB) activation. In addition to inhibiting NFkB, propionic acid also activates the peroxisome proliferator-activated receptor gamma, which may enhance insulin sensitivity (Segain et al. 2000). Additionally, SCFA affects at least two molecular signalling pathways histone deacetylation (HDAC) and G-protein-coupled receptors (GPCRs), which have broad regulatory effects throughout the body. SCFA are organic GPCR activators and inhibitors of histone deacetylases (Tan et al. 2014). The primary pathway for carbohydrate digestion and absorption in enterocytes was impaired in biopsies of autistic children with GI complaints, showing a deficit of genes encoding disaccharidases and hexose transport enzymes (Williams et al. 2011). This finding suggests that bacterial dysbiosis results from an underlying impairment of digestion and absorption. A unique genus of aerobic gram-negative rods called *Sutterella* was found in the ileal biopsies of autistic children with GI complaints in a later report, but not in any GI complaints in children who were not autistic (Williams et al. 2012). One of the main areas of research in autism today is the role of the gut microbiome and its metabolites in autism spectrum disorder (ASD) but the results do not yet allow for a single coherent theory on which to base therapeutic choices (Mulle et al. 2013; Stoner et al. 2014). The roots of autism may be found in utero, according to a recent article in the *New England Journal of Medicine* that describes structural brain abnormalities in autistic children that started during prenatal brain development (Hsiao et al. 2013). Perhaps the maternal gestational microbiome should receive more attention. Maternal immune activation (MIA) of pregnant mice can cause behavioural changes in their offspring that are like ASD. Bacteroides fragilis, a single probiotic, improves communication and stereotypes while reducing excessive gut permeability and gut microbial composition (Hsiao et al. 2013).

D-Lactic Acid D-Lactate, which is a byproduct of the microbial fermentation of carbohydrates, is typically produced in excess when a small bowel resection enables the delivery of a significant amount of carbohydrates to the colon. Increased intestinal permeability and bacterial translocation across the intestinal mucosal barrier may also result in an increase in D-lactate in plasma following other types of abdominal surgery (Qiao et al. 2009). Prebiotics and probiotics can reduce the amount of D-lactic acid that is produced in the gut, but they must be carefully chosen (Munakata et al. 2010). Some *Lactobacillus* species can produce D-lactate. Beta-glucan (found in oats and barley) in high doses can also increase intestinal permeability (Ewaschuk et al. 2012). A combination of *Bifidobacterium breve* Yakult and *Lactobacillus casei* Shirota as probiotics and galacto-oligosaccharide as a prebiotic successfully prevented repeated neurotoxicity in a man with recurrent D-lactic acidosis caused by short bowel syndrome who had become resistant to antibiotics and dietary restriction. By inhibiting the growth of D-lactate-producing bacteria and promoting intestinal motility, the combination, known as a symbiotic, enabled reduction in colonic absorption of D-lactate (Takahashi et al. 2013).

Ammonia A well-known neurotoxin called ammonia is created in intestinal tract from urea by bacterial ureases. The liver absorbs ammonia from the gut and uses it in the urea cycle. Hepatic encephalopathy's (HE) pathogenesis is aided by ammonia. Ammonia alters the blood-brain barrier's ability to function, which results in abnormal neurotransmitters being produced in addition to direct neurotoxic injury (Qureshi et al. 2014). Cirrhotic patients with or without HE has a different gut microbiome composition, which has been linked to cognitive dysfunction (Bajaj et al. 2012). Amounts of urease-producing microbes are positively associated with cognitive problems in cirrhotic patients (Zhang et al. 2013). When the non-absorbed antibacterial drug rifaximin is added to standard lactulose treatment, the rate of total inversion of HE increases from 51% to 76% and the fatality rate decreases from 49.1% to 23.8%, highlighting the importance of gut microflora in pathogenesis (Sharma et al. 2013). It has also been shown that using symbiotic to change the gut microbiome can help patients with cirrhosis who suffer from cognitive dysfunction.

15.5 Neuroendocrine Mechanisms

Bacterial hormone and neurotransmitter syntheses and reactions are both possible. *Lactobacillus* species produce acetylcholine and gamma-aminobutyrate (GABA); *Bifidobacterium* species produce GABA; *Escherichia* species produce norepinephrine, serotonin and dopamine; and *Bacillus* species produce norepinephrine and dopamine (Cryan and Dinan 2012). Human hormones and neurotransmitters influence the growth and virulence of these organisms. Host stressors may cause massive changes in gut microbes' composition as well as specific effects on potential pathogens (Lyte 2004).

Immune modulation is a function of lactic acid-producing bacteria that may affect the overall composition of the gut microbiome. Humans appear to react similarly to, if not entirely differently than, laboratory animals when exposed to psychosocial stress. Because gut microbiota alters stress reactions in animal models, probiotics have been used in several human clinical trials to investigate their effects on stress reactivity and mood (Messaoudi et al. 2011). Tillisch et al. studied a group of healthy women who were given a fermented dairy product containing *Bifidobacterium animalis* subsp. *lactis*, *Streptococcus thermophilus*, *Lactobacillus bulgaricus* and *Lactococcus lactis* subsp. *lactis* for 4 weeks. Patients attended functional magnetic resonance imaging (MRI) both prior to and following the test to measure resting brain activity and reaction to an emotional reactivity test. A control group was given the same dairy product minus the probiotics. Consumption of a probiotic beverage has been linked to changes in midbrain interconnection and a decrease in work response in areas of the brain responsible for processing emotions and visceral sensations (Tillisch et al. 2013). Thus, much research on the neuroendocrine impacts of gut microbes takes a pharmacodynamic approach rather than an ecological one: a particular intervention is implemented, and specific outcomes are tested. Unlike pharmaceuticals, however, gut microbes arise in a network of interconnected, highly organized living groups. A probiotic accomplishes a lot more than introducing an unfamiliar species of bacteria that may or may not be capable of identifying a market segment in the community.

It might alter community function, but it might also change community structure in unexpected ways. According to human studies, people's gut microbial composition varies significantly depending on their age, genetic background, physiological state, communication with other microbes, environment and diet (Filippo et al. 2010). Furthermore, the ileum microbiota is dominated by various bacterial phyla and is both easier and a little less steady than that of the colonic fecal matter microflora (Booijink et al. 2010). Because of this complication, clinical and laboratory studies on the health effects of modifying the microbiome must be conducted with specific patient characteristics in mind (Galland 2006) (Fig. 15.1).

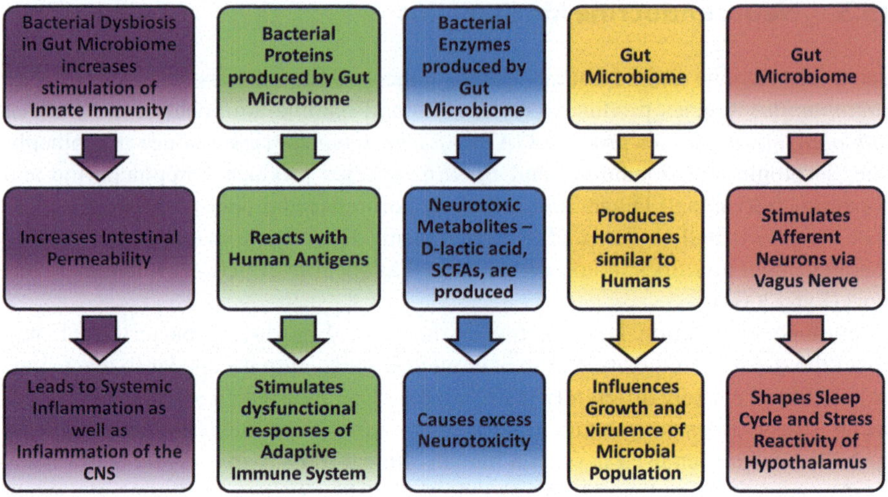

Fig. 15.1 Mechanisms of gut microbiome affecting brain functionality

15.6 Parkinson's Disease and Its Association with the Gut Microbiota

With a varied diversity in the gut microbiota, it has become an emerging point of research to detect different diseases, and one of the intriguing discoveries is the role of the gut-brain axis in neurodegenerative diseases. Among these neurodegenerative diseases, Parkinson's disease has been found to be one of the most severely affected by gut microbiome dysbiosis (Keshavarzian et al. 2020). Affecting over 4 million individuals in the year 2005, and suspected to affect up to 8.7 individuals by 2030, Parkinson's disease is the second most common neurodegenerative disease, and it is constantly worsening without any permanent cure and affecting multiple systems of the body (Dorsey et al. 2018; Poewe et al. 2017; Zhu et al. 2022). It is usually seen to be occurring amongst the middle aged to elderly population (Lajoie et al. 2021).

Parkinson's disease is specified by abnormalities in the motor functioning, which is characterized by tremors even while resting, bradykinesia, stiffness, rigidity, no integrity in the body structure and issues with walking, and also by non-motor signs like anosmia and hyposmia, fast blinking of the eye, troubles with sleeping, anxiety or depression, not proper expression of the cognitive functions, dysautonomia, impairment in the autonomic and enteric nervous system and disorders pertaining to the gastro-intestinal tract, like constipation (Kalia and Lang 2015; Goldman and Postuma 2014). One of the major symptoms of Parkinson's disease pathogenesis is the decrease in the number of dopaminergic neurons located in the substantia nigra pars compacta, as well as the formation of Lewy bodies, which are formed by the aggregation of α-synuclein (Abeliovich and Gitler 2016). The α-synuclein protein found in Lewy bodies is toxic to the body and plays a key role in the onset of

Parkinson's disease. The development of the disease involves multiple mechanisms like genetic aspects, aggregation of α-synuclein protein and difficulty in its removal, improper functioning of mitochondria, ferroptosis, inflammation in the neural pathways and oxidative stress (Wang et al. 2022a; Yildirim et al. 2022; Segura-Aguilar et al. 2014).

With an increase in the detection of symptoms, disorders of the gastrointestinal tract including constant constipation for over a decade and gastroparesis seems to be one of the earliest biomarkers to detect Parkinson's (Wallen et al. 2022; Qian et al. 2020). The involvement of the infections of the GI tract made gut-brain axis a point of interest for researchers who found that the axis shares a two-directional bridge for the connection between enteric nervous system and the brain and is also responsible for the distribution of α-synuclein due to the caudo-rostral gradient (Bullich et al. 2019). The involvement of the gastrointestinal tract has also led to the research of gut dysbiosis in relation to pathogenesis of Parkinson's disease, and it is found that the development of PD can be aggravated by the influence of the microorganisms in the gut on the permeability of the intestine and inflammation of the colon, which, in turn raises neuronal inflammation, increases the rate of deposition of α-synuclein, increases radicals leading to oxidative stress and reduces the production of neurotransmitter (Zhu et al. 2022). Owing to evidence of dysbiosis in the gut microbiome in the Parkinson's disease patient, multiple scientists have researched in this field, and even studies on animal models have been carried out to detect the origin of Parkinson's from the gut and how the microorganisms affect the pathogenesis (Sampson et al. 2016).

Though there has been no conclusion that could be reached, there have been multiple discoveries made on the way that show us the intricate involvement of multiple gut microorganisms in the pathogenesis of PD. The meta-analysis of the available 16S microbiome datasets revealed that there was a change in the gut microbiome across the different patients affected by Parkinson's disease. There was an overexpression of the *Lactobacillus*, *Akkermansia* and *Bifidobacterium* genera and Verrucomicrobiaceae family, and the under expression of the Lachnospiraceae family and the *Faecalibacterium* genus (Romano et al. 2021). All the mentioned bacteria are involved in the synthesis of short-chain fatty acids; however the change in their numbers lead to a difference in the final metabolites being produced and a reduction in the concentration of SCFAs, which causes difference in gut permeability and leads to inflammation (Hu et al. 2020). Other studies show that change in the number of Lactobacillus and Enterococcus faecalis are responsible for the change in metabolism of Levodopa, by altering the levels of tyrosine decarboxylase. This tyrosine decarboxylase is known to further disrupt the environment of gut due to gut-brain axis by direct influence or by stimulating deep brain signals (Maini Rekdal et al. 2019; Lubomski et al. 2021). Research has shown that the entry of certain gram-negative bacteria in the gut triggers the immune mechanism in a way which leads to the neuronal degradation and issues in motor functions. Also, curli, which is an amyloidogenic protein, if produced in excess by *Escherichia coli* initiates the aggregation of α-synuclein and increases the rate of neurodegeneration of the brain (Sampson et al. 2020; Chen et al. 2016). Research has also shown that individuals

Table 15.1 Gut microbiome dysbiosis that leads to pathogenesis of Parkinson's disease

Change in gut	Effect on PD
Lactobacillus↑ *Akkermansia*↑ *Bifidobacterium*↑ *Lachnospiraceae*↓ *Faecalibacterium*↓ *Proteus*↑ *Bilophila*↑ *Roseburia*↑	Reduced secretion of short-chain fatty acids leads to increase in gut permeability
Lactobacillus↑ *Enterococcus*↑	Changes level of tyrosine decarboxylase, affecting gut-brain axis
Escherichia coli↑	Increases the production of alpha-synuclein
Helicobacter pylori infection	Degrades motor functions severely
Prevotellaceae↓	Leads to increase in gut permeability and increase in the production of endotoxins by the bacteria
Ralstonia↑ *Proteobacteria*↑ *Enterococcaceae*↑ *Coprococcus*↓ *Roseburia*↓ *Faecalibacterium*↓	Increase in the production of pro-inflammatory cytokine
Desulfovibrionaceae↑ *Bifidobacterium*↑*Bilophila*↑ *Akkermansia*↑ *Lachnospiraceae*↓ *Faecalibacterium*↓	Fermentation of carbohydrates is reduced, butyrate production is lowered, amino acid metabolism increases

↑ indicates increase in number of bacteria
↓ indicates decrease in number of bacteria

who have suffered from an infection by *Helicobacter pylori* are at an added risk of Parkinson's and the motor functions get degraded severely (Romano et al. 2021).

It is being concluded that in a patient with Parkinson's disease, there is a significant change in the composition of gut microbiome which results in change of the metabolic products produced by the gut—there is an error in the β-glucuronate and tryptophan degradation pathways, and neuroprotective compounds like SCFAs, ubiquinones and salicylate, along with neurodegenerative compounds like sphingosine, ceramides and trimethylamine N-oxide, see a definite change in their levels in the patient. This constant relation of gut microbiota and Parkinson's disease establishes that any change in the concentration of the gut microbiome leads to a difference in the symptoms of Parkinson's including the length of the disease, the rate of depletion of motor functions and the development of one or more non-motor characteristics (López-Otín et al. 2013; Kierdorf et al. 2019).

Owing to the difference in gut microbiome in PD patients, the gut microorganism environment is being considered as a biomarker for the diagnosis and evaluation of the progression of Parkinson's disease (Rowland et al. 2018). There has been proof of many other bacteria, the change in whose numbers leads to different symptoms in PD, and they have been described in Table 15.1 (Fig. 15.2).

15.7 Influence of Gut Microbiota Metabolites on Brain Ageing

In recent years, changes in the gut microbiome have been linked to the onset of age-related illnesses such as Alzheimer's and Parkinson's disease, as well as other brain conditions such as anxiety, depression and schizophrenia. For a variety of reasons,

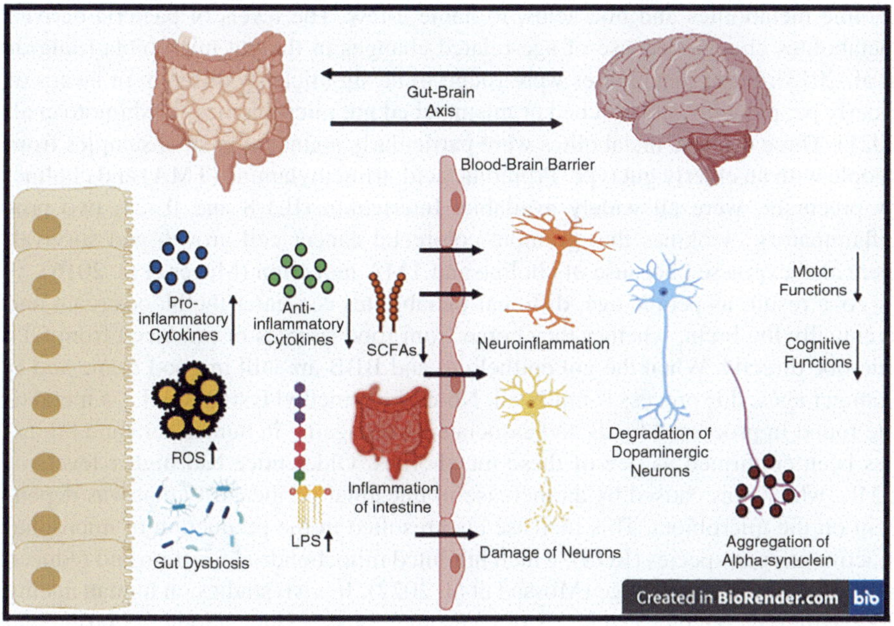

Fig. 15.2 Changes in the gut that affects the brain leading to Parkinson's disorder

the bacterial population of the host's gastrointestinal tract can change. Aside from dietary changes and regular antibiotic use, ageing has a significant impact on the microbiome. Older people with age-related diseases and cognitive impairments have higher levels of metabolites derived from the microbiota when compared to younger, healthy age groups (López-Otín et al. 2013). Choline and trimethylamine are two metabolites that have been identified as having higher concentrations in elderly hosts and are also known risk factors for age-related diseases. These metabolites have been shown to activate innate immunity inside the central nervous system by affecting the improvement and activation status of brain-resident macrophages, even though the underlying mechanisms and pathways remain largely unknown. Non-parenchymal macrophages and parenchymal microglia, which make up the brain's macrophage population, live in the perivascular spaces, meninges and choroid plexus, respectively (Kierdorf et al. 2019).

15.7.1 Gut Metabolites in Ageing

Bacteria in the GIT contribute significantly to the breakdown of indigestible substances by producing a variety of enzymes (Rowland et al. 2018). Gut microbiotas produce a wide range of metabolites in addition to aiding digestion. These bioactive substances have significant effects on the host, including the regulation of metabolic pathways via feedback mechanisms, nutrient absorption and the composition of the microbiota itself. Bacterial metabolites include short-chain fatty acids (SCFAs), choline metabolites and bile acids, to name a few. The levels of bacteria-derived metabolites change because of age-related changes in the gut microbiota (Salazar et al. 2013). Eight metabolites were found to be significantly different in faeces of elderly people with age-matched or mismatched gut microbiomes (Yoshimoto et al. 2021). Three of these metabolites were particularly abundant in stool samples from people with an elderly gut type. Propionic acid, trimethylamine (TMA) and choline, its precursor, were all widely available. Interleukin (IL) 8 and IL-21, two pro-inflammatory cytokines that promote colorectal cancer cell growth and survival, were also expressed because of choline and TMA induction (Mager et al. 2016).

As a result, as people age, different metabolites can enter the bloodstream and eventually the brain, whether they come from food sources or produced from GIT microbe directly. When the gut epithelium and BBB are still in good shape and at younger ages, this process is inhibited. N6-carboxymethyl lysine (CML), a metabolite found in processed foods and associated with ageing in human sera and brains, has been confirmed as one of these metabolites. Older mice had higher levels of CML, which was caused by an increase in intestinal permeability that was dependent on the microbiota. This increase also resulted in the production of microglial reactive oxygen species (ROS), which inhibited mitochondrial function and reduced ATP production and storage (Mossad et al. 2022). In vivo studies on human ageing are conflicting, but elevated levels of some age-associated metabolites contribute to increased gut permeability in vitro. More recent in vivo and ex vivo human studies found no discernible differences in the permeability of the small intestine, colon or entire gut between young and old people (Wilms et al. 2020).

15.7.2 Parenchymal Microglia in Ageing

Microglia are thought to play a role in several CNS disorders during neurodevelopment and neurodegeneration (Sierra et al. 2019). They play crucial roles in the development, homeostasis and several CNS pathologies in the brain. Rapidly reacting to pathogen invasion and neuronal injury, microglia regulate neuronal cell survival, apoptosis, synaptic pruning, synaptogenesis and myelination (Li and Barres 2018).

Through constant motion, protrusion and retraction of the processes, microglia can scour large areas of their microenvironment for damage- and pathogen-associated molecular patterns (DAMPs, PAMPs). Furthermore, cell-cell interactions between microglia and vascular system cells, astrocytes and neurons are influenced by their processes (Colonna and Butovsky 2017). When microglia come into contact with DAMPs or PAMPs as a result of infection, trauma or neurodegenerative pathologies, they undergo morphological change into an amoeboid structure, allowing migration and phagocytosis at the site of injury (Nimmerjahn et al. 2005). MHC II antigens were found to be expressed more frequently because of altered gene expression as well as morphological changes. Furthermore, activated microglia release pro-inflammatory cytokines, amplifying the inflammatory response (Kim and Joh 2006). Increased levels of pro-inflammatory cytokines, such as those from the interleukin (IL) family, such as IL-1 and IL-6 and tumour necrosis factor (TNF) can harm glial cells and neurons as well as the CNS cells they are supposed to protect. As a result, chronically activated microglia or an imbalance in the release of pro- and anti-inflammatory cytokines are thought to play a role in the onset and progression of neurodegenerative diseases (Smith et al. 2012). Recent research has also revealed that the aged brain's microglia release higher baseline levels of pro-inflammatory cytokines such as TNF, IL-1, IL-6 and IL-12b, as well as anti-inflammatory mediators such as TGF-1 and IL-10, which balance one another to maintain a steady state. Similar results have been observed in human brains with Alzheimer's disease (AD) (Miller and Streit 2007).

15.8 Treatment Methods: Altering the Gut Microbiome to Alter Parkinson's Disease

Since the gut microbiome is very intricately related to pathogenesis of Parkinson's disease (PD), it was thought to be a good approach for treating PD. All methods—use of probiotics, prebiotics and antibiotics; performing faecal microbiota transplantation (FMT) and modifying diet; focus on changing the gut microbiome to make it suitable to relieve the symptoms of PD. They focus on increasing the production of neurotransmitters, altering host metabolites and increasing neuroprotective actions of different pathways.

Probiotics: Probiotics is defined as "live microorganisms which when administered in adequate amounts confer a health benefit on the host" by the International Scientific Association for Probiotics and Prebiotics (ISAPP). They know to be extremely good at handling disorders of the gastrointestinal tracts, and can be usually used without the stress of side effects (Gibson et al. 2017; Hill et al. 2014). Most of these tests have been carried out using 1-methyl-4-phenyl-1,2,3,6-tetrahydropyridine (MPTP) and rotenone toxin-induced Parkinson's disease mouse models to check for the effects of the different probiotics on human hosts (Sun et al. 2018; Zhao et al. 2021). Strains like *Bifidobacterium*, *Streptococcus* and *Lactobacilli* are known to reverse the symptoms of Parkinson's to some extent (Georgescu et al. 2016). The constant consumption of milk shows an improvement in bowel movement and prevents the growth for *Staphylococci* in the gut, due to the presence of *Lactobacillus casei* Shirota (Cassani et al. 2011). A *Bacillus* species taken as probiotic has the capacity to convert L-tyrosine into L-DOPA, which can substitute the function of dopamine in patients with Parkinson's. Also, addition of certain bacteria of the *Enterococcus* species in the intestine produces tyrosine decarboxylases, and can change levodopa to dopamine (Surwase and Jadhav 2011).

Also, use of probiotics containing a mixture of *B. animalis lactis*, *L. rhamnosus GG* and *L. acidophilus* acts as a protector to the neurons (Srivastav et al. 2019). The combination of multiple bacteria in the probiotics like *L. acidophilus* and *B. infantis* benefit in treating of bloating and reduction of abdominal pain in the PD patients (Hill et al. 2014). Also, another combination of *L. reuteri*, *L. acidophilus*, *B. bifidum* and *L. fermentum* can decrease issues related to motor functions (Tamtaji et al. 2019). On the other hand, supplementing *L. plantarum* PS128 for about 3 months with regular medicines for PD increases the UPDRS motor score and QOL of the patients suffering from PD (Lu et al. 2021). It can also prevent hyperactivation of glial cells, maintain MPTP-induced oxidative-stress and swelling of the neurons, reduces deficiency in motor functions and neurotoxicity and even increases the concentration of serotonin and dopamine to control anxiety-related behaviour, by managing the gut microbiome and increasing motility and production of mucin in the intestine (Liao et al. 2020; Chen et al. 2021).

There is also a probiotic mixture made up of *L. acidophilus* and *B. animalis lactis* that is known to increase the amount of butyrate in the body. This, in turn, protects nigral dopaminergic neurons from MPTP and neurotoxicity due to rotenone (Srivastav et al. 2019). Also, oral intake of probiotic *B. breve* strain A1 can increase the memory extinction in hippocampus by restoring induction rate in neurons (Ishii et al. 2021).

Also, *Clostridium butyricum* in the form of probiotic increases motor functions, reduces neuron loss and issues at the synapse and activates the microglia using gut microbiome through the GLP-1 pathway (Sun et al. 2021). Other than these, another mixture comprising *S. thermophilus* CRL 807, *L. plantarum* CRL 2130 and CRL 808 is known to facilitate motor behaviours and decrease the rate of neuroinflammation.

Even other strains like *L. salivarius* AP-32 and *L. rhamnosus* HA-114 are under research as they have been noted to show properties to increase antioxidant activities and also improve hippocampal activities (Perez Visñuk et al. 2020).

Thus, it can be said that in PD patients, the constant administration of probiotics is known to show neuroprotective actions on the dopaminergic neurons. Despite most of these probiotics showing eminent results in animal models, before using them in clinical trials the determination of exact dosage and duration of administration must be determined to avoid mishaps (Xie and Prasad 2020). Apart from direct probiotics, there are different classes amongst them that get used in the management of PD. Postbiotics are metabolites produced by bacteria in the form of probiotics which are functionally bioactive. They have immunomodulatory, anti-inflammatory, anti-obesogenic, anti-proliferative, anti-hypertensive, hypocholesterolemia and antioxidant activities (Liu et al. 2020). One such compound is butyrate, which can act towards the revival of motor functions, decrease dopamine deficiency and prevent inflammation of neurons. Also, it can stimulate the production of colonic glucagon which shows neuroprotective actions (Cuevas-González et al. 2020). Another class of compounds are the psychobiotics that usually act as an adjunct to other treatment methodologies and provide psychological relief in case of neurodegenerative disorder in patients. *L. plantarum* DP189 is a tested psychobiotic that prevents accumulation of α-synuclein by managing oxidative stress and inflammation (Wang et al. 2022b). Live biotherapeutic products (LBP) are another class of medicines that may or may not be classified under probiotics. LBPs not only include food supplements but also go on to include drugs and medicines. *Parabacteroides distasonis* (MRX0005) and *Megasphaera massiliensis* (MRX0029) have shown anti-inflammatory and antioxidant effects in glioblastoma and astrocytoma cells (Ahmed et al. 2019).

Thus, it can be said that in PD patients, the constant administration of probiotics is known to show neuroprotective actions on the dopaminergic neurons. Despite most of these probiotics showing eminent results in animal models, before using them in clinical trials the determination of exact dosage and duration of administration must be determined to avoid mishaps.

Prebiotics They are fibres taken in with the diet which are easily digested. They can increase the activity of different microorganisms present in the gut to the benefit of an individual (Klaenhammer et al. 2012). The first-time use of prebiotics in the treatment of PD came into concept when sodium oligomannate was able to successfully result in the mitigation of Alzheimer's. Ever since, multiple clinical trials have been carried out which have shown a resounding decrease of symptoms developed in Parkinson's (Wang et al. 2019; Xiao et al. 2021). Supplementation of prebiotics in the diet has shown reduction of non-motor symptoms and decrease in calprotectin, the inflammatory marker. It has also shown an increase in butyrate in PD patients, which can thus cure the issues cropping up owing to reduction of SCFAs in butyrate-producing bacteria (Becker et al. 2022). This will also help in facilitating the function of gut-brain axis and improve the integrity of blood-brain connection and inhibit histone deacetylase which is responsible for the cell death of neurons (Dalile et al. 2019). Butyrate-producing bacteria can also initiate Atg5- and PI3K/Akt/mTOR-related autophagy which is known to degrade alpha-synuclein, and increase striatal dopamine (DA) concentration (Qiao et al. 2020). Also, prebiotics can be useful in the maintenance of the CNS, as β-galacto-oligosaccharides and fructooligosaccharide play a key role in the maintenance of synapse plasticity and coordination of memory (Savignac et al. 2013).

Synbiotics Synbiotics are composed of probiotics and prebiotics combined to produce a synergistic effect, where the prebiotic acts to facilitate the metabolism of a probiotic microorganism. Polymannuronic acid and *Lacticaseibacillus rhamnosus GG* when supplied in combination shows better results than they do individually in preventing loss of dopaminergic neurons, improving locomotive functionality and increasing production of tyrosine hydroxylase (Liu et al. 2022). They also reduce inflammation in the neurons, fortify the blood-brain barrier and increase the production of neurotrophic factor and thus prevent cell death in the striatum. However, the problem lies in the fact that identifying a pair of prebiotic and probiotic that correspond to show synergistic effect is a challenge and hence the application is limited (Lin et al. 2023).

Antibiotics Antibiotics are compounds derived from microorganisms that can kill or inhibit the growth of other microorganisms at low concentrations. For PD induced due to pathogenic infection in the GI tract, antibiotics have shown remarkable results in treatment. *Helicobacter pylori* in known to infect the small intestine which disbalances the concentration of other bacteria leading to a great impact on the initiation of Parkinson's; however the use of antibiotics to prevent the multiplication of *H. pylori* showed a great improvement in the characteristics of PD (Bjarnason et al.

2005). Combinations of different antibiotics are known to prevent neurotoxicity due to MPTP of the dopamine neurons and striatum (Koutzoumis et al. 2020). There were pronounced results even with broad-spectrum antibodies like vancomycin, tetracycline, minocycline, rifampicin and neomycin in preventing loss of dopamine neurons, increasing motor functions and reducing inflammatory reactions (Bortolanza et al. 2018; Cankaya et al. 2019; Burak et al. 2014). Certain antibodies are known to decrease amounts of *Firmicutes* and increase amounts of *Proteobacteria*, *Verrucomicrobia*, *Bacteroidetes* and *Cyanobacteria*, which in turn decrease Interleukins and tumour-necrosis factor levels to prevent inflammatory reactions (Koutzoumis et al. 2020; Pu et al. 2019).

15.9 Faecal Microbiota Transplantation (FMT)

It is a process in which to re-establish the normal gut microbiome, stool from a healthy individual is placed in the GI tract of the unhealthy patient (Fan et al. 2022). In the procedure, firstly the required microorganism is recognized, and then the stool sample is homogenized, filtered out and resuspended, then colonoscopy and enema are performed in the unhealthy individual and finally the lyophilized capsule is administered into the patient via orogastric tube or as an oral pill (Biagi et al. 2013). FMT over time has been used as a treatment method for diseases like bowel syndrome, type II diabetes, ulcerative colitis and various neurodegenerative disorders. In PD, FMT helps in curing GI-related symptoms, and in establishing communication with the vagus nerve (Glass et al. 2010). FMT allows the change in gut metabolites that affect neural pathways and activates the production of neuroactive compounds while preventing the production of neurotoxic compounds, strengthens the gut barrier and initiates the immune system. Recent studies have shown that FMT treatment in PD patients can increase the *Blautia* and *Prevotella* species and show a decrease in *Bacteroidetes* phylum, thus reducing constipation and alleviating non-motor symptoms of the disease (XiaoYi et al. 2021). Also, faecal transplantation can increase the number of *Proteobacteria*, *Turicibacterales* and *Enterobacteriales* and decrease the number of *Firmicutes* and *Clostridiales* in the gut thus bettering the gut dysbiosis, reducing comorbidities in GI, increasing the DA and 5-HT concentration, reducing the initiation of microglia and astrocytes in substantia nigra and improving the motor impairments. It also subdues the TLR4/TNF-α signalling pathway thus preventing the gut and brain from swelling (Sun et al. 2018). FMT can also reduce trembling in PD patients, relieves physical dysfunction and improves anxiety or depression (Huang et al. 2019). However, FMT sometimes may cause worse dyskinesia, reducing the numbers of *Lachnospiraceae* and *Ruminococcaceae* if the concentration of stool samples is not well estimated, and so during administration there has to be extra care taken (Keshavarzian et al. 2015).

Diet Modifications: Researchers have revealed that diet plays a key role in affecting the onset, development and curing of Parkinson's disease. Usually, consumption of a balanced quantity of carbohydrates, proteins and fats, along with vegetables, fruits and omega-3 fatty acids, with high quantities of fibres, flavonoids and polyphenols, like seen in a Mediterranean diet, has great neuroprotective actions thus preventing PD (Uyar and Yildiran 2019; Maraki et al. 2019). Usually, it is also recorded that depending on the calorie taken in by a patient, there's a change in the severity of symptoms experienced by the patients (Barichella et al. 2017). Initially it was thought that the diet helps by mitigating inflammatory reactions; however, recent discoveries show that diet brings changes to the gut microbiome which in turn affects PD. Diet high in carbohydrates or containing fibres allows the release of SCFAs by the gut microorganisms that in turn are beneficial in PD. Diet primarily based on protein, with low carbohydrates, known as ketogenic diet is known to improve the motor functions in PD patients, by increasing SCFAs and decreasing γ-glutamyl amino acid (Kraeuter et al. 2019; Li et al. 2022). Polyunsaturated ω-3 fatty acid also can affect the gut microbiome to reduce oxidative stress and the aggregation of α-synuclein (Perez-Pardo et al. 2018). Also, diet based on calorie limitation and timely fasting increases *Lactobacillus* species in the gut and increases the amount of β-hydroxybutyrate, fibroblast growth factor and ghrelin levels by altering the gut microbiome to increase neuroprotective actions in PD patients (Bayliss et al. 2016; Srivastava and Haigis 2011). Dietary treatment though looks to be a promising venture for treating PD, but it requires to be personalized making it difficult to be an assured method.

15.10 Conclusion and Future Perspective

The gut microbiota can have an impact on human health in both positive and negative ways. There is mounting evidence that the gut microbiome resembles an astonishingly diverse and densely populated microbial community that is essential to both human health preservation and disease pathogenesis. The gut microbiome is the genome of all the microorganisms that live in the gastrointestinal (GI) tract. Researchers have been able to describe the gut microbiome with unprecedented precision thanks to advancements in DNA sequencing technology and new bioinformatics tools. Modern brain-disorder treatments frequently focus on restoring dysregulated neurotransmission in the affected brain regions. Intestinal microbiota and brain communication research have peaked in recent years, but it has also revealed several mechanisms by which the human host responds to commensal and pathogenic bacteria. The communication between the brain and the microbiota is mediated by epithelial receptor-mediated signalling, immune regulation and stimulation of enteric neurons by bacterial metabolites.

The industry has made significant efforts to combat diseases affecting the elderly, such as metabolites produced in the gut that can travel to the brain and affect brain macrophages. As a result, understanding the underlying mechanisms of age-related dysbiosis, which alters metabolites derived from the gut and thus affects the CNS as well as the host's immune and endocrine responses, is critical. According to several studies, microbial metabolites can influence the morphology and function of brain macrophages. Two of these modifications are polarization and phagocytic capacity, which in turn regulate behaviour and emotional processes. A fundamental issue that must be addressed is the notion that age-related changes in gut microbial ecology and function, as well as the involvement of specific bacterial species, may be predictive of relevant clinical issues. The moderate and long-term impact on mental as well as cognitive performance is a barrier in current microbe scientific research. Parkinson's disease and other neurodegenerative problems in older people, and autism and attentiondeficit/hyperactivity disorder affecting younger people, have been a major region of research over the past decade, and therapies are being developed for the same, however the result currently is depressing. Future research will reveal whether microbes can be used to produce therapeutic neurotransmitters for the treatment of psychiatric disorders. For therapy to be effective, any potential side effects, such as those caused by the presence of receptors or epigenetic processes in tissues other than the brain, must be investigated.

Finally, GI microbiome regulation or modification via diet may offer significant benefits for preventing and treating brain-related disorders, prompting a few experts to propose specific microbiota developments for use as potential psychotropic therapies. Because there are numerous ways to combine prebiotics and probiotics with other nutritional compounds, future mechanistic studies are required to establish the true potential of such psychotropic therapies to deliver the anticipated benefits in the targeted populations. Treatments based on probiotics and microbiota may take months or years to influence neuropsychiatric conditions, but the microbiome's impact on host coagulation can be felt fairly quickly. In this regard, integrating multiple metabolomic, metagenomic, meta transcriptomic and proteomic methods will be a promising approach to enable a more detailed portrayal of the multifaceted microbial ecosystem and key metabolites to validate their therapeutic potential as adjuvants in the treatment of age-related gut-brain axis pathologies. Further research is needed to explore how specific microbial metabolites influence brain macrophage morphology, function, and their role in regulating behaviour and emotional processes. Investigating whether alterations in specific gut bacterial species can predict the development of age-related brain disorders holds promise for early intervention and preventative strategies. Exploring the potential of manipulating the gut microbiome to produce therapeutic neurotransmitters for treating psychiatric disorders is a fascinating avenue for future research. However, evaluating potential side effects and targeting specific brain regions are crucial considerations. The potential of dietary modulation to regulate or modify the gut microbiome offers significant promise for preventing and treating brain-related disorders. Future mechanistic studies are needed to determine the most effective prebiotic and probiotic combinations and their long-term benefits for specific patient populations. Developing psychotropic therapies based on probiotics and microbiota manipulation requires further investigation, considering the potential delayed effects on neuropsychiatric conditions. Integrating metabolomic, metagenomic, metatranscriptomic, and proteomic methods holds promise for comprehensively characterizing the gut microbiome ecosystem and identifying key metabolites with therapeutic potential for age-related gut-brain axis pathologies.

References

Abeliovich A, Gitler AD (2016) Defects in trafficking bridge Parkinson's disease pathology and genetics. Nature 539:207–216. https://doi.org/10.1038/nature20414

Ahmed S, Busetti A, Fotiadou P, Vincy Jose N, Reid S, Georgieva M et al (2019) In vitro characterization of gut microbiota-derived bacterial strains with neuroprotective properties. Front Cell Neurosci 13:402

Arumugam M, Raes J, Pelletier E, Le Paslier D, Yamada T, Mende DR, Fernandes GR, Tap J, Bruls T, Batto JM et al (2011) Enterotypes of the human gut microbiome. Nature 473:174–180

Bajaj JS, Ridlon JM, Hylemon PB et al (2012) Linkage of gut microbiome with cognition in hepatic encephalopathy. Am J Physiol Gastrointest Liver Physiol 302:G168–G175

Barichella M, Cereda E, Cassani E, Pinelli G, Iorio L, Ferri V et al (2017) Dietary habits and neurological features of Parkinson's disease patients: implications for practice. Clin Nutr 364:1054–1061. https://doi.org/10.1016/j.clnu.2016.06.020

Bayliss JA, Lemus MB, Stark R, Santos VV, Thompson A, Rees DJ et al (2016) Ghrelin-AMPK signalling mediates the neuroprotective effects of calorie restriction in Parkinson's disease. J Neurosci 3610:3049–3063. https://doi.org/10.1523/jneurosci.4373-15.2016

Becker A, Pierre Schmartz G, Gröger L, Grammes N, Galata V, Philippeit H, et al (2022) Effects of resistant starch on symptoms, faecal markers and gut microbiota in Parkinson's disease—the RESISTA-PD trial. Dev Reprod Biol S1672–0229(21)(2021)00245-X

Berer K, Krishnamoorthy G (2012) Commensal gut flora and brain autoimmunity: a love or hate affair? Acta Neuropathol 123:639–651

Biagi E, Candela M, Turroni S, Garagnani P, Franceschi C, Brigidi P (2013) Ageing and gut microbes: perspectives for health maintenance and longevity. Pharmacol Res 69:11–20. https://doi.org/10.1016/j.phrs.2012.10.005

Bjarnason I, Bjarnason I, Charlett A et al (2005) Role of chronic infection and inflammation in the gastrointestinal tract in the aetiology and pathogenesis of idiopathic parkinsonism. Part 2: response of facets of clinical idiopathic parkinsonism to *Helicobacter pylori* eradication. A randomized, double-blind, placebo-controlled efficacy study. Helicobacter 104:276–287

Booijink CC, El-Aidy S, Rajilić-Stojanović M et al (2010) High temporal and inter-individual variation detected in the human ileal microbiota. Environ Microbiol 12:3213–3227

Bortolanza M, Nascimento G, Socias S et al (2018) Tetracycline repurposing in neurodegeneration: focus on Parkinson's disease. J Neural Trans 12510:1403–1415

Bullich C, Keshavarzian A, Garssen J, Kraneveld A, Perez-Pardo P (2019) Gut vibes in Parkinson's disease: the microbiota-gut-brain axis. Mov Disord Clin Pract 6:639–651. https://doi.org/10.1002/mdc3.12840

Burak Y, Lütfü H, Ertugrul K, Rüdiger SW (2014) RIFAMPICIN: an antibiotic with brain protective function. Brain Res Bull 107:37–42

Cai W, Zhang K, Zhu L, Xu J, Yang B, Hu X et al (2017) Dysfunction of the neurovascular unit in ischemic stroke and neurodegenerative diseases: an ageing effect. Ageing Res Rev 72:77–87. https://doi.org/10.1016/j.arr.2016.09.006

Cankaya S, Cankaya B, Kilic U, Kilic E, Yulug B (2019) The therapeutic role of minocycline in Parkinson's disease. Drugs Context 8:212553

Caracciolo B, Xu W, Collins S et al (2014) Cognitive decline, dietary factors and gut–brain interactions. Mech Ageing Dev 136–137:59–69

Cassani E, Privitera G, Pezzoli G, Pusani C, Madio C, Iorio L et al (2011) Use of probiotics for the treatment of constipation in Parkinson's disease patients. Minerva gastroenterologica e Dietologica 57:117–121. https://doi.org/10.1109/IFOST.2011.6021087

Cattaneo A, Cattane N, Galluzzi S et al (2017) Association of brain amyloidosis with pro-inflammatory gut bacterial taxa and peripheral inflammation markers in cognitively impaired elderly. Neurobiol Aging 49:60–68

Chen SG et al (2016) Exposure to the functional bacterial amyloid protein curli enhances alpha-synuclein aggregation in aged Fischer 344 rats and Caenorhabditis elegans. Sci Rep 6:34477

Chen CM, Wu CC, Huang CL, Chang MY, Cheng SH, Lin CT et al (2021) Lactobacillus plantarum PS128 promotes intestinal motility, mucin production, and serotonin signalling in mice. Probiotics Antimicrob Proteins 143:535–545. https://doi.org/10.1007/s12602-021-09814-3
Colonna M, Butovsky O (2017) Microglia function in the central nervous system during health and neurodegeneration. Annu Rev Immunol 35:441–468. https://doi.org/10.1146/annurev-immunol-051116-052358
Cryan JF, Dinan TG (2012) Mind-altering microorganisms: the impact of the gut microbiota on brain and behaviour. Nat Rev Neurosci 13:701–712
Cuevas-González PF, Liceaga AM, Aguilar-Toalá JE (2020) Postbiotics and paraprobiotics: from concepts to applications. Food Res Int 136:109502. https://doi.org/10.1016/j.foodres.2020.109502
Dalile B, Van Oudenhove L, Vervliet B, Verbeke K (2019) The role of short-chain fatty acids in microbiota-gut-brain communication. Nat Rev Gastroenterol Hepatol 16(8):461–478. https://doi.org/10.1038/s41575-019-0157-3. PMID: 31123355
Daneman R, Prat A (2015) The blood–brain barrier. Cold Spring Harb Perspect Biol 7:a020412. https://doi.org/10.1101/cshperspect.a020412
David LA, Maurice CF, Carmody RN, Gootenberg DB, Button JE, Wolfe BE et al (2014) Diet rapidly and reproducibly alters the human gut microbiome. Nature 505:559–563. https://doi.org/10.1038/nature12820
De Preter V, Geboes KP, Bulteel V, Vandermeulen G (2011) Kinetics of butyrate metabolism in the normal colon and in ulcerative colitis: the effects of substrate concentration and carnitine on the β-oxidation pathway. Aliment Pharmacol Ther 34:526–532
Dorsey ER, Sherer T, Okun MS, Bloem BR (2018) The emerging evidence of the Parkinson pandemic. J Parkinsons Dis 8:S3–S8
Duerkop BA, Vaishnava S, Hooper LV (2009) Immune responses to the microbiota at the intestinal mucosal surface. Immunity 31:368–376
Engelhardt B, Liebner S (2014) Novel insights into the development and maintenance of the blood-brain barrier. Cell Tissue Res 355:687–699. https://doi.org/10.1007/s00441-014-1811-2
Ewaschuk JB, Johnson IR, Madsen KL et al (2012) Barley-derived β-glucans increase gut permeability, *ex vivo* epithelial cell binding to *E. coli*, and naive T-cell proportions in weanling pigs. J Anim Sci 90:2652–2662
Fan H-X, Sheng S, Zhang F (2022) New hope for Parkinson's disease treatment: targeting gut microbiota. CNS Neurosci Ther 28:1675–1688. https://doi.org/10.1111/cns.13916
Filippo C, Cavalieri D, Di Paola M, Ramazzotti M, Poullet JB, Massart S, Collini S, Pieraccini G, Lionetti P (2010) Impact of diet in shaping gut microbiota revealed by a comparative study in children from Europe and rural Africa. Proc Natl Acad Sci U S A 107:14691–14696
Fried LP, Tangen CM, Walston J et al (2001) Frailty in older adults: evidence for a phenotype. J Gerontol Biol Sci Med Sci 56:M146–M157
Galland L (2006) Patient-centred care: antecedents, triggers, and mediators. Altern Ther Health Med 12:62–70
Georgescu D, Ancusa OE, Georgescu LA, Ionita I, Reisz D (2016) Nonmotor gastrointestinal disorders in older patients with Parkinson's disease: is there hope? Clin Interv Aging 11:1601–1608. https://doi.org/10.2147/cia.S106284
Gérard C, Vidal H (2019) Impact of gut microbiota on host glycemic control. Front Endocrinol Lausanne 10:29. https://doi.org/10.3389/fendo.2019.00029
Gibson GR, Hutkins R, Sanders ME et al (2017) Expert consensus document: The International Scientific Association for Probiotics and Prebiotics ISAPP consensus statement on the definition and scope of prebiotics. Nat Rev Gastroenterol Hepatol 148:491–502
Glass CK, Saijo K, Winner B, Marchetto MC, Gage FH (2010) Mechanisms underlying inflammation in neurodegeneration. Cell 140:918–934. https://doi.org/10.1016/j.cell.2010.02.016
Goldman JG, Postuma R (2014) Premotor nonmotor features Parkinson's disease. Curr Opin Neurol 274:434–441. https://doi.org/10.1097/wco.0000000000000112
Heijtz RD, Wang S, Anuar F, Qian Y, Björkholm B, Samuelsson A et al (2011) Normal gut microbiota modulates brain development and behaviour. Proc Natl Acad Sci U S A 108:3047–3052. https://doi.org/10.1073/pnas.1010529108

Heumann D, Barras C, Severin A, Glauser MP, Tomasz A (1994) Gram-positive cell walls stimulate synthesis of tumour necrosis factor alpha and interleukin-6 by human monocytes. Infect Immun 62:2715–2272

Hill C, Guarner F, Reid G et al (2014) Expert consensus document. The International Scientific Association for Probiotics and Prebiotics consensus statement on the scope and appropriate use of the term probiotic. Nat Rev Gastroenterol Hepatol 118:506–514

Hsiao EY, McBride SW, Hsien S et al (2013) Microbiota modulate behavioural and physiological abnormalities associated with neurodevelopmental disorders. Cell 155:1451–1463

Hu L, Dong MX, Huang YL, Lu CQ, Qian Q, Zhang CC et al (2020) Integrated metabolomics and proteomics analysis reveals plasma lipid metabolic disturbance in patients with Parkinson's disease. Front Mol Neurosci 13:80. https://doi.org/10.3389/fnmol.2020.00080

Huang H, Xu H, Luo Q, He J, Li M, Chen H et al (2019) Fecal microbiota transplantation to treat Parkinson's disease with constipation: a case report. Medicine (Baltimore) 9826:e16163. https://doi.org/10.1097/md.0000000000016163

Hugon P, Dufour J-C, Colson P, Fournier P-E, Sallah K, Raoult D (2015) A comprehensive repertoire of prokaryotic species identified in human beings. Lancet Infect Dis 15:1211–1219. https://doi.org/10.1016/S1473-30991500293-5

Ishii T, Furuoka H, Kaya M, Kuhara T (2021) Oral administration of probiotic bifidobacterium breve improves facilitation of hippocampal memory extinction *via* restoration of aberrant higher induction of neuropsin in an MPTP-induced mouse model of Parkinson's disease. Biomedicines 92:167. https://doi.org/10.3390/biomedicines9020167

Kalia LV, Lang AE (2015) Parkinson's disease. Lancet 386:896–912. https://doi.org/10.1016/S0140-67361461393-3

Keshavarzian A, Green SJ, Engen PA, Voigt RM, Naqib A, Forsyth CB et al (2015) Colonic bacterial composition in Parkinson's disease. Mov Disord 30:1351–1360. https://doi.org/10.1002/mds.26307

Keshavarzian A, Engen P, Bonvegna S, Cilia R (2020) The gut microbiome in Parkinson's disease: a culprit or a bystander? Prog Brain Res 252:357–450. https://doi.org/10.1016/bs.pbr.2020.01.004. Epub 2020 Mar 5. PMID: 32247371

Kierdorf K, Masuda T, Jordão MJC, Prinz M (2019) Macrophages at CNS interfaces: ontogeny and function in health and disease. Nat Rev Neurosci 20:547–562. https://doi.org/10.1038/s41583-019-0201-x

Kim YS, Joh TH (2006) Microglia, major player in brain inflammation: their roles in the pathogenesis of Parkinson's disease. Exp Mol Med 38:333–347. https://doi.org/10.1038/emm.2006.40

Klaenhammer TR, Kleerebezem M, Kopp MV, Rescigno M (2012) The impact of probiotics and prebiotics on the immune system. Nat Rev Immunol 1210:728–734. https://doi.org/10.1038/nri3312

Koutzoumis D, Vergara M, Pino J et al (2020) Alterations of the gut microbiota with antibiotics protects dopamine neuron loss and improves motor deficits in a pharmacological rodent model of Parkinson's disease. Exp Neurol 325:113159

Kraeuter AK, Guest PC, Sarnyai Z (2019) The therapeutic potential of ketogenic diet throughout life: focus on metabolic, neurodevelopmental and neurodegenerative disorders. Adv Exp Med Biol 1178:77–101. https://doi.org/10.1007/978-3-030-25650-0_5

Lajoie AC, Lafontaine AL, Kaminska M (2021) The spectrum of sleep disorders in Parkinson disease: a review. Chest 1592:818–827. https://doi.org/10.1016/j.chest.2020.09.099

Laterza L, Rizzatti G, Gaetani E, Chiusolo P, Gasbarrini A (2016) The gut microbiota and immune system relationship in human graft-versus-host disease. Mediterr J Hematol Infect Dis 8:e2016025

Leclercq S, Matamoros S, Cani PD et al (2014) Intestinal permeability, gut-bacterial dysbiosis, and behavioural markers of alcohol-dependence severity. Proc Natl Acad Sci U S A 111:E4485–E4493

Ley RE, Turnbaugh PJ, Klein S, Gordon JI (2006) Microbial ecology: human gut microbes associated with obesity. Nature 444:1022–1023

Li Q, Barres BA (2018) Microglia and macrophages in brain homeostasis and disease. Nat Rev Immunol 18:225–242. https://doi.org/10.1038/nri.2017.125

Li J, Jia H, Cai X, Zhong H, Feng Q, Sunagawa S et al (2014) An integrated catalogue of reference genes in the human gut microbiome. Nat Biotechnol 32:834–841. https://doi.org/10.1038/nbt.2942

Li P, Zhang Y, Xu Y, Cao H, Li L (2022) Characteristics of CD8+ and CD4+ tissue-resident memory lymphocytes in the gastrointestinal tract. Adv Gut Microbiome Res 2022:9157455. https://doi.org/10.1155/2022/9157455

Liao JF, Cheng YF, You ST, Kuo WC, Huang CW, Chiou JJ et al (2020) Lactobacillus plantarum PS128 alleviates neurodegenerative progression in 1-methyl-4-phenyl-1,2,3,6-tetrahydropyridine-induced mouse models of Parkinson's disease. Brain Behav Immun 90:26–46. https://doi.org/10.1016/j.bbi.2020.07.036

Lin C-H, Lai H-C, Wu M-S (2023) Gut-oriented disease modifying therapy for Parkinson's disease. J Formosan Med Assoc 122(1):9–18. https://doi.org/10.1016/j.jfma.2022.09.010. ISSN 0929-6646

Liu J, Xu F, Nie Z, Shao L (2020) Gut microbiota approach—a new strategy to treat Parkinson's disease. Front Cell Infect Microbiol. 10:570658. https://doi.org/10.3389/fcimb.2020.570658. PMID: 33194809; PMCID: PMC7643014

Liu X, Du ZR, Wang X, Sun XR, Zhao Q, Zhao F et al (2022) Polymannuronic acid prebiotic plus Lacticaseibacillus rhamnosus GG probiotic as a novel synbiotic promoted their separate neuroprotection against Parkinson's disease. Food Res Int 155:111067

Lobo A, Launer LJ, Fratiglioni L et al (2000) Prevalence of dementia and major subtypes in Europe: a collaborative study of population-based cohorts. neurologic diseases in the elderly research group. Neurology 5411(Suppl 5):S4–S9

Long-Smith C, O'Riordan KJ, Clarke G, Stanton C, Dinan TG, Cryan JF (2020) Microbiota-gut-brain axis: new therapeutic opportunities. Annu Rev Pharmacol Toxicol 60:477–502. https://doi.org/10.1146/annurev-pharmtox-010919-023628

López-Otín C, Blasco MA, Partridge L, Serrano M, Kroemer G (2013) The hallmarks of the aging Europe PMC funders group. Cell 153:1194–1217. https://doi.org/10.1016/j.cell.2013.05.039

Lu CS, Chang HC, Weng YH, Chen CC, Kuo YS, Tsai YC (2021) The add-on effect of Lactobacillus plantarum PS128 in patients with Parkinson's disease: a pilot study. Front Nutr 8:650053. https://doi.org/10.3389/fnut.2021.650053

Lubomski M, Xu X, Holmes AJ, Yang JYH, Sue CM, Davis RL (2021) The impact of device-assisted therapies on the gut microbiome in Parkinson's disease. J Neurol 269:780–795. https://doi.org/10.1007/s00415-021-10657-9

Lyte M (2004) Microbial endocrinology and infectious disease in the 21st century. Trends Microbiol 12:14–20

Macfarlane GT, Macfarlane S (2011) Fermentation in the human large intestine: its physiologic consequences and the potential contribution of prebiotics. J Clin Gastroenterol 45(Suppl):S120–S127

Macpherson AJ, Uhr T (2004) Induction of protective IgA by intestinal dendritic cells carrying commensal bacteria. Science 303:1662–1665

Maes M, Mihaylova I, Leunis JC (2007) Increased serum IgA and IgM against LPS of enterobacteria in chronic fatigue syndrome CFS: indication for the involvement of gram-negative enterobacteria in the aetiology of CFS and for the presence of an increased gut-intestinal permeability. J Affect Disord 99:237–240

Mager LF, Wasmer MH, Rau TT, Krebs P (2016) Cytokine-induced modulation of colorectal cancer. Front Oncol 6:96. https://doi.org/10.3389/fonc.2016.00096

Maini Rekdal V, Bess EN, Bisanz JE, Turnbaugh PJ, Balskus EP (2019) Discovery and inhibition of an interspecies gut bacterial pathway for Levodopa metabolism. Science 364:6445. https://doi.org/10.1126/science.aau6323

Maraki MI, Yannakoulia M, Stamelou M, Stefanis L, Xiromerisiou G, Kosmidis MH et al (2019) Mediterranean diet adherence is related to reduced probability of prodromal Parkinson's disease. Mov Disord 341:48–57. https://doi.org/10.1002/mds.27489

Mart'yanov SV, Botchkova EA, Plakunov VK, Gannesen AV (2021) The impact of norepinephrine on mono-species and dual-species staphylococcal biofilms. Microorganisms 9:1–10. https://doi.org/10.3390/microorganisms9040820

Messaoudi M, Lalonde R, Violle N et al (2011) Assessment of psychotropic-like properties of a probiotic formulation (*Lactobacillus helveticus* R0052 and *Bifidobacterium longum* R0175) in rats and human subjects. Br J Nutr 105:755–764
Miller KR, Streit WJ (2007) The effects of ageing, injury and disease on microglial function: a case for cellular senescence. Neuron Glia Biol 3:245–253. https://doi.org/10.1017/S1740925X08000136
Mizrahi-Man O, Davenport ER, Gilad Y (2013) Taxonomic classification of bacterial 16S rRNA genes using short sequencing reads: evaluation of effective study designs. PLoS One 8:e53608
Moore WE, Holdeman LV (1974) Human faecal flora: the normal flora of 20 Japanese-Hawaiians. Appl Microbiol 27:961–979
Mossad O, Batut B, Yilmaz B, Dokalis N, Mezö C, Nent E et al (2022) Gut microbiota drive age-related oxidative stress and mitochondrial damage in microglia via the metabolite N6-carboxymethyl lysine. Nat Neurosci 25:295–305. https://doi.org/10.1038/s41593-022-01027-3
Mulle JG, Sharp WG, Cubells JF (2013) The gut microbiome: a new frontier in autism research. Curr Psychiatry Rep 15:337
Munakata S, Arakawa C, Kohira R et al (2010) A case of D-lactic acid encephalopathy associated with use of probiotics. Brain Dev 32:691–694
Nimmerjahn A, Kirchhoff F, Helmchen F (2005) Resting microglial cells are highly dynamic surveillants of brain parenchyma in vivo. Neuroforum 308:1314–1318. https://doi.org/10.1126/science.1110647
O'Toole PW (2012) Changes in the intestinal microbiota from adulthood through to old age. Clin Microbiol Infect 184:44–46
O'Toole PW, Jeffery IB (2015) Gut microbiota and ageing. Science 350:1214–1215. https://doi.org/10.1126/science.aac8469
Perez Visñuk D, Savoy De Giori G, Leblanc JG, de Moreno De Leblanc A (2020) Neuroprotective effects associated with immune modulation by selected lactic acid bacteria in a Parkinson's disease model. Nutrition 79–80:110995. https://doi.org/10.1016/j.nut.2020.110995
Perez-Pardo P, Dodiya HB, Broersen LM, Douna H, Van Wijk N, Lopes Da Silva S et al (2018) Gut-brain and brain-gut axis in Parkinson's disease models: effects of a uridine and fish oil diet. Nutr Neurosci 216:391–402. https://doi.org/10.1080/1028415x.2017.1294555
Peterson DA, McNulty NP, Guruge JL, Gordon JI (2007) IgA response to symbiotic bacteria as a mediator of gut homeostasis. Cell Host Microbe 2(5):328–339. https://doi.org/10.1016/j.chom.2007.09.013
Poewe W, Seppi K, Tanner CM, Halliday GM, Brundin P, Volkmann J et al (2017) Parkinson disease. Nat Rev Dis Primers 3:17013. https://doi.org/10.1038/nrdp.2017.13
Poretsky R, Rodriguez-R LM, Luo C, Tsementzi D, Konstantinidis KT (2014) Strengths and limitations of 16S rRNA gene amplicon sequencing in revealing temporal microbial community dynamics. PLoS One 9:e93827
Pu Y, Chang L, Qu Y, Wang S, Zhang K, Hashimoto K (2019) Antibiotic-induced microbiome depletion protects against MPTP-induced dopaminergic neurotoxicity in the brain. Ageing Albany NY 11:6915. https://doi.org/10.18632/aging.102221
Qian Y, Yang X, Xu S, Huang P, Li B, Du J et al (2020) Gut metagenomics-derived genes as potential biomarkers of Parkinson's disease. Brain 143:2474–2489. https://doi.org/10.1093/brain/awaa201
Qiao Z, Li Z, Li J et al (2009) Bacterial translocation and change in intestinal permeability in patients after abdominal surgery. J Huazhong Univ Sci Technol Med Sci 29:486–491
Qiao C-M, Sun M-F, Jia X-B, Shi Y, Zhang B-P, Zhou Z-L et al (2020) Sodium butyrate causes α-synuclein degradation by an Atg5-dependent and PI3K/Akt/mTOR-related autophagy pathway. Exp Cell Res 387:111772. https://doi.org/10.1016/j.yexcr.2019.111772
Qureshi MO, Khokhar N, Shafqat F (2014) Ammonia levels and the severity of hepatic encephalopathy. J Coll Physicians Surg Pak 24:160–163
Rinninella E, Raoul P, Cintoni M, Franceschi F, Miggiano GAD, Gasbarrini A, Mele MC (2019) What is the healthy gut microbiota composition? A changing ecosystem across age, environment, diet, and diseases. Microorganisms 71:14. https://doi.org/10.3390/microorganisms7010014

Romano S, Savva GM, Bedarf JR, Charles IG, Hildebrand F, Narbad A (2021) Meta-analysis of the Parkinson's disease gut microbiome suggests alterations linked to intestinal inflammation. NPJ Parkinsons Dis 7:27. https://doi.org/10.1038/s41531-021-00156-z

Rowland I, Gibson G, Heinken A, Scott K, Swann J, Thiele I et al (2018) Gut microbiota functions: metabolism of nutrients and other food components. Eur J Nutr 57:1–24. https://doi.org/10.1007/s00394-017-1445-8

Salazar N, López P, Valdés L, Margolles A, Suárez A, Patterson ÁM et al (2013) Microbial targets for the development of functional foods accordingly with nutritional and immune parameters altered in the elderly. J Am Coll Nutr 32:399–406. https://doi.org/10.1080/07315724.2013.827047

Sampson TR et al (2016) Gut microbiota regulate motor deficits and neuroinflammation in a model of Parkinson's disease. Cell 167:1469–1480.e12

Sampson TR et al (2020) A gut bacterial amyloid promotes alpha-synuclein aggregation and motor impairment in mice. Elife 9:e53111

Savignac HM, Corona G, Mills H, Chen L, Spencer JP, Tzortzis G et al (2013) Prebiotic feeding elevates central brain derived neurotrophic factor, N-methyl-D-aspartate receptor subunits and D-serine. Neurochem Int 638:756–764. https://doi.org/10.1016/j.neuint.2013.10.006

Segain JP, Raingeard de la Blétière D, Bourreille A et al (2000) Butyrate inhibits inflammatory responses through NF kappa B inhibition: implications for Crohn's disease. Gut 47:397–403

Segura-Aguilar J, Paris I, Muñoz P, Ferrari E, Zecca L, Zucca FA (2014) Protective and toxic roles of dopamine in Parkinson's disease. J Neurochem 1296:898–915. https://doi.org/10.1111/jnc.12686

Sharma BC, Sharma P, Lunia MK et al (2013) A randomised, double-blind, controlled trial comparing rifaximin plus lactulose with lactulose alone in treatment of overt hepatic encephalopathy. Am J Gastroenterol 108:1458–1463

Sierra A, Paolicelli RC, Kettenmann H (2019) Cien Años de Microglía: milestones in a century of microglial research. Trends Neurosci 42:778–792. https://doi.org/10.1016/j.tins.2019.09.004

Smith JA, Das A, Ray SK, Banik NL (2012) Role of pro-inflammatory cytokines released from microglia in neurodegenerative diseases. Brain Res Bull 87:10–20. https://doi.org/10.1016/j.brainresbull.2011.10.004

Srivastav S, Neupane S, Bhurtel S, Katila N, Maharjan S, Choi H et al (2019) Probiotics mixture increases butyrate, and subsequently rescues the nigral dopaminergic neurons from MPTP and rotenone-induced neurotoxicity. J Nutr Biochem 69:73–86. https://doi.org/10.1016/j.jnutbio.2019.03.021

Srivastava S, Haigis MC (2011) Role of sirtuins and calorie restriction in neuroprotection: implications in Alzheimer's and Parkinson's diseases. Curr Pharm Des 1731:3418–3433. https://doi.org/10.2174/138161211798072526

Stoner R, Chow ML, Boyle MP et al (2014) Patches of disorganisation in the neocortex of children with autism. N Engl J Med 370:1209–1219

Sudo N, Chida Y, Aiba Y, Sonoda J, Oyama N, Yu XN et al (2004) Postnatal microbial colonisation programs the hypothalamic-pituitary-adrenal system for stress response in mice. J Physiol 558:263–275. https://doi.org/10.1113/jphysiol.2004.063388

Sun MF, Zhu YL, Zhou ZL, Jia XB, Xu YD, Yang Q et al (2018) Neuroprotective effects of fecal microbiota transplantation on MPTP-induced Parkinson's disease mice: gut microbiota, glial reaction and TLR4/TNF-alpha signalling pathway. Brain Behav Immun 70:48–60

Sun J, Li H, Jin Y, Yu J, Mao S, Su KP et al (2021) Probiotic Clostridium butyricum ameliorated motor deficits in a mouse model of Parkinson's disease *via* gut microbiota-GLP-1 pathway. Brain Behav Immun 91:703–715. https://doi.org/10.1016/j.bbi.2020.10.014

Surwase SN, Jadhav JP (2011) Bioconversion of L-tyrosine to L-DOPA by a novel bacterium Bacillus sp. JPJ. Amino Acids 41:495–506. https://doi.org/10.1007/s00726-010-0768-z

Takahashi K, Terashima H, Kohno K, Ohkouchi N (2013) A stand-alone synbiotic treatment for the prevention of D-lactic acidosis in short bowel syndrome. Int Surg 98:110–113

Tamtaji OR, Taghizadeh M, Daneshvar Kakhaki R, Kouchaki E, Bahmani F, Borzabadi S et al (2019) Clinical and metabolic response to probiotic administration in people with Parkinson's disease: a randomised, double-blind, placebo-controlled trial. Clin Nutr 383:1031–1035. https://doi.org/10.1016/j.clnu.2018.05.018

Tan J, McKenzie C, Potamitis M et al (2014) The role of short-chain fatty acids in health and disease. Adv Immunol 121:91–119

Thursby E, Juge N (2017) Introduction to the human gut microbiota. Biochem J 474:1823–1836

Tillisch K, Labus J, Kilpatrick L et al (2013) Consumption of fermented milk product with probiotic modulates brain activity. Gastroenterology 144:1394–1401

Uyar G, Yildiran H (2019) A nutritional approach to microbiota in Parkinson's disease. Biosci Microbiota Food Health 384:115–127. https://doi.org/10.12938/bmfh.19-002

Wallen ZD, Demirkan A, Twa G et al (2022) Metagenomics of Parkinson's disease implicates the gut microbiome in multiple disease mechanisms. Nat Commun 13:6958. https://doi.org/10.1038/s41467-022-34667-x

Wang X, Sun G, Feng T, Zhang J, Huang X, Wang T et al (2019) Sodium oligomannate therapeutically remodels gut microbiota and suppresses gut bacterial amino acids-shaped neuroinflammation to inhibit Alzheimer's disease progression. Cell Res 2910:787–803. https://doi.org/10.1038/s41422-019-0216-x

Wang ZL, Yuan L, Li W, Li JY (2022a) Ferroptosis in Parkinson's disease: glia-neuron crosstalk. Trends Mol Med 284:258–269. https://doi.org/10.1016/j.molmed.2022.02.003

Wang L, Zhao Z, Zhao L, Zhao Y, Yang G, Wang C et al (2022b) Lactobacillus plantarum DP189 reduces α-syn aggravation in MPTP-induced Parkinson's disease mice *via* regulating oxidative damage, inflammation, and gut microbiota disorder. J Agric Food Chem 704:1163–1173. https://doi.org/10.1021/acs.jafc.1c07711

Williams BL, Hornig M, Buie T et al (2011) Impaired carbohydrate digestion and transport and mucosal dysbiosis in the intestines of children with autism and gastrointestinal disturbances. PLoS One 6:e24585

Williams BL, Hornig M, Parekh T, Lipkin WI (2012) Application of novel PCR-based methods for detection, quantitation, and phylogenetic characterization of Sutterella species in intestinal biopsy samples from children with autism and gastrointestinal disturbances. mBio 3:e00261-11

Wilms E, Troost FJ, Elizalde M, Winkens B, de Vos P, Mujagic Z et al (2020) Intestinal barrier function is maintained with aging—a comprehensive study in healthy subjects and irritable bowel syndrome patients. Sci Rep 10:1–10. https://doi.org/10.1038/s41598-019-57106-2

Xiao S, Chan P, Wang T, Hong Z, Wang S, Kuang W et al (2021) A 36-week multicenter, randomised, double-blind, placebo-controlled, parallel-group, phase 3 clinical trial of sodium oligomannate for mild-to-moderate Alzheimer's dementia. Alzheimers Res Ther 131:62. https://doi.org/10.1186/s13195-021-00795-7

XiaoYi K, XiaoHan Y, LiJuan X et al (2021) Evaluation of faecal microbiota transplantation in Parkinson's disease patients with constipation. Microb Cell Fact 201:98

Xie C, Prasad AA (2020) Probiotics treatment improves hippocampal dependent cognition in a rodent model of Parkinson's disease. Microorganisms 811:1661. https://doi.org/10.3390/microorganisms8111661

Yildirim S, Ozkan A, Aytac G, Agar A, Tanriover G (2022) Role of melatonin in TLR4-mediated inflammatory pathway in the MPTP-induced mouse model. Neurotoxicology 88:168–177. https://doi.org/10.1016/j.neuro.2021.11.011

Yoshimoto S, Mitsuyama E, Yoshida K, Odamaki T, Xiao JZ (2021) Enriched metabolites that potentially promote age-associated diseases in subjects with an elderly-type gut microbiota. Gut Microbes 13:1–11. https://doi.org/10.1080/19490976.2020.1865705

Zhang Z, Zhai H, Geng J et al (2013) Large-scale survey of gut microbiota associated with MHE Via 16S rRNA-based pyrosequencing. Am J Gastroenterol 108:1601–1611

Zhao Z, Ning J, Bao XQ, Shang M, Ma J, Li G, Zhang D (2021) Fecal microbiota transplantation protects rotenone-induced Parkinson's disease mice via suppressing inflammation mediated by the lipopolysaccharide-TLR4 signaling pathway through the microbiota-gut-brain axis. Microbiome 9(1):226. https://doi.org/10.1186/s40168-021-01107-9. PMID: 34784980; PMCID: PMC8597301

Zhu M, Liu X, Ye Y, Yan X, Cheng Y, Zhao L, Chen F, Ling Z (2022) Gut microbiota: a novel therapeutic target for Parkinson's disease. Front Immunol 13:937555. https://doi.org/10.3389/fimmu.2022.937555. PMID: 35812394; PMCID: PMC9263276

Gut Microbiome: A Key Factor in Ageing and an Aim for Anti-senescence Intervention

16

Hanaa R. Abbas, Mohamed Y. Zaky,
and Manal Abdul-Hamid

Abstract

The ageing process is increasingly being attributed to the gut microbiome. It has been hypothesized that age-related gut dysbiosis is influenced by the immune system deterioration in the ageing intestinal epithelium. There is evidence that the microbiome structure may have an impact on how quickly people age. These changes take place over time gradually. The age-related changes, including nutrition, frailty, and inflammation have a major impact on the breakdown of the symbiotic relationship between the host and the gut microbiome. Reduced microbiota diversity has demonstrated a key sign of age-onset dysbiosis, which may cause certain bacterial groups to proliferate. A healthy gut microbiota is essential for controlling metabolism, fighting off infection and inflammation, preventing cancer and autoimmune diseases, and regulating the brain-gut axis. A healthy and non-pathological ageing process may be facilitated by gut microbiota manipulation, which may also serve to prevent the development of degenerative pathways. In this chapter, we describe and explain how the gut microbiome affects human cells in health and disease, its involvement in cellular senescence in ageing, and the link between the gut microbiome and cellular senescence. We also discuss how to use this knowledge to develop microbiome-based anti-ageing and pro-longevity treatments.

H. R. Abbas · M. Abdul-Hamid (✉)
Cell Biology, Histology and Genetics Division, Department of Zoology, Faculty of Science, Beni-Suef University, Beni-Suef, Egypt
e-mail: manal.mohamed3@science.bsu.edu.eg

M. Y. Zaky
Molecular Physiology Division, Department of Zoology, Faculty of Science, Beni-Suef University, Beni-Suef, Egypt

S. Pathak, A. Banerjee (eds.), *Gut Microbiome and Brain Ageing*,
https://doi.org/10.1007/978-981-99-8803-7_16

Keywords

Ageing · Gut microbiome · Cellular senescence · Interventions · Anti-senescence

16.1 Introduction

The genetic makeup of the diverse ecosystem of all microorganisms living in the human gastrointestinal tract is known as the gut microbiome (GMB) (Thursby and Juge 2017). The gut microbiota is composed of more than 1500 types of microorganisms, including bacteria, protozoa, viruses, fungi, yeast, and archaea. It comprises more than 3 million genes and more than 100 trillion of these microbes (Dekaboruah et al. 2020). The metabolites that these genes produce, which number in the millions, modify or replace numerous human processes (Gilbert et al. 2018). The gut microbiome helps maintain homeostasis throughout life. Numerous inflammatory infections and diseases have been related to the pathophysiology of dysbiosis, or altered gut microbial composition (Vaiserman et al. 2017).

The four bacterial phyla Firmicutes, Bacteroides, Proteobacteria, and Actinobacteria constitute around 98% of these microorganisms (Alam et al. 2020). The extensive, multifactorial process of ageing, which is primarily brought on by the development of DNA damage, comprises a variety of physiological, epigenomic, metabolic, and immunological alterations that result in functional degradation (Conway and Duggal 2021). The host's gut microbiota play crucial roles in digestion and mineral absorption, lipid storage and metabolism, vitamin and amino acid biosynthesis, the suppression of pathogenic microbial species, carbohydrate fermentation, maintenance of intestinal barrier integrity, induction of an immune response, and the development of the central nervous system (Rowland et al. 2018).

The results of recent studies involving model organisms suggest that age-related intestinal dysbiosis may be a factor in unhealthy ageing and shorter lifespan. Through a variety of proteins, food signalling-independent pathways, and epigenetic mechanisms, the gut microbiota interacts with the host (Kim and Jazwinski 2018). It has been determined that senescent cells (SCs) are responsible for organismal ageing and also they seriously affect the gut microbiome (Sharma 2022). They increase with ageing and their senescence-associated secretory phenotype promotes the chronic, pro-inflammatory systemic condition (inflammaging) that characterizes ageing, decreasing stem cells' ability for regeneration and increasing the chance of developing age-related diseases (Ferrucci and Fabbri 2018). In this chapter, we delve deeper into the intricate relationship between the gut microbiome and the aging process. We explore how age-related changes in the gut microbiota may contribute to senescence and age-related disorders, and conversely, how manipulating the gut microbiome might offer promising avenues for preventing or mitigating these effects.

16.2 The Relevance of the Gut Microbiome in Health and Disease

The gut microbiota interacts with the host organism from conception to senescence, helping to control a number of metabolic pathways by producing substances that can impact the epigenetic changes linked to critical cell activities (Rooks and Garrett 2016).

The gut microbiome equilibrium is essential for maintaining the homeostatic balance of epithelial cell proliferation, metabolism, inflammation, and immune system stimulation. Furthermore, it regulates gut-brain communication, which affects mental and neurological health (Kho and Lal 2018). The human gut microbiota also plays a number of significant roles, such as controlling insulin resistance and having an effect on insulin secretion. These roles include aiding pathogen defence by colonizing mucosal surfaces and producing various antimicrobial compounds (Caricilli and Saad 2013). Multiple sclerosis, depression, allergies, and anxiety have all been related to the disruption of this ecosystem brought on by either flaws in the integrity of the gut barrier or an increase in pro-inflammatory cytokines produced by the gut microbiome (Durack and Lynch 2019).

16.3 Ageing and Gut Microbiome

Ageing is a normal, time-dependent physiological process or a complicated interaction between hereditary and environmental factors, and it causes changes in many aspects of life, including biology, the environment, behaviour, and society (Shilpa et al. 2018). The overall functions also start to decline as a result. Genomic instability, epigenetic changes, telomere attrition, and loss of proteostasis are the main molecular and cellular markers of ageing (Lidzbarsky et al. 2018).

These factors trigger compensatory mechanisms like disturbed nutrient sensing, cellular senescence, and mitochondrial dysfunction, which then lead to stem cell exhaustion and altered intercellular communication, which are the main causes of functional decline in ageing. Because it regulates immune system response and its dysregulation may perpetuate pro-inflammatory states, the gut microbiome contributes to age-associated alterations and longevity (Tan et al. 2021).

As a result, low-grade systemic inflammation (also known as "inflammaging") increases a person's vulnerability to chronic illnesses and disabilities, such as metabolic disease, cognitive decline, cardiovascular disease, frailty, and death. Age-related alterations in the gut microbiota have been observed as a rise in pro-inflammatory symbionts and a decline in helpful microorganisms that may shed light on the genetic and biological origins of ageing and age-related disorders (Xu et al. 2021). It has been determined that the gut microbiome and ageing are correlated, as evidenced by the decline in physiological function and deterioration of tissue structure that occur with ageing, both of which are accompanied by changes to the gut microbiota that affect the host-immune system's maturation, regulation, function, and senescence (Nagpal et al. 2018).

16.4 The Structure of the Microbiota Changes with Age and External Factors

The composition of the gut microbiota is influenced by a number of important factors, including lifestyle, nutrition, diseases, alcohol usage, antibiotic use, probiotic use, immune response induction, and the amount of IgA produced by B cells (Hasan and Yang 2019). When comparing young and elderly people, the taxonomic composition reflects their current and historical health status (Hou et al. 2022).

Ruminococcus, *Prevotella*, and related genera had lower relative abundance in the gut microbiota of elderly persons than in young people, and the relative proportion of *Bacteroides* spp. was larger (53%) than that of *Clostridium* species (Rinninella et al. 2019). A low-grade inflammatory condition that affects the immune system is brought on by age-related abnormalities in the gut microbiota, which result in a combination of fewer beneficial bacteria and more pro-inflammatory organisms (Wu et al. 2021a).

Adults have a gut microbiota that is more flexible, sophisticated, and able to withstand exogenous invasions like stress, but it is still susceptible to changes brought on by the environment (Gilbert et al. 2018). A unique core microbiota that is dominated by the *Bacteridaceae*, *Ruminococcaceae*, and *Lachnospiraceae* families and that diminishes with age was reported in studies on Italian centenarians (99–104 years old) and semi-supercentenarians (105–109 years old) (Ragonnaud and Biragyn 2021).

16.5 How the Gut Microbiome Affects Human Ageing!

In contrast to chronological age, some changes in the composition and diversity of the gut microbiota are related to biological or functional age. The gut microbiome affects inflammation and perhaps even an elderly person's longevity (Badal et al. 2020). Young adults have a varied population of gut flora colonizing the lumen and, in particular, the mucin layer of their intestines. These microbes coexist with the host in a symbiotic relationship (Thursby and Juge 2017). *Akkermansia muciniphila* in particular is a member of the Verrucomicrobia phylum that supports gut barrier integrity, preventing leakage and the accompanying production of inflammation. Due to an increase in potentially pro-inflammatory bacteria and a decrease in beneficial microbes, such as members of the *Verrucomicrobia* genus, the composition of the gut flora and microbial diversity in the elderly are altered (Geerlings et al. 2018). Systemic inflammation that results from this promotes ageing-related morbidities and early death. It also causes gut leakiness. Despite the microbiota of centenarians changing, it still exhibits diversity and beneficial microbiota, which helps to reduce overt inflammation and promote good ageing (Al Bander and Nitert 2020).

The microbiome is alone in charge of producing short-chain fatty acids (SCFAs) because it breaks down indigestible fibres. These have a variety of roles in the body, including defending and maintaining intestinal integrity, battling infections, supplying bacteria and colonocytes with energy, regulating microbial functions,

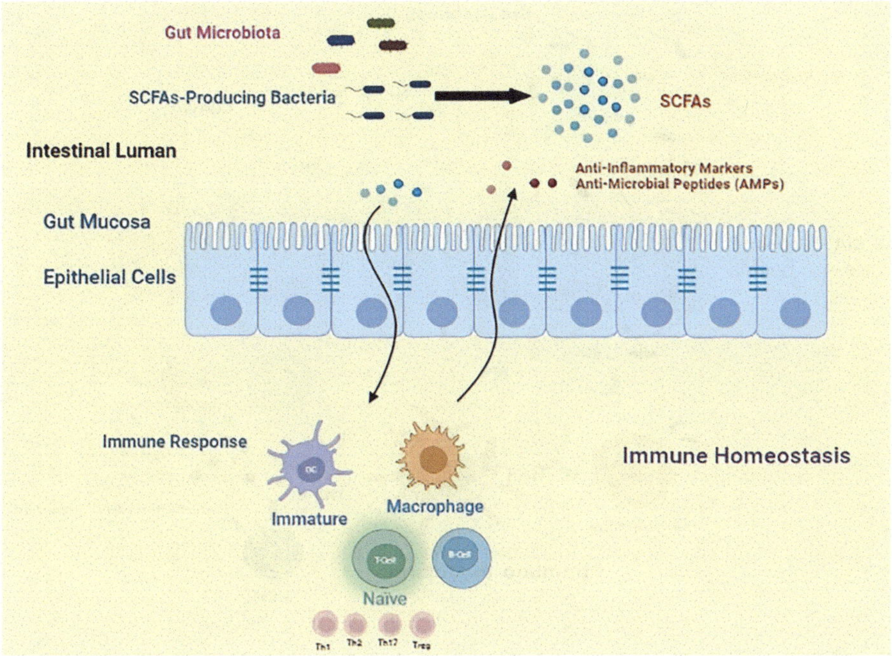

Fig. 16.1 Gut homeostasis characteristics in healthy ageing. In these circumstances, the gut microbiota controls the generation of anti-microbial peptides (AMPs) and improves the gut barrier integrity by creating short-chain fatty acids (SCFAs). Immune cells contribute to host defence including, dendritic cells (DCs) and macrophages induce the activation and differentiation of naïve T cells for production of anti-inflammatory markers by lymphoid cells (e.g. Th1, Th17) and B cells that induce anti-bodies from plasma cells to bind commensal microbes and soluble antigens

modulating immune cell development, and activating CD8+ T lymphocytes (Morrison and Preston 2016). Th17 cells, ILC2 and IgA-secreting B cells, regulatory T cells (Tregs), IL-10, and Transforming growth factor (TGF)-producing cells are all induced by the gut microbiota (Pandiyan et al. 2019). This is providing an important role in the regulation of immunological homeostasis (Fig. 16.1). Study on aged mice showed a rise in pro-inflammatory microbes and a fall in good gut flora such as SCFA-producing species of *Clostridium* and *A. muciniphila* (Ragonnaud and Biragyn 2021).

16.6 Microbiome-Based Ageing Interventions

Age-related inflammation is accompanied with immune system remodelling (immunosenescence) that causes autoimmunity, immune incompetence, increased susceptibility to bacterial and viral infections, and subpar vaccination responses (inflammaging) (Ray and Yung 2018).

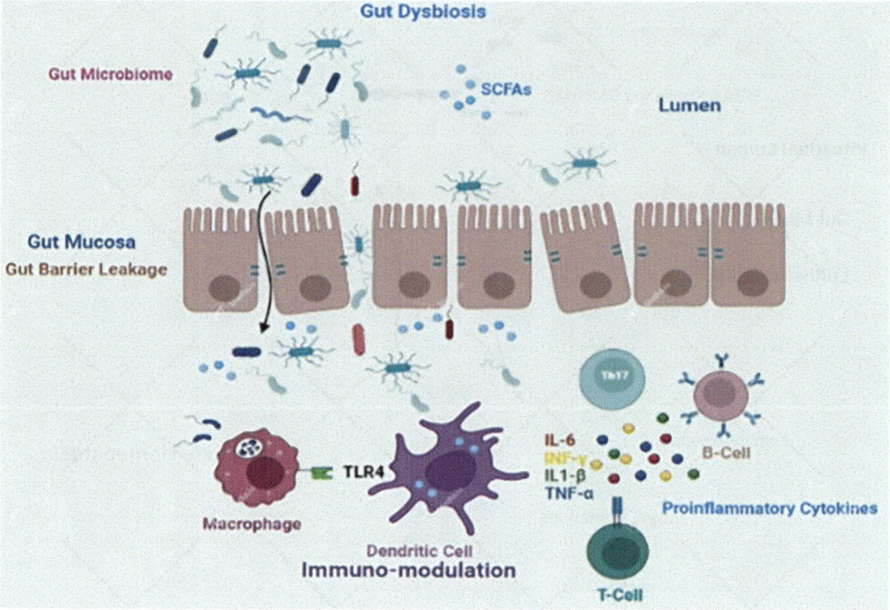

Fig. 16.2 Schematic illustrations of dysbiotic gut microbiome with the loss of barrier integrity and breach in the intestinal epithelial cell barrier in ageing. Impaired intestinal barrier integrity causes microbe translocation, microbial partial expansion of certain bacteria, and metabolic instability. Intestinal immune system is activated through TLR activation. Inflammatory cytokines are produced as a result of the hyperactivation of T helper cells (e.g. Th17), which sets off an inflammatory cascade

The immune system, including the maturation of antigen-presenting cells, the establishment of germinal centres for B-cell maturation, and the microbicidal activities of macrophages and neutrophils, is affected by the gut microbiome, which has the capacity to determine healthy ageing. The microbiome-based interventions may possibly reduce immune ageing and ameliorate several age-related diseases due to altering the gut microbiome composition (Fig. 16.2). The gut epithelium is damaged resulting in reduced intestinal integrity because risk of malnutrition increases with advancing age (Walrath et al. 2021).

16.7 Anti-ageing Interventions and Intestinal Microbiota

Due to the fact that the gut microbiota may be affected by a various anti-ageing factors, involving nutrition, medication, and exercise, it has emerged as a crucial component of the anti-ageing process (Hasan and Yang 2019). A balanced diet and regular exercise are linked to a longer life expectancy and good ageing. Prebiotics are indigestible fibre substances like inulin and galacto-oligosaccharides that specifically promote the development of a small number of the gut microbes. Probiotics are live bacteria that make up a healthy gut microbiota, such as lactobacilli and *bifidobacteria* (Holscher 2017). They also have immune-modulatory effects, cause the synthesis of anti-inflammatory cytokines (IL-12 and IL-10), improve granulocyte phagocytosis, and stimulate intestinal IgA production. By boosting butyrate generation and promoting the expansion of SCFA-producing bacteria in the gut, they have beneficial immunomodulatory effects (Raheem et al. 2021).

Clinical studies have demonstrated that prebiotics with probiotics, often known as synbiotics, can treat malnutrition in older adults by reducing the release of pro-inflammatory mediators (TNF-, IL-1, and CRP) to a larger extent than even probiotics alone. Consuming probiotics may cause changes in the distribution of T cell subsets, including a decrease in the frequency of memory T cells and senescent CD28 (Markowiak and Śliżewska 2017).

16.8 Diet Intervention and Gut Microbiome

Natural foods and healthy eating habits can modify the functioning of the intestinal barrier and the makeup of the gut microbiota, which has anti-ageing benefits and plays a significant part in the aetiology of various age-related disorders (Zeppa et al. 2022). Vegetables, fruits, and beverages should make up a large portion of a healthy diet as these whole foods are highly connected with overall health due to the presence of phenolic compounds and fibres (Du et al. 2021).

16.9 Calorie Restriction (CR) and Intermittent Fasting (IF)

Calorie restriction is a type of dietary restriction that limits caloric consumption without causing malnutrition or reducing intake of vital nutrients, which may induce alterations in intestinal microbiota (Kökten et al. 2021). Intermittent fasting, a type of periodic CR, has been shown to improve metabolism by increasing insulin sensitivity, lowering the blood glucose levels, and decreasing fat mass and overall weight. Animals can develop a healthy gut microbiota that provides health improvements, extends lifespan, and protects against age-related diseases, as shown by the impact of calorie restriction on the microbiome (Patterson et al. 2015).

By lowering the phyla negatively linked to longevity, CR therapy alters the microbiota structure (including *Lactobacillus* and *Bifidobacterium*), reduces diet-associated metabolic problems, and extends and promotes a healthy lifespan in normal animals. Intermittent fasting (IF) is a successful and healthy weight-control strategy because it results in microbiota reconfiguration. IF, which is viewed as a sort of periodic CR that can alter and expand the gut microbiota, leads to a higher abundance of the families *Bacteroidaceae*, *Lactobacillaceae*, and *Prevotellacea* (Du et al. 2021).

16.10 Dietary Patterns and the Gut Microbiome, as well as Healthy Foods

Thanks to developments in nutritional science, we now realize more about how nourishing food stimulates the balance and activity of the gut microbiota. A healthy diet with fibre and phenolic substances may be able to influence the microbiome (Singh et al. 2017). Gut microbiota can be altered by healthy eating patterns like the Mediterranean diet (MD), which emphasizes whole grains, vegetables, fruits, nuts, and olive oil. Dietary fibre can affect the gut microbiota (increased number of SCFA producers) and metabolic control since they create metabolites like SCFAs (Nagpal et al. 2019).

A number of beneficial intestinal microorganisms, including *Bifidobacterium*, *Lactobacillus*, *Verrucomicrobia*, *Akkermansia*, and *Christensenellaceae*, have been shown to proliferate when exposed to polyphenols. Lemon polyphenols have shown to decrease the gut microbiota that may be related to ageing. Black and green tea polyphenols reduced body mass, which caused *Firmicutes* to fall and *Bacteroidetes* to increase (Kumar Singh et al. 2019). Red wine polyphenols considerably reduced undesired bacteria such as *Enterobacter cloacae* and *Escherichia coli* while significantly increasing butyrate-producing bacteria like *Faecalibacterium prausnitzii* and *Roseburia* (Yang et al. 2020).

16.11 Exercise Effect on the Intestinal Microbiota and Ageing

Exercise is a healthy habit that lowers the risk of developing chronic conditions including diabetes, cancer, and cardiovascular disease. It also increases the number of bacteria that produce butyric acid, which alters the metabolism of fat storage and prevents obesity (Zhang et al. 2022). Numerous studies have demonstrated that exercising raises the number of butyrate-producing taxa in the gastrointestinal tract, including Bacteroidales S24-7, *Faecalibacterium prausnitzii*, *Clostridiaceae*, and *Roseburia hominis*. Exercise improves the richness of gut microbiota and beneficial bacteria, which in turn reduces age-related problems and delays ageing (Cataldi and Poli 2022).

Improvements in exercise tolerance, insulin sensitivity, adipose tissue inflammation, and systolic blood pressure have all been linked to exercise-induced changes in the gut microbiota (Verheggen et al. 2021). The impact of exercise on the gut microbiota in elderly adults was revealed using data from the American Gut Project. The study demonstrated that older adults who were overweight had a higher diversity of gut microbiota than older adults who were of normal weight (Ramos et al. 2022).

16.12 Gut Microbiota-Drug Link and Ageing

The gut microbiota can both directly and indirectly affect how well a host can metabolize drugs, despite the fact that many medications undergo microbial biotransformation (Pant et al. 2022). Aspirin, resveratrol, metformin, and rapamycin are examples of oral medications that must first undergo intestinal metabolization before being absorbed. Drugs can alter the makeup of the gut microbiota through primarily two routes of action. The first mode is when medications stimulate microbial translocation from other bodily parts to the intestine, which results in an unbalanced microbial ecology. As a result of proton pump inhibitors, which weaken the stomach's acid barrier, oral bacteria, for instance, can enter the intestine through the stomach (Walsh et al. 2018).

Drugs that alter the gut microbes and directly affect bacterial growth are likely to be the primary component of the second approach. Some anti-ageing drugs as well as some antibacterial medications have the ability to control the proliferation of gut microbes. Besides promoting the development of intestinal bacteria that produce SCFAs, metformin has a therapeutic effect on lowering insulin resistance and glucose homeostasis (Weersma et al. 2020).

Metformin is introduced as an anti-ageing and health-improving drug because it can change the intestinal microbiota composition of obese mice, rats, and T2D patients. The natural macrolide molecule rapamycin, which was obtained from bacteria, has the ability to slow down the ageing process and successfully treat age-related disorders. It lowers the concentration of Alphaproteobacteria, a class of microorganisms that has previously been linked to a drop in senior mortality (Zhang and Hu 2020).

16.13 The Gut Microbiota and Human Ageing Diseases

An inflammatory response in the gut that has grown with age is brought on by changes in the gut flora of elderly persons. The intestinal barrier's integrity has declined as a result of the age-related imbalance in the gut microbiota (Walrath et al. 2021). Therefore, microbial product leakage may induce a permanent low-grade inflammatory state that had been linked to cognitive decline and dementia by up-regulating the amounts of circulating pro-inflammatory substances such as interferons, interleukin-6, interleukin-1, and TNF-α (Ferrucci and Fabbri 2018).

16.14 Gut Microbiota in Gut-Muscle Axis and the Function in Ageing

Preclinical and in vitro studies have established a connection between gut bacteria and muscle mass. The gut microbiota can influence the host's muscle mass and function through regulating the systemic inflammatory response and immunology, mitochondrial biogenesis, endocrine, and insulin sensitivity (Zhao et al. 2021).

Probiotics or butyrate treatment of muscle-wasting animal models has been associated with increased muscle mass, as well as better fermentative capacity and distinctive microbiota composition in mice with lower muscle mass. A gut-muscle axis may be critical for skeletal muscle function and maintenance as well as for preventing the beginning and progression of sarcopenia and age-related physical weakness (Chen et al. 2022).

A link between grip strength and the microbiota makeup, which includes significant taxa like Faecalibacterium and Bifidobacterium, has been demonstrated in a few small human investigations. Similar studies on parkinsonism patients have revealed that those who walk more slowly may have a different microbiome composition (Ticinesi et al. 2019).

The potential significance of microbiome in the pathogenesis of sarcopenia and physical frailty is getting lots and lots of research interest. The development of osteosarcopenic obesity may have a pathophysiological link with ageing microbiota. In particular, changes in the makeup of the gut microbiota may promote anabolic resistance and chronic inflammation, which could lead to weaker muscles, decreased physical performance, and worse clinical consequences (Daily and Park 2022).

16.15 Ageing Microbiota-Gut-Liver Axis

Functionally, the liver and gut microbiome are linked; in healthy individuals, the liver clears the intestines from bacteria, antigens, and inflammatory metabolites while regulating immune system homeostasis and the activity of inflammatory cells (Gebrayel et al. 2022). The gut microbiota also influences the production of bile acids, oxidative stress, pro-inflammatory cytokines, and hepatic lipogenesis, which further modifies the microbiota in the gut. Because intestinal bacteria and their metabolites can enter the liver through the portal venous system, the intestinal barrier function is disrupted, which further impairs the liver defence mechanisms. Then, Kupffer and stellate cells of the liver are activated (Kanmani et al. 2020).

Gut microbial dysregulation may stimulate the production of bacterial DNA, deoxycholic acids, and lipopolysaccharide (LPS), which can pass through the portal vein and cause chronic liver inflammation (Wang et al. 2022). Numerous inflammatory factors are generated as a result of their action, and these factors ultimately promote cirrhosis and liver fibrosis by causing necrotic inflammation and steatosis in the liver cells as well as aggravating intestinal mucosal damage. Clinical studies have shown that the gut-liver axis affects alcoholic liver disease (ALD), non-alcoholic fatty liver disease (NAFLD), and chronic hepatitis B (Roehlen et al. 2020).

16.16 Ageing Microbiota-Gut-Pancreas Axis

Curiously, both type 1 diabetes (T1DM) and type 2 diabetes mellitus (T2DM) sufferers reported higher *C. albicans* colonization compared to healthy controls. The diversity of fungi is considerably larger in T1DM patients. These results suggest that the pathophysiology of diabetes may include intestinal fungi (Wu et al. 2021a). Additionally, Nod1 ligands from intestinal commensal bacteria (functioning as signal molecules) are necessary for insulin transport in pancreatic beta cells. In the pancreas of patients with pancreatic ductal adenocarcinoma (PDA), bacteria and fungi are significantly more prevalent. In both humans and mice, *Malassezia* spp. is substantially enriched in PDA, and gut fungus move from the stomach to the pancreas perhaps via the sphincter of Oddi (Zhang et al. 2019).

In particular, recombinant C3a (rC3a) increases pancreatic tumour volume while mannose binding lectin (MBL)-deleted animals have minimal tumour pathogenesis, suggesting that the glycan of fungal cell wall-MBL-complement cascade pathway plays a significant role in pancreatic disorders (Wu et al. 2021b).

16.17 Ageing Microbiota-Gut-Heart Axis

Studies carried out over the last 10 years indicated that the gut microbiome may be involved in the aetiology of atherosclerosis, cardiovascular disease, and heart failure (Rahman et al. 2022). Growing evidence implicates the gut microbiome (GMB) in the development of immune-mediated subtypes of cardiomyopathy and heart failure (HF) (myocarditis and anthracycline-induced cardiotoxicity). This is a response to certain cardiovascular medications, as well as HF-related comorbidities such, chronic kidney diseases, cardiorenal syndrome, insulin resistance, malnutrition, and cardiac cachexia (Mamic et al. 2021).

The onset of gut ischemia in HF is a result of both elevated gut venous pressures and decreased blood flow in the splanchnic arteries. A decrease in the anti-inflammatory bacteria and an increase in potentially harmful microorganisms are observed in chronic HF (such as *Shigella*, *Salmonella*, *Candida*, etc.). These microorganisms colonize chronic heart failure (HF) patients more frequently and abundantly than healthy people, and the degree of colonization is correlated with the severity of HF (Gallo et al. 2022). The pathophysiology of HF is mechanistically linked to the gut microbial metabolites of short-chain fatty acids, trimethylamine *N*-oxide (TMAO), amino acid metabolites, and bile acids, according to recent study. As a result, these metabolites may be useful targets for HF treatment (Kazemian et al. 2020).

16.18 Ageing Microbiota-Gut-Skin Axis

The disruption in cutaneous homeostasis results from an imbalance caused by abnormally high amounts of commensal bacteria, which can also cause systemic inflammation (Toor et al. 2019). According to the suggested mechanism, an imbalance in the gut causes T-cell activation while also altering the immunosuppressive cytokines and Treg cells necessary to maintain microbial tolerance. As a result, there is a pattern of persistent inflammation in the skin and gut that cannot be self-regulated by the normal immune response (Rocamora-Reverte et al. 2020).

Atopic dermatitis (AD) is a long-lasting inflammatory skin disorder that is characterized by immune dysregulation, compromised skin barrier function, and skin and gut microbial dysbiosis. Less Bifidobacterium and more *Enterobacteriaceae*, *Clostridiums*, and *Staphylococcus* colonization in AD patients compared to healthy controls result in a reduced variety of the gut microbiota in this disease (Kim and Kim 2019).

16.19 Ageing Microbiota-Gut-Bone Axis

The gut microbiota maintains skeletal homeostasis by affecting host metabolism, immunological function, and the hormone release (Behera et al. 2020). It has been determined that the mechanisms underlying probiotics or prebiotics facilitating mineral utilization have become the larger absorption surface, greater expression of calcium-binding proteins, and improved solubility of minerals created by enhanced short-chain fatty acids (SCFAs) (Whisner and Castillo 2018).

The gut microbiome has a significant impact on how vitamin D, B, and K are metabolized. In ovariectomized (OVX) Sprague-Dawley (SD) rats, probiotics have been found to improve the physical and biomechanical properties of the femur, whereas long-term antibiotic exposure harms the gut microbiomes and impairs bone material properties in mice, most notably a decrease in whole bone-bending strength (Tu et al. 2021).

16.20 Gut-Brain Axis with Ageing Microbiota

A network of bidirectional communication between the host and the microbiome is known as the microbiota-gut-brain axis (MGBA) (Rutsch et al. 2020). Dysbiosis results in defective MGBA signalling, which controls the growth and operation of the host's immunological, metabolic, and neurological systems. Age-related and stroke-related dysfunction of the MGBA caused by dysbiosis is associated with the emergence of well-known stroke risk factors such as obesity, diabetes, and atherosclerosis (Honarpisheh et al. 2022).

Healthy gut microbiome colonisation is vital in brain development and behaviour because it improves the blood-brain barrier (BBB), changes synaptic plasticity, and results in social interaction deficits when absent (Tang and Zhu 2020). The gut microbiome may influence the central nervous system (CNS) via communicating with the brain and vice versa through the vagus nerve (VN) in both health and disease. Dysbiotic microbiota has been associated with both IL-17 expression in the brain and an increase in CD4+-IL-17 cells in the gut (Ma et al. 2019).

In addition to the central neuroinflammation seen in Alzheimer's disease (AD) and Parkinson's disease (PD), there is compelling evidence linking changes in the gut microbiota to peripheral immune response (Shandilya et al. 2022). Clinical investigations on stroke patients showed a rise in biodiversity, a marginal decline in *Bacteroidetes* members, and a rise in *Lactobacillus ruminis* (from the phylum *Firmicutes*) (Chidambaram et al. 2022).

Comparatively, studies using faeces from AD patients revealed less diversity and more Bacteroidetes, which was supported by studies using animal models. Patients' PD microbiomes are characterized by dramatically less *Prevotellaceae*, a member of the phylum *Bacteroidetes*, and dramatically more *Akkermansia*, *Bifidobacteriaceae*, *Enterobacteriaceae*, and *Lactobacillaceae* (Fang et al. 2020).

16.21 Cellular Senescence Links Biological Ageing and the Aim of Gut Microbiome for Anti-senescence Intervention

Cellular senescence, which is assumed to be a pre-encoded cancer suppressor mechanism, is characterized by cell cycle arrest, a lack of proliferative potential, global cell growth, and noticeably distorted nuclei. Additionally, the presence of chromatin foci that exhibit a prolonged DNA damage response enhanced nuclear factor-B (NF-B) signalling, and resistance to apoptosis are implicated (Ferrucci and Fabbri 2018). It is conceivable to think of a two-way interaction between the gut microbiome and cellular senescence, in which the secretory metabolites of gut microbiota undergo biotransformation in one direction (Sharma 2022).

The accumulation of senescent cells (epithelial/fibroblasts) in the intestine and senescence-associated secretory phenotype (SASP) may have an impact on cellular processes and cause immune reactions in the gut (Fig. 16.3). Dietary factors can have a direct impact on the ability of intestinal cells to experience cellular senescence (Naylor et al. 2013). It has been demonstrated that metabolites produced by a probiotic called *Lactobacillus fermentum*, which was isolated from human faeces, can reduce a number of senescence-related characteristics. These involved cellular hypertrophy, senescence-associated galactosidase (SA-gal) activity, induction of SASP, DNA damage response, Akt/mTOR pathway, and activation of cell cycle inhibitor signalling (Di Micco and Krizhanovsky 2021).

Additional study reported that a *Sphingomonas hydrophobicum* extract could delay skin senescence by lowering the activation of p21, p16Ink4a cell cycle inhibitors, and SA-gal activity, which would have skin remodelling and psychobiological impacts. Persistent SASP produced by senescent enterocytes may induce an inflammatory state and/or an oncogenic mutation, which may have a detrimental effect on

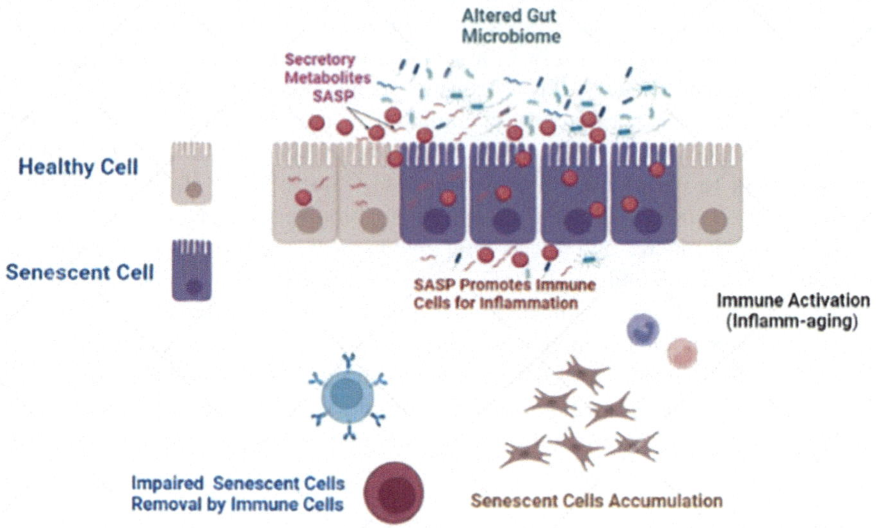

Fig. 16.3 Senescent cells can accumulate and develop more quickly in many organs when immune system activities are compromised. Persistent senescent cells produce SASP, which influences neighbouring healthy cells via paracrine processes. This promotes an inflammatory and tumorigenic setting, which ultimately compromises organ functions and accelerates the ageing pattern

gut permeability, immune activation, and the structure of the gut microbiome (Sharma 2022).

Recent researches at the cellular and molecular levels have shown that *Lactobacillus* can improve the ability of cells, tissues, and organs to fend off ageing and diseases associated with it. Studies showed that the supplementation of *Lactobacillus* strains, either alone or in combination, greatly had improved numerous age-related physical and physiological aspects of health in experimental animals or even people (Samtiya et al. 2022).

16.22 Conclusion and Future Perspectives

It seems acceptable to investigate human biological systems again from the aspect of gut microbiome and to detect the particular microbial dysbiosis as signs of cellular senescence and ageing. The gut microbiota significantly has effects on many characteristics of overall health, and also due to the abnormalities in diversity and functionality of the microbiota which are linked to a variety of inflammatory diseases. With emerging evidence of gut microbiome role in cellular senescence as a central dogma of ageing and age-related pathologies, the demand for novel antisenescence therapies depending on gut microbiome has increased to develop novel approaches for healthier ageing. Developing personalized interventions based on individual gut microbiome profiles holds promise for optimizing anti-senescence strategies and minimizing side effects. This could involve targeted manipulation of the gut microbiota using prebiotics, probiotics, or faecal microbiota transplantation (FMT). Assessing the long-term safety and efficacy of these microbiome-based interventions for promoting healthy ageing and preventing age-related diseases remains vital. Conducting well-designed clinical trials with appropriate control groups is essential to translate promising preclinical findings into effective therapeutic strategies for humans. Identifying specific microbial signatures or metabolites associated with cellular senescence could serve as valuable biomarkers for early detection and monitoring of age-related decline. By continuing to explore the intricate relationship between the gut microbiome and cellular senescence, researchers can potentially unlock novel avenues for promoting healthy aging and delaying the onset of age-related diseases. This exciting field of research has the potential to significantly improve lifespan and enhance health throughout life.

References

Al Bander Z, Nitert MD (2020) The gut microbiota and inflammation: an overview. Int J Environ Res Public Health 17(20):7618. https://doi.org/10.3390/ijerph17207618

Alam MT, Amos G, Murphy A, Murch S, Wellington E, Arasaradnam R (2020) Microbial imbalance in inflammatory bowel disease patients at different taxonomic levels. Gut Pathog 12(1):1. https://doi.org/10.1186/s13099-019-0341-6

Badal V, Vaccariello E, Murray E, Yu K, Knight R, Jeste D, Nguyen T (2020) The gut microbiome, aging, and longevity: a systematic review. Nutrients 12(12):3759. https://doi.org/10.3390/nu12123759

Behera J, Ison J, Tyagi SC, Tyagi N (2020) The role of gut microbiota in bone homeostasis. Bone 135:115317. https://doi.org/10.1016/j.bone.2020.115317

Caricilli AM, Saad MJ (2013) The role of gut microbiota on insulin resistance. Nutrients 5(3):829–851. https://doi.org/10.3390/nu5030829

Cataldi S, Poli L (2022) The effects of physical activity on the gut microbiota and the gut-brain axis in preclinical and human models: a narrative review. Nutrients 14(16):3293. https://doi.org/10.3390/nu14163293

Chen LH, Chang SS, Chang HY, Wu CH, Pan CH, Chang CC, Chan CH, Huang HY (2022) Probiotic supplementation attenuates age-related sarcopenia via the gut-muscle axis in SAMP8 mice. J Cachexia Sarcopenia Muscle 13(1):515–531. https://doi.org/10.1002/jcsm.12849

Chidambaram SB, Rathipriya AG, Mahalakshmi AM, Sharma S, Hediyal TA, Ray B, Sunanda T, Rungratanawanich W, Kashyap RS, Qoronfleh MW, Essa MM, Song B-J, Monaghan TM (2022) The influence of gut dysbiosis in the pathogenesis and management of ischemic stroke. Cell J 11(7):1239

Conway J, Duggal NA (2021) Ageing of the gut microbiome: potential influences on immune senescence and inflammageing. Ageing Res Rev 68:101323

Daily JW, Park S (2022) Sarcopenia is a cause and consequence of metabolic dysregulation in aging humans: effects of gut dysbiosis, glucose dysregulation, diet and lifestyle. Cell J 11(3):338. https://doi.org/10.3390/cells11030338

Dekaboruah E, Suryavanshi MV, Chettri D, Verma AK (2020) Human microbiome: an academic update on human body site specific surveillance and its possible role. Arch Microbiol 202(8):2147–2167. https://doi.org/10.1007/s00203-020-01931-x

Di Micco R, Krizhanovsky V (2021) Cellular senescence in ageing: from mechanisms to therapeutic opportunities. Nat Rev Mol Cell Biol 22(2):75–95. https://doi.org/10.1038/s41580-020-00314-w

Du Y, Gao Y, Zeng B, Fan X, Yang D, Yang M (2021) Effects of anti-aging interventions on intestinal microbiota. Gut Microbes 13(1):1994835

Durack J, Lynch SV (2019) The gut microbiome: relationships with disease and opportunities for therapy. J Exp Med 216(1):20–40. https://doi.org/10.1084/jem.20180448

Fang P, Kazmi SA, Jameson KG, Hsiao EY (2020) The microbiome as a modifier of neurodegenerative disease risk. Cell Host Microbe 28(2):201–222. https://doi.org/10.1016/j.chom.2020.06.008

Ferrucci L, Fabbri E (2018) Inflammageing: chronic inflammation in ageing, cardiovascular disease, and frailty. Nat Rev Cardiol 15(9):505–522. https://doi.org/10.1038/s41569-018-0064-2

Gallo A, Macerola N, Favuzzi AM, Nicolazzi MA, Gasbarrini A, Montalto M (2022) The gut in heart failure: current knowledge and novel frontiers. Med Princ Pract 31(3):203–214. https://doi.org/10.1159/000522284

Gebrayel P, Nicco C, Al Khodor S, Bilinski J, Caselli E, Comelli EM, Egert M, Giaroni C, Karpinski TM, Loniewski I, Mulak A, Reygner J, Samczuk P, Serino M, Sikora M, Terranegra A, Ufnal M, Villeger R, Pichon C, Konturek P, Edeas M (2022) Microbiota medicine: towards clinical revolution. J Transl Med 20(1):111. https://doi.org/10.1186/s12967-022-03296-9

Geerlings SY, Kostopoulos I, de Vos WM, Belzer C (2018) Akkermansia muciniphila in the human gastrointestinal tract: when, where, and how? Microorganisms 6(3):75. https://doi.org/10.3390/microorganisms6030075

Gilbert JA, Blaser MJ, Caporaso JG, Jansson JK, Lynch SV, Knight R (2018) Current understanding of the human microbiome. Nat Med 24(4):392–400. https://doi.org/10.1038/nm.4517

Hasan N, Yang H (2019) Factors affecting the composition of the gut microbiota, and its modulation. PeerJ 7:e7502. https://doi.org/10.7717/peerj.7502

Holscher HD (2017) Dietary fiber and prebiotics and the gastrointestinal microbiota. Gut Microbes 8(2):172–184. https://doi.org/10.1080/19490976.2017.1290756

Honarpisheh P, Bryan RM, McCullough LD (2022) Aging microbiota-gut-brain axis in stroke risk and outcome. Circ Res 130(8):1112–1144

Hou K, Wu Z-X, Chen X-Y, Wang J-Q, Zhang D, Xiao C, Zhu D, Koya JB, Wei L, Li J, Chen Z-S (2022) Microbiota in health and diseases. Signal Transduct Target Ther 7(1):135. https://doi.org/10.1038/s41392-022-00974-4

Kanmani P, Suganya K, Kim H (2020) The gut microbiota: how does it influence the development and progression of liver diseases. Biomedicine 8(11):501

Kazemian N, Mahmoudi M, Halperin F, Wu JC, Pakpour S (2020) Gut microbiota and cardiovascular disease: opportunities and challenges. Microbiome 8(1):36. https://doi.org/10.1186/s40168-020-00821-0

Kho ZY, Lal SK (2018) The human gut microbiome—a potential controller of wellness and disease. Front Microbiol 9:1835. https://doi.org/10.3389/fmicb.2018.01835

Kim S, Jazwinski SM (2018) The gut microbiota and healthy aging: a mini-review. J Gerontol 64(6):513–520

Kim JE, Kim HS (2019) Microbiome of the skin and gut in atopic dermatitis (AD): understanding the pathophysiology and finding novel management strategies. J Clin Med 8(4):444. https://doi.org/10.3390/jcm8040444

Kökten T, Hansmannel F, Ndiaye NC, Heba AC, Quilliot D, Dreumont N (2021) Calorie restriction as a new treatment of inflammatory diseases. Adv Nutr 12(4):1558–1570. https://doi.org/10.1093/advances/nmaa179

Kumar Singh A, Cabral C, Kumar R, Ganguly R, Kumar Rana H, Gupta A, Rosaria Lauro M, Carbone C, Reis F, Pandey AK (2019) Beneficial effects of dietary polyphenols on gut microbiota and strategies to improve delivery efficiency. Nutrients 11(9):2216. https://doi.org/10.3390/nu11092216

Lidzbarsky G, Gutman D, Shekhidem HA, Sharvit L, Atzmon G (2018) Genomic instabilities, cellular senescence, and aging: in vitro, in vivo and aging-like human syndromes. Front Med 17(5):104. https://doi.org/10.3389/fmed.2018.00104

Ma Q, Xing C, Long W, Wang HY, Liu Q, Wang R-F (2019) Impact of microbiota on central nervous system and neurological diseases: the gut-brain axis. J Neuroinflammation 16(1):53. https://doi.org/10.1186/s12974-019-1434-3

Mamic P, Chaikijurajai T, Tang WHW (2021) Gut microbiome—a potential mediator of pathogenesis in heart failure and its comorbidities: state-of-the-art review. J Mol Cell Cardiol 152:105–117. https://doi.org/10.1016/j.yjmcc.2020.12.001

Markowiak P, Śliżewska K (2017) Effects of probiotics, prebiotics, and synbiotics on human health. Nutrients 9(9):1021. https://doi.org/10.3390/nu9091021

Morrison DJ, Preston T (2016) Formation of short chain fatty acids by the gut microbiota and their impact on human metabolism. Gut Microbes 7(3):189–200. https://doi.org/10.1080/19490976.2015.1134082

Nagpal R, Mainali R, Ahmadi S, Wang S, Singh R, Kavanagh K, Kitzman DW, Kushugulova A, Marotta F, Yadav H (2018) Gut microbiome and aging: physiological and mechanistic insights. Nutr Healthy Aging 4(4):267–285. https://doi.org/10.3233/nha-170030

Nagpal R, Shively CA, Register TC, Craft S, Yadav H (2019) Gut microbiome-Mediterranean diet interactions in improving host health. F1000Res 8:699

Naylor RM, Baker DJ, van Deursen JM (2013) Senescent cells: a novel therapeutic target for aging and age-related diseases. Clin Pharmacol Ther 93(1):105–116. https://doi.org/10.1038/clpt.2012.193

Pandiyan P, Bhaskaran N, Zou M, Schneider E, Jayaraman S, Huehn J (2019) Microbiome dependent regulation of T(regs) and Th17 cells in mucosa. Front Immunol 10:426. https://doi.org/10.3389/fimmu.2019.00426

Pant A, Maiti TK, Mahajan D, Das B (2022) Human gut microbiota and drug metabolism. Microb Ecol 86:97. https://doi.org/10.1007/s00248-022-02081-x

Patterson RE, Laughlin GA, LaCroix AZ, Hartman SJ, Natarajan L, Senger CM, Martínez ME, Villaseñor A, Sears DD, Marinac CR, Gallo LC (2015) Intermittent fasting and human metabolic health. J Acad Nutr Diet 115(8):1203–1212. https://doi.org/10.1016/j.jand.2015.02.018

Ragonnaud E, Biragyn A (2021) Gut microbiota as the key controllers of "healthy" aging of elderly people. Immun Ageing 18(1):2. https://doi.org/10.1186/s12979-020-00213-w

Raheem A, Liang L, Zhang G, Cui S (2021) Modulatory effects of probiotics during pathogenic infections with emphasis on immune regulation. Front Immunol 12:616713

Rahman MM, Islam F, Or-Rashid MH, Mamun AA, Rahaman MS, Islam MM, Meem AFK, Sutradhar PR, Mitra S, Mimi AA, Emran TB, Fatimawali Idroes R, Tallei TE, Ahmed M, Cavalu S (2022) The gut microbiota (microbiome) in cardiovascular disease and its therapeutic regulation. Front Cell Infect Microbiol 12:903570. https://doi.org/10.3389/fcimb.2022.903570

Ramos C, Gibson GR, Walton GE, Magistro D, Kinnear W, Hunter K (2022) Systematic review of the effects of exercise and physical activity on the gut microbiome of older adults. Nutrients 14(3):674

Ray D, Yung R (2018) Immune senescence, epigenetics and autoimmunity. Clin Immunol 196:59–63. https://doi.org/10.1016/j.clim.2018.04.002

Rinninella E, Raoul P, Cintoni M, Franceschi F, Miggiano GAD, Gasbarrini A, Mele MC (2019) What is the healthy gut microbiota composition? A changing ecosystem across age, environment, diet, and diseases. Microorganisms 7(1):14. https://doi.org/10.3390/microorganisms7010014

Rocamora-Reverte L, Melzer FL, Würzner R, Weinberger B (2020) The complex role of regulatory T cells in immunity and aging. Front Immunol 11:616949. https://doi.org/10.3389/fimmu.2020.616949

Roehlen N, Crouchet E, Baumert TF (2020) Liver fibrosis: mechanistic concepts and therapeutic perspectives. J Cell 9(4):875. https://doi.org/10.3390/cells9040875

Rooks MG, Garrett WS (2016) Gut microbiota, metabolites and host immunity. Nat Rev Immunol 16(6):341–352. https://doi.org/10.1038/nri.2016.42

Rowland I, Gibson G, Heinken A, Scott K, Swann J, Thiele I, Tuohy K (2018) Gut microbiota functions: metabolism of nutrients and other food components. Eur J Nutr 57(1):1–24. https://doi.org/10.1007/s00394-017-1445-8

Rutsch A, Kantsjö JB, Ronchi F (2020) The gut-brain axis: how microbiota and host inflammasome influence brain physiology and pathology. Front Immunol 11:604179

Samtiya M, Puniya AK, Puniya M, Shah NP, Dhewa T, Vemuri R (2022) Probiotic regulation to modulate aging gut and brain health: a concise review. Bacteria 1(4):250–265

Shandilya S, Kumar S, Kumar Jha N, Kumar Kesari K, Ruokolainen J (2022) Interplay of gut microbiota and oxidative stress: perspective on neurodegeneration and neuroprotection. J Adv Res 38:223–244. https://doi.org/10.1016/j.jare.2021.09.005

Sharma R (2022) Emerging interrelationship between the gut microbiome and cellular senescence in the context of aging and disease: perspectives and therapeutic opportunities. Probiotics Antimicrob Proteins 14(4):648–663. https://doi.org/10.1007/s12602-021-09903-3

Shilpa A, Kalyani S, Manisha S (2018) Ageing process and physiological changes. In: Grazia DO, Antonio G, Daniele S (eds) Journal of gerontology (Ch. 1). IntechOpen, Rijeka

Singh RK, Chang H-W, Yan D, Lee KM, Ucmak D, Wong K, Abrouk M, Farahnik B, Nakamura M, Zhu TH, Bhutani T, Liao W (2017) Influence of diet on the gut microbiome and implications for human health. J Transl Med 15(1):73. https://doi.org/10.1186/s12967-017-1175-y

Tan Q, Liang N, Zhang X, Li J (2021) Dynamic aging: channeled through microenvironment. Front Physiol 12:702276. https://doi.org/10.3389/fphys.2021.702276

Tang W, Zhu H (2020) The impact of gut microbiota disorders on the blood-brain barrier. Infect Drug Resist 13:3351–3363. https://doi.org/10.2147/idr.s254403

Thursby E, Juge N (2017) Introduction to the human gut microbiota. Biochem J 474(11):1823–1836. https://doi.org/10.1042/bcj20160510

Ticinesi A, Nouvenne A, Cerundolo N, Catania P, Prati B, Tana C, Meschi T (2019) Gut microbiota, muscle mass and function in aging: a focus on physical frailty and sarcopenia. Nutrients 11(7):1633

Toor D, Wsson MK, Kumar P, Karthikeyan G, Kaushik NK, Goel C, Singh S, Kumar A, Prakash H (2019) Dysbiosis disrupts gut immune homeostasis and promotes gastric diseases. Int J Mol Sci 20(10):2432. https://doi.org/10.3390/ijms20102432

Tu Y, Yang R, Xu X, Zhou X (2021) The microbiota-gut-bone axis and bone health. J Leukoc Biol 110(3):525–537

Vaiserman AM, Koliada AK, Marotta F (2017) Gut microbiota: a player in aging and a target for anti-aging intervention. Ageing Res Rev 35:36–45

Verheggen R, Konstanti P, Smidt H, Hermus A, Thijssen DHJ, Hopman MTE (2021) Eight-week exercise training in humans with obesity: marked improvements in insulin sensitivity and modest changes in gut microbiome. Obesity (Silver Spring) 29(10):1615–1624. https://doi.org/10.1002/oby.23252

Walrath T, Dyamenahalli KU, Hulsebus HJ, McCullough RL, Idrovo JP, Boe DM (2021) Age-related changes in intestinal immunity and the microbiome. J Leukoc Biol 109(6):1045–1061. https://doi.org/10.1002/jlb.3ri0620-405rr

Walsh J, Griffin BT, Clarke G, Hyland NP (2018) Drug-gut microbiota interactions: implications for neuropharmacology. Br J Pharmacol 175(24):4415–4429. https://doi.org/10.1111/bph.14366

Wang L, Cao ZM, Zhang LL, Li JM, Lv WL (2022) The role of gut microbiota in some liver diseases: from an immunological perspective. Front Immunol 13:923599. https://doi.org/10.3389/fimmu.2022.923599

Weersma RK, Zhernakova A, Fu J (2020) Interaction between drugs and the gut microbiome. Gut 69(8):1510–1519. https://doi.org/10.1136/gutjnl-2019-320204

Whisner CM, Castillo LF (2018) Prebiotics, bone and mineral metabolism. Calcif Tissue Int 102(4):443–479. https://doi.org/10.1007/s00223-017-0339-3

Wu X, Xia Y, He F, Zhu C, Ren W (2021a) Intestinal mycobiota in health and diseases: from a disrupted equilibrium to clinical opportunities. Microbiome 9(1):1–18

Wu YL, Xu J, Rong XY, Wang F, Wang HJ, Zhao C (2021b) Gut microbiota alterations and health status in aging adults: from correlation to causation. Aging Med (Milton) 4(3):206–213. https://doi.org/10.1002/agm2.12167

Xu Y, Liu X, Liu X, Chen D, Wang M, Jiang X, Xiong Z (2021) The roles of the gut microbiota and chronic low-grade inflammation in older adults with frailty. Front Cell Infect Microbiol 11:675414. https://doi.org/10.3389/fcimb.2021.675414

Yang Q, Liang Q, Balakrishnan B, Belobrajdic DP, Feng Q-J, Zhang W (2020) Role of dietary nutrients in the modulation of gut microbiota: a narrative review. Nutrients 12(2):381

Zeppa SD, Agostini D, Ferrini F, Gervasi M, Barbieri E, Bartolacci A, Piccoli G, Saltarelli R, Sestili P, Stocchi V (2022) Interventions on gut microbiota for healthy aging. J Cell 12(1):34

Zhang Q, Hu N (2020) Effects of metformin on the gut microbiota in obesity and type 2 diabetes mellitus. Diabetes Metab Syndr Obes 13:5003–5014. https://doi.org/10.2147/dmso.s286430

Zhang Q, Pan Y, Zeng B, Zheng X, Wang H, Shen X, Li H, Jiang Q, Zhao J, Meng Z-X, Li P, Chen Z, Wei H, Liu Z (2019) Intestinal lysozyme liberates Nod1 ligands from microbes to direct insulin trafficking in pancreatic beta cells. Cell Res 29(7):516–532. https://doi.org/10.1038/s41422-019-0190-3

Zhang L, Liu Y, Sun Y, Zhang X (2022) Combined physical exercise and diet: regulation of gut microbiota to prevent and treat of metabolic disease: a review. Nutrients 14(22):4774

Zhao J, Huang Y, Yu X (2021) A narrative review of gut-muscle axis and sarcopenia: the potential role of gut microbiota. Int J Gen Med 14:1263–1273. https://doi.org/10.2147/ijgm.s301141

The manufacturer's authorised representative in the EU is Springer Nature Customer Service Centre GmbH, Europaplatz 3, 69115 Heidelberg, Germany. If you have any concerns regarding our products, please contact ProductSafety@springernature.com

Printed and bound by CPI Group (UK) Ltd, Croydon, CR0 4YY

01/10/2026

02225896-0015